Sustainable Utilization of Metals

Sustainable Utilization of Metals—Processing, Recovery and Recycling

Special Issue Editor

Bernd Friedrich

MDPI • Basel • Beijing • Wuhan • Barcelona • Belgrade • Manchester • Tokyo • Cluj • Tianjin

Special Issue Editor
Bernd Friedrich
RWTH Aachen University
Germany

Editorial Office
MDPI
St. Alban-Anlage 66
4052 Basel, Switzerland

This is a reprint of articles from the Special Issue published online in the open access journal *Metals* (ISSN 2075-4701) (available at: https://www.mdpi.com/journal/metals/special_issues/Sustainable_Recycling).

For citation purposes, cite each article independently as indicated on the article page online and as indicated below:

LastName, A.A.; LastName, B.B.; LastName, C.C. Article Title. *Journal Name* **Year**, *Article Number*, Page Range.

ISBN 978-3-03928-885-4 (Pbk)
ISBN 978-3-03928-886-1 (PDF)

Cover image courtesy of Tom Gertjegerdes.

© 2020 by the authors. Articles in this book are Open Access and distributed under the Creative Commons Attribution (CC BY) license, which allows users to download, copy and build upon published articles, as long as the author and publisher are properly credited, which ensures maximum dissemination and a wider impact of our publications.

The book as a whole is distributed by MDPI under the terms and conditions of the Creative Commons license CC BY-NC-ND.

Contents

About the Special Issue Editor ix

Bernd Friedrich
Sustainable Utilization of Metals-Processing, Recovery and Recycling
Reprinted from: *Metals* **2019**, *9*, 769, doi:10.3390/met9070769 1

Benedikt Flerus, Thomas Swiontek, Katrin Bokelmann, Rudolf Stauber and Bernd Friedrich
Thermochemical Modelling and Experimental Validation of In Situ Indium Volatilization by Released Halides during Pyrolysis of Smartphone Displays
Reprinted from: *Metals* **2018**, *8*, 1040, doi:10.3390/met8121040 5

Chiara Bonomi, Alexandra Alexandri, Johannes Vind, Angeliki Panagiotopoulou, Petros Tsakiridis and Dimitrios Panias
Scandium and Titanium Recovery from Bauxite Residue by Direct Leaching with a Brønsted Acidic Ionic Liquid
Reprinted from: *Metals* **2018**, *8*, 834, doi:10.3390/met8100834 17

Şerif Kaya, Edward Michael Peters, Kerstin Forsberg, Carsten Dittrich, Srecko Stopic and Bernd Friedrich
Scandium Recovery from an Ammonium Fluoride Strip Liquor by Anti-Solvent Crystallization
Reprinted from: *Metals* **2018**, *8*, 767, doi:10.3390/met8100767 35

Tamara Ebner, Stefan Luidold, Matthias Honner, Helmut Antrekowitsch and Christoph Czettl
Conditioning of Spent Stripping Solution for the Recovery of Metals
Reprinted from: *Metals* **2018**, *8*, 757, doi:10.3390/met8100757 45

Johannes Vind, Alexandra Alexandri, Vicky Vassiliadou and Dimitrios Panias
Distribution of Selected Trace Elements in the Bayer Process
Reprinted from: *Metals* **2018**, *8*, 327, doi:10.3390/met8050327 59

Wei-Sheng Chen and Hsing-Jung Ho
Recovery of Valuable Metals from Lithium-Ion Batteries NMC Cathode Waste Materials by Hydrometallurgical Methods
Reprinted from: *Metals* **2018**, *8*, 321, doi:10.3390/met8050321 81

Şerif Kaya, Carsten Dittrich, Srecko Stopic and Bernd Friedrich
Concentration and Separation of Scandium from Ni Laterite Ore Processing Streams
Reprinted from: *Metals* **2017**, *7*, 557, doi:10.3390/met7120557 97

Chenna Rao Borra, Thijs J. H. Vlugt, Jeroen Spooren, Peter Nielsen, Yongxiang Yang and S. Erik Offerman
Characterization and Feasibility Studies on Complete Recovery of Rare Earths from Glass Polishing Waste
Reprinted from: *Metals* **2019**, *9*, 278, doi:10.3390/met9030278 105

Evangelos Bourbos, Antonis Karantonis, Labrini Sygellou, Ioannis Paspaliaris and Dimitrios Panias
Study of Nd Electrodeposition from the Aprotic Organic Solvent Dimethyl Sulfoxide
Reprinted from: *Metals* **2018**, *8*, 803, doi:10.3390/met8100803 117

Bo Zhang, Yong Fan, Chengjun Liu, Yun Ye and Maofa Jiang
Reduction Characteristics of Carbon-Containing REE–Nb–Fe Ore Pellets
Reprinted from: Metals **2018**, *8*, 204, doi:10.3390/met8040204 131

Chenna Rao Borra, Thijs J. H. Vlugt, Yongxiang Yang and S. Erik Offerman
Recovery of Cerium from Glass Polishing Waste: A Critical Review
Reprinted from: Metals **2018**, *8*, 801, doi:10.3390/met8100801 147

Andrei Shishkin, Viktors Mironovs, Hong Vu, Pavel Novak, Janis Baronins, Alexandr Polyakov and Jurijs Ozolins
Cavitation-Dispersion Method for Copper Cementation from Wastewater by Iron Powder
Reprinted from: Metals **2018**, *8*, 920, doi:10.3390/met8110920 163

Maximilian V. Reimer, Heike Y. Schenk-Mathes, Matthias F. Hoffmann and Tobias Elwert
Recycling Decisions in 2020, 2030, and 2040—When Can Substantial NdFeB Extraction be Expected in the EU?
Reprinted from: Metals **2018**, *8*, 867, doi:10.3390/met8110867 175

Marina Gnatko, Cong Li, Alexander Arnold and Bernd Friedrich
Purification of Aluminium Cast Alloy Melts through Precipitation of Fe-Containing Intermetallic Compounds
Reprinted from: Metals **2018**, *8*, 796, doi:10.3390/met8100796 191

Anton Andersson, Amanda Gullberg, Adeline Kullerstedt, Erik Sandberg, Mats Andersson, Hesham Ahmed, Lena Sundqvist-Ökvist and Bo Björkman
A Holistic and Experimentally-Based View on Recycling of Off-Gas Dust within the Integrated Steel Plant
Reprinted from: Metals **2018**, *8*, 760, doi:10.3390/met8100760 203

Lei Guo, Xiaochun Wen, Qipeng Bao and Zhancheng Guo
Removal of Tramp Elements within 7075 Alloy by Super-Gravity Aided Rheorefining Method
Reprinted from: Metals **2018**, *8*, 701, doi:10.3390/met8090701 221

Fabian Diaz, Yufengnan Wang, Tamilselvan Moorthy and Bernd Friedrich
Degradation Mechanism of Nickel-Cobalt-Aluminum (NCA) Cathode Material from Spent Lithium-Ion Batteries in Microwave-Assisted Pyrolysis
Reprinted from: Metals **2018**, *8*, 565, doi:10.3390/met8080565 233

Piotr Palimąka, Stanisław Pietrzyk, Michał Stępień, Katarzyna Ciećko and Ilona Nejman
Zinc Recovery from Steelmaking Dust by Hydrometallurgical Methods
Reprinted from: Metals **2018**, *8*, 547, doi:10.3390/met8070547 249

Alicia Gauffin and Petrus Christiaan Pistorius
The Scrap Collection per Industry Sector and the Circulation Times of Steel in the U.S. between 1900 and 2016, Calculated Based on the Volume Correlation Model
Reprinted from: Metals **2018**, *8*, 338, doi:10.3390/met8050338 263

Stanisław Małecki and Krzysztof Gargul
Low-Waste Recycling of Spent CuO-ZnO-Al_2O_3 Catalysts
Reprinted from: Metals **2018**, *8*, 177, doi:10.3390/met8030177 279

Daniel Fernández-González, José Sancho-Gorostiaga, Juan Piñuela-Noval and Luis Felipe Verdeja González
Anodic Lodes and Scrapings as a Source of Electrolytic Manganese
Reprinted from: Metals **2018**, *8*, 162, doi:10.3390/met8030162 287

Seifeldin R. Mohamed, Semiramis Friedrich and Bernd Friedrich
Refining Principles and Technical Methodologies to Produce Ultra-Pure Magnesium for High-Tech Applications
Reprinted from: *Metals* **2019**, *9*, 85, doi:10.3390/met9010085 303

Jil Schosseler, Anna Trentmann, Bernd Friedrich, Klaus Hahn and Hermann Wotruba
Kinetic Investigation of Silver Recycling by Leaching from Mechanical Pre-Treated Oxygen-Depolarized Cathodes Containing PTFE and Nickel
Reprinted from: *Metals* **2019**, *9*, 187, doi:10.3390/met9020187 315

Xingbang Wan, Jani Fellman, Ari Jokilaakso, Lassi Klemettinen and Miikka Marjakoski
Behavior of Waste Printed Circuit Board (WPCB) Materials in the Copper Matte Smelting Process
Reprinted from: *Metals* **2018**, *8*, 887, doi:10.3390/met8110887 327

Minna Rämä, Samu Nurmi, Ari Jokilaakso, Lassi Klemettinen, Pekka Taskinen and Justin Salminen
Thermal Processing of Jarosite Leach Residue for a Safe Disposable Slag and Valuable Metals Recovery
Reprinted from: *Metals* **2018**, *8*, 744, doi:10.3390/met8100744 337

Željko Kamberović, Milisav Ranitović, Marija Korać, Zoran Andjić, Nataša Gajić, Jovana Djokić and Sanja Jevtić
Hydrometallurgical Process for Selective Metals Recovery from Waste-Printed Circuit Boards
Reprinted from: *Metals* **2018**, *8*, 441, doi:10.3390/met8060441 347

Stefan Steinlechner and Jürgen Antrekowitsch
Thermodynamic Considerations for a Pyrometallurgical Extraction of Indium and Silver from a Jarosite Residue
Reprinted from: *Metals* **2018**, *8*, 335, doi:10.3390/met8050335 367

About the Special Issue Editor

Bernd Friedrich born in 1958, studied metallurgical engineering and electrometallurgy at RWTH Aachen University, where he received his doctorate in 1988. After many years in management positions at GfE Nürnberg and VARTA Batterie AG, he returned to RWTH Aachen University in 1999 as a Full Professor of Metallurgical Process Engineering and Metal Recycling. His institute (IME) covers the topic of "circular economy" at RWTH, as well as in EIT RawMaterials for NRW. As a result of the steadily growing interest in battery recycling from the public, industry and political spheres, the IME has been systematically addressing recycling concepts since 2002 in an aim to develop optimal industry-oriented recycling strategies for all common battery systems. Lead–acid batteries, nickel–cadmium batteries, nickel–metal hydride batteries, primary batteries, Li-ion portable batteries and Li-ion traction batteries have all been tested down to the demo scale, and some of them have been industrially implemented as BAT (Best Available Technology). Prof. Friedrich has authored more than 75 publications, including his contributions to international conferences. His work was awarded in 2008 with Europe's highest endowed industry prize of the German Business Association, Dusseldorf, and in 2012 with the "German Resource Efficiency Prize" of the Federal Ministry of Economics and Energy as an outstanding example of raw material and material-efficient products, processes or services and application-oriented research results.

Editorial

Sustainable Utilization of Metals-Processing, Recovery and Recycling

Bernd Friedrich

IME Process Metallurgy and Metal Recycling Department, RWTH Aachen University, 52056 Aachen, Germany; BFriedrich@metallurgie.rwth-aachen.de

Received: 1 July 2019; Accepted: 2 July 2019; Published: 10 July 2019

1. Introduction and Scope

Our modern everyday life and thus our technical progress is based on a variety of metals. For example, computer chips and smartphones contain up to 60 different metals in various compounds and concentrations. The secure supply of our industry with these raw materials has become a strategic element of global politics and always leads to conflicts of interest between economics and ecology, as well as social needs.

Georesources, i.e., metal-containing primary raw materials, will continue to provide the bulk of supply for a long time, but deposits will become poorer and more complex. In the wake of Europe's demand for a "circular economy", the reuse of metal-containing materials or their elements themselves is becoming increasingly important. However, recycling in a growing demand environment, especially in the Asian region, can never close the gap even with theoretically complete recirculation. In this respect, the sustainable use of our metals is very important. However, that does not mean recycling at any price, as there is always an optimum balance between resource use (energy, materials, personnel, land, water, etc.) and the resource proceeds of every metal extraction process.

The high demand on advanced metallic materials raises the need for an extensive recycling of metals and a more sustainable use of raw materials. Advanced materials are crucial for technological applications, coexisting with an increasing scarcity of natural resources. This Special Issue, "Sustainable Utilization of Metals - Processing, Recovery and Recycling", is dedicated to the latest scientific achievements in efficient production of metals, purposing a sustainable resource use. Research centers from three continents present in 25 research papers and two review papers the results of their work in recent years on this topic.

These also include primary raw materials directly, waste from past mining and processing operations, metallurgical slags and end-of-life products such as Waste Electric and Electronic Equipment (WEEE) or batteries. Depending on the country situation, the implementation of new processes or the use of new substances in existing plants requires adapted technologies, in particular to create or maintain jobs in less industrialized regions, thus ensuring social peace.

2. Contributions

The idea of circular economy is the point of origin for contributions, aiming on the recirculation of metal-rich waste streams—such as WEEE, multi-metal alloys and composite materials—back into metal production. This topic goes along with pursuing the holistic use of input materials, resulting in the avoidance of waste by-products. In order to minimize material losses and energy consumption, this issue explores concepts for the optimization concerning the interface between mechanical and thermal pre-treatment and metallurgical processes. Furthermore, the direct re-use of complex alloys and composite materials without splitting them up into their single constituents is taken into account.

Papers in this issue are also engaged with the question of how the properties of indispensable advanced materials and alloys can be preserved by a more responsible input or even avoidance of particular constituents. In this regard, new approaches in material design, structural engineering and substitution are provided.

Considering both principal aspects—circular economy and material design—the recovery and the use of minor metals play an essential role, since their importance for technological applications often goes along with a lack of supply on the world market. Additionally, their ignoble character, as well as their low concentration in recycling materials cause a low recycling rate of these metals, awarding them the status of "critical metals". The research of this increasingly important material group will be discussed in this Special Issue in seven research papers [1–7].

Also classified as critical metals but included in a separate category is the group of Rare Earth Elements (REE). Recovery of these elements and thus securing of supply of raw materials independently of non-European market is still the focus of research today. Four papers deal with the recovery of REE [8–11].

Base and precious metals will be more and more in the focus of future research, as primary deposits will become poorer and more complex and thus winning is not that easy. Consumer and production wastes show excellent recovery opportunities for this group of metals. That is why eleven papers deal with the recovery of base metals [12–22] and one of precious metals [23].

Also, the recycling from complex systems like WEEE or batteries is in focus in this Special Issue. In these systems selective extraction of several metals in one step is not feasible or even goal of research. Numerous groups of metals are extracted simultaneously and treated in further steps to be separated optimally from multicomponent systems. For this reason, four papers show an overlap over multiple metal groups, base, precious and critical metals [24–27].

Conflicts of Interest: The author declare no conflict of interest.

References

1. Flerus, B.; Swiontek, T.; Bokelmann, K.; Stauber, R.; Friedrich, B. Thermochemical Modelling and Experimental Validation of In Situ Indium Volatilization by Released Halides during Pyrolysis of Smartphone Displays. *Metals* **2018**, *8*, 1040. [CrossRef]
2. Bonomi, C.; Alexandri, A.; Vind, J.; Panagiotopoulou, A.; Tsakiridis, P.; Panias, D. Scandium and Titanium Recovery from Bauxite Residue by Direct Leaching with a Brønsted Acidic Ionic Liquid. *Metals* **2018**, *8*, 834. [CrossRef]
3. Kaya, Ş.; Peters, E.M.; Forsberg, K.; Dittrich, C.; Stopic, S.; Friedrich, B. Scandium Recovery from an Ammonium Fluoride Strip Liquor by Anti-Solvent Crystallization. *Metals* **2018**, *8*, 767. [CrossRef]
4. Ebner, T.; Luidold, S.; Honner, M.; Antrekowitsch, H.; Czettl, C. Conditioning of Spent Stripping Solution for the Recovery of Metals. *Metals* **2018**, *8*, 757. [CrossRef]
5. Vind, J.; Alexandri, A.; Vassiliadou, V.; Panias, D. Distribution of Selected Trace Elements in the Bayer Process. *Metals* **2018**, *8*, 327. [CrossRef]
6. Chen, W.-S.; Ho, H.-J. Recovery of Valuable Metals from Lithium-Ion Batteries NMC Cathode Waste Materials by Hydrometallurgical Methods. *Metals* **2018**, *8*, 321. [CrossRef]
7. Kaya, Ş.; Dittrich, C.; Stopic, S.; Friedrich, B. Concentration and Separation of Scandium from Ni Laterite Ore Processing Streams. *Metals* **2017**, *7*, 557. [CrossRef]
8. Borra, C.R.; Vlugt, T.J.H.; Spooren, J.; Nielsen, P.; Yang, Y.; Offermann, S.E. Characterization and Feasibility Studies on Complete Recovery of Rare Earths from Glass Polishing Waste. *Metals* **2019**, *9*, 278. [CrossRef]
9. Bourbos, E.; Karanronis, A.; Sygellou, L.; Paspaliaris, I.; Panias, D. Study of Nd Electrodeposition from the Aprotic Organic Solvent Dimethyl Sulfoxide. *Metals* **2018**, *8*, 803. [CrossRef]
10. Zhang, B.; Fan, Y.; Liu, C.; Ye, Y.; Jiang, M. Reduction Characteristics of Carbon-Containing REE–Nb–Fe Ore Pellets. *Metals* **2018**, *8*, 204. [CrossRef]
11. Borra, C.R.; Vlugt, T.J.H.; Yang, Y.; Offerman, S.E. Recovery of Cerium from Glass Polishing Waste: A Critical Review. *Metals* **2018**, *8*, 801. [CrossRef]
12. Shishkin, A.; Mironovs, V.; Vu, H.; Novak, P.; Baronins, J.; Polyakov, A.; Ozolins, J. Cavitation-Dispersion Method for Copper Cementation from Wastewater by Iron Powder. *Metals* **2018**, *8*, 920. [CrossRef]

13. Reimer, M.V.; Schenk-Mathes, H.Y.; Hoffmann, M.F.; Elwert, T. Recycling Decisions in 2020, 2030, and 2040—When Can Substantial NdFeB Extraction be Expected in the EU? *Metals* **2018**, *8*, 867. [CrossRef]
14. Gnatko, M.; Li, C.; Arnold, A.; Freidrich, B. Purification of Aluminium Cast Alloy Melts through Precipitation of Fe-Containing Intermetallic Compounds. *Metals* **2018**, *8*, 796. [CrossRef]
15. Andersson, A.; Gullberg, A.; Kullerstedt, A.; Sandberg, E.; Andersson, M.; Ahmed, H.; Sundqvist-Ökvist, L.; Björkman, B. A Holistic and Experimentally-Based View on Recycling of Off-Gas Dust within the Integrated Steel Plant. *Metals* **2018**, *8*, 760. [CrossRef]
16. Guo, L.; Wen, X.; Bao, Q.; Guo, Z. Removal of Tramp Elements with 7075 Alloy by Super-Gravity Aided Rheorefining Method. *Metals* **2018**, *8*, 701. [CrossRef]
17. Diaz, F.; Wang, Y.; Moorthy, T.; Friedrich, B. Degradation Mechanism of Nickel-Cobalt-Aluminum (NCA) Cathode Material from Spent Lithium-Ion Batteries in Microwave-Assisted Pyrolysis. *Metals* **2018**, *8*, 565. [CrossRef]
18. Palimąka, P.; Pietrzyk, S.; Stępień, M.; Ciećko, K.; Nejman, I. Zinc Recovery from Steelmaking Dust by Hydrometallurgical Methods. *Metals* **2018**, *8*, 547. [CrossRef]
19. Gauffin, A.; Pistorius, P.C. The Scrap Collection per Industry Sector and the Circulation Times of Steel in the U.S. between 1900 and 2016, Calculated Based on the Volume Correlation Model. *Metals* **2018**, *8*, 338. [CrossRef]
20. Małecki, S.; Gargul, K. Low-Waste Recycling of Spent CuO-ZnO-Al_2O_3 Catalysts. *Metals* **2018**, *8*, 177. [CrossRef]
21. Fernández-González, D.; Sancho-Gorostiaga, J.; Piñuela-Noval, J.; Verdeja González, L.F. Anodic Lodes and Scrapings as a Source of Electrolytic Manganese. *Metals* **2018**, *8*, 162. [CrossRef]
22. Mohamed, S.R.; Friedrich, S.; Friedrich, B. Refining Principles and Technical Methodologies to Produce Ultra-Pure Magnesium for High-Tech Applications. *Metals* **2019**, *9*, 85. [CrossRef]
23. Schosseler, J.; Trentmann, A.; Friedrich, B.; Hahn, K.; Wotruba, H. Kinetic Investigation of Silver Recycling by Leaching from Mechanical Pre-Treated Oxygen-Depolarized Cathodes Containing PTFE and Nickel. *Metals* **2019**, *9*, 187. [CrossRef]
24. Wang, X.; Fellmann, J.; Jokilaakso, A.; Klemettinen, L.; Marjakoski, M. Behavior of Waste Printed Circuit Board (WPCB) Materials in the Copper Matte Smelting Process. *Metals* **2018**, *8*, 887.
25. Rämä, M.; Nurmi, S.; Jokilaakso, A.; Klemettinen, L.; Taskinen, P.; Salminen, J. Thermal Processing of Jarosite Leach Residue for a Safe Disposable Slag and Valuable Metals Recovery. *Metals* **2018**, *8*, 744. [CrossRef]
26. Kamberović, Ž.; Ranitović, M.; Korać, M.; Andjić, Z.; Gajić, N.; Djokić, J.; Jevtić, S. Hydrometallurgical Process for Selective Metals Recovery from Waste-Printed Circuit Boards. *Metals* **2018**, *8*, 441. [CrossRef]
27. Steinlechner, S.; Antrekowitsch, J. Thermodynamic Considerations for a Pyrometallurgical Extraction of Indium and Silver from a Jarosite Residue. *Metals* **2018**, *8*, 335. [CrossRef]

© 2019 by the author. Licensee MDPI, Basel, Switzerland. This article is an open access article distributed under the terms and conditions of the Creative Commons Attribution (CC BY) license (http://creativecommons.org/licenses/by/4.0/).

Article

Thermochemical Modelling and Experimental Validation of In Situ Indium Volatilization by Released Halides during Pyrolysis of Smartphone Displays

Benedikt Flerus [1,2,*], Thomas Swiontek [1], Katrin Bokelmann [2], Rudolf Stauber [2] and Bernd Friedrich [1]

1. Institute of Process Metallurgy and Metal Recycling IME, RWTH Aachen University, Intzestraße 3, 52056 Aachen, Germany; thomas.swiontek@accurec.de (T.S.); bfriedrich@ime-aachen.de (B.F.)
2. Project Group, Materials Recycling and Resource Strategies IWKS, Fraunhofer Institute for Silicate Research ISC, Brentanostraße 2, 63755 Alzenau, Germany; katrin.bokelmann@isc.fraunhofer.de (K.B.); rudolf.stauber@isc.fraunhofer.de (R.S.)
* Correspondence: bflerus@ime-aachen.de; Tel.: +49-(0)241-80-95856

Received: 22 November 2018; Accepted: 5 December 2018; Published: 8 December 2018

Abstract: The present study focuses on the pyrolysis of discarded smartphone displays in order to investigate if a halogenation and volatilization of indium is possible without a supplementary halogenation agent. After the conduction of several pyrolysis experiments it was found that the indium evaporation is highly temperature-dependent. At temperatures of 750 °C or higher the indium concentration in the pyrolysis residue was pushed below the detection limit of 20 ppm, which proved that a complete indium volatilization by using only the halides originating from the plastic fraction of the displays is possible. A continuous analysis of the pyrolysis gas via FTIR showed that the amounts of HBr, HCl and CO increase strongly at elevated temperatures. The subsequent thermodynamic consideration by means of FactSage confirmed the synergetic effect of CO on the halogenation of indium oxide. Furthermore, HBr is predicted to be a stronger halogenation agent compared to HCl.

Keywords: pyrolysis; smartphone; displays; halogenation; indium; volatilization; thermodynamics; recycling

1. Introduction

During the last decade, the role of waste electric and electronic equipment (WEEE) as a feedstock has become increasingly important for European metal refineries. Both its significant domestic supply and its high metal content compared to primary resources have made it an attractive raw material for the recovery of valuable metals—especially copper and precious metals (Au, Ag). Moreover, WEEE contains a broad variety of several other metals, ranging from base metals (Fe, Zn, Sn, Al, Pb) to special metals (Ga, Ge, In, Ta, rare earth elements), whereby the content and the actual occurrence of the individual metals depend on the particular kind of WEEE. The metal content of a smartphone, for instance, differs strongly from that of a washing machine, resulting in a high heterogeneity of the total WEEE stream. Following the example of the smartphone, which represents a contemporary, widely distributed type of electronic consumer product, it can be stated that this kind of device exhibits a high complexity—not only in terms of various metals, but also regarding other sorts of materials, like glass, ceramics and plastics. This complexity makes holistic recycling and metal recovery a big challenge. Currently, WEEE with a high content of copper and precious metals is introduced into the pyrometallurgical copper route which is able to handle larger amounts of feedstock. However, the chance of an easy recovery of precious metals is hindered by the high diversity of additional materials and elements, which leads to an increased input of impurities into the copper phase and requires an extensive slag design during the smelting process. Simultaneously, less noble trace metals, such as indium, tantalum and gallium, are lost in the slag and cannot be recovered [1,2]. From the environmental point of view, the processing of WEEE in copper smelters makes high demands of the off-gas treatment because the combustion of adhering plastics releases harmful substances such as dioxins and halides. The regard of all these aspects induces the consideration of appropriate techniques that can be applied prior to the smelting process in order to remove unwanted substances or separate particular elements which cannot be recovered in the smelter. Existing mechanical dissembling and sorting processes reach their limit when they are faced with composite materials and miniaturized components, i.e., smartphones and printed circuit boards. At this time, a thermal pretreatment of the electronic scrap via pyrolysis is a promising way to overcome these barriers. In this context, the work of Diaz et al. [3] shows that the pyrolysis of adhering plastics has several benefits:

- breakup of plastic-metal composites, so that a mechanical separation of a concentrated metal fraction is possible;
- removal of harmful organic substances and corrosive halides;
- production of a high-caloric pyrolysis gas that can be used as fuel or reduction agent.

Additional to these aspects, the process of pyrolysis enables not only the volatilization of hydrocarbons and halides, but also specific metals because of their affinity to form volatile metal halides. Thus, this work focusses on the case of indium in terms of a complete indium separation during the pyrolysis of indium-containing WEEE.

1.1. ITO Displays

As soon as electronic devices equipped with a flat panel display are discarded, indium is introduced into the stream of WEEE. Incorporated into the structure of indium tin oxide (ITO), indium is an indispensable compound to realize the functionality of different types of flat displays which are used for televisions, PCs, tablets and smartphones [4]. In 2017, the European Commission confirmed indium's status as critical because of its insecure import reliance and a recycling rate of 0% [5]. Regarding the latter, and keeping in mind that there is a domestic indium supply by way of discarded flat panel displays, an uncomplicated and cost-efficient way to include a pyrolytical indium separation step into the pyrometallurgical recycling route for WEEE by keeping all the other mentioned benefits of pyrolysis should be considered.

The most widespread technology for flat panel displays in electronic consumer products (i.e., televisions) is LCD technology (liquid crystal display). However, with respect to other more modern devices such as smartphones, OLED technology (organic light emitting diode) is becoming more established as a standard. In both cases, the displays are composed of several glass and polymer layers: lid (glass), polarizer film (polymer), active layer (LCD or OLED), ITO film (glass or polymer substrate), thin film transistors (glass substrate) and optical layers (polymer) [6]. A visualization of this sandwich construction is shown in Figure 1. The exact number and combination of layers depend on the technology, as well as the manufacturer, and will not be examined further. Nevertheless, the organic polymer films, and especially the brominated flame retardants (BFR) enclosed in the polymer structure, play a fundamental role for this work, which is why their chemical composition is taken into account. Due to their high transparency and their good mechanical and electrical properties, polyethylene terephthalate (PET), polycarbonate (PC), polyethylene naphthalate (PEN) and polymethyl methacrylate (PMMA) are appropriate materials to be used in flat displays [7,8]. For the polarizer film, a polymer based on cellulose triacetate (CTA), is state of the art [9]. With regard to PC, PET and PEN, it has been reported that their structure allows the application of BFR [10]. Coming back to the idea of the pyrolytic pretreatment of WEEE, the question of using bromine, originating from the BFR, as a halogenation agent for the formation of volatile indium bromide remains.

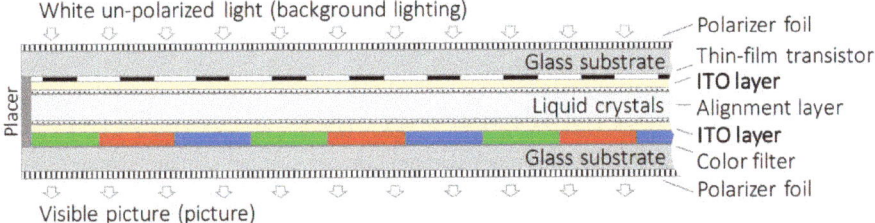

Figure 1. Sandwich construction of a liquid crystal display (LCD) [9].

1.2. Halogenation of Metals

Generally, there are several types of halogenation reactions, depending on the number and nature of reactants in the system. Nonetheless, all have the transformation of a metal oxide to a metal halide in common, which can be seen in the following equations, where the chlorination of a bivalent metal oxide is carried out [11]:

$$MeO + Cl_2 \Longleftrightarrow MeCl_2 + \frac{1}{2}O_2 \quad (1)$$

$$MeO + C + Cl_2 \Longleftrightarrow MeCl_2 + CO \quad (2)$$

$$MeO + 2HCl \Longleftrightarrow 2MeCl_2 + H_2O \quad (3)$$

As can be seen from Equations (1)–(3), both the use of pure chlorine (halogenation) or gaseous HCl (hydrohalogenation) is possible. Furthermore, chlorine can be replaced by bromine without changing the stoichiometry of the reaction agents, however, it has to be taken into account that the thermodynamics of the reactions are slightly different. As it will be explained later, the halogenation reactions are strongly affected by temperature and other gas components occurring in the system. In this context, the formation of so-called subhalides plays an important role.

1.3. Current Research on Indium Volatilization from Displays through Chlorination

So far, there have been several scientific works on the halogenation and volatilization of indium from flat screens. All of these were conducted by using HCl as the halogenation agent. Ma et al. [12] examined the vacuum chlorination of LCD glass powder, whereby HCl was generated by the thermal decomposition of NH_4Cl, which was blended with the glass powder. Prior to the chlorination process, the polymers from the displays were removed via a pyrolysis step. The results of the chlorination showed that the indium recovery increased with the applied temperature due to the higher vapor pressure of $InCl_3$. As a second significant influencing factor, a smaller particle size of the LCD powder was identified to increase the indium recovery.

Similar investigations were carried out by Terakado et al. [13], who also used NH_4Cl and concluded that high temperatures support the chlorination/volatilization. For this work, the input material was synthesized by covering soda glass with ITO, so that no polymers were involved. The temperature was adjusted between 400 and 800 °C under normal pressure. Additionally, Terakado et al. found that the addition of small amounts of carbon powder had a beneficial effect on the indium recovery. In contrast to Ma et al., a longer milling time of the glass and thus a smaller particle size lowered the rate of indium recovery.

Another concept of HCl generation for the chlorination was followed by Kameda et al. [14], who used the products of the thermal decomposition of polyvinyl chloride (PVC) to create a HCl-rich atmosphere. The experiments of Kameda et al. were conducted under air as well as under an inert nitrogen atmosphere, resulting in a higher indium volatilization when nitrogen was applied. Furthermore, the In recovery from pure In_2O_3 was lower than using LCD powder as input material.

Similar to Me et al., Takahashi et al. [15] removed the polymer fraction from the milled LCD displays via incineration. Afterwards, the LCD powder was moistened with aqueous HCl and dried so that water was removed and HCl remained in the solid material. The thermal process was executed between 400 and 700 °C. Like the previous research, high temperatures and a nitrogen-rich atmosphere promoted the formation and evaporation of $InCl_3$.

1.4. Motivation and Innovative Approach of this Work

In each of the papers presented above, a supplementary chlorination agent was used without consideration of the polymer fraction from the LCD displays. However, this polymer fraction probably contains BFR, which can influence the whole process and act as a halogenation agent in situ unless the polymers are not removed. In one of our preceding papers, we investigated the temperature dependence of the gaseous and solid pyrolysis products during the thermal decomposition of printed circuit boards (PCB) [16]. It was found that elevated temperatures and a high heating rate have a considerable influence on the decomposition mechanism of the organics and thus on the composition of the pyrolysis gas. One important aspect was the formation of HBr at 700 °C during the decomposition of BFR in the epoxy resin. Hence, it is expected that this HBr generation will also occur during the pyrolysis of LCDs.

Therefore, this work focuses on the possibility of an in situ halogenation of indium oxide by HBr. A special point of interest is the influence of other gaseous pyrolysis products on the halogenation reaction. At this point, carbon monoxide (CO) is a matter of particular interest because its formation increases with higher temperatures. According to Peek [11] and Grabda et al. [17], the presence of CO has a synergetic effect on the thermodynamics of the halogenation reaction.

2. Materials and Methods

2.1. Materials

The input materials for the experimental work were displays from 20 discarded smartphones of different brands. First, the devices were dismantled manually so that the multi-layer displays, having an overall mass of 458 g, could be removed. To obtain a bulk material with a homogenous chemical composition, the displays were processed in a Fritsch Pulverisette 25/19 cutting mill, which resulted in a maximum particle size of 4 mm. The final comminution was completed via cryogenic grinding in a SPEX 6870D Freezer/Mill. Concerning the chemical analysis, the concentrations of most elements were measured by means of an ICP–OES (Perkin Elmer Optima 8300, Waltham, MA, USA), whereas Br and Cl were analyzed via RFA (PANalytical Axios, Malvern Panalytical, Almelo, The Netherlands). The total carbon (TC) was measured in a LECO R612 (LECO, St. Joseph, CT, USA) multiphase carbon determinator. Table 1 contains the results of the elemental analysis. Concerning the results for bromine and chlorine, a certain discrepancy has to be accepted because no appropriate standard material is available to fulfil an appropriate calibration and thus a precise analysis.

Due to the high share of polymer films, the comminuted product was not a flowable powder but a dense, rubbery fluff. In order to ensure smooth feeding, the material was compressed to pellets with a diameter of 6 mm and a length of 20 mm. For the second test series, pure ammonium chloride was mixed with the display fluff, adjusting to a ratio of 1:100, which means that 10 g display fluff was mixed with 0.1 g NH_4Cl.

Table 1. Chemical analysis of milled smartphone displays.

Element	conc./ppm	conc./wt.-%	Method
In	130	0.013	ICP–OES
Sn	470	0.047	ICP–OES
Ba	2000	0.2	ICP–OES
Ca	18,200	1.82	ICP–OES
K	5500	0.55	ICP–OES
Mg	11,800	1.18	ICP–OES
Na	35,500	3.55	ICP–OES
Sr	1400	0.14	ICP–OES
Al	41,100	4.11	ICP–OES
Si	17,400	17.4	ICP–OES
Br	50	0.005	RFA
Cl	100	0.010	RFA
C	231,000	23.100	TC

2.2. Experimental Setup and Procedure

The pyrolysis experiments were performed in a closed 1.5 L stainless steel reactor, which was placed in an electric resistance furnace, as can be seen in Figure 2. To prepare for the insertion of further equipment (thermocouple, charging tube and probe head for the off-gas analyzer) the lid was equipped with four gastight lead-throughs. Additionally, the lid was water-cooled to avoid any damage of the rubber seals by thermal impact.

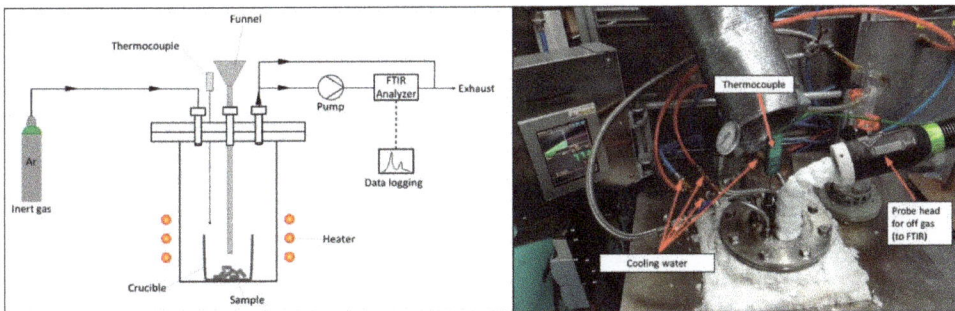

Figure 2. Schematic figure (**left**) and photo (**right**) of the experimental setup.

For the temperature measurement, a type K thermocouple was chosen. In order to maintain an oxygen-free atmosphere, the reactor was purged with a constant argon stream of 3 L/min. The test runs were conducted at various temperatures between 300 and 800 °C, raising the temperature in steps of 50 °C with each trial. After heating the reactor up to the desired temperature, 10 g of display pellets were charged into the hot reactor using a densely sintered alumina pipe and a funnel. On the bottom of the reactor, an alumina crucible was placed to collect the pellets. Subsequent to feeding, the pipe and the funnel were removed and the lead-through was closed with a steel plug to avoid any leakage of pyrolysis gas. For each temperature step, the above outlined procedure was repeated twice. Following the pyrolysis, it was necessary to let the reactor cool down slowly, before the solid residue could be removed from the crucible and homogenized in a ball mill. Finally, the concentrations of indium and tin were measured via ICP–OES. One separate series of trials was performed to measure the composition of the pyrolysis gas at 300, 500 and 700 °C. For this purpose, a gas pump, manufactured by Ansyco, was used to extract an off-gas volume of 2.0 L/min from the reactor. Downstream from the pump, the gas flew through a Gasmet DX4000 FTIR (Gasmet Technologies Oy, Helsinki, Finland) (Fourier transform infrared) gas analyzer which allowed the continuous detection and quantification of various compounds in the pyrolysis gas. Compounds of special interest were HBr, HCl and CO. The overrun gas flow was released from the reactor through a bypass.

3. Results and Discussion

In contrast to earlier studies on the pyrolytic volatilization of indium, the experimental setup used in this work did not allow the characterization of volatilized indium compounds. The water cooling of the lid resulted in the condensation of pyrolysis oil on its underside. Hence, in any case of indium halide vaporization, indium must occur in the condensed oil. However, there currently exists no successful procedure for an elemental analysis of the oil. Regarding a structural analysis of the solid pyrolysis residue via XRD, no satisfying results were obtained due to the high amount of carbon and glass, as well as the low content of indium compounds, in the powder. Therefore, the following results refer to the elemental analysis of the solid residue (via ICP–OES) and the off-gas analysis.

3.1. Mass Loss and Volatilization of ITO

The mass loss of the pyrolyzed materials can be seen in Figure 3. Apart from slight discrepancies at the beginning and the end, both graphs show an almost identical trend, reaching their maximum at 650 °C (35% for pure display powder) and 800 °C (39% for NH4Cl-addition).

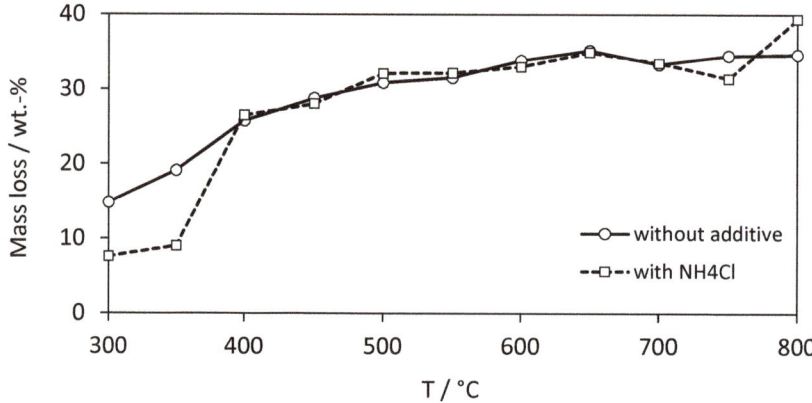

Figure 3. Mass loss during the pyrolysis of ground smartphone displays.

Figure 4 shows the concentration of indium and tin in the solid residue at several pyrolysis temperatures. The first value on the abscissa, which is labelled "NP", represents the indium content of the non-pyrolyzed material. It should also be taken into account that the detection limit of indium in the ICP–OES was 20 ppm. In the case of five samples (between 700 and 800 °C), the analysis resulted in indium concentrations below this limit. To illustrate these points in the diagram, the values were set to zero but it must be assumed that the actual indium concentrations are located somewhere between 0 and 19 ppm. Apart from that aspect, it can be seen clearly that the processing temperature has a strong influence on the volatilization of indium from the material whether or not NH_4Cl was added as a supportive chlorination agent. Looking at both graphs, no decrease of indium in the material can be observed up to 350 °C. After the application of higher temperatures, an increasing amount of indium left the material. The increase before can be due to the mass loss referring to the pyrolysis of polymers. Both graphs exhibit a similar course, where the graph showing the experiments with NH_4Cl addition keeps an average distance of 57 ppm below the graph without any additive. Therefore, the detection limit of 20 ppm was already reached at 700 °C, whereas the samples without NH_4Cl had to be processed at least at 750 °C to bring the indium content to the same level. Obviously, a complete volatilization of indium without any extra halide-providing substance is possible and it follows that there must be sufficient halides in the polymer fraction of the displays. According to Figure 4, a significant indium vaporization starts between 350 and 400 °C because the enrichment of indium by the act of pyrolysis of polymers is balanced.

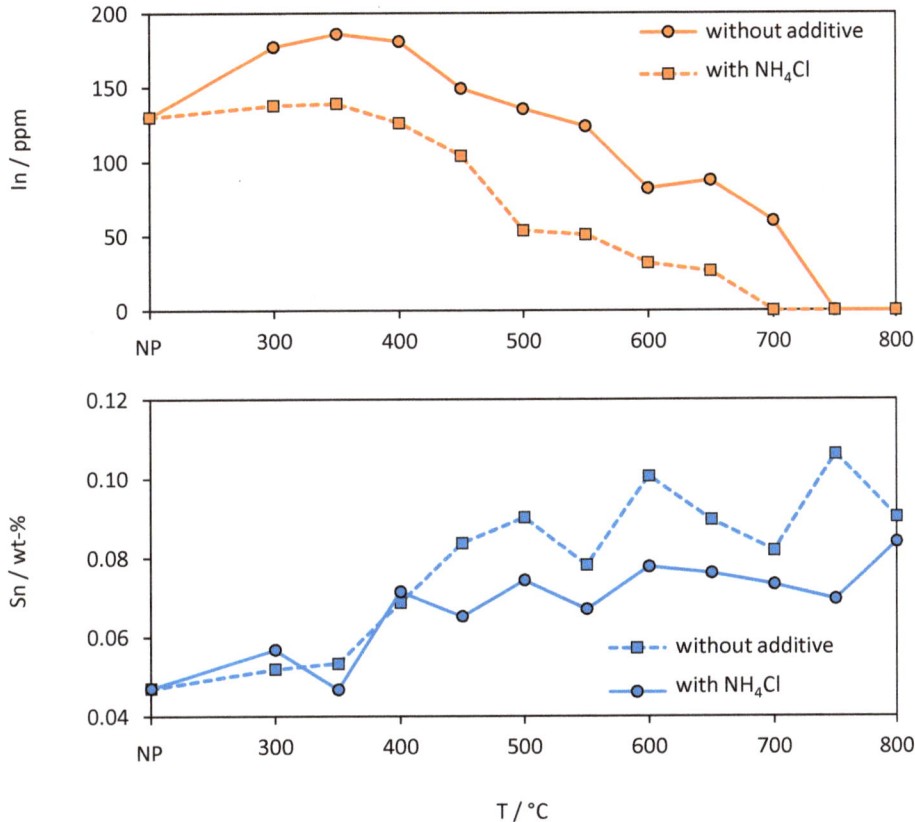

Figure 4. Concentration of In and Sn in the solid residue at different pyrolysis temperatures.

Concerning the analysis results of tin, the graphs in the second diagram of Figure 3 show an unsteady performance which is contrary to the results of indium. All in all, tin tends to be enriched in the material and no volatilization seems to happen. Comparing the tin concentration of the non-pyrolyzed material and the residue from the 800-°C trial, there is an increase of more than 90%, which is consistent with the mass loss of 35% (pure display powder) and 39% (NH_4Cl-addition) as can be obtained from Figure 2.

3.2. Off-Gas Analysis

The results of the off-gas analysis for CO, HBr and HCl during the pyrolysis of display powder at 300, 500 and 700 °C are depicted in Figure 5. While the off gas contained numerous other gaseous species—especially organic substances—for this work, the three named compounds appear to be the most important ones. The concentrations of each component are normalized to one gram of charged material. By comparing the three diagrams, the changing scale of both axes has to be respected. Basically, the course of the curves resembles the results from our previous work where printed circuit boards were pyrolyzed (Diaz et al. [2]). At 300 °C, little gas formation could be measured, at which no detection of HCl and HBr happened. The oscillating CO concentration around 8 ppm proves that there is a slow thermal decomposition which lasts for at least 50 min. With increasing the temperature to 500 °C and 700 °C, the amount of all gas components rose rapidly, particularly in the case of CO. Different to Diaz et al., a significant release of HBr was already observed at 500 °C but did not increase at 700 °C. This is different for HCl, which exhibits a strong increase from 500 °C to 700 °C. Concerning this temperature increase, the period from the first time of gas evolution to the last point of significant concentrations was reduced to a third. However, the graphs for HBr reveal some challenges regarding its detection in the gas phase. It must be kept in mind that the detection of HBr using FTIR is possible but is strongly affected by the sampling. From practice it is known that HBr tends to adhere to the inside surface of the sampling equipment so that the molecules reach the FTIR detector with a certain delay. Thus, its concentration–time record, as it is shown in Figure 5, is drawn-out as against the records for CO and HCl. It was also observed that the concentration of HBr continued to oscillate between 0 and 2 ppm although the formation of all other compounds, and therefore the pyrolysis itself, were finished.

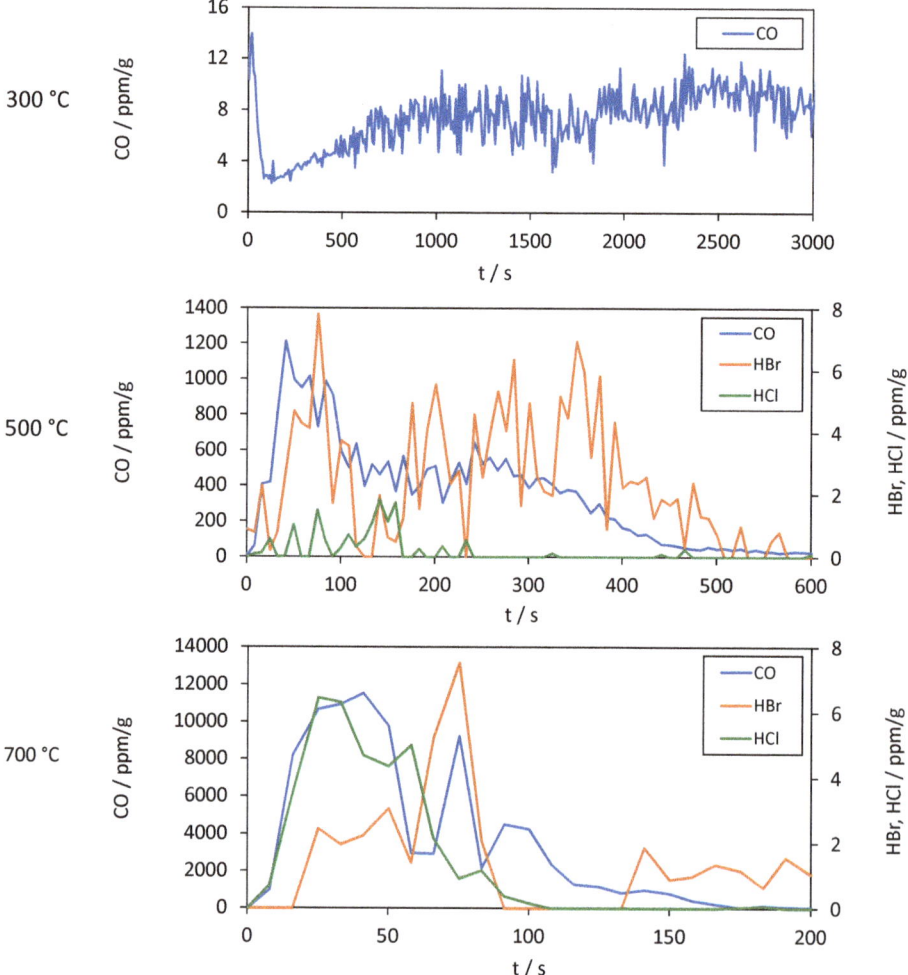

Figure 5. Concentrations of CO, HBr and HCl during the pyrolysis of display powder at 300, 500 and 700 °C.

Despite the limited HBr measurement, it can be stated that higher pyrolysis temperatures favor the formation of CO, HBr and HCl. On the one hand, this aspect confirms the results of the indium decline with increasing temperature (see Figure 4), but on the other hand, it is not known yet which compound—HBr or HCl—acted as the major halogenation agent. Although the pyrolysis reaction at 700 °C is three times faster compared to 500 °C, there seems to be no limitation in terms of kinetics of the volatilization. In other words, the halogenation reaction and evaporation are fast enough and do not require residence times of HBr/HCl and CO longer than 100 s.

3.3. Thermochemical Modelling

For a more detailed investigation on the halogenation reactions occurring during the pyrolysis of powdered smartphone displays, several thermochemical calculations were executed using the software FactSage™ 7.0 [18], which is provided by GTT Technologies. The background of these calculations are considerations of Ma et al. [3] concerning the formation of indium sub chlorides, which may depend on the presence of a reducing agent.

All calculations in FactSage™ were done by assuming ideal conditions, which means that there are no molecular interactions in the gas phase. The required thermodynamic data was received from the SGPS database for pure substances. As a result, two series of Gibbs free energy lines were obtained: one series for the hydrobromination and the other for the hydrochlorination of one mole In_2O_3 (see Figure 6). In fact, a calculation involving ITO would represent the real system rather than only In_2O_3. However, the available data bases of FactSage™ do not provide any data for ITO so a compromise had to be made.

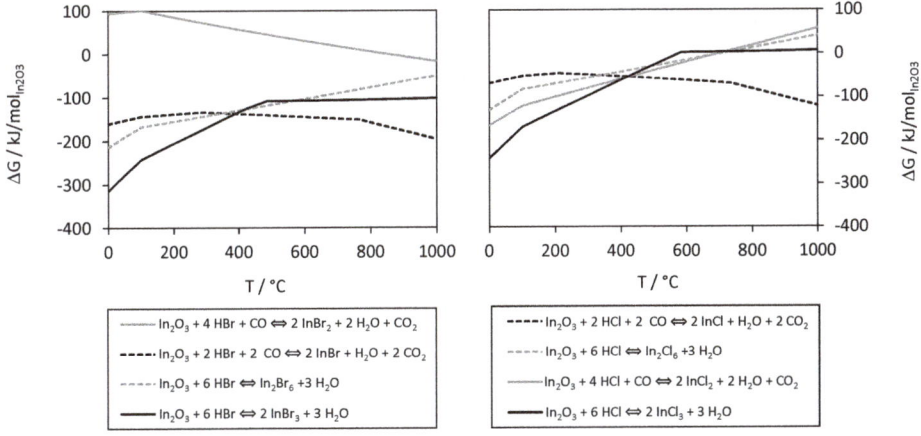

Figure 6. Gibbs free energy for hydrobromination (**left**) and hydrochlorination (**right**) of indium oxide.

All the graphs exhibit kinks at 100 °C, which indicates a phase change due to the evaporation of water at this temperature. Inflection points at higher temperatures represent the formation of halides in the gaseous state. In the cases of InCl and InBr, the melting point at 210 and 285 °C can be observed. Comparing the curves of the reactions, there is only a small difference between HBr and HCl, leading to the conclusion that HBr might be a slightly better halogenation agent due to the lower Gibbs free energy. As temperature increases, there are only three halogenation reactions, which become more favorable from the thermochemical point of view—as long as CO is added to the system in order to act as a reducing agent. Thus, if temperatures above 400 °C and sufficient CO is present, the generation of indium in the form of InBr and InCl seems to be the predominant mechanism. However, at 750 °C the formation of $InBr_3$ and In_2Br_6 is predicted to be also possible. As seen in Figure 3, 750 °C is the temperature where a complete volatilization of indium from the display powder without any additional NH$_4$Cl was achieved. The different courses of the solid grey lines, which represent the formation of InBr$_2$ and InCl$_2$, create doubt that the quality of the thermodynamic data is appropriate to reach a conclusion regarding the temperature-dependent formation of these particular compounds.

As a further step, the temperature dependence of the evaporation of the formed halides from Figure 6 will be investigated, because all halides can occur in different aggregate states and the success of the practical work is based on their evaporation. Therefore, the logarithmic vapor pressure lines of the different gaseous indium bromides and chlorides are shown in Figure 7.

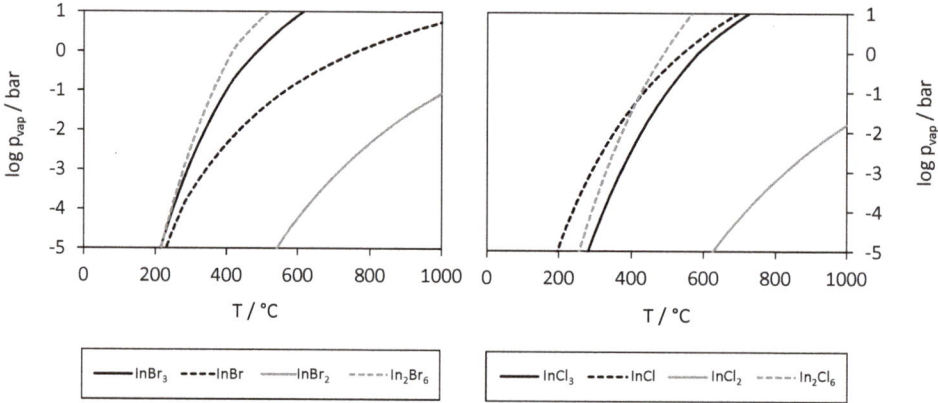

Figure 7. Vapor pressure curves of pure indium bromides (**left**) and indium chlorides (**right**).

If the curves are compared to the Gibbs free energy lines (see Figure 6), it becomes clear that there is a competition between the Gibbs free energy and the vapor pressures of the single components, which means that a halogenation reaction with a low Gibbs free energy forms a halide with a low vapor pressure (except InBr$_2$ and InCl$_2$). Therefore, the process of evaporation is another barrier which is strongly influenced by temperature. Assuming that mostly InBr is generated (according to Figure 6) it still requires sufficiently high temperatures to bring it to the gas phase.

To complete the contemplation of thermodynamics, we will have a closer look at some equilibrium calculations—especially with respect to the amount of supplied CO and the resulting proportion of the gaseous indium halides. In other words, the results from Figures 6 and 7 are combined in one diagram. For this purpose, a system was defined consisting of 1 mole of In$_2$O$_3$ and an excess amount of 6 moles each of HBr and HCl. The quantity of CO was varied from 0 to 5 moles in steps of 0.1 mole and the whole procedure was executed for 700 °C and 800 °C. Finally, Figure 8 illustrates the resulting data from the equilibrium calculations. Also, here attention must be paid to the different scale of the axes. Due to the great difference in their amount, bromides and chlorides are shown in two different diagrams, although they exist in the same system.

As already predicted by the Gibbs free enthalpy lines in Figure 6, the occurrence of indium chlorides in the gas phase is inhibited by the preferred formation of indium bromide and for all amounts of CO, both In$_2$Br$_6$ and In$_2$Cl$_6$ play only a minor role. In contrast, the proportion of InBr/InBr$_3$ shifts to higher values as the supply of CO increases. Raising the temperature from 700 °C (broken lines) to 800 °C (solid lines) intensifies this effect. This fact can be also observed concerning the formation of chlorides, but in a much lower scale as was previously mentioned. Again, it is theoretically proven that a CO-rich (reducing) atmosphere supports the formation of gaseous subhalides as well as the halogenation process. Referring to the results from the off-gas measurement (see Figure 5), a preferred formation of InBr seems to be the most probable mechanism at 700 °C because there is an enormous excess of CO compared to HBr and HCl.

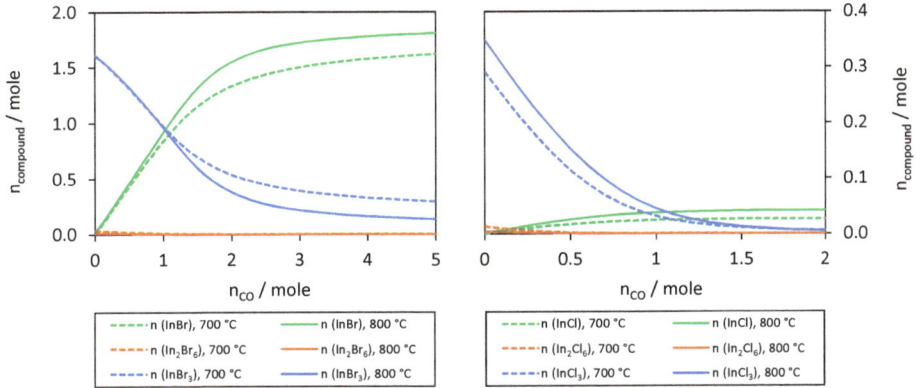

Figure 8. Proportion of formed indium halides by HBr (**left**) and HCl (**right**) depending on the amount of CO.

4. Conclusions

The aim of this work was to assess whether it is possible to volatilize indium from the display material of discarded smartphones without using an additional halogenation agent. This requirement was accomplished at a minimum processing temperature of 750 °C. At this temperature, significant amounts of HBr/HCl and CO were released from the plastic fraction to react with indium oxide. The theoretical study on thermodynamics confirmed the idea of a synergetic effect of CO on the halogenation at which In tends to be evaporated in the form of InBr/InCl rather than $InBr_3$/$InCl_3$ at low CO concentrations. Compared to HCl, HBr is slightly preferred in order to act as a halogenation agent from the thermochemical point of view. Due to the successful decrease of indium and the recorded concentrations of HCl and HBr in the off gas it can be concluded that the amount of halide in the displays' plastics is sufficient for a complete indium halogenation so that no additive is required. Of course, the indium volatilization can also be realized at lower temperatures by adding an excess of NH_4Cl, as was demonstrated in other papers [12–15], as well as in this work. However, both thermochemical calculations and experiments have shown that even though a successful halogenation would have happened, at least 700 °C is necessary to achieve a complete evaporation of the indium halides from the display material. As a final result, it can be stated that the success of a complete indium volatilization during pyrolysis depends on a chain of several mechanisms which are all promoted by high temperatures:

- thermal decomposition of plastics and the supply of sufficient HBr/HCl and CO;
- halogenation of indium oxide and the formation of (sub) halides depending on temperature and supply of CO;
- evaporation of formed (sub) halides.

Although no product material was collected, the process appears to be selective because the concentration of tin in the pyrolysis residue increased (due to of the mass loss of the material) which means that tin was not volatilized significantly. In this paper, the basic idea is to include the process of indium volatilization into a current recycling concept for WEEE with as little effort as possible. The main requirement for this, indeed, is the application of a pyrolysis step in the WEEE processing route which provides several other—probably more substantial—advantages [3,16]. Based on this fact, however, a simultaneous indium separation would be a great feature.

Author Contributions: B.F. (Benedikt Flerus) and T.S. conceived and designed the experiments; T.S. performed the experiments; B.F. (Benedikt Flerus) analyzed the data and wrote the paper with contributions of K.B.; R.S. and B.F. (Bernd Friedrich) supervised the work and contributed reagents, materials and analysis tools.

Conflicts of Interest: The authors declare no conflict of interest.

References

1. Forsén, O.; Aromaa, J.; Lundström, M. Primary Copper Smelter and Refinery as a Recycling Plant—A System Integrated Approach to Estimate Secondary Raw Material Tolerance. *Recycling* **2017**, *2*, 19. [CrossRef]
2. Lennartsson, A.; Engström, F.; Samuelsson, C.; Björkman, B.; Pettersson, J. Large-Scale WEEE Recycling Integrated in an Ore-Based Cu-Extraction System. *J. Sustain. Metall.* **2018**, *4*, 222–232. [CrossRef]
3. Diaz, F.; Florez, S.; Friedrich, B. High recovery recycling route of WEEE: The potential of pyrolysis. In Proceedings of the EMC European Metallurgical Conference, Düsseldorf, Germany, 14–17 June 2015; GDMB Verlag GmbH: Clausthal-Zellerfeld, Germany, 2015.
4. USGS National Minerals Information Center INDIUM. *Mineral Commodity Summaries*; U.S. Geological Survey: Reston, VA, USA, 2012.
5. European Commission. *The 2017 List of Critical Raw Materials for the EU*; European Commission: Brussels, Belgium, 2017.
6. Guenther, B.D.; Steel, D. *Encyclopedia of Modern Optics*, 2nd ed.; Elsevier Science & Technology: San Diego, CA, USA, 2018.
7. Aleksandrova, M. Specifics and Challenges to Flexible Organic Light-Emitting Devices. *Adv. Mater. Sci. Eng.* **2016**. [CrossRef]
8. Salhofer, S.; Spitzbart, M.; Maurer, K. Recycling of LCD Screens in Europe—State of the Art and Challenges. In Proceedings of the 18th CIRP International Conference on Life Cycle Engineering, Braunschweig, Germany, 2–4 May 2011. [CrossRef]
9. Ueberschaar, M.; Schlummer, M.; Jalalpoor, D.; Kaup, N.; Rotter, V. Potential and Recycling Strategies for LCD Panels from WEEE. *Recycl. Metals* **2017**, *2*, 7. [CrossRef]
10. Lassen, C.; Astrup Jensen, A.; Crookes, M.; Christensen, F.; Nyander Jeppesen, C.; Clausen, N.J.A.J.; Mikkelsen, S.H.M. *Survey of Brominated Flame Retardants*; The Danish Environmental Protection Agency: Copenhagen, Denmark, 2014.
11. Peek, E.M.L. Chloride Pyrohydrolysis. Lixiviant Regeneration and Metal Separation. Ph.D. Thesis, Techn. Univ., Delft, The Netherlands, 1996.
12. Ma, E.; Lu, R.; Xu, Z. An efficient rough vacuum-chlorinated separation method for the recovery of indium from waste liquid crystal display panels. *Green Chem.* **2012**, *14*, 3395. [CrossRef]
13. Terakado, O.; Iwaki, D.; Murayama, K.; Hirasawa, M. Indium Recovery from Indium Tin Oxide, ITO, Thin Film Deposited on Glass Plate by Chlorination Treatment with Ammonium Chloride. *Mater. Trans.* **2011**, *52*, 1655–1660. [CrossRef]
14. Kameda, T.; Park, K.-S.; Sato, W.; Grause, G.; Yoshioka, T. Recovery of indium from In_2O_3 and liquid crystal display powder via a chloride volatilization process using polyvinyl chloride. *Thermochim. Acta* **2009**, *493*, 105–108. [CrossRef]
15. Takahashi, K.; Sasaki, A.; Dodbiba, G.; Sadaki, J.; Sato, N.; Fujita, T. Recovering Indium from the Liquid Crystal Display of Discarded Cellular Phones by Means of Chloride-Induced Vaporization at Relatively Low Temperature. *Metall. Mater. Trans. A* **2009**, *40*, 891–900. [CrossRef]
16. Diaz, F.; Flerus, B.; Nagraj, S.; Bokelmann, K.; Stauber, R.; Friedrich, B. Comparative Analysis About Degradation Mechanisms of Printed Circuit Boards (PCBs) in Slow and Fast Pyrolysis: The Influence of Heating Speed. *J. Sustain. Metall.* **2018**, *4*, 205–221. [CrossRef]
17. Grabda, M.; Oleszek, S.; Shibata, E.; Nakamura, T. Distribution of inorganic bromine and metals during co-combustion of polycarbonate (BrPC) and high-impact polystyrene (BrHIPS) wastes containing brominated flame retardants (BFRs) with metallurgical dust. *J. Mater. Cycles Waste Manag.* **2018**, *20*, 201–213. [CrossRef]
18. Bale, C.W.; Bélisle, E.; Chartrand, P.; Decterov, S.A.; Eriksson, G.; Gheribi, A.E.; Hack, K.; Jung, I.H.; Kang, Y.B.; Melançon, J.; et al. FactSage Thermochemical Software and Databases -2010–2016. *Calphad* **2016**, *54*, 35–53. [CrossRef]

© 2018 by the authors. Licensee MDPI, Basel, Switzerland. This article is an open access article distributed under the terms and conditions of the Creative Commons Attribution (CC BY) license (http://creativecommons.org/licenses/by/4.0/).

Article

Scandium and Titanium Recovery from Bauxite Residue by Direct Leaching with a Brønsted Acidic Ionic Liquid

Chiara Bonomi [1,*], Alexandra Alexandri [1], Johannes Vind [1,2], Angeliki Panagiotopoulou [3,4], Petros Tsakiridis [1] and Dimitrios Panias [1,*]

1. School of Mining and Metallurgical Engineering, National Technical University of Athens, Iroon Polytechniou 9, Zografou Campus, 15780 Athens, Greece; aalexandri@metal.ntua.gr (A.A.); jvind@metal.ntua.gr (J.V.); ptsakiri@central.ntua.gr (P.T.)
2. Department of Continuous Improvement and System Management, Aluminium of Greece Plant, Metallurgy Business Unit, Mytilineos S.A., Agios Nikolaos, 32003 Viotia, Greece
3. Institute of Biosciences & Applications, National Centre for Scientific Research "Demokritos", Patr. Gregoriou E & 27 Neapoleos Str, Agia Paraskevi, 15310 Athens, Greece; apanagio@bio.demokritos.gr
4. Institute of Biosciences & Applications, National Centre for Scientific Research "Demokritos", Neapoleos 10, Agia Paraskevi, 15310 Athens, Greece
* Correspondence: bonomich@metal.ntua.gr (C.B.); panias@metal.ntua.gr (D.P.); Tel.: +30-210-7724054 (C.B. & D.P.)

Received: 20 September 2018; Accepted: 15 October 2018; Published: 17 October 2018

Abstract: In this study, bauxite residue was directly leached using the Brønsted acidic ionic liquid 1-ethyl-3-methylimidazolium hydrogensulfate. Stirring rate, retention time, temperature, and pulp density have been studied in detail as the parameters that affect the leaching process. Their optimized combination has shown high recovery yields of Sc, nearly 80%, and Ti (90%), almost total dissolution of Fe, while Al and Na were partially extracted in the range of 30–40%. Si and rare earth element (REEs) dissolutions were found to be negligible, whereas Ca was dissolved and reprecipitated as $CaSO_4$. The solid residue after leaching was fully characterized, providing explanations for the destiny of REEs that remain undissolved during the leaching process. The solid residue produced after dissolution can be further treated to extract REEs, while the leachate can be subjected to metal recovery processes (i.e., liquid–liquid extraction) to extract metals and regenerate ionic liquid.

Keywords: bauxite residue; red mud; ionic liquids; scandium recovery; titanium recovery

1. Introduction

Bauxite residue (BR), also known as red mud, is the major byproduct of the Bayer process for alumina production, produced by the alkali leaching of bauxite. On average, for each metric ton of alumina, 1–1.5 metric tons of BR are generated [1,2], which leads to a global production of over 150 million metric tons per year [1,3].

BR composition can differ depending on the type of bauxite ore from which alumina are produced and Bayer processing techniques [4,5]. During the Bayer process, valuable base and trace elements like iron (Fe), some aluminum (Al), titanium (Ti), and rare earth elements (REEs) remain in the bauxite residue. As a consequence, REEs are enriched with a factor of about 2 in BR comparing to the initial ore [6,7]. Particularly interesting is the case of scandium (Sc), as its concentration in BR (in Greek BR accounts to 130 ppm on average) is much higher than in the Earth's crust (22 ppm on average) [8]; that means a notable enrichment of Sc in BR. Due to the high market price (Sc_2O_3—4600 US$/kg, 99.99% purity, in 2017) [9], Sc may represent 95% of the economic value of rare earths in BR [10]. It has also been listed as a critical raw material by the European Commission due to its high economic importance and supply risk [11]. In fact, Sc is mainly produced as a byproduct during the processing of various ores, from titanium and REEs ores (China), uranium ore (Kazakhstan and Ukraine), and apatite ore (Russia). It can also be recovered from previously processed tailings or residues [9,12,13]. For these reasons, BR can be accounted as a secondary raw material source [14], and the recovery of Sc could represent a high economic interest.

BR can also be considered a secondary source for Ti, which is a photocatalyst and it is applied in the white pigment industry [15]. Since the availabilities and qualities of Ti ores are decreasing [16], it is important to find methods for extracting Ti from secondary sources.

Many studies, patents, and pilot scale implementations have been carried out for Sc and Ti recovery from BR, mainly by investigating hydrometallurgical or combined pyro-hydrometallurgical processes [5,12,16–23], but none of them has reached an industrial scale. Nowadays, the impact of the zero-waste valorization policy motivates the research community on finding innovative, greener, and economical viable routes for metal extraction from complex polymetallic matrices, such as the bauxite residue [24].

Ionometallurgical approach can be exploited as an alternative to conventional hydrometallurgical processing. The term ionometallurgy indicates the use of ionic liquids (ILs) as solvents in metals processing. ILs are liquid at room temperature and consist solely of ions; generally an organic cation and inorganic/organic anion. ILs have superior properties against conventional organic solvents, such as nonflammability, a wide electrochemical window, high thermal stability, negligible vapor pressure, and low volatility [25]. For these reasons and thanks to the vast number of combinations of the cation and the anion during synthesis, ILs have potential for many applications, such as solvent extraction [26,27], catalytic reactions [28,29], and electrodeposition of metals [30,31]. In the past few decades, ILs have been used also as lixiviants for metals dissolution [25,32–35]. Applying ionic liquid leaching on secondary raw material resources eventually improves efficiency yields, reduce waste effluent, and increases selectivity.

The aim of this work is to investigate the direct leaching of bauxite residue by using a Brønsted acidic ionic liquid, achieving high Ti and Sc recovery yields. To optimize the process, several parameters were studied. Moreover, solid residue after leaching was fully characterized and explanations of the destiny of REEs were given.

2. Materials and Methods

Bauxite residue was provided by Aluminium of Greece (Mytilineos S.A.), dried at 100 °C overnight, homogenized, and split in order to take a representative sample that was next crushed and ground. The sample was then subjected to chemical, mineralogical and physical characterization. Chemical composition was analyzed after complete dissolution of the sample via fusion method: 0.1 g of BR was mixed with 1.5 g of $Li_2B_4O_7$ and 0.1 g of KNO_3 and then fused at 1000 °C for 1 h, followed by dissolution in HNO_3 10% v/v. The main elements were identified by a Perkin Elmer 2100 Atomic Absorption Spectrometer (AAS) (Waltham, MA, USA), while minor elements were analyzed by a Thermo Fisher Scientific™ X-series 2 Inductively Coupled Plasma Mass Spectrometer (ICP-MS) (Waltham, MA, USA) and a Perkin Elmer Optima 8000 Inductively Coupled Plasma Atomic Emission Spectrometer (ICP-OES) (Waltham, MA, USA). The calcium oxide content was measured in the solid sample with a Spectro Xepos Energy Dispersive X-ray fluorescence spectroscopy (SPECTRO, Kleve, Germany) (ED-XRF). Mineralogical characterization was performed with a Bruker D8 focus X-ray powder diffractometer (XRD) (Bruker, Billerica, MA, USA) with nickel-filtered CuKa radiation, and quantitative evaluation was done via profile fitting by using XDB Powder Diffraction Phase Analytical System version 3.107 that targets specifically bauxite and bauxite residue [36,37]. Particle size analysis was carried out by a Malvern Mastersizer TM Laser particle size analyzer (Malvern Instruments, Malvern, UK).

The ionic liquid 1-ethyl-3-methylimidazolium hydrogensulfate ([Emim][HSO_4]) was supplied by Iolitec (Iolitec Ionic Liquids Technologies, Heilbronn, Germany) with >98% purity and characterized.

Infrared measurements were conducted with a Perkin Elmer FTIR spectrum 100 (Waltham, MA, USA). Viscosity analysis was performed with a Brookfield viscometer DV-I + LV supported by a Brookfield Thermosel accessory (Brookfield Ametek, Harlow, UK). Nuclear Magnetic Resonance (NMR) spectra were obtained in DMSO-d_6 at 25 °C on a Bruker Avance DRX 500 MHz (Bruker Biospin, Germany) (^{1}H at 500.13 MHz and ^{13}C at 125.77 MHz) equipped with a 5 mm multi nuclear broad band inverse detection probe.

Batch leaching experiments were performed in a 50 mL Trallero and Schlee mini reactor (Trallero and Schlee, Barcelona, Spain) combined with a mechanical stirrer, a vapor condenser, and a temperature controller, by adding BR to the IL when the set temperature was reached. Vacuum filtration was executed by cooling the system at 120 °C and adding a nonviscous/volatile solvent (dimethyl sulfoxide, further denoted as DMSO) to the leachates, to decrease viscosity and ease the process. After filtration, pregnant leaching solutions (PLS) were digested through acidic treatment (HNO_3 65% v/v and aqua regia) to oxidize and destroy the organics and then analyze with AAS, ICP-OES and ICP-MS. Solid residues were characterized via fusion method (already described above) and XRD. Microstructural characterization was carried out by a JEOL 6380 LV Scanning Electron Microscope (JEOL, Tokyo, Japan) coupled with Energy Dispersive System (SEM-EDS) and a JEOL 2100 HR (JEOL, Tokyo, Japan) 200 kV Transmission Electron Microscope (TEM) in order to detect and locate REEs.

3. Results and Discussion

3.1. Bauxite Residue Characterization

The main component of BR was found to be Fe_2O_3, accounting for 42.34 wt.%, followed by Al_2O_3 with 16.25 wt.%, while TiO_2 was 4.27 wt.%, and total rare earth oxides (REO) assessed to 0.19 wt.%, as it is shown in Table 1.

Table 1. Bauxite residue chemical analysis. Note: REO, rare earth oxides; LOI, loss of ignition.

Unit	Fe_2O_3	Al_2O_3	SiO_2	TiO_2	CaO	Na_2O	REO	LOI	Others	Sum
wt.%	42.34	16.25	6.97	4.27	11.64	3.83	0.19	12.66	1.85	100.00

In particular, cerium (Ce) was found to be the main rare earth element in concentration (402.2 mg/kg), followed by lanthanum (La) (145 mg/kg), scandium (Sc) (134 mg/kg), neodymium (Nd) (127.1 mg/kg), and yttrium (Y) (112 mg/kg).

Identification and quantification of mineralogical phases (Table 2) denoted hematite as the main mineral in BR with 30 wt.%, while Ti-containing phases were perovskite, anatase and rutile with 4.5, 0.5 and 0.5 wt.% respectively.

Table 2. Bauxite residue mineralogical phases and quantification.

Mineralogical Phase	Formula	wt.%
Hematite	Fe_2O_3	30
Calcium aluminum iron silicate hydroxide	$Ca_3AlFe(SiO_4)(OH)_8$	17
Cancrinite	$Na_6Ca_2(AlSiO_4)_6(CO_3)_2$	15
Diaspore	α-AlOOH	9
Goethite	$Fe_2O_3 \cdot H_2O$	9
Perovskite	$CaTiO_3$	4.5
Chamosite	$(Fe^{2+},Mg)_5Al(AlSi_3O_{10})(OH)_8$	4
Calcite	$CaCO_3$	4
Boehmite	γ-AlOOH	3
Gibbsite	$Al(OH)_3$	2
Rutile	TiO_2	0.5
Anatase	TiO_2	0.5
Sum		98.5

From particle size distribution analysis, it was found that 50% of the particles were below 1.87 µm, while 90% were smaller than 42.87 µm.

3.2. Ionic Liquid Characterization

1-ethyl-3-methylimidazolium hydrogensulfate ([Emim][HSO$_4$]) is a Brønsted acidic ionic liquid whose molecular weight is 208.24 g/mol and density (ρ) at room temperature is 1367.9 kg/m^3. The molecular structure of the IL is shown in Figure 1.

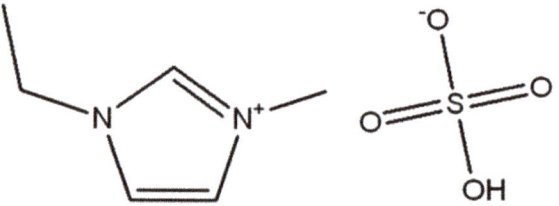

Figure 1. [Emim][HSO$_4$] molecular structure.

Viscosity measurements (Figure 2) have revealed that even though [Emim][HSO$_4$] is very viscous at room temperature (1642 mPa·s), by increasing temperature its viscosity dramatically decreases, reaching 221 mPa·s at 60 °C and 33 mPa·s at 120 °C.

Midinfrared spectrum have shown bands (cm^{-1}) at 3452 (OH), 3151 (aromatic/imidazole CH), 3106 (imidazole ring), 2985 (CH), 2944 (CH), 2881 ((CH$_2$)$_n$–CH$_3$), 2583–2497 (HOSO•••HOSO), 1636 (OH), 1572 (C=C, C=N, C–N), 1454 (CH$_3$), 1431 (S=O$_2$), 1389 (CH$_3$), 1211 (S=O$_2$), 1160 (S–O attached to C$_2$H$_5$), 1089 (HSO$_4^-$), 1023 (C–N–C), 960 (O–S–O), 832 (imidazole ring), 757 (CH of imidazole ring) and 701 (C–H–C). ^{1}H NMR (500 MHz, DMSO) δ (ppm): 1.36 (t, 3H, CH$_3$), 3.85 (s, 3H, CH$_3$), 4.19 (q, 2H, CH$_2$), 7.71 (s, 1H, CH=CH), 7.78 (s, 1H, CH=CH), 9.19 (s, 1H, N–CH–N).

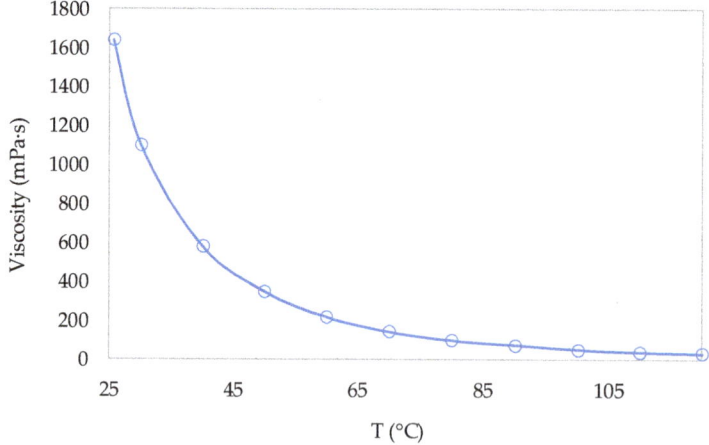

Figure 2. Viscosity measurements of [Emim][HSO$_4$] versus temperature.

Reaction Mechanism

To investigate the mechanism of the reaction that takes place, two monometallic solutions of 11 g/L of Sc and 11 g/L of Al were prepared, by dissolving Sc$_2$O$_3$ and Al$_2$O$_3$ in [Emim][HSO$_4$].

The two monometallic solutions were then analyzed with ^{1}H and ^{13}C NMR. Assignment of ^{1}H and ^{13}C chemical shifts was based on the combined analyses of a series of ^{1}H–^{1}H and ^{1}H–^{13}C correlation experiments recorded using standard pulse sequences from the Bruker library.

From the results (Appendix A, Figures A1 and A2), it could be concluded that there is no significant rearrangement in the carbon chain after the dissolution procedure in all three metal cases.

^{1}H and ^{13}C NMR spectra did not indicate any notable differences in the chemical shifts depending on the leached metal. The similar chemical shifts for the protons and the carbons localized in between the two nitrogen atoms indicate metal interaction through the anion of the IL.

There is not any steric effect of electron clouds changing of electrostatic interactions between ionic charges.

The results obtained from NMR analysis of two monometallic leachates have led to the following proposed reaction:

$$Me_2O_n + n[Emim][HSO_4] = Me_2(SO_4)_n + nOH^- + n[Emim]^+ \quad (1)$$

where Me is the metal and n is the oxidation state of the metal.

3.3. Leaching Process Optimization: Parameters Affecting the System

In order to optimize the process, stirring rate, retention time, temperature, and pulp density were investigated. Each parameter was studied separately, keeping the others constant, and choosing the combination that gave the best results. In each case, Ca and Si in leachates were below the detection limit and Ce, Nd, Y and La recovery was lower than 1%.

3.3.1. Stirring Rate

Initially, experiments were carried out by examining four different stirring rates, 100, 200, 400 and 600 revolutions per minute (rpm), while keeping constant all the other parameters at 150 °C, 5% w/v pulp density and 24 h. Results are given in Figure 3.

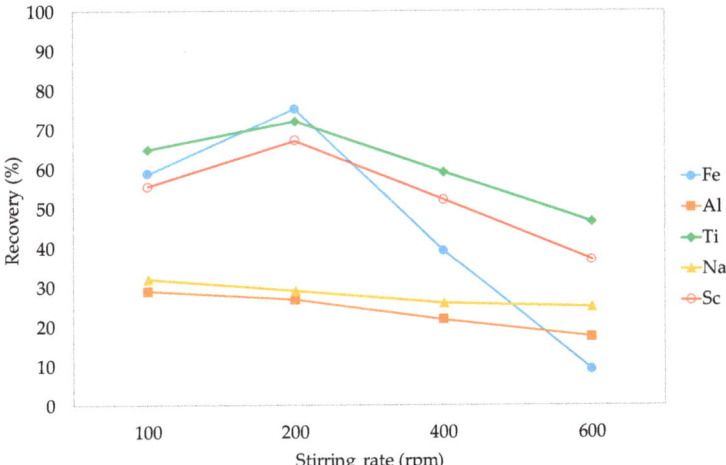

Figure 3. Investigation of the stirring rate effect on metals dissolution by leaching BR with [Emim][HSO$_4$] at 150 °C, 5% w/v pulp density for 24 h.

At these conditions, it is possible to observe an increase of Fe, Ti, and Sc extraction when the stirring rate increases from 100 to 200 rpm (from 55% of Sc, 57% of Fe, and 65% of Ti at 100 rpm to 67% of Sc, 75% of Fe and 72% of Ti at 200 rpm). On the other hand, as the stirring rate increases from 200 to 600 rpm, Fe, Ti, and Sc extraction is observed to linearly decrease (from 67% of Sc, 75% of Fe, and 72% of Ti at 200 rpm to 37% of Sc, 46% of Ti, and only 9% for Fe at 600 rpm). Na and Al recovery were slightly affected by the stirring rate as they remained almost stable in a range of 17–32% of recovery. This effect of stirring rate on metal recovery is typical in hydrometallurgy. Under low stirring rates, a thick boundary layer was developed on the surface of the solid particles, making the diffusion of chemical species from and to the solid particles surface inefficient. Therefore, at stirring rates lower than 200 rpm, the leaching process is slowed down and metal recovery decreases, as it is seen in Figure 3. At stirring rates higher than 200 rpm, the thickness of the boundary layer is substantially decreased, but the high convective mass transfer of reactants from the surface of the particles, makes the surface reactions again inefficient and thus the recovery yields are diminishing, as it is seen in Figure 3. Therefore, a compromise is always found under intermediate stirring rates which, for this system, is around 200 rpm. At this stirring rate, Fe, Ti, and Sc have the highest recovery yields (75%, 72% and 67% respectively).

3.3.2. Kinetic Studies

Several sets of kinetic have been performed at 200 rpm stirring rate, 5% w/v pulp density, analyzing the behavior of the system at three different temperatures: 150, 175, and 200 °C.

In Figure 4 it is observed that at low temperature (150 °C), all metals show the same trend in the first twelve hours; an initial metal dissolution occurred in the first six hours, whilst in the following six hours, metals dissolution is decreased, reaching their lowest concentration at 12 h retention time. This unusual behavior can be attributed to the precipitation of Ca as $CaSO_4$ that massively occurs within the first 6 h (Appendix A, Figure A3), while Fe dissolution is low. In the latter 6 h, adsorption phenomena were more important and faster than dissolution and, being in contact with anhydrite, metals are removed from the leachates, attaining the minimum at 12 h. Then, metals continue their dissolution and as the anhydrite precipitation has been completed they are gradually desorbed, increasing their concentration in solution and reaching the equilibrium at 24 h retention time, with the exception of iron that continues to be dissolved but at a substantially lower rate.

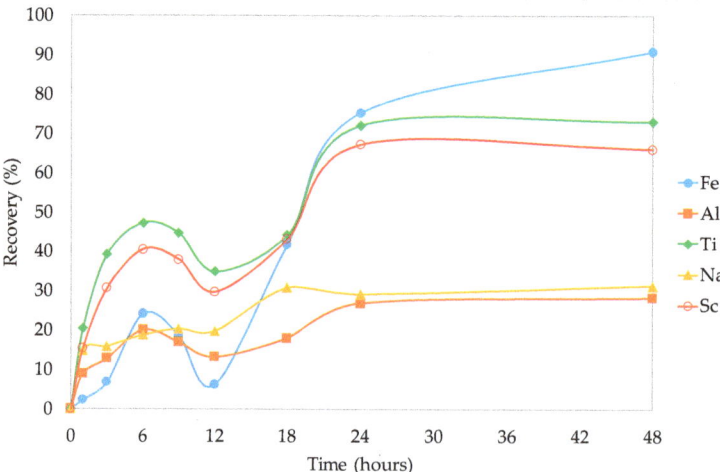

Figure 4. Kinetic curves for metals dissolution by leaching BR with [Emim][HSO$_4$] at 1, 3, 6, 12, 18, 24 and 48 h, 200 rpm, 150 °C and 5% w/v pulp density.

Kinetic studies have been carried out at 175 °C (Figure 5), in this case the unusual dissolution phenomenon observed at 150 °C was not seen and the plateau has been reached faster, after 12 h, achieving 90% of Fe, 70% of Ti and Sc dissolution and again moderate Al and Na recovery (30%). After 1 h, more than 35% of Sc has been dissolved, this, as mentioned, is due to the fact that goethite is totally dissolved and hematite starts to be leached as well. The equilibrium has been reached at 70% of Sc and 90% of Fe recovery, which is in agreement with Vind et al. studies, as the main mineralogical Sc containing phases in bauxite residue are hematite and goethite (55% and 25% on average, respectively) [38].

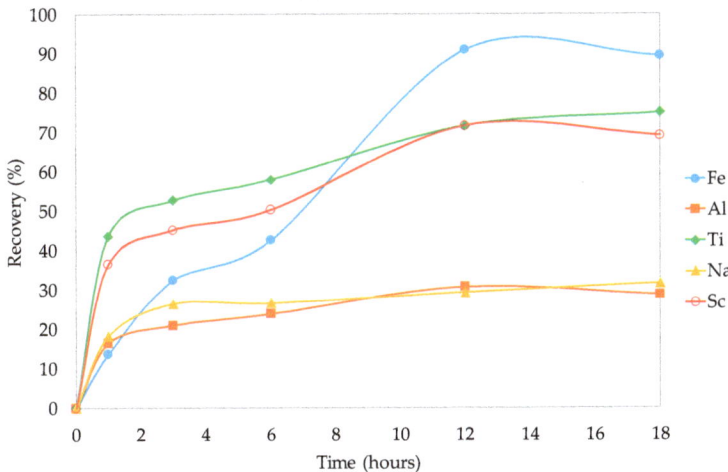

Figure 5. Kinetic curves for metals dissolution by leaching BR with [Emim][HSO$_4$] at 1, 3, 6, 12, and 18 h, 200 rpm, 175 °C and 5% w/v pulp density.

At 200 °C (Figure 6), Fe, Ti and Sc are considerably leached even after 1 h (60–74%). The maximum extraction of these metals has been reached after 12 h, where Fe was almost totally dissolved, Ti recovery was over 90% and Sc reached nearly 80%. Al and Na dissolution remained stable along the kinetic curve in a range of 30–40%.

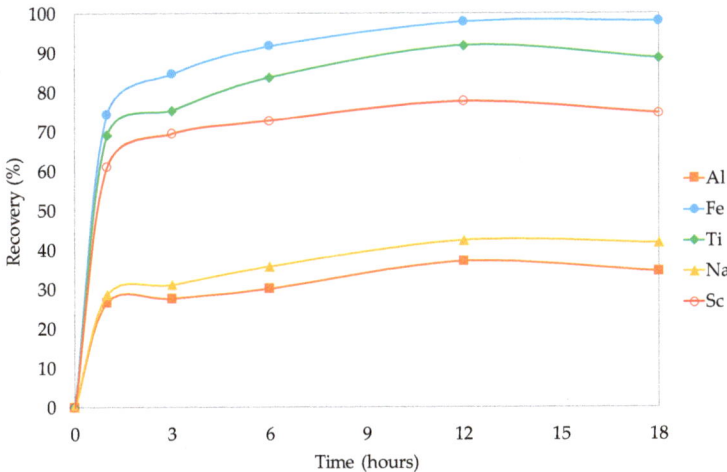

Figure 6. Kinetic curves for metals dissolution by leaching BR with [Emim][HSO$_4$] at 1, 3, 6, 12, and 18 h, 200 rpm, 200 °C, and 5% w/v pulp density.

Extraction of Sc at these high recovery yields (nearly 80%), when Fe was also almost totally dissolved, again confirms that Sc was found to be hosted mainly in hematite and goethite mineralogical phases in bauxite residue [38], as already mentioned. It was hypothesized before that in the same experimental setup as given here, the unrecovered proportion of Sc (about 20%) may be associated mainly with the chemically durable zirconium orthosilicate (ZrSiO$_4$), that contains around 10% of the total Sc in bauxite residue, but also with other undissolved (or partially dissolved) phases as boehmite, diaspore, and titanium-containing phases, which have been determined to be carriers of Sc in Greek BR [38].

3.3.3. Pulp Density

Four experiments have been conducted to investigate the effect of pulp density on the system, at 2.5%, 5%, 10%, and 14.3% w/v pulp density, under constant temperature, time and stirring rate (200 °C, 12 h, and 200 rpm). Results are shown in Figure 7.

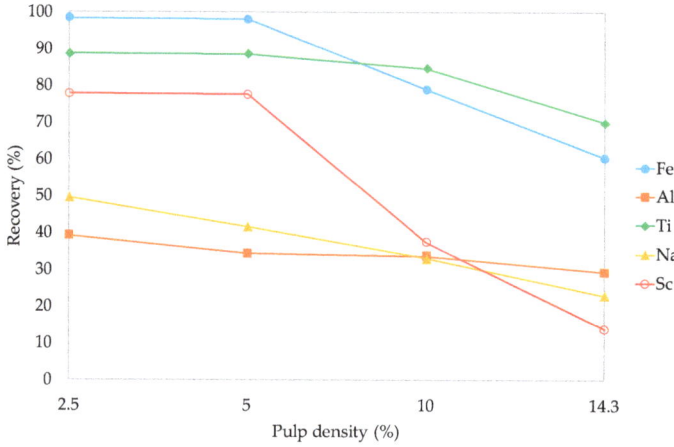

Figure 7. Study on the system behavior for metals dissolution by changing pulp density when leaching BR with [Emim][HSO$_4$] at 12 h, 200 rpm, 200 °C.

Fe, Ti, and Sc present constant recovery in the area of 2.5–5% w/v pulp density (almost total dissolution for Fe, 88% for Ti, and 78% for Sc). This behavior can be explained by the extremely high ionic liquid excess and the relatively low viscosity of the system due to the low concentration of dissolved metals. By increasing pulp density, the ionic liquid excess decreases and viscosity substantially rises, due to the increase of the number of suspended BR particles as well as the dissolved metal concentrations affecting the ions mobility phenomena and the thickness of the boundary layer. This results in a sharp and linear decreasing recovery, reaching the minimum at 14.3% w/v (60% of Fe, 70% Ti, and 14% of Sc). Experiments at pulp density higher than 14.3% w/v were not carried out due to the high viscosity, which prevented filtration and caused serious problems during the leaching process.

3.4. Characterization of the Solid Residue after Leaching

The solid residue collected after leaching bauxite residue at optimum conditions (200 rpm, 200 °C, 12 h and 5% w/v pulp density) was characterized via fusion method, XRD, SEM, and TEM analyses. The resulting residue was found to be 48% of the weight of the initial BR mass.

As it can be seen from chemical analysis shown in Table 3, solid residue after leaching is high in aluminum, calcium, and silicon, while it is depleted in iron and titanium. REEs remain in the solid residue (with the exception of scandium) and can be leached afterwards.

Table 3. Chemical analysis of the residue after leaching BR at optimum conditions.

Metal Oxide	Fe$_2$O$_3$	Al$_2$O$_3$	SiO$_2$	TiO$_2$	CaO	Na$_2$O	REO	SO$_3$	LOI	Others
wt.%	3.71	27.44	14.51	1.46	24.73	2.75	0.19	18.69	6.00	0.52

From the comparison of the XRD spectra of bauxite residue and residue generated after leaching (Appendix A, Figure A4), it is possible to observe that peaks attributed to hematite, goethite, calcium aluminum iron silicate hydroxide, gibbsite, and perovskite, which are present in bauxite residue, disappear after leaching. Aluminum phases like diaspore and boehmite remain relatively intact after leaching, as well as cancrinite, chamosite, and calcite. On the other hand, a new mineralogical phase calcium sulfate anhydrite (causing the consumption of about 2 wt.% of the IL), which was formed due to the interaction between calcium and the anion of the ionic liquid, is created during leaching. The above observations explain well the behavior of Al and Na during leaching as their main minerals in BR such as diaspore, boehmite and cancrinite remain insoluble, leading to low to moderate recoveries. On the other hand, Fe and Ti bearing minerals were depleted in leaching residue thus confirming their observed high recoveries. Regarding Ca leaching, phases like calcium aluminum iron silicate hydroxide are substantially soluble, while phases such as cancrinite and chamosite resist dissolution. Calcite is partially dissolved in IL solution and in the presence of HSO$_4^-$ anions, undertakes a transition to anhydrite, which is a secondary precipitated phase during the leaching process.

SEM-EDS analysis of the solid residue after leaching confirmed the findings of chemical and XRD analyses (Figure 8). The matrix, which is mainly composed of Al, Ca, and Na silicates, surrounds phases transitioning from CaCO$_3$ to CaSO$_4$.

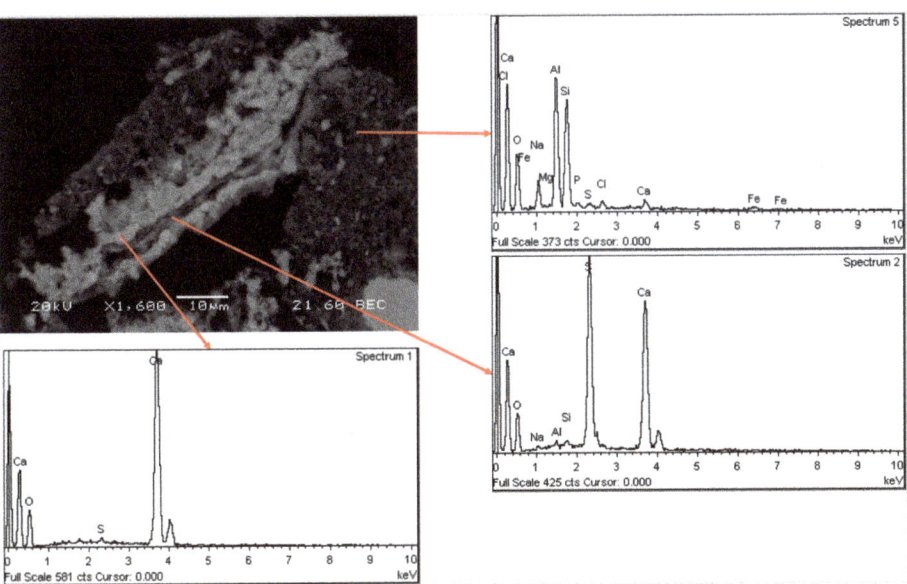

Figure 8. Scanning electron microscope (SEM image of the matrix of the residue after leaching).

3.5. REEs in the Solid Residue

Small REEs-containing particles (about 10 μm) were detected in SEM-EDS, in particular YPO$_4$ particles including heavy rare earths like gadolinium and dysprosium (Figure 9 left). This is consistent with Vind et al. studies of raw bauxite residue [39] where the presence of heavy rare earth phosphates with the major constituent being yttrium and containing other heavy REEs like gadolinium, dysprosium, and erbium is reported. This is an indication that these grains endure the [Emim][HSO$_4$] leaching process without being subjected to any dissolution and thus explaining negligible heavy REEs recoveries.

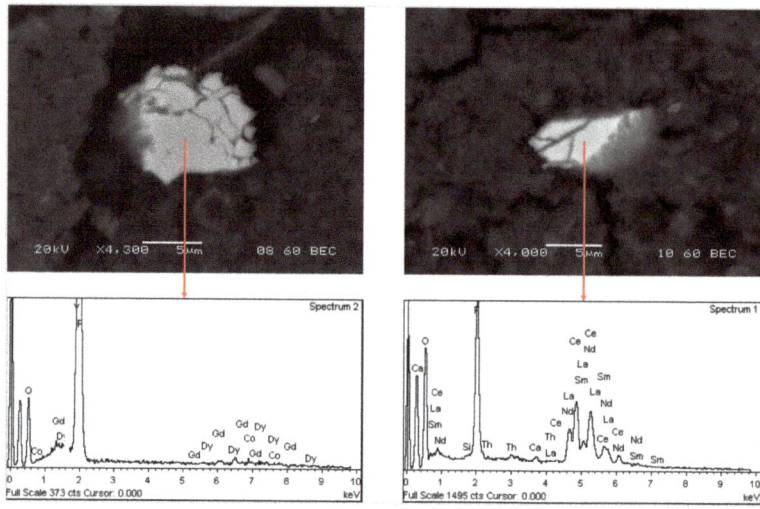

Figure 9. SEM image of a YPO$_4$ particle (**left**) and a CePO$_4$ particle (**right**).

Small mixed calcium–cerium phosphate particles were also identified; in this case, grains included light rare earths like neodymium, lanthanum, and samarium (Figure 9 right). Vind et al. reported the presence of light rare earths as calcium-containing phosphate phases in bauxite residue [39]. In the case of the solid residue after leaching, grains containing light REEs (LREEs) phosphates are also present. This may indicate a partial dissolution of calcium from the mixed Ca-LREEs phases, leaving behind smaller phosphate particles which are beneficiated in LREEs. This was also implied by TEM analysis, detecting very fine (<500 nm) particles of Al-containing CePO$_4$ (Figure 10).

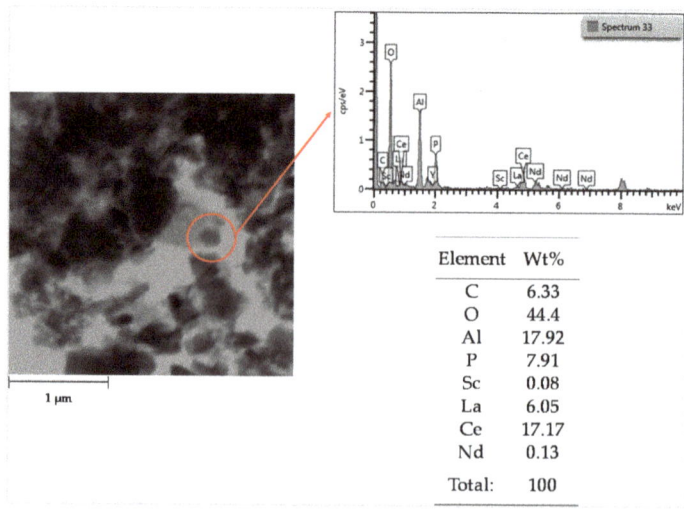

Element	Wt%
C	6.33
O	44.4
Al	17.92
P	7.91
Sc	0.08
La	6.05
Ce	17.17
Nd	0.13
Total:	100

Figure 10. TEM image of a CePO$_4$, Al containing, particle.

4. Conclusions

In this study, Brønsted acidic ionic liquid 1-ethyl-3-methylimidazolium hydrogensulfate was used to directly leach bauxite residue. Experiments were carried out in a closed mini reactor, equipped with a condenser and a temperature controller. Stirring rate, time, temperature and pulp density were thoroughly examined to find the optimum conditions for Sc (nearly 80%) and Fe (almost totally dissolved) high recovery yields. This outcome confirms that Sc is mainly hosted in hematite and goethite mineralogical phases (55% and 25%, respectively) in bauxite residue, in accordance to the work of Vind et al. [38]. The undissolved Sc content might be attributed to $ZrSiO_4$, containing around 10% of the total Sc in bauxite residue, but also to other phases, such as boehmite, diaspore, and titanium-containing phases that host Sc in Greek bauxite residue [38].

At the optimum conditions, 90% of Ti was dissolved, while Al and Na were partially extracted (in a range of 30–40%). Si and REEs dissolutions were found to be negligible, whereas Ca was partially dissolved and precipitated as $CaSO_4$ consuming about 2 wt.% of the ionic liquid.

Solid residue after leaching was fully characterized and found to be rich in Al, Ca, and Si, with the main minerals present being anhydrite, diaspore, and cancrinite; it could be further treated to extract REEs. SEM and TEM analyses of the solid residues provided explanations for the destiny of REEs, which remain undissolved enduring the leaching process.

It can be concluded that [Emim][HSO_4] ionic liquid is a good leaching agent for dissolving metals from bauxite residue and, since it is not selective against iron, high recovery yields of Sc can be achieved, reaching up to 80% of extraction.

Author Contributions: C.B. and D.P. conceived and designed the experiments; C.B. performed the experiments; A.P. performed NMR analysis and analyzed the spectra; A.A. performed chemical analyses; J.V. and C.B. performed the SEM-EDS analysis and analyzed the data; P.T. performed TEM analysis; C.B. analyzed the data and wrote the paper; D.P. contributed to the analysis of the data and the writing of the paper.

Acknowledgments: The research leading to these results has received funding from the European Community's Horizon 2020 Program (H2020/2014—2019) under Grant Agreement no. 636876 (MSCA–ETN REDMUD).

Conflicts of Interest: The authors declare no conflict of interest. The funding sponsors had no role in the design of the study; in the collection, analyses, or interpretation of data; in the writing of the manuscript, and in the decision to publish the results.

Appendix A

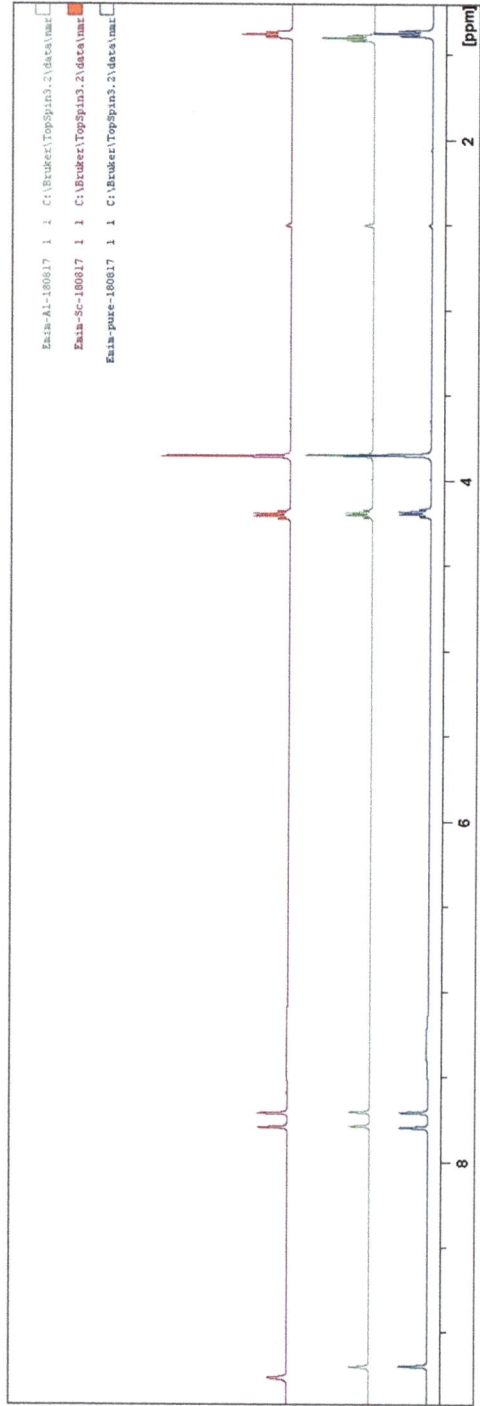

Figure A1. ^{1}H Nuclear Magnetic Resonance (NMR) comparison between [Emim][HSO$_4$] (blue), [Emim][HSO$_4$] after leaching Al$_2$O$_3$ (green), [Emim][HSO$_4$] after leaching Sc$_2$O$_3$ (red).

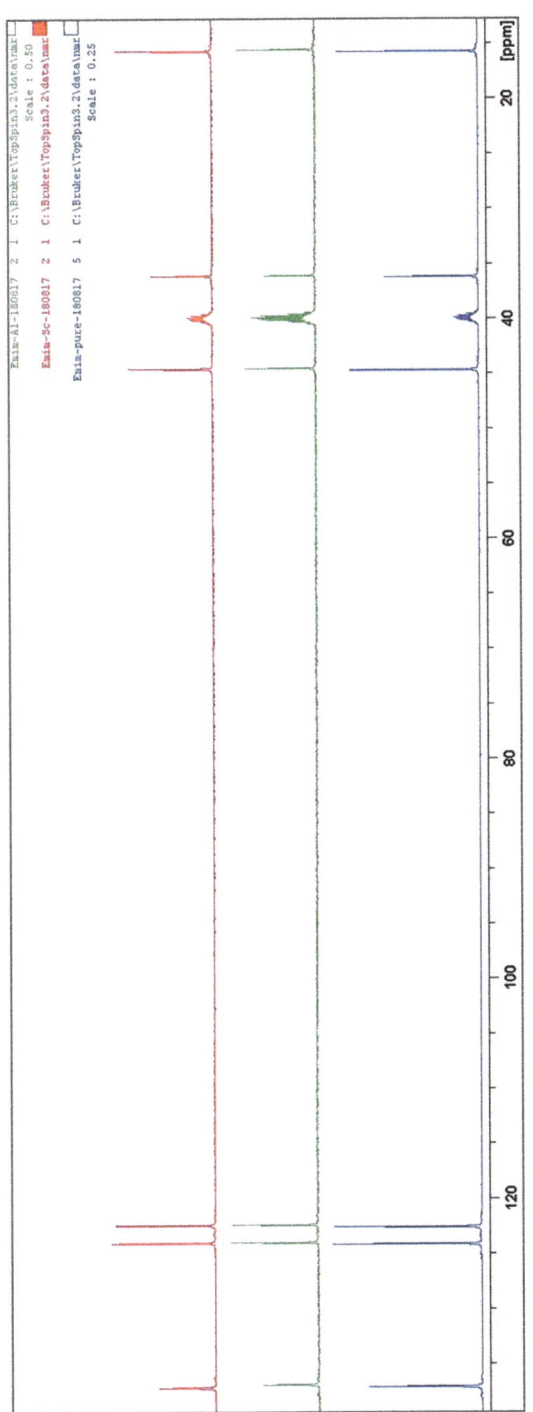

Figure A2. ^{13}C NMR comparison between [Emim][HSO$_4$] (blue), [Emim][HSO$_4$] after leaching Al$_2$O$_3$ (green), [Emim][HSO$_4$] after leaching Sc$_2$O$_3$ (red).

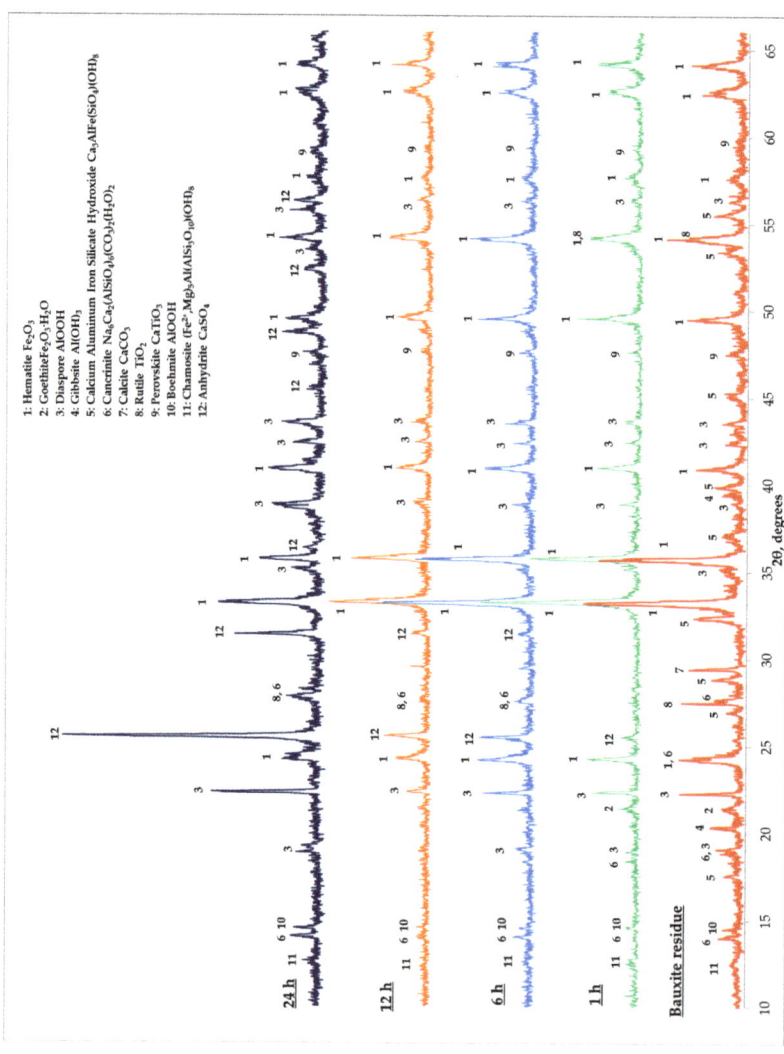

Figure A3. Comparison between X-ray powder diffractometer (XRD) of bauxite residue and solid residue after leaching bauxite residue at 150 °C, 200 rpm, 5% w/v pulp density for 1, 6, 12, and 24 h.

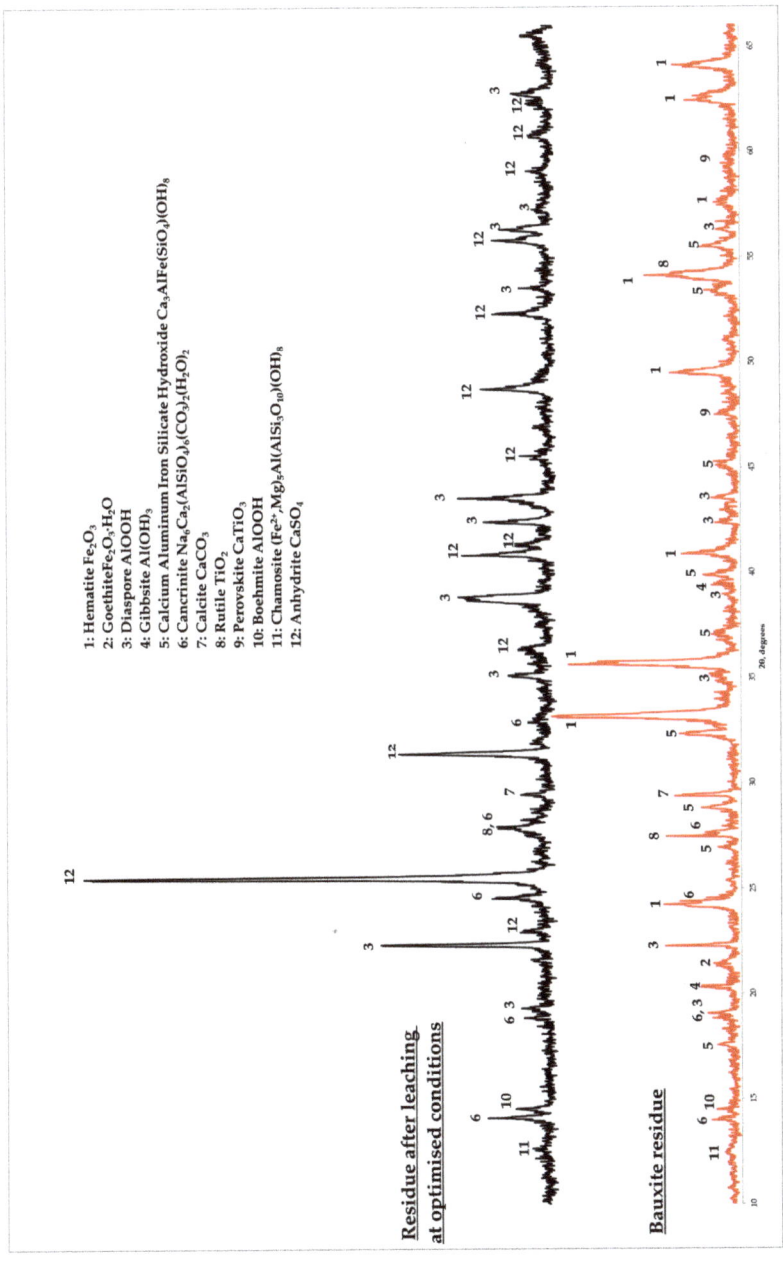

Figure A4. Comparison of the XRD spectra of bauxite residue and solid residue after leaching at optimum conditions.

References

1. Evans, K. The History, Challenges, and New Developments in the Management and Use of Bauxite Residue. *J. Sustain. Metall.* **2016**, *2*, 316–331. [CrossRef]
2. Zhang, R.; Zheng, S.; Ma, S.; Zhang, Y. Recovery of Alumina and Alkali in Bayer Red Mud by the Formation of Andradite-Grossular Hydrogarnet in Hydrothermal Process. *J. Hazard. Mater.* **2011**, *189*, 827–835. [CrossRef] [PubMed]
3. Klauber, C.; Gräfe, M.; Power, G. Bauxite Residue Issues: II. Options for Residue Utilization. *Hydrometallurgy* **2011**, *108*, 11–32. [CrossRef]
4. Gräfe, M.; Power, G.; Klauber, C. Bauxite Residue Issues: III. Alkalinity and Associated Chemistry. *Hydrometallurgy* **2011**, *108*, 60–79. [CrossRef]
5. Liu, Y.; Naidu, R. Hidden Values in Bauxite Residue (Red Mud): Recovery of Metals. *Waste Manag.* **2014**, *34*, 2662–2673. [CrossRef] [PubMed]
6. Ochsenkühn-Petropulu, M.; Lyberopulu, T.; Parissakis, G. Direct Determination of Landthanides, Yttrium and Scandium in Bauxites and Red Mud from Alumina Production. *Anal. Chim. Acta* **1994**, *296*, 305–313. [CrossRef]
7. Vind, J.; Alexandri, A.; Vassiliadou, V.; Panias, D. Distribution of Selected Trace Elements in the Bayer Process. *Metals (Basel)* **2018**, *8*, 327. [CrossRef]
8. Rudnick, R.L.; Gao, S. Composition of the Continental Crust. In *Treatise on Geochemistry*; Elsevier: Amsterdam, The Netherlands, 2003; pp. 1–64.
9. Gambogi, J. *Mineral Commodity Summaries, Scandium*; U.S. Geological Survey: Reston, VA, USA, 2017; pp. 146–147.
10. Balomenos, E.; Davris, P.; Pontikes, Y.; Panias, D. Mud2Metal: Lessons Learned on the Path for Complete Utilization of Bauxite Residue Through Industrial Symbiosis. *J. Sustain. Metall.* **2017**, *3*, 551–560. [CrossRef]
11. Deloitte Sustainability; British Geological Survey; Bureau de Recherches Géologiques et Minières; TNO. *Study on the Review of the List of Critical Raw Materials*; EU Law and Publications: Luxembourg, 2017; ISBN 978-92-79-47937-3.
12. Borra, C.R.; Blanpain, B.; Pontikes, Y.; Binnemans, K.; Van Gerven, T. Recovery of Rare Earths and Other Valuable Metals From Bauxite Residue (Red Mud): A Review. *J. Sustain. Metall.* **2016**, *2*, 365–386. [CrossRef]
13. Ochsenkühn-Petropulu, M.; Lyberopulu, T.; Parissakis, G. Selective Separation and Determination of Scandium from Yttrium and Lanthanides in Red Mud by a Combined Ion Exchange/Solvent Extraction Method. *Anal. Chim. Acta* **1995**, *315*, 231–237. [CrossRef]
14. Binnemans, K.; Jones, P.T.; Blanpain, B.; Van Gerven, T.; Pontikes, Y. Towards Zero-Waste Valorisation of Rare-Earth-Containing Industrial Process Residues: A Critical Review. *J. Clean. Prod.* **2015**, *99*, 17–38. [CrossRef]
15. Bonomi, C.; Cardenia, C.; Yin, P.T.W.; Panias, D. Review of Technologies in the Recovery of Iron, Aluminium, Titanium and Rare Earth Elements from Bauxite Residue (Red Mud). In Proceedings of the International Symposium on Enhanced Landfill Mining, Lisboa, Portugal, 8–10 February 2016; pp. 259–276.
16. Alkan, G.; Schier, C.; Gronen, L.; Stopic, S.; Friedrich, B. A Mineralogical Assessment on Residues after Acidic Leaching of Bauxite Residue (Red Mud) for Titanium Recovery. *Metals* **2017**, *7*, 458. [CrossRef]
17. Agatzini-Leonardou, S.; Oustadakis, P.; Tsakiridis, P.E.; Markopoulos, C. Titanium Leaching from Red Mud by Diluted Sulfuric Acid at Atmospheric Pressure. *J. Hazard. Mater.* **2008**, *157*, 579–586. [CrossRef] [PubMed]
18. Alkan, G.; Yagmurlu, B.; Cakmakoglu, S.; Hertel, T.; Kaya, Ş.; Gronen, L.; Stopic, S.; Friedrich, B. Novel Approach for Enhanced Scandium and Titanium Leaching Efficiency from Bauxite Residue with Suppressed Silica Gel Formation. *Sci. Rep.* **2018**, *8*, 5676. [CrossRef] [PubMed]
19. Liu, Z.; Li, H. Metallurgical Process for Valuable Elements Recovery from Red Mud—A Review. *Hydrometallurgy* **2015**, *155*, 29–43. [CrossRef]
20. Ochsenkühn-Petropoulou, M.T.; Hatzilyberis, K.S.; Mendrinos, L.N.; Salmas, C.E. Pilot-Plant Investigation of the Leaching Process for the Recovery of Scandium from Red Mud. *Ind. Eng. Chem. Res.* **2002**, *41*, 5794–5801. [CrossRef]
21. Ochsenkühn-Petropulu, M.; Lyberopulu, T.; Ochsenkühn, K.M.; Parissakis, G. Recovery of Lanthanides and Yttrium from Red Mud by Selective Leaching. *Anal. Chim. Acta* **1996**, *319*, 249–254. [CrossRef]

22. Borra, C.R.; Blanpain, B.; Pontikes, Y.; Binnemans, K.; Van Gerven, T. Smelting of Bauxite Residue (Red Mud) in View of Iron and Selective Rare Earths Recovery. *J. Sustain. Metall.* **2016**, *2*, 28–37. [CrossRef]
23. Borra, C.R.; Pontikes, Y.; Binnemans, K.; Van Gerven, T. Leaching of Rare Earths from Bauxite Residue (Red Mud). *Miner. Eng.* **2015**, *76*, 20–27. [CrossRef]
24. Bonomi, C.; Davris, P.; Balomenos, E.; Giannopoulou, I.; Panias, D. Ionometallurgical Leaching Process of Bauxite Residue: A Comparison between Hydrophilic and Hydrophobic Ionic Liquids. In Proceedings of the 35th International ICSOBA Conference, Hamburg, Germany, 2–5 October 2017; pp. 557–564.
25. Abbott, A.P.; Frisch, G.; Hartley, J.; Ryder, K.S. Processing of Metals and Metal Oxides Using Ionic Liquids. *Green Chem.* **2011**, *13*, 471. [CrossRef]
26. Vander Hoogerstraete, T.; Onghena, B.; Binnemans, K. Homogeneous Liquid–Liquid Extraction of Rare Earths with the Betaine—Betainium Bis(Trifluoromethylsulfonyl)Imide Ionic Liquid System. *Int. J. Mol. Sci.* **2013**, *14*, 21353–21377. [CrossRef] [PubMed]
27. Wellens, S.; Thijs, B.; Möller, C.; Binnemans, K. Separation of Cobalt and Nickel by Solvent Extraction with Two Mutually Immiscible Ionic Liquids. *Phys. Chem. Chem. Phys.* **2013**, *15*, 9663. [CrossRef] [PubMed]
28. Wasserscheid, P.; Keim, W. Ionic Liquids—New "Solutions" for Transition Metal Catalysis. *Angew. Chemie* **2000**, *39*, 3772–3789. [CrossRef]
29. Welton, T. Room-Temperature Ionic Liquids. Solvents for Synthesis and Catalysis. *Chem. Rev.* **1999**, *99*, 2071–2084. [CrossRef] [PubMed]
30. Bourbos, E.; Giannopoulou, I.; Karantonis, A.; Paspaliaris, I.; Panias, D. Electrodeposition of Rare Earth Metals from Ionic Liquids. In *Rare Earths Industry*; Elsevier: Amsterdam, The Netherlands, 2016; pp. 199–207.
31. Abbott, A.P.; McKenzie, K.J. Application of Ionic Liquids to the Electrodeposition of Metals. *Phys. Chem. Chem. Phys.* **2006**, *8*, 4265. [CrossRef] [PubMed]
32. Reddy, R.G. Emerging Technologies in Extraction and Processing of Metals. *Metall. Mater. Trans. B* **2003**, *34*, 137–152. [CrossRef]
33. Abbott, A.P.; Capper, G.; Davies, D.L.; Shikotra, P. Processing Metal Oxides Using Ionic Liquids. *Miner. Process. Extr. Metall.* **2006**, *115*, 15–18. [CrossRef]
34. Davris, P.; Balomenos, E.; Panias, D.; Paspaliaris, I. Selective Leaching of Rare Earth Elements from Bauxite Residue (Red Mud), Using a Functionalized Hydrophobic Ionic Liquid. *Hydrometallurgy* **2016**, *164*, 125–135. [CrossRef]
35. Binnemans, K.; Jones, P.T. Solvometallurgy: An Emerging Branch of Extractive Metallurgy. *J. Sustain. Metall.* **2017**, *3*, 570–600. [CrossRef]
36. Sajó, I.E. X-Ray Diffraction Quantitative Phase Analysis of Bayer Process Solids. In Proceedings of the 10th International Conference of ICSOBA, Bhubaneshwar, India, 23–30 November 2008; pp. 71–76.
37. Sajò, I.E. *XDB Powder Diffraction Phase Analytical System*, Version 3.107; Computer Software: Budapest, Hungary, 2005.
38. Vind, J.; Malfliet, A.; Bonomi, C.; Paiste, P.; Sajó, I.E.; Blanpain, B.; Tkaczyk, A.H.; Vassiliadou, V.; Panias, D. Modes of Occurrences of Scandium in Greek Bauxite and Bauxite Residue. *Miner. Eng.* **2018**, *123*, 35–48. [CrossRef]
39. Vind, J.; Malfliet, A.; Blanpain, B.; Tsakiridis, P.; Tkaczyk, A.; Vassiliadou, V.; Panias, D. Rare Earth Element Phases in Bauxite Residue. *Minerals* **2018**, *8*, 77. [CrossRef]

© 2018 by the authors. Licensee MDPI, Basel, Switzerland. This article is an open access article distributed under the terms and conditions of the Creative Commons Attribution (CC BY) license (http://creativecommons.org/licenses/by/4.0/).

Article

Scandium Recovery from an Ammonium Fluoride Strip Liquor by Anti-Solvent Crystallization

Şerif Kaya [1,*], Edward Michael Peters [2,*], Kerstin Forsberg [2,*], Carsten Dittrich [3], Srecko Stopic [4] and Bernd Friedrich [4]

1. Mining Engineering Department, Middle East Technical University, Ankara 06800, Turkey
2. KTH Royal Institute of Technology, Department of Chemical Engineering, Teknikringen 42, SE-100 44 Stockholm, Sweden
3. MEAB Chemie Technik GmbH, 52068 Aachen, Germany; carsten@meab-mx.com
4. IME Institute of Process Metallurgy and Metal Recycling, RWTH Aachen University, 52056 Aachen, Germany; sstopic@ime-aachen.de (S.S.); bfriedrich@ime-aachen.de (B.F.)
* Corresponding author: serifkaya@gmail.com (Ş.K.); edwpet@kth.se (E.M.P.); kerstino@ket.kth.se (K.F.); Tel.: +90-537-513-0949 (Ş.K.); +46-72-565-9653 (E.M.P.); +46-8-790-6404 (K.F.)

Received: 6 September 2018; Accepted: 22 September 2018; Published: 26 September 2018

Abstract: In this study, the crystallization of scandium from ammonium fluoride strip liquor, obtained by solvent extraction, was investigated using an anti-solvent crystallization technique. Acetone, ethanol, methanol and isopropanol were added individually to the strip liquor as the anti-solvent and scandium was precipitated and obtained in the form of $(NH_4)_3ScF_6$ crystals. The results show that scandium can be effectively crystallized from the strip liquor to obtain an intermediate, marketable scandium product. Yields greater than 98% were obtained using an anti-solvent to strip liquor volumetric ratio of 0.8. Acetone had the least performance at lower anti-solvent to strip liquor volumetric ratios, possibly due to its limited H bonding capability with water molecules when compared to alcohols.

Keywords: scandium; anti-solvent crystallization; solvent extraction; precipitation; ammonium scandium hexafluoride; chemical equilibrium diagram

1. Introduction—Previous Studies and State of the Art

Scandium was identified as a critical raw material (CRM) to the European Union in 2017 and in the US draft list of critical minerals in 2018 [1,2]. Scandium is used in solid oxide fuel cells (SOFCs), which is a rapidly growing market [3,4]. Currently more than 90% of the annual global production of Sc is used for the production of SOFCs [3]. Scandium also finds use in, e.g., laser garnets, as phosphors for light emitting diodes and as an alloying element for aluminium [5–7]. The aluminium scandium alloys have high strength, are corrosion resistant and allow welding without loss in strength, which makes them attractive for the aerospace and automotive industries [3,7]. Viable sources of scandium include nickel-cobalt laterites, uranium processing wastes, residues from titanium oxide production and bauxite residues (so called red mud) [8–13]. Scandium is produced in a few countries worldwide with 66% of the production in China, 26% in Russia and 7% in Ukraine [1].

In the conventional processing to extract scandium, the last steps often consist of the precipitation of scandium oxalate or scandium hydroxide from a purified solution [8]. The scandium salts can then be calcined to obtain pure scandium oxide [14]. The scandium oxide can then be fluorinated with HF (Hydrofluoric acid) to produce ScF_3 [15]. The latter is the precursor used in scandium metal production. This is a long procedure with many stages, driving the cost and environmental footprint of scandium fluoride and scandium metal. According to sale statistics in the US, the estimated price of Sc_2O_3 (99.99% purity) is 4.60 US dollars/g, that of ScF_3 (99.9% purity) is 277 US dollars/g, whilst that of Sc metal ingot is 132 US dollars/g in 2018 [16]. An alternative approach is to pre-concentrate scandium from waste streams by leaching, solvent extraction, and then stripping scandium from the organic phase using ammonium fluoride. A pure scandium ammonium fluoride solid phase can then be obtained by crystallization. Anti-solvent crystallization is a technique that involves adding a solvent that is soluble in a solution so as to lower the solubility of the desired salt, thereby generating supersaturation in the resultant mixture [17].

There is information about the crystallization of scandium fluoride phases in literature. Scandium can be precipitated from an ammonium fluoride solution as $ScF_3 \cdot (0-0.25)H_2O$, NH_4ScF_4, $(NH_4)_3ScF_6$, $(NH_4)_5Sc_3F_{14}$, $NH_4Sc_3F_{10}$ or $(NH_4)_2Sc_3F_{11}$, depending on the composition of the solution and temperature [18–20]. Scandium could also be precipitated as sodium or potassium fluoride and ammonium fluoride salts [19]. Scandium has been crystallized as scandium trifluoride on an industrial scale after the stripping of Ti(IV), Th(IV) and Sc(III) from dodecyl phosphoric acid in kerosene using hydrofluoric acid in uranium processing [13]. Scandium and thorium were precipitated while titanium was left in the solution. The solids were further processed after dissolution in a sodium hydroxide solution. Scandium has also been stripped from loaded organic phases of D2EHPA by using sodium fluoride solution from which Na_3ScF_6 has been obtained [14]. The precipitate, $(NH_4)_2NaScF_6$, has also been reported to form from a Sc containing strip liquor using NaOH solution at pH 9. After calcination of this intermediate product, a cryolite type phase ($NaScF_4$-Na_3ScF_6) can be obtained for use in Al electrolysis in place of Na_3AlF_6 to obtain Al-Sc alloys [10].

In the present work, a novel approach is suggested where scandium is precipitated as an ammonium scandium hexafluoride $(NH_4)_3ScF_6$ salt by addition of an alcohol to an NH_4F strip liquor containing Sc. The alcohol acts as an anti-solvent, which effectively reduces the solubility of the scandium salt in the resultant mixture. The precipitated ammonium scandium hexafluoride salt can directly be used for scandium metal production by eliminating the use of environmentally undesirable hydrofluoric acid during the conversion of Sc_2O_3 into ScF_3.

2. Experimental Procedure

2.1. Thermodynamic Modeling

A thermodynamic model of the Sc-F system was constructed using MEDUSA (Make Equilibrium Diagrams Using Sophisticated Algorithms) software, version 2017-Jan-27, developed at KTH Royal Institute of Technology, Stockholm, Sweden [21]. The software is based on algorithms for the computation of multicomponent, multiphase solution equilibrium [22–24]. The pH was varied between 1 and 12, and the logarithmic total fluoride concentration, $\log[F^-]_{TOT}$, was varied between -2 and 2, corresponding to a total F^- concentration of 0.01 to 100 mol/L at 25 °C. All models were conducted for a total Sc concentration of 67 mmol/L (3000 mg/L) at 25 °C. The software predicts the solution speciation based on the stability constants and solubility products of various complexes and salts, respectively. The overall stability constants of the Sc-F and Sc-OH complexes used as well as the solubility products of the respective solids are shown in Table 1. The overall stability constant of ScF_6^{3-} was extrapolated and approximated from the constants of the lower Sc-F complexes, since they had a perfect quadratic fit. The existence of the pentafluoride complex, ScF_5^{2-}, was only mentioned briefly [20] with no adequate information available in other sources. There is also evidence of the polynuclear complex, $Sc_2F_3^{3+}$ [25]. The software computed the ionic strength as the fluoride concentration, and the pH was varied. The first thermodynamic model was conducted by considering soluble complexes only, without considering the formation of any solids; the second model considered the formation of solids; and the third model was conducted by varying the fluoride activity, $\log\{F^-\}$, between -8 and 2, instead of the total fluoride concentration, as was the case in the first and second models.

Table 1. Stability constants of Sc complexes and solubility products of Sc salts.

Complex	pK$_n$	Ref.	Complex	pK$_n$	Ref.
ScF^{2+} (aq)	7.08		ScO(OH)	−9.4	
ScF$_2{}^+$ (aq)	12.89		Sc(OH)$_3$	−29.7	
ScF$_3{}^0$ (aq)	17.36	[26]	Sc(OH)$^{2+}$ (aq)	−4.3	
ScF$_4{}^-$ (aq)	20.21		Sc(OH)$_2{}^+$ (aq)	−9.7	[27]
ScF$_6{}^{3-}$ (aq)	22.00 *		Sc(OH)$_3$ (aq)	−16.1	
Sc$_2$F$_3{}^{3+}$ (aq)	20.7	[28]	Sc(OH)$_4{}^-$ (aq)	−26	
ScF$_3$ (s)	−11.5	[29]	Sc$_2$(OH)$_2{}^{4+}$ (aq)	−6	
Sc$_2$O$_3$ (s)	−36.3	[27]	Sc$_3$(OH)$_5{}^{4+}$ (aq)	−16.34	

* denotes extrapolated.

The model served to determine the dominant Sc-F complexes under the experimental conditions.

2.2. Experimental Procedure

The scandium containing strip liquor was prepared through the solvent extraction of a synthetic scandium sulphate solution followed by stripping since this procedure mimics that employed to recover scandium by the envisaged process. The organic solvent that was used contained D2EHPA purchased from Lanxess, Germany; the actual composition is withheld for proprietary reasons. This solvent was mixed with a scandium sulphate solution containing 3671 mg/L Sc at room temperature in a beaker for 10 min at an organic to aqueous phase volumetric ratio of O/A: 1/1. The synthetic scandium sulphate solution was prepared by dissolving 99.9% scandium sulphate pentahydrate obtained from Richest Group, China. After mixing the organic extractant and the scandium-containing aqueous phase, the loaded organic was separated from the raffinate via a separation funnel, and the raffinate phase was analyzed for scandium content to calculate the % extraction. No purification steps were conducted on the reagents used during the solvent extraction tests. The stripping of scandium from the loaded organic phase was conducted at room temperature by using a 3 mol/L reagent grade ammonium fluoride solution with an aqueous to organic phase ratio of A/O: 1/1. After stripping, the stripped organic phase was separated from the scandium-containing strip liquor via a separation funnel and the scandium content of the ammonium scandium fluoride strip liquor was analyzed to determine the stripping efficiency. A synthetic strip liquor was then obtained for use in the anti-solvent crystallization experiments.

After stripping, reagent grade acetone, ethanol, methanol and isopropanol were added to the ammonium scandium hexafluoride strip liquor separately at room temperature as anti-solvents. The solutions were mixed for 10 min whereby precipitation occurred. The precipitated crystals were separated from the solution by means of filtration using a 0.22 μm membrane. The obtained crystals were then washed with the same anti-solvent, dried overnight at 60 °C, and analyzed by a "Rigaku Ultima-IV" model powder X-ray diffractometer (Rigaku, San Antonio, TX, USA) with a Cu-Kα X-ray tube working under 40 kV and 40 mA to identify the crystal structure of the precipitates. A Spectro Arcos ICP-OES analyzer (SPECTRO Analytical Instruments GmbH, Kleve, North Rhine-Westphalia, Germany) capable of true-axial and true-radial plasma observations was used to analyze the total concentration of the liquid samples. The samples were diluted by a factor of 100, such that the organic concentration in the samples analyzed was below 0.5% v/v. A JEOL JSM-6490LV SEM (Scanning Electron Microscope) (JEOL USA, Peabody, MA, USA) was used to determine the crystal morphology and size of the precipitates.

3. Results and Discussion

3.1. Thermodynamic Model

Figure 1A,B shows the thermodynamic models conducted for a total Sc concentration of 67 mmol/L (3000 mg/L) at 25 °C to determine the stability of soluble complexes with varying pH and total fluoride concentration, as well as a consideration of solids precipitation, respectively. Figure 1C shows the model (67 mmol/L Sc at 25 °C) with variation of fluoride activity and pH.

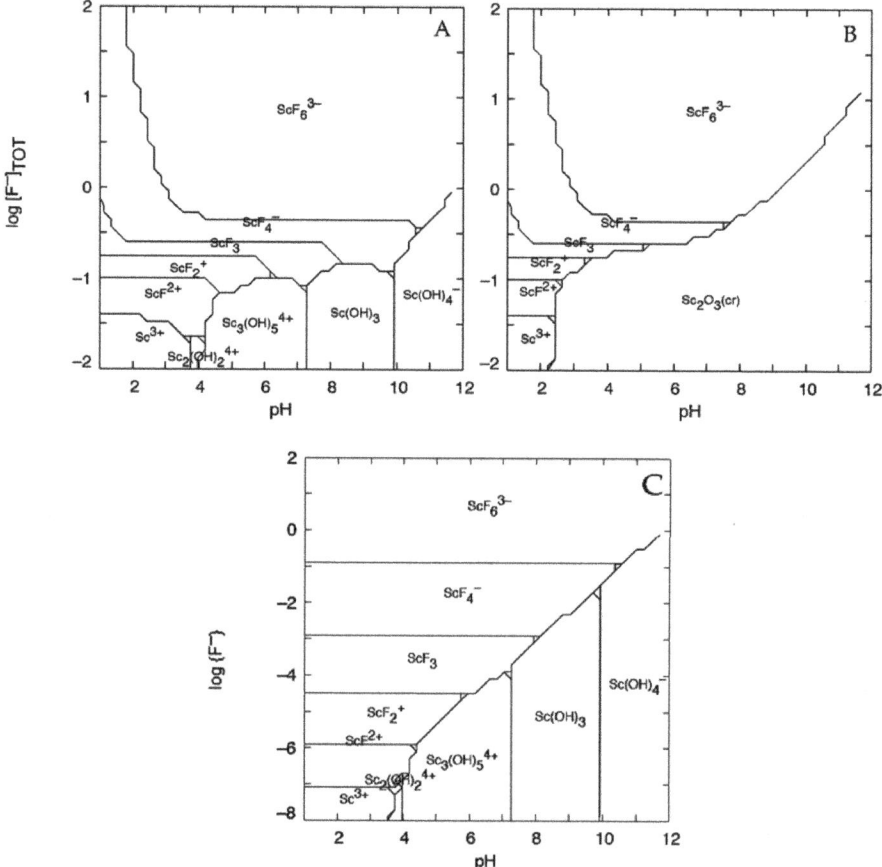

Figure 1. Predominance area diagram of Sc species in fluoride medium at 67 mmol/L (3000 mg/L) Sc at 25 °C. (**A**) Soluble complexes only with respect to logarithm of total fluoride ion concentration and pH; (**B**) Soluble complexes and solids formation with respect to logarithm of total fluoride concentration and pH; (**C**) Soluble complexes only with respect to logarithm of fluoride activity and pH.

The diagrams show that higher Sc-F complexes become more stable with an increase in the F^- ion concentration. The fluoride complexes, $[ScF_n]^{3-n}$, for values of $n = 0$ to 4, are more stable at a total fluoride concentration below 0.5 mol/L, corresponding to $\log[F^-]_{TOT}$ of -0.3; above that, the ScF_6^{3-} complex becomes more dominant. The Sc-OH complexes are more dominant in alkaline media at low fluoride concentrations. The tetra- and hexafluoride complexes are also stable over the entire range of pH values at fluoride concentrations above $\log[F^-]_{TOT}$ -0.6. The strip liquor used in this study had a pH of about 5.6 and a fluoride concentration of about 3 mol/L, corresponding to $\log[F^-]_{TOT}$ of

0.48 mol/L. The region of operation is the one in which ScF_6^{3-} is more stable; therefore $(NH_4)_3ScF_6$ is likely to be precipitated under the experimental conditions. However, Figure 1B shows that Sc_2O_3 is likely to precipitate out as the pH increases above 2.5 within a certain region of lower fluoride concentrations; dominated by hydroxide complexes in Figure 1A and extending to higher fluoride concentrations above pH 8. Presumably, it is actually the hydroxide, $Sc(OH)_3$, that precipitates as the pH increases, but the software most certainly predicts the oxide since it is the most insoluble. The results of this model are not universal and should be used with care; otherwise a new model is required for each different system.

A similar model, depicted in Figure 1C, was conducted by varying the logarithm of fluoride activities; the result matched the one in the literature [30] up to $\log\{F^-\}$ of -3 over the entire range of pH 1–12. The difference between the two models is that the new model considers higher scandium fluoride complexes (ScF_4^- and ScF_6^{3-}) at higher activities $\log\{F^-\}$ between -3 and 2, as well as some polynuclear Sc-OH complexes. Note that $\log\{F^-\}$ is used herein to refer to logarithm of activity and $\log[F^-]_{TOT}$ refers to logarithm of total concentration as shown in Figure 1A–C.

3.2. Strip Liquor Preparation

The chemical analysis of the raffinate phase with O/A: 1/1 showed that after mixing scandium sulphate solution containing 3671 mg/L Sc with the organic phase, almost all of the scandium present in the aqueous phase was extracted to the organic phase with a raffinate scandium concentration of <1 mg/L and an extraction efficiency of >99.9%. The extracted scandium was then stripped from the organic phase by 3 mol/L NH_4F solution. The chemical analysis result of the strip liquor obtained with A/O: 1/1 ratio showed that the strip liquor contained 3660 mg/L scandium with a 98% stripping efficiency.

The high stripping efficiency of scandium with 3 mol/L NH_4F solution is due to the complexation of scandium and fluoride ions in high concentration of ammonium fluoride media [20,31–34] according to the following extraction and stripping reactions Equations (1) and (2), in which the organic extractant is represented as R:

$$Sc^{3+} + 3R \cdot H = R_3Sc + 3H^+ \text{ (during extraction)} \quad (1)$$

$$R_3Sc + 6NH_4^+ + 6F^- = 3NH_4^+ + ScF_6^{3-} + 3(R \cdot NH_4^+) \text{ (during stripping)} \quad (2)$$

The scandium will exist with fluoride and ammonium ions in complexes with different stoichiometry, as shown in Figure 1 with the ScF_6^{3-} complex being dominant at F^- concentrations above 0.5 mol/L.

3.3. Anti-Solvent Crystallization

After the stripping of scandium during the solvent extraction step, a synthetic stock solution of 3660 mg/L scandium-containing ammonium fluoride strip liquor was obtained and was used for anti-solvent crystallization experiments. In order to decrease the solubility of scandium, the direct addition of ethanol, methanol, isopropanol and acetone was tested. The results from these experiments are presented in Figure 2. Figure 3 shows the results from another experiment conducted using ethanol as the anti-solvent at higher ethanol to strip liquor ratios. In these figures, the vol.% reagent added is expressed as a fraction of the strip liquor volume, not the total solution mixture volume. It can be seen that the solubility of scandium decreases as the volume of anti-solvent added to the initial strip liquor increases. This is attributed to the reduction in the solubility of $(NH_4)_3ScF_6$ in the resultant solvent mixture.

According to the information given in Figure 2, acetone was found to be the least effective anti-solvent in precipitating out a Sc phase from the strip liquor, compared to the other solvents. This could be due to differences in the molecular structure between a ketone and an alcohol. The ketone has a limited hydrogen-bonding capability compared with the alcohols, implying that it interacts weakly with water molecules in solution. The lower molecular weight ketones are capable of forming hydrogen bonds with water as hydrogen acceptors due to the lone pair of electrons on the oxygen atom [35], whereas alcohols and water are both hydrogen acceptors and donors. To further understand the degree of interaction of solvents, the solubility parameters of various solvents were computed [36] taking into account the contributions of the dispersion (London) forces, polar forces and hydrogen bonding. This parameter is used to determine the degree of interaction of solvents; the closer the values of the solubility parameters of two solvents, the greater the level of interaction. The solubility parameters of water, methanol, ethanol and acetone have been reported as 23.5, 14.28, 12.92 and 9.77 $cal^{1/2} \cdot cm^{-3/2}$, implying the decreasing solubility of the organic solvents in water in that order. By using 25 vol.% and 50 vol.% of ethanol, methanol or isopropanol, >88% and >98% of Sc present in the strip liquor can be precipitated and separated out from the strip liquor, respectively.

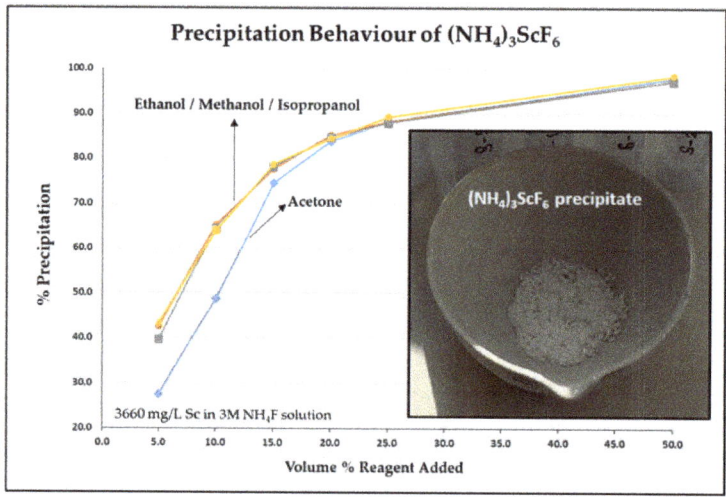

Figure 2. Effect of different anti-solvent addition on the crystallization behavior of scandium.

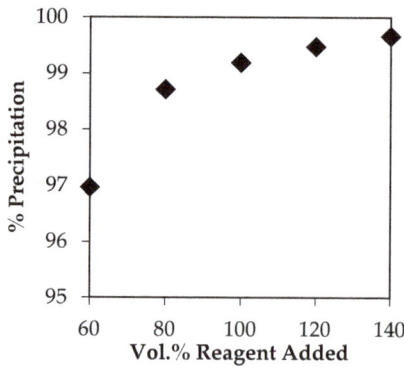

Figure 3. Effect of ethanol addition at higher ratios (Adopted from [37]).

Figures 2 and 3 are in close agreement and the latter shows that the extent of the precipitation becomes asymptotical at a ratio of 0.8 with yields greater than 98.5%. This means that the amount of anti-solvent added could be optimised within the ratio 0.8–1.0 since the further addition of anti-solvent has a minor precipitation effect, hence unnecessary cost and process overloading. After adding the anti-solvent, scandium immediately started to precipitate from the strip liquor in the form of fine crystals, which settled down to the bottom of the beaker when stirring was stopped. The crystals were easily separated by filtration and analysed by powder XRD (X-Ray Diffraction), the result of which is given in Figure 4. According to the XRD data, the precipitated crystals were found to be in the form of $(NH_4)_3ScF_6$ [18].

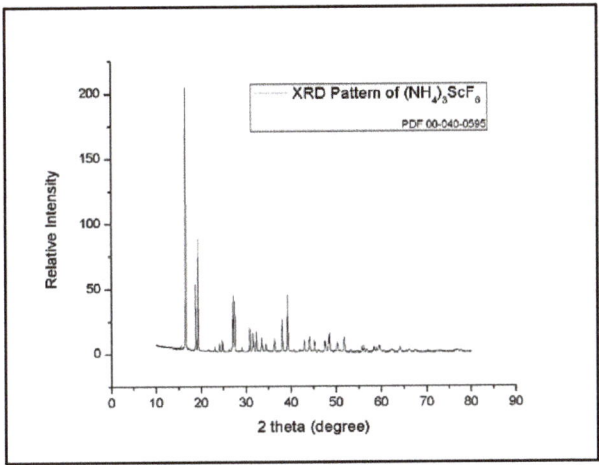

Figure 4. XRD pattern of the obtained $(NH_4)_3ScF_6$ crystals.

At the end, fine $(NH_4)_3ScF_6$ crystals can be obtained, and since this compound is a water soluble form of scandium, it may be marketed to the industry in this form for various applications. Alternatively, the obtained $(NH_4)_3ScF_6$ crystals may be calcined in an atmosphere controlled furnace (to capture HF gas) according to the following chemical reactions Equations (3) and (4) [38] to obtain ScF_3, which is the precursor for the production of Sc metal or alloys by metallothermic or electrometallurgical processing techniques.

$$(NH_4)_3ScF_6 \rightarrow NH_4ScF_4 + 2NH_{3(gas)} + 2HF_{(gas)} \quad (260\text{--}290\ ^\circ C) \tag{3}$$

$$NH_4ScF_4 \rightarrow ScF_3 + NH_{3(gas)} + HF_{(gas)} \quad (340\text{--}350\ ^\circ C) \tag{4}$$

A detailed techno-economic viability study would be required in the future for a comparison of this process with the traditional fluorination route. In the traditional process, scandium was crystallized as the hydroxide or oxalate which were then calcined at temperatures of about 700–800 °C to obtain scandium oxide [13] followed by fluorination using HF acid to obtain ScF_3. In the current processing route, scandium is precipitated as $(NH_4)_3ScF_6$ from an NH_4F solution. This precipitate is then calcined at 350 °C to obtain ScF_3. Whilst large alcohol quantities are required for this process, it is possible to recover the alcohol by distillation for re-use in the crystallization process. This implies higher equipment costs, but a substantial reduction in operating costs could be realised since the alcohol can be re-used, the calcination temperature is much lower and the process eliminates the use of HF acid. On the other hand, the main disadvantage of the envisaged process is the use of alcohols, which are highly flammable and pose a safety hazard in an industrial scale operation. However, using equipment that minimizes the risk of fire, along with adherence to safety standards and regulations can minimize this risk.

The morphology of the $(NH_4)_3ScF_6$ crystals was analysed by SEM as shown in Figure 5. Discrete regular shaped crystals were obtained and the bulk of the crystals had sizes in the range ca. 1–3 µm. There were no significant discrepancies in morphology and crystal sizes observed amongst the different anti-solvents used. All experiments were conducted by the addition of the anti-solvent in bulk without control of supersaturation. It is expected that a better crystal product quality could be obtained by controlling the supersaturation through the controlled addition of the anti-solvent.

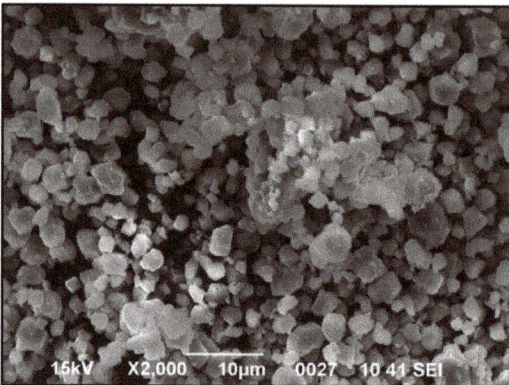

Figure 5. SEM image of $(NH_4)_3ScF_6$ crystals obtained using ethanol at a ratio of 0.2.

4. Conclusions

In this study, scandium was successfully crystallized and recovered from ammonium fluoride strip liquor in the form of fine $(NH_4)_3ScF_6$ crystals by means of an anti-solvent crystallization technique. The use of acetone, ethanol, methanol and isopropanol was predicted to decrease the solubility of $(NH_4)_3ScF_6$, and it was proven that upon adding the above-mentioned anti-solvent reagents, scandium immediately started to crystallize out of the solution in the form of fine $(NH_4)_3ScF_6$ within just 10 min of mixing. Very high precipitate yields above 98% could be obtained with an anti-solvent to strip liquor ratio of 0.8, after which it levelled off. The crystal product was regular shaped with sizes in the range of ca. 1–3 μm, and no discrepancies were observed in the product quality amongst the different anti-solvents employed. Acetone had the least effectiveness at very low ratios, which was ascribed to its limited H-bonding capability in comparison to the alcohols, and hence its weaker solvent-solvent molecular interactions in solution.

Author Contributions: Ş.K. and E.M.P. conceived, designed and performed the experiments; all of the authors analysed the data; C.D. contributed reagents/materials; S.S. and B.F. contributed analysis tools; Ş.K., E.M.P. and K.F. wrote the paper.

Funding: This research was funded by the European Union's Horizon 2020 research and innovation program under the Grant Agreements No. 730105-SCALE.

Acknowledgments: MEAB Chemie Technik GmbH and RWTH Aachen University-IME Department of Process Metallurgy and Metal Recycling are gratefully acknowledged for providing the necessary chemicals and conducting the chemical analyses, respectively.

Conflicts of Interest: The authors declare no conflict of interest.

References

1. European Commission. *Communication from the Commission to the European Parliament, the Council, the European Economic and Social Committee and the Committee of the Regions on the 2017 List of Critical Raw Materials for the EU*; European Commission: Brussels, Belgium, 2017.
2. Magyar, M.J.; Petty, T.R. Draft list of critical minerals. *Fed. Regist.* **2018**, *83*, 33.
3. Binnemans, K.; Jones, P.T.; Müller, T.; Yurramendi, L. Rare Earths and the Balance Problem: How to Deal with Changing Markets? *J. Sustain. Metall.* **2018**, *4*, 126–146. [CrossRef]

4. Laguna-Bercero, M.A.; Kinadjan, N.; Sayers, R.; El Shinawi, H.; Greaves, C.; Skinner, S.J. Performance of La2-xSrxCo0.5Ni0.5O4±δ as an Oxygen Electrode for Solid Oxide Reversible Cells. *Fuel Cells* **2011**, *11*, 102–107. [CrossRef]
5. Boulon, G. Fifty years of advances in solid-state laser materials. *Opt. Mater.* **2012**, *34*, 499–512. [CrossRef]
6. Xia, Z.; Liu, Q. Progress in discovery and structural design of color conversion phosphors for LEDs. *Prog. Mater. Sci.* **2016**, *84*, 59–117. [CrossRef]
7. Riva, S.; Yusenko, K.V.; Lavery, N.P.; Jarvis, D.J.; Brown, S.G.R. The scandium effect in multicomponent alloys. *Int. Mater. Rev.* **2016**, *61*, 203–228. [CrossRef]
8. Wang, W.; Pranolo, Y.; Cheng, C.Y. Metallurgical processes for scandium recovery from various resources: A review. *Hydrometallurgy* **2011**, *108*, 100–108. [CrossRef]
9. Smirnov, D.I.; Molchanova, T.V. The investigation of sulphuric acid sorption recovery of scandium and uranium from the red mud of alumina production. *Hydrometallurgy* **1997**, *45*, 249–259. [CrossRef]
10. Kaya, Ş.; Dittrich, C.; Stopic, S.; Friedrich, B. Concentration and Separation of Scandium from Ni Laterite Ore Processing Streams. *Metals* **2017**, *7*, 557. [CrossRef]
11. Li, D.; Wang, C. Solvent extraction of Scandium(III) by Cyanex 923 and Cyanex 925. *Hydrometallurgy* **1998**, *48*, 301–312. [CrossRef]
12. Hatzilyberis, K.; Lymperopoulou, T.; Tsakanika, L.A.; Ochsenkühn, K.M.; Georgiou, P.; Defteraios, N.; Tsopelas, F.; Ochsenkühn-Petropoulou, M. Process Design Aspects for Scandium-Selective Leaching of Bauxite Residue with Sulfuric Acid. *Minerals* **2018**, *8*, 79. [CrossRef]
13. Lash, L.D.; Ross, J.R. Vitro chemical recovers costly scandium from uranium solutions. *Min. Eng.* **1961**, *220*, 967.
14. Xu, S.Q.; Li, S.Q. Review of the extractive metallurgy of scandium in China (1978–1991). *Hydrometallurgy* **1996**, *42*, 337–343.
15. Martinez, A.M.; Osen, K.S.; Gudbrandsen, H.; Sommerseth, C.; Wang, Z.; Darell, O. Direct Method for Producing Scandium Metal and Scandium-Aluminium Intermetallic Compounds from the Oxides. In *Light Metals 2018*; Martin, O., Ed.; The Minerals, Metals and Materials Society: Pittsburgh, PA, USA, 2018; pp. 1559–1564.
16. Gambogi, J. *Mineral Commodity Summaries: Scandium*; Document No. (703) 648-7718. U.S. Geological Survey: Reston, VA, USA, 2018.
17. Moldoveanu, G.A.; Demopoulos, G.P. Organic solvent-assisted crystallization of inorganic salts from acidic media. *J. Chem. Technol. Biotechnol.* **2015**, *90*, 686–692. [CrossRef]
18. Sviridova, T.A.; Sokolova, Y.V. Pirozhenko, K.Y. Crystal structure of $(NH_4)_5Sc_3F_{14}$. *Crystallogr. Rep.* **2013**, *58*, 220–225. [CrossRef]
19. Sokolova, Y.V.; Cherepanin, R.N. Preparation and examination of the properties of complex scandium fluorides. *Russ. J. Appl. Chem.* **2011**, *84*, 1319–1323. [CrossRef]
20. Mioduski, T.; Gumiński, C.; Zeng, D. IUPAC-NIST Solubility Data Series. 100. Rare Earth Metal Fluorides in Water and Aqueous Systems. Part 1. Scandium Group (Sc, Y, La). *J. Phys. Chem. Ref. Data* **2014**, *43*, 013105. [CrossRef]
21. Puigdomenech, I. Windows software for the graphical presentation of chemical speciation. In Proceedings of the 219th American Chemical Society National Meeting, San Francisco, CA, USA, 26–30 March 2000; American Chemical Society: Washington, DC, USA, 2000.
22. Eriksson, G. An algorithm for the computation of aqueous multicomponent, multiphase equilibria. *Anal. Chim. Acta* **1979**, *112*, 375–383. [CrossRef]
23. Ingri, N.; Kakolowicz, W.; Sillén, L.G.; Warnqvist, B. Errata: High-speed computers as a supplement to graphical methods—V: Haltafall, a general program for calculating the composition of equilibrium mixtures. *Talanta* **1968**, *15*, xi–xii.
24. Ingri, N.; Kakolowicz, W.; Sillén, L.G.; Warnqvist, B. High-speed computers as a supplement to graphical methods—V: Haltafall, a general program for calculating the composition of equilibrium mixtures. *Talanta* **1967**, *14*, 1261–1286. [CrossRef]
25. Constable, E.C. Scandium. *Coord. Chem. Rev.* **1984**, *57*, 229–236. [CrossRef]
26. Kury, J.W.; Paul, A.D.; Hepler, L.G.; Connick, R.E. The Fluoride Complexing of Scandium(III) in Aqueous Solution: Free Energies, Heats and Entropies. *J. Am. Chem. Soc.* **1959**, *81*, 4185–4189. [CrossRef]

27. Baes, C.F.; Mesmer, R.E. *The Hydrolysis of Cations*; Wiley-VCH Verlag GmbH & Co. KGaA: Weinheim, Germany, 1976.
28. Burgess, D. *Standard Reference Data NIST46—NIST Critically Selected Stability Constants of Metal Complexes: Version 8.0*; NIST: Gaithersburg, MD, USA, 2013.
29. Itoh, H.; Hachiya, H.; Tsuchiya, M.; Suzuki, Y.; Asano, Y. Determination of solubility products of rare earth fluorides by fluoride ion-selective electrode. *Bull. Chem. Soc. Jpn.* **1984**, *57*, 1689–1690. [CrossRef]
30. Wood, S.A.; Samson, I.M. The aqueous geochemistry of gallium, germanium, indium and scandium. *Ore Geol. Rev.* **2006**, *28*, 57–102. [CrossRef]
31. Mackay, K.M.; Mackay, R.A.; Henderson, W. *Introduction to Modern Inorganic Chemistry*, 6th ed.; Nelson Thornes Ltd.: Cheltenham, UK, 2002.
32. Watanabe, M.; Nishimura, S. Process for producing fluorides of metals. U.S. Patent 4,741,893, 3 May 1988.
33. Stevenson, P.C.; Nervik, W.E. *The Radiochemistry of the Rare Earths, Scandium, Yttrium and Actinium*; USAEC Technical Information Center: Oak Ridge, TN, USA, 1961.
34. Vickery, R.C. Some Reactions of Scandium. *J. Chem. Soc.* **1956**, 3113–3120. [CrossRef]
35. Ouellette, R.J.; Rawn, J.D. *Principles of Organic Chemistry*; Elsevier: Amsterdam, The Netherlands, 2015.
36. Hansen, C.M. The Three Dimensional Solubility Parameter and Solvent Diffusion Coefficient. Their Importance in Surface Coating Formulation. Ph.D. Thesis, Polytechnic Lrereanstalt, Danmarks Tekniske Højskole, Copenhagen, Denmark, August 1967.
37. Peters, E.; Kaya, Ş.; Dittrich, C.; Forsberg, K. Recovery of scandium by crystallization techniques. In Proceedings of the 2nd International Bauxite Residue Valorization and Best Practices Conference, Athens, Greece, 7–10 May 2018; pp. 401–408.
38. Rakov, E.G.; Mel'nichenko, E.I. The Properties and Reactions of Ammonium Fluorides. *Russ. Chem. Rev.* **1984**, *53*, 851–869. [CrossRef]

© 2018 by the authors. Licensee MDPI, Basel, Switzerland. This article is an open access article distributed under the terms and conditions of the Creative Commons Attribution (CC BY) license (http://creativecommons.org/licenses/by/4.0/).

Article

Conditioning of Spent Stripping Solution for the Recovery of Metals

Tamara Ebner [1,*], Stefan Luidold [1], Matthias Honner [1], Helmut Antrekowitsch [1] and Christoph Czettl [2]

1. Chair of Nonferrous Metallurgy, Montanuniversitaet Leoben, Leoben 8700, Austria; stefan.luidold@unileoben.ac.at (S.L.); matthias.honner@unileoben.ac.at (M.H.); helmut.antrekowitsch@unileoben.ac.at (H.A.)
2. CERATIZIT Austria GmbH, Reutte 6600, Austria; christoph.czettl@ceratizit.com
* Correspondence: tamara.ebner@unileoben.ac.at; Tel.: +43-3842-402-5236

Received: 26 June 2018; Accepted: 21 September 2018; Published: 25 September 2018

Abstract: The objective of this study was to develop an eco-friendly method for processing spent stripping solutions, which originate from the wet chemical decoating of metal cutting tools, to generate a product that represents a useful basis for the recovery of valuable components. These liquids contain, for example, considerable quantities of Ti, Co, and W. Hence, the treatment of these solutions, especially because of the dissolved Co, is essential. The process is based on the precipitation of an insoluble compound with the use of a Ca source. The thermal treatment of the precipitate enables its reuse in the procedure, which leads to a minimum amount of solid process waste. The suggested method, which can be readily controlled by pH adjustment, results in a reduction of hazardous substances and an enrichment of valuable compounds in the solid product. Therefore, this process represents an effective preliminary step in the recovery of concentrated metals, such as Ti.

Keywords: closed-loop circulation; environmentally friendly process; enrichment of Ti; preparation for recovery; reduction of Co; precipitation; thermal treatment; hydrometallurgy

1. Introduction

The wet chemical decoating of metal cutting tools allows for the removal of abraded or faulty coatings and prepares them for subsequent reuse. Therefore, a complete and substrate-sensitive technique is necessary to do this. Previous studies [1–4] have indicated that a mixture of 3.0 mol/L ammonia, 4.7 mol/L hydrogen peroxide, and 0.1 mol/L citric acid at the appropriate process parameters has led to the best results with respect to these requirements. In former investigations, commercially available indexable inserts, which consisted of WC-MX-Co cemented carbides coated with hard materials, were treated. These inserts involved TiAlN-PVD, TiAlTaN-PVD, and TiB_2-CVD coatings as well as a TiN-bonding layer in each case to improve the wear resistance of the cutting tools. Because cemented carbides principally represent a composite material of tungsten carbide and cobalt [5,6] and because there is inevitable degradation of the substrate during the decoating procedure, the spent etchants contain considerable amounts of Co and W. Furthermore, these solutions exhibit other metal ions, such as Ti or Ta, which originate from the dissolved hard coatings. The great number of different substances in the used stripping agents requires processing to reduce the hazardous components, especially Co.

Spent stripping solutions, which have arisen from the previously mentioned studies [1–4], have chemical compositions that are not common by-products of any known processes. Therefore, these liquids represent an exceptional residue for which a novel approach must first be established in the laboratory, because no standardised reconditioning step is currently available. Hence, this study was performed to develop an environmentally friendly and low-waste technique to process these alkaline solutions comprising spent ammonia, hydrogen peroxide, and citric acid and containing Co, Ti, and other elements. However, another question that has arisen is whether the production of a secondary, solid raw material is possible within the process; such a material would constitute a useful basis for the recovery of valuable components, such as Ti.

As mentioned before, one of the main components of these liquids is citric acid. This colourless, odourless, and crystalline solid has the molecular formula $C_6H_8O_7$. Its monohydrate and anhydrous compounds melt at 100 °C and 153 °C, respectively. Citric acid is a weak acid that dissociates stepwise in water, as given in Equations (1)–(4), whereby the distribution of the citric ions is dependent on the dissociation constants and the pH value. H_3AOH represents citric acid, and A corresponds with $C_6H_4O_6$ [7–10].

$$H_3AOH \leftrightarrow H_2AOH^- + H^+ \qquad (1)$$

$$H_2AOH^- \leftrightarrow HAOH^{2-} + H^+ \qquad (2)$$

$$HAOH^{2-} \leftrightarrow AOH^{3-} + H^+ \qquad (3)$$

$$AOH^{3-} \leftrightarrow AO^{4-} + H^+ \qquad (4)$$

According to Al-Khaldi et al. (2007), Figure 1 represents the distribution of citric acid ions depending on the equilibrium pH value. As can be seen, AOH^{3-} respectively Cit^{3-} constitutes the predominant citric acid species in the alkaline milieu, which is the focus of the current study [10].

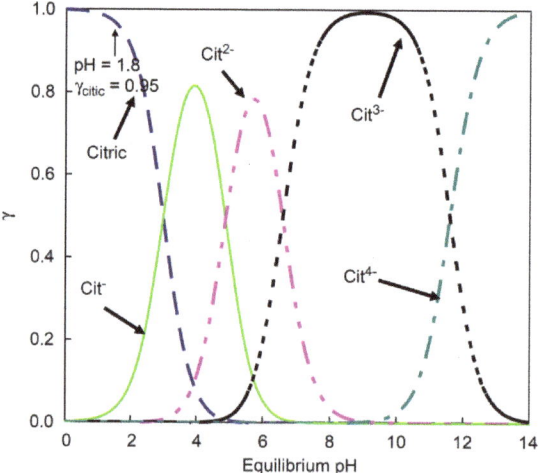

Figure 1. Molar fraction (γ) of different citric acid species depending on the equilibrium pH according to Al-Khaldi et al. (2007) at a temperature of 25 °C and a pressure of 1 atm. Reproduced from [10], with copyright permission from Elsevier, 2007.

Citric acid forms thermodynamically stable chelate complexes with various metal ions, such as Co, Ni, Cu, Fe, and W. In doing so, bonds between the carboxyl or hydroxyl groups and the metal ion occur. As stated by Wyrzykowsky and Chmurzyński (2010), Co^{2+}, Ni^{2+}, Mn^{2+}, or Zn^{2+} form 1:1 complexes at a pH of 6 at 25 °C, because the equilibrium lies at Cit^{2-}, which can also be seen in Figure 1. In some cases, more than one citric acid molecule can be involved in this reaction. Co and citrate ions, for example, can form different complex species in aqueous solutions, such as $[Co(CitH)]$, $[Co(Cit)]^-$, $[Co(Cit)_2]^{4-}$, $[Co(CitH_{-1})]^{2-}$, or $[Co_2(CitH_{-1})_2]^{4-}$. Their distribution depends on the pH value, as well as the metal/ligand ratio in the liquid. This formation of stable complexes impedes the precipitation of the solid metal-containing compounds, leading to certain challenges with respect to the removal of the hazardous components [9–12].

This problem can be solved by the addition of a Ca source and its dissolution into the spent stripping solutions. In the presence of calcium ions, citric acid combines, for example, to form CaHCit, CaH_2Cit^+, or Ca_3Cit_2. The latter exhibits low solubility in water, which decreases with the rising temperature [13,14].

Because the citrate is mainly present as a trivalent Cit^{3-} in the alkaline spent stripping solutions, which have been processed in this study, and tricalcium dicitrate is a known calcium salt of citric acid, the following chemical equation was assumed to take place during the precipitation:

$$3\,Ca^{2+} + 2Cit^{3-} \leftrightarrow Ca_3Cit_2 \qquad (5)$$

Taking advantage of these circumstances enables the processing of the spent stripping agents and the diminution of the metal ions.

2. Materials and Methods

Our suggested method for processing alkaline, metal-containing spent stripping solutions in a laboratory model, therefore, is based on the precipitation of an insoluble compound through the use of a Ca source. The following flow chart, given in Figure 2, shows the steps of the developed treatment.

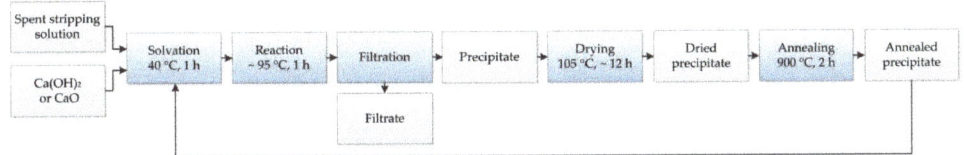

Figure 2. Principal flow chart for the processing of used stripping solutions [3].

First, $Ca(OH)_2$ or CaO (Carl Roth GmbH + Co. KG, Karlsruhe, Germany) was added to the liquid as a Ca^{2+} source in a certain excess. While heating the suspension, the citrate reacted with the dissolved calcium ions to form a poorly soluble calcium citrate compound, which precipitated as a white, voluminous solid. The solid reaction product was separated from the solution by filtration. After drying, the filter cake had to be annealed in air for 2 h at 900 °C. This temperature appeared appropriate for the thermal treatment, because Ca citrates decompose stepwise in CaO through the emission of CO_2, H_2O, and other volatile constituents. Furthermore, the excess of the undissolved $Ca(OH)_2$ reacted to CaO. Figure 3 depicts the thermogravimetric analyses of pure $Ca(OH)_2$ and Ca_3Cit_2, as well as of two precipitates from preliminary investigations, in which $Ca(OH)_2$ served as the precipitating agent. The graph shows the weight losses that occurred in the course of heating these substances to 1000 °C in air. A stepwise decomposition of the materials took place until a constant weight was reached. The product of the thermal procedure, calcium oxide, represented a suitable input material, as a Ca^{2+} source for a renewed use as a precipitating agent for citrates in stripping solutions. This enabled a closed-loop circulation of the solid residue in the form of a low-waste process, in which the losses needed to be compensated for with a fresh material. Aside from the diminution of the citrates, a decline of the metal ion concentrations of the solution occurred, which resulted in an enrichment of certain components in the precipitate. Hence, this method represents an effective preliminary step for the recovery of concentrated compounds from such a solid.

Two different series of tests were performed. Series 1 was meant to provide information regarding the effectiveness of the developed technique and to clarify if the reuse of the thermally treated residue was possible without adding a fresh Ca source. Series 2 was applied with CaO as the starting material in constant excess.

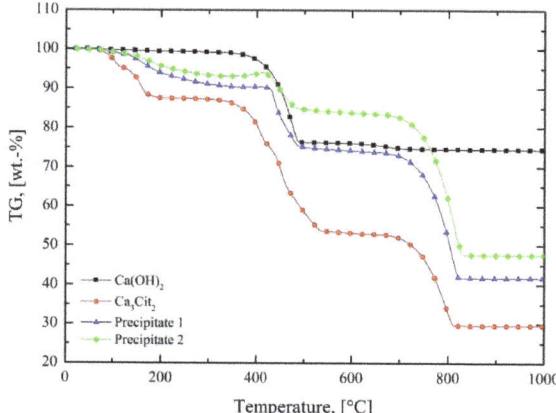

Figure 3. Thermogravimetric analyses in air of pure $Ca(OH)_2$ and Ca_3Cit_2, as well as two precipitates from the aforementioned preliminary investigations, in which $Ca(OH)_2$ served as the precipitating agent.

The experimental procedure was based on the processing method depicted in Figure 2. First, 200 mL of the starting solution was heated to 40 °C while being stirred in a beaker. After reaching this temperature, the addition of a fresh Ca source occurred in the first experiment of each series of the thermally treated precipitate for the subsequent tests. The temperature remained constant for 1 h. Meanwhile, the soluble amount of the solid dissolved. During the ensuing boiling of the suspension for another hour, a calcium citrate compound precipitated in the form of a white and voluminous solid. As required, the evaporated water was compensated for with the addition of deionized water. After cooling the suspension to the ambient temperature, the suspended particles were separated by filtration. The drying of the filter cake took place at 105 °C until its weight remained constant. Then, it was annealed in air for 2 h at 900 °C. The thermally treated residue served as a precipitating agent for the removal of the citrates in the following tests. Depending on the aim of the series, fresh $Ca(OH)_2$ (series 1) or CaO (series 2) was added to the solid. Deionized water was added to the filtrate, which originated from the solid–liquid separation, until the solution reached 200 mL. The solution was analysed with ion chromatography (IC, Ion Chromatography System ICS-2000, Dionex Corporation, Sunnyvale, CA, USA) and inductively coupled plasma mass spectrometry (ICP-MS, Agilent 7500 CE, Agilent Technologies, Santa Clara, CA, USA) to measure the citrate and the metal content of the liquid, respectively.

In series 1 (test designations T1–T6, Figure 4a), 200 mL of two different spent stripping solutions, S1 and S2, were processed by adding Ca^{2+} ions in the form of $Ca(OH)_2$ in triple excess in relation to the citrate concentration present in each solution (T1). The resulting thermally treated residues served as starting materials for the respective, subsequent tests (T2–T5), which were used without the addition of a fresh Ca source. In the last test of the first series (T6), $Ca(OH)_2$ was added to the remaining solid, which ensured a high excess of Ca^{2+} ions. Over the course of the first 5 tests (T1–T5) of series 1, the number of Ca^{2+} ions declined because of losses based on, for example, the significant Ca concentrations in the processed solutions, the filtration, and the handling. The annealed residues, which resulted from the last tests (T6), were analysed by X-ray fluorescence (XRF, XRF WDXRF Axios Panalytical, Malvern Panalytical, Malvern, UK; Almelo, Netherlands) to determine its chemical composition and draw conclusions concerning the metal content of the solid.

Series 2 comprised of 6 experiments (test designations T7–T12, Figure 4b) with 200 mL of two different initial solutions, S3 and S4, and was meant to provide information regarding whether the use of CaO as a precipitation agent was possible. After each test (T7–T11), the Ca losses were compensated for with the addition of fresh calcium oxide, which theoretically led to an approximately constant threefold excess of Ca^{2+} ions over the course of the tests. Therefore, the simplifying assumption implicated that the solid resulting from the previous precipitation only consisted of CaO after the annealing step. The thermally treated solid residue of the last tests (T12) were analysed by XRF, as given in Section 3.

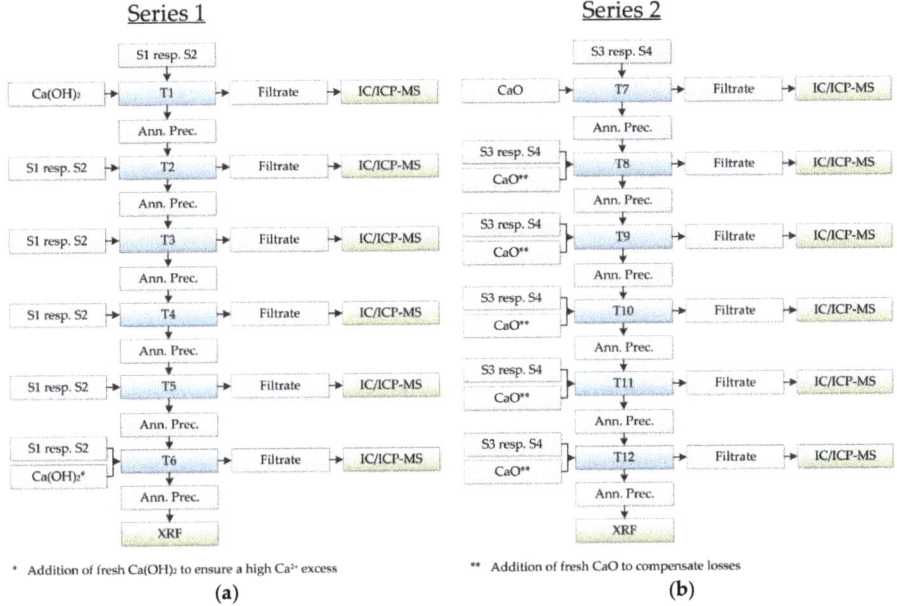

Figure 4. Test procedure of series 1 (**a**) and series 2 (**b**) Ann. Prec., annealed precipitate; IC, ion chromatography; ICP-MS, inductively coupled plasma mass spectrometry; and XRF, X-ray fluorescence.

Initial Solutions

Table 1 summarises the results of the IC and ICP-MS analyses of the 4 different, transparent, and solid-free initial solutions, which were treated in the course of this study. This investigation was performed with a focus on the dissolved metal ions Al, Co, Nb, Ta, Ti, and W, which represent the most important constituents of the applied spent stripping solutions. The liquids S1 and S2 were processed in series 1 (T1–T6), whereas S3 and S4 were examined in series 2 (T7–T12). The pH values of S1, S2, S3, and S4 corresponded with 10.3, 10.4, 10.0, and 9.6, respectively.

Table 1. Concentrations of the citrates and the different metal ions of the initial solutions S1, S2, S3, and S4 (b.a.l., below the analytical limit of determination) measured by IC and ICP-MS [3].

Component	Concentration, [mg/L]			
	Series 1		Series 2	
	S1	S2	S3	S4
Citrate	17,000	16,100	21,000	21,000
Al	1.89	13.0	3.26	2.58
Co	34.7	220	53,1	20.9
Nb	8.19	2.00	b.a.l.	b.a.l.
Ta	13.0	3.99	0.0820	0.0680
Ti	65.3	31.5	490	360
W	160	120	110	69.2

3. Results

Table 2 shows the multiple of the stoichiometrically required amount of Ca^{2+}, calculated by Equation (5), in relation to the citrate concentration present in each spent stripping agent. In the first tests of each series (T1 and T7), a fresh Ca source was applied as a precipitating agent. In series 1, using S1 and S2 as the initial solutions, the thermally treated residues were reused in the subsequent tests without the need of the further addition of a Ca source, as mentioned before. The relative amount declined from T1 to T5 because of losses concerning, for example, the dissolved Ca in the filtrates or the handling. In T5 for the processing of S1, the number of Ca^{2+} ions fell under the stoichiometrically required quantity, so it can be hypothesised that it is not possible to remove all the present citrates. In the last tests (T6) of series 1, fresh $Ca(OH)_2$ was added to the solid for the precipitation, which led to a 5.63- and 5.56-fold excess of Ca^{2+} in S1 and S2, respectively, which should have improved the effectiveness of the process to a greater extent. In series 2 (T7–T12), the Ca losses over the course of the experiments were compensated for with the addition of fresh CaO, which resulted in a theoretically constant threefold excess.

Table 2. The multiple of the stoichiometrically required amount of Ca^{2+} ions in relation to the citrate concentration in the initial solutions S1, S2, S3, and S4 [3].

Series 1			Series 2		
Test Designation	S1	S2	Test Designation	S3	S4
T1	3.00	3.00	T7	3.00	3.00
T2	2.14	2.76	T8	3.00	3.00
T3	1.34	2.38	T9	3.00	3.00
T4	1.32	2.20	T10	3.00	3.00
T5	0.93	1.89	T11	3.00	3.00
T6	5.63	5.56	T12	3.00	3.00

The following graphs in Figure 5 compare the pH value before the experiment (start) with the pH value after the removal of the solid and the cooling of the filtrate of series 1 (Figure 5a) and 2 (Figure 5b). The pH of the filtrates resulting from the processing of S1 (Figure 5a) surpassed the starting value in T1 and T6 only; in T2–T5, it declined constantly until the lowest point of 5.6 was reached in T5. All the final pH values of S2 (Figure 5a) were in the alkaline range. In S2, the lowest pH was also reached in T5 and corresponded to about 8.5. The pH decreased over the course of the tests due to the inevitable losses of the solid. It was possible to counteract this trend by adding a fresh Ca source, as shown in T6. Based on the compensation for the losses in series 2 (Figure 5b), the pH of the filtrates should have been more alkaline than the initial solutions S3 and S4. This assumption can only be confirmed for T7–T10. The reason for the lower pH values in T11 and T12 is discussed in the analysis of the XRF results below.

Figure 6 compares the citrate concentrations of the four initial solutions to the resulting filtrates from series 1 (Figure 6a) and 2 (Figure 6b). By analysing, for example, the outcome of T1 (Figure 6a) and T7 (Figure 6b), it can be concluded that a successful diminution of the citrates was achieved by the use of $Ca(OH)_2$ (series 1) and CaO (series 2), because 97.8% and 98.2% of the citrates in S1 and S2 of series 1, respectively, and 99.7% and 98.2% of the present citrates in S3 and S4 of series 2, respectively, were removed in the first step (Table 3). Furthermore, we can conclude that the reuse of the thermally treated precipitate, according to the proposed process in Figure 2, is possible. The effectiveness of the citrate reduction was associated with the presence of the Ca^{2+} ions and the pH value. Hence, the contents of the citrates in series 1 (Figure 6a) increased during the tests, because the amounts of the provided Ca^{2+} and the pH declined, as given in Table 2 and Figure 5, especially in the T5 for the processing of S1 and S2. The low citrate concentrations of the series 2 filtrates (Figure 6b) showed that the continuous addition of fresh CaO resulted in a satisfying reduction of the citrates, because the citrate concentrations constantly remained below 1 g/L. The comparison of the citrate contents and the pH values allowed us to conclude that the best results were achieved in an alkaline milieu (pH value > 8).

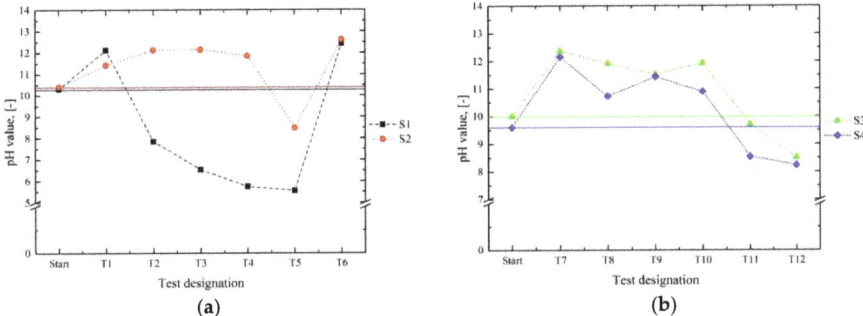

Figure 5. pH value before (start, continuous line) and after the tests (T1–T6) of series 1 (**a**), as well as the tests (T7–T12) of series 2 (**b**) [3].

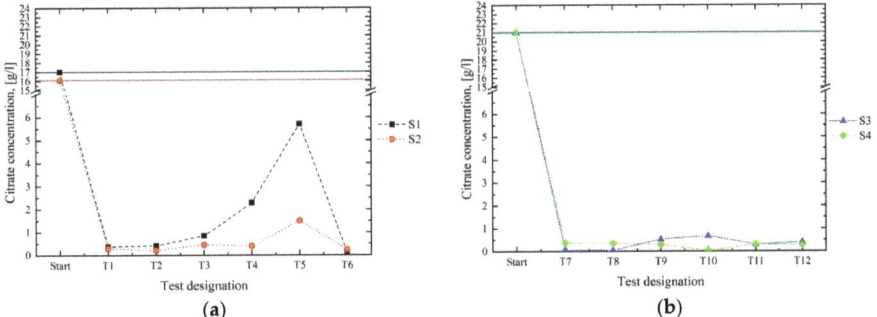

Figure 6. Concentration of the citrates in the initial solutions (start, continuous line) and the filtrates of series 1 (**a**, T1–T6) and series 2 (**b**, T7–T12) [3].

Table 3. Concentrations of the citrates and different metal ions of the treated solutions S1, S2, S3, and S4 after the first tests (T1 and T7, respectively) of series 1 and 2 (b.a.l., below the analytical limit of determination) measured by IC and ICP-MS and the reduction (Red.) in percentage with respect to the initial values given in Table 1 [3].

Component	Series 1				Series 2			
	S1		S2		S3		S4	
	T1, [mg/L]	Red., [%]	T1, [mg/L]	Red., [%]	T7, [mg/L]	Red., [%]	T7, [mg/L]	Red., [%]
Citrate	370	97.8	290	98.2	60.0	99.7	370	98.2
Al	0.110	94.2	0.190	98.5	0.340	89.6	0.130	95.0
Co	1.24	96.4	1.70	99.2	0.620	98.8	0.590	97.2
Nb	b.a.l.	-	b.a.l.	-	-	-	-	-
Ta	0.0260	99.8	0.0240	99.4	b.a.l.	-	b.a.l.	-
Ti	0.0840	99.9	0.0520	99.8	0.160	100	0.0600	100
W	120	25.0	48.3	59.8	12.0	89.1	12.0	82.7

During the precipitation of the calcium citrates, a simultaneous diminution of most other dissolved metal ions occurred. As an example, Table 3 shows the concentrations of the metal ions in the filtrates after the first tests of each series (T1 and T7). Furthermore, Table 3 highlights the reduction in the concentrations in relation to the initial values, given in Table 1. As can be seen, the contents of most of the metal ions, except for W in series 1, were significantly decreased because of the sufficient excess of the added Ca source. For example, a diminution of 96.4% and 99.2% was observed for Co in S1 and S2 in series 1, respectively, and a decrease of 98.8% and 97.2% occurred in S3 and S4 in series 2, respectively, which is of particular importance for this hazardous element. Figures 7–11 compare the initial value (start, Table 1) of an ion in the spent stripping solution to its concentration in the treated filtrates (T1–T6 and T7–T12). The values of the filtrates of T1 and T7 of each figure correspond with the values given in Table 3.

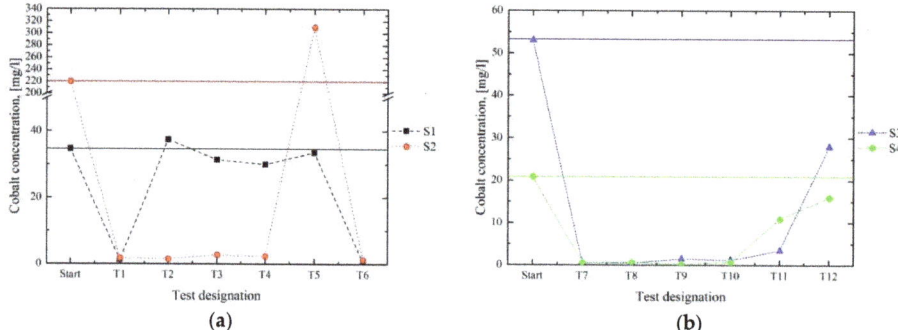

Figure 7. Concentration of cobalt in the initial solutions (start, continuous line) and in the filtrates of series 1 (**a**, T1–T6) and series 2 (**b**, T7–T12) [3].

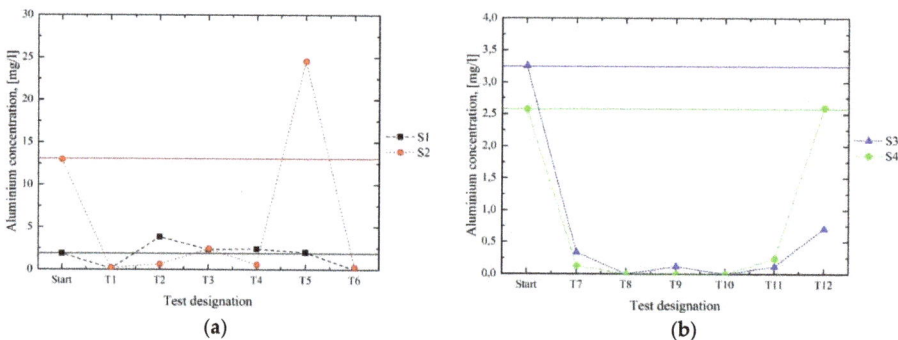

Figure 8. Concentration of aluminium in the initial solutions (start, continuous line) and in the filtrates of series 1 (**a**, T1–T6) and series 2 (**b**, T7–T12) [3].

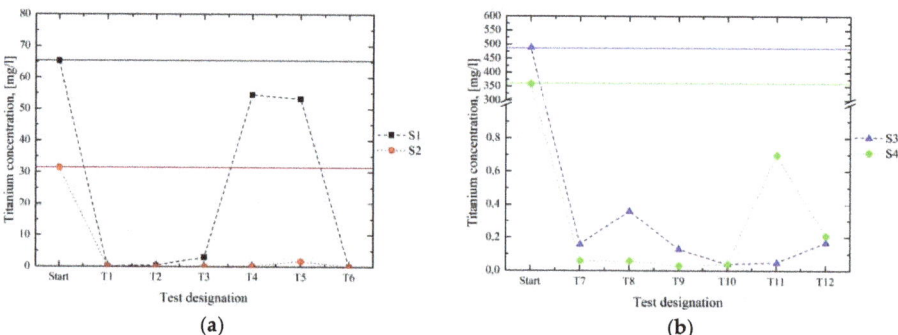

Figure 9. Concentration of titanium in the initial solutions (start, continuous line) and in the filtrates of series 1 (**a**, T1–T6) and series 2 (**b**, T7–T12) [3].

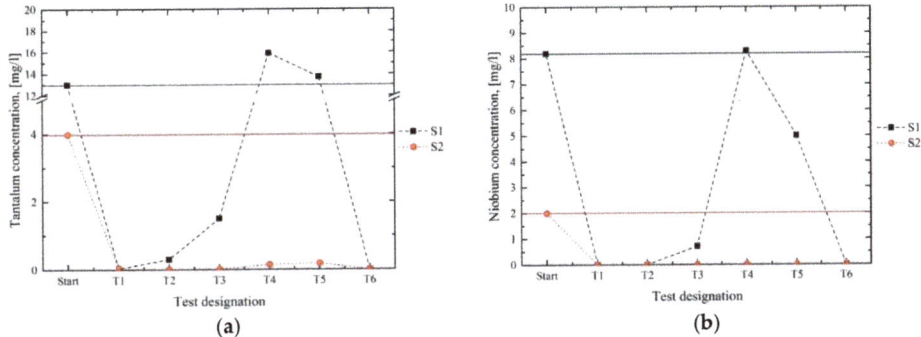

Figure 10. Concentration of tantalum (**a**) and niobium (**b**) in the initial solutions (start, continuous line) and in the filtrates of series 1 [3].

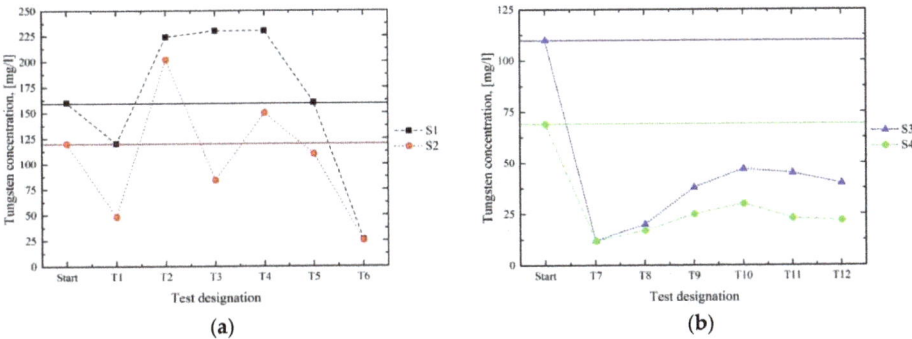

Figure 11. Concentration of tungsten in the initial solutions (start, continuous line) and in the filtrates of series 1 (**a**, T1–T6) and series 2 (**b**, T7–T12) [3].

Figures 7 and 8 present the results for the Co and Al contents of the filtrates. In series 1 (a), the best results were achieved in the first and the last tests (T1 and T6) for both S1 and S2. Throughout the further processing of S1 in T2–T5, hardly any reduction of Co and Al occurred. During the processing of S2 in series 1, most of the Co and Al contents were removed, except for in T5. The results of series 2 (b) show satisfying results in T7–T10. These results indicated that the separation of Co and Al occurred most successfully in a strongly alkaline milieu at pH 10–12. In principle, the metal concentrations in the filtrate—which exceeded the initial contents in the initial solutions, such as in T5 in the processing of S2 for Co and Al—may be explained by the dissolution of an already-separated metal compound in the reused precipitate.

The graphs, given in Figures 9 and 10a, present the removal of Ti and Ta compared with the initial concentrations. The evaluation of the graphs indicated that in series 1 (Figure 9a, Figure 10a) the best results were achieved in tests T1 and T6, whereas the amount of both elements increased in T2–T5. Series 2 exhibited a satisfying reduction of Ti (Figure 9b) and Ta in each test run. Because the concentrations of Ta in the solutions of T7–T12 were located below the analytical limit, no graph is presented for those tests. Regarding the influence of the pH value, we have concluded that a satisfying diminution of both elements is possible in the alkaline range (pH value > 8).

Figure 10b displays the Nb concentrations before and after the tests in series 1. As seen in Table 1, the amount of Nb in S3 and S4 was located below the analytical limit; hence, the results of series 2 are not presented here. For series 1, no Nb was detected in T1, T2, and T6 for S1. In T3–T5 only a partial diminution was achieved. In the course of the treatment of S2, no Nb was detected in any filtrate. Thus, a pH value > 8 led to the successful reduction of Nb.

Figure 11 shows the W content of the filtrates after the tests and compares them with the initial values. The concentrations in series 1 (Figure 11a) were only diminished significantly in T1 and T6. With respect to series 2 (Figure 11b), the lowest contents of W were reached in the first test T7. These results indicated that it is possible to remove certain amounts of W; nevertheless, this element tended to remain dissolved, which implied that no complete precipitation can be achieved.

Table 4 represents the main components and trace elements of the annealed residues after series 1 and 2, which were analysed by XRF. All four solids comprised calcium oxide as a main component, with 92.9, 88.3, 65.4, and 72.7 wt.%, respectively. CaO was formed by annealing the filter cake at 900 °C in air via stepwise decomposition through the separation of H_2O, CO_2, and other volatile components. Because the solid was subsequently heated up to 1000 °C prior to the XRF analysis, an ignition loss occurred based on the removal of other volatile substances. Concerning the most important compounds formed by the substrate and coating elements, TiO_2 and Al_2O_3 were especially noteworthy. The content of Al_2O_3 accounted for 0.07, 0.28, 0.40, and 0.41 wt.% in the residues of S1, S2, S3, and S4, respectively. A clear difference between the two series can be seen regarding the TiO_2 contents, with 1.03 and 0.81 wt.% in S1 and S2 of series 1, respectively, and 21.0 and 17.0 wt.% in S3 and S4 of series 2, respectively. This can be explained by the enhanced removal of Ti in series 2 due to the theoretically constant excess of Ca^{2+} ions and the higher Ti concentration in the initial solutions, as given in Table 1. Other elements, such as Co, W, Ta, and Nb separated to certain extents from the liquid, but their enrichment in the solid merely remained in the ppm range.

Table 4. Main compounds and trace elements of the annealed precipitates at the end of each test series processing the initial solutions S1, S2, S3, and S4, analysed by X-ray fluorescence (XRF) (I.L., ignition loss) [3].

Component	Main Compounds, (wt.%)			
	Series 1		Series 2	
	S1	S2	S3	S4
Al_2O_3	0.07	0.28	0.40	0.41
CaO	92.9	88.3	65.4	72.7
Fe_2O_3	0.14	0.14	0.18	0.21
I.L.	2.19	3.46	5.16	3.46
MgO	0.86	0.91	1.26	0.44
MnO	0.02	0.02	0.09	0.10
P_2O_5	0.03	0.56	1.46	2.33
SiO_2	0.19	0.27	0.53	0.61
SO_3	0.12	0.33	0.16	0.21
TiO_2	1.03	0.81	21.0	17.0
Σ	97.6	95.1	95.6	97.4
Component	Trace Elements, (ppm)			
	Series 1		Series 2	
	S1	S2	S3	S4
Co	1506	25,430	11,000	4720
Nb	438	224	22.0	17.00
Ta	980	400	b.a.l.	b.a.l.
W	6800	530	14,300	6510

4. Discussion

The results of this study allowed us to conclude that a removal of citrates from spent stripping solutions based on the proposed method, stated in Figure 2, is possible. The findings of series 1 indicated that the separation rate can be enhanced by the addition of a fresh Ca compound. Hence, a refreshing of the annealed solid, for example, with $Ca(OH)_2$ or CaO appears reasonable. This presumption was confirmed by the outcome of series 2, because the reduction of the pH values in both of the final tests had no negative influence on the removal of the citrates. This follows the fact that the supply of Ca^{2+} ions due to the addition of a fresh Ca source was sufficient for the formation of calcium citrate.

Parallel to the citrate removal over the course of precipitation, a diminution of various metal ions also occurred. As shown in Table 3, the removal of a high proportion of most of the dissolved ions was observed, especially with a sufficient excess of a Ca source. The decisive factor for this diminution was the pH value. A decline of the pH to 8 (series 2) had no negative influence on the separation rate of the citrates; nevertheless, it led to an increase of several metal ions in the filtrates, as shown by the trend of Co in series 2, as shown in Figure 7. The decrease of the pH in T11 and T12 of series 2 (Figure 5b) can be explained by the aforementioned separation of the metal ions, such as Ti. Therefore, the annealed residue not only consisted of CaO, but also of a certain amount of other metal oxides, such as TiO_2. Because an accumulation of these oxides occurred in series 2, the decreasing amount of CaO and the subsequent decline of the pH value became especially apparent in the last tests T11 and T12. This was confirmed by the XRF analyses given in Table 4, because the annealed solid arising from the processing of S3 only consisted of 65.4% CaO; the remaining part comprised various other oxides. As a consequence of this continuous enrichment, the aspired threefold excess from the addition of a fresh Ca source was not reached in the real system. Hence, the theoretical excess resembled the real excess at the beginning of series 2 but declined in further tests. This circumstance could be counteracted by analysing the residue after every test, which would give a better understanding of the oxidic content. The required amount of input material for the XRF measurement needed to be compensated for by adding fresh CaO in the precipitating agent, which constituted a source of error in the subsequent experiments. Because the necessary amount of solid for the analysis approximately corresponds with the mass of the annealed solid for processing 200 mL of the spent stripping solution, this analysis procedure seems to be disadvantageous.

The results of series 1 and 2 for Co and Al (Figures 7 and 8) indicated that a successful removal of these elements is possible, especially in an alkaline medium. The best results were achieved at pH values around 12. For further relevant constituents, such as Ti, Ta, and Nb (Figures 9 and 10), high separation rates were obtained as well, even at a comparatively lower pH value of > 8.

The removal of W (Figure 11) presented a certain challenge, because the solubility of this element strongly depends on the pH value. The results of series 1 indicated that a high excess of Ca^{2+} ions and the resulting pH values of 12.4 and 12.6 in T6 led to a reduction rate of 83.6% and 78.8% in S3 and S4, respectively. Consequently, no satisfying separation of W, just a certain diminution, was achieved. Nevertheless, these results allowed us to conclude that, at least, the removal of a high proportion of W occurs in a strongly alkaline milieu (pH value > 12).

With respect to the composition of the annealed solids (Table 4) resulting from series 1 (S1 and S2), only a slight enrichment of the separated elements occurred. The content of 6800 ppm W in the annealed precipitate of S1 was attributed to the considerable concentration of 160 mg/L W in the initial solution (Table 1). In contrast, the residue of S2 contained 25,430 ppm Co as a result of the high Co concentration of 220 mg/L in the spent stripping solution. Due to the enhancement in series 2 regarding the addition of a fresh Ca source, an enrichment of 21.0 and 17.0 wt.% TiO_2 was achieved. This provides a useful starting base for further investigations concerning the recovery of valuable components. The optimised procedure in series 2, furthermore, led to higher concentrations of W (14,300 and 6510 ppm in S3 and S4, respectively) and Co (11,100 and 4720 ppm in S3 and S4, respectively) in the annealed solid.

5. Conclusions

During the wet chemical decoating of metal cutting tools, as shown in previous studies [1–4], spent stripping solutions arise which contain, for example, considerable quantities of citrates, Ti, Co, and W. Based on their compositions, these liquids represent a novel type of process residue, which do not resemble any other common by-products of any currently known processes. Therefore, this study was conducted to develop a new, environmentally friendly and low-waste method in a laboratory model to process these spent stripping solutions. A further objective was to transfer the dissolved metal ions into a solid to provide the basis for the recovery of these valuable components.

Finally, the results of the tests have allowed us to conclude that the proposed process (stated in Figure 2) represents a successful, pH-controlled method for processing spent stripping solutions. The treatment of the stripping media with a minimum threefold excess of a Ca source in relation to the citrate ions in the solution enabled the precipitation of a poorly soluble calcium citrate compound in the form of a white, voluminous solid, which can be readily separated. Particularly noteworthy is the fact that this led to the simultaneous reduction of the majority of the dissolved metal ions in the strongly alkaline milieu (pH value > 12). This phenomenon is especially important with respect to hazardous substances, such as Co. The thermal reconditioning step allowed for the reuse of the annealed solid as a precipitating agent for the citrates in the stripping solutions. As a result of this closed-loop process, a minimum amount of process waste was produced, and the enrichment of certain valuable components in the solid was achieved. This method constitutes a suitable, as well as a low-waste, basis for the processing of spent stripping solutions and, furthermore, offers the opportunity to recover valuable constituents due to their accumulation in the annealed solid.

6. Prospects

Because the noteworthy enrichment of several valuable components in the solid product was achieved, based on the closed-loop process, this method represents an effective preliminary step for the recovery of such valuable components. As concentrations of 21.0 and 17.0 wt.% TiO_2 in the solids of series 2 were obtained, the focus of future studies will be on the recovery of this constituent. A combination of various pyro- and hydrometallurgical processes should lead to a successful separation of titanium and titanium compounds, respectively. After calcination, the selective leaching of the associated components with different acids at suitable process parameters, including pH, concentration, and temperature, seems feasible. Another possibility may be to use a chloride process to separate undesirable compounds by chlorination. The resulting $TiCl_4$ could be converted into the pigment TiO_2 in a further step.

In conclusion, the suggested method leads to the removal of hazardous substances from spent stripping solutions, as well as enabling the satisfying enrichment of the valuable components in the solid product. Therefore, this technique represents a useful basis for the recovery of, for example, titanium or titanium compounds by various pyro- and hydrometallurgical processes.

Author Contributions: Conceptualization, investigation, and methodology, T.E. and M.H.; Formal analysis, T.E., S.L., M.H., H.A. and C.C.; Project administration, S.L.; Resources, C.C.; Supervision, H.A.; Validation, T.E., S.L., H.A. and C.C.; Visualization and original draft preparation, T.E.; and Review and editing of the final manuscript, S.L., H.A. and C.C.

Funding: This research was funded by Austrian Research Promotion Agency (FFG), the grant number was 851894.

Acknowledgments: The authors would like to thank the Austrian Research Promotion Agency and the Austrian Ministry for Transport, Innovation, and Technology, as well as the Federal Ministry of Science, Research, and Economy for supporting this project (FFG project number: 851894).

Conflicts of Interest: The authors declare no conflicts of interest.

References

1. Ebner, T.; Kücher, G.; Luidold, S.; Schnideritsch, H.; Czettl, C.; Storf, C.; Antrekowitsch, H. Reuse of metal cutting tools by wet chemical decoating. In Proceedings of the EMC 2015—European Metallurgical Conference, Düsseldorf, Germany, 14–17 June 2015; GDMB Verlag GmbH: Clausthal-Zellerfeld, Germany, 2015; Volume 2, pp. 725–738.
2. Ebner, T.; Kücher, G. Wet chemical decoating of metal cutting tools. In Proceedings of the EMC 2015—European Metallurgical Conference, Düsseldorf, Germany, 14–17 June 2015; GDMB Verlag GmbH: Clausthal-Zellerfeld, Germany, 2015; Volume 2, pp. 1104–1105.

3. Ebner, T. Concepts for the Optimization of Cemented Carbide Recycling. Ph.D. Thesis, Montanuniversität Leoben, Leoben, Austria, 2016.
4. Kücher, G. Chemical Stripping of Cemented Carbides. Master's Thesis, Montanuniversität Leoben, Leoben, Austria, 2014.
5. Lassner, E.; Schubert, W.D. Tungsten in Hardmetals. In *Tungsten: Properties, Chemistry, Technology of the Element, Alloys, and Chemical Compounds*; Kluwer Academic/Plenum Publishers: New York, NY, USA, 1999; pp. 321–324. ISBN 0-306-45053-4.
6. Schedler, W. Was ist Hartmetall—Was sind Hartstoffe. In *Hartmetall Für Den Praktiker: Aufbau, Herstellung, Eigenschaften und Industrielle Anwendung Einer Modernen Werkstoffgruppe*; VDI-Verlag: Düsseldorf, Germany, 1988; pp. 1–20, ISBN 9783184008031.
7. Sicherheitsdatenblatt Citronensäure ≥ 99.5%, p.a., ACS, Wasserfrei. Available online: https://www.carlroth.com/downloads/sdb/de/X/SDB_X863_DE_DE.pdf (accessed on 12 January 2016).
8. Sicherheitsdatenblatt Citronensäure Monohydrat zum Entkalken. Available online: https://www.carlroth.com/downloads/sdb/de/1/SDB_1818_DE_DE.pdf (accessed on 12 January 2016).
9. Verhoff, F.H. Citric Acid. In *Ullmann's Encyclopedia of Industrial Chemistry, Volume A 7, Chlorophenols to Copper Compounds*, 5th ed.; Gerhartz, W., Yamamoto, Y.S., Campbell, F.T., Pfefferkorn, R., Rounsaville, J.F., Eds.; VCH Verlagsgesellschaft mbH: Weinheim, Germany, 1986; pp. 103–108. ISBN 3-527-20107-6.
10. Al-Khaldi, M.H.; Nasr-El-Din, H.A.; Mehta, S.; Al-Aamri, A.D. Reaction of citric acid with calcite. *Chem. Eng. Sci.* **2007**, *62*, 5880–5896. [CrossRef]
11. Frank, A.C.; Sumodjo, P.T.A. Electrodeposition of cobalt from citrate containing baths. *Electrochim. Acta* **2014**, *132*, 75–82. [CrossRef]
12. Wyrzykowski, D.; Chmurzyński, L. Thermodynamics of citrate complexation with Mn^{2+}, Co^{2+}, Ni^{2+} and Zn^{2+} ions. *J. Therm. Anal. Calorim.* **2010**, *102*, 61–64. [CrossRef]
13. Burgos, G.; Birch, G.; Buijse, M. Acid Fracturing with Encapsulated Citric Acid. In Proceedings of the SPE International Symposium and Exhibition on Formation Damage Control, Lafayette, LA, USA, 18–20 February 2004.
14. Apelblat, A. Solubilities of organic salts of magnesium, calcium and iron in water. *J. Chem. Thermodyn.* **1993**, *25*, 1443–1445. [CrossRef]

© 2018 by the authors. Licensee MDPI, Basel, Switzerland. This article is an open access article distributed under the terms and conditions of the Creative Commons Attribution (CC BY) license (http://creativecommons.org/licenses/by/4.0/).

Article

Distribution of Selected Trace Elements in the Bayer Process †

Johannes Vind [1,2,*], Alexandra Alexandri [2], Vicky Vassiliadou [1] and Dimitrios Panias [2,*]

1. Department of Continuous Improvement and Systems Management, Aluminium of Greece Plant, Metallurgy Business Unit, Mytilineos S.A., Agios Nikolaos, 32003 Viotia, Greece; vicky.vassiliadou@alhellas.gr
2. School of Mining and Metallurgical Engineering, National Technical University of Athens, Iroon Polytechniou 9, Zografou Campus, 15780 Athens, Greece; aalexandri@metal.ntua.gr
* Correspondence: johannes.vind@alhellas.gr (J.V.); panias@metal.ntua.gr (D.P.); Tel.: +30-210-7722184 (J.V.); +30-210-772-2276 (D.P.)
† This paper is the written, extended and updated version of the oral presentation entitled as "Distribution of Trace Elements Through the Bayer Process and its By-Products", presented in the 35th International Conference and Exhibition ICSOBA-2017 which was held in Hamburg, Germany from 2–5 October 2017; Vind, J.; Vassiliadou, V.; Panias, D. Distribution of trace elements through the Bayer process and its by-products. In Proceedings, Travaux 46; Hamburg, Germany, 2–5 October 2017; pp. 255–267; https://icsoba.org/sites/default/files/2017papers/Alumina%20Papers/AA14%20-%20Distribution%20of%20Trace%20Elements%20Through%20the%20Bayer%20Process%20and%20its%20By%20Products.pdf (accessed on 23 March 2018).

Received: 9 April 2018; Accepted: 3 May 2018; Published: 8 May 2018

Abstract: The aim of this work was to achieve an understanding of the distribution of selected bauxite trace elements (gallium (Ga), vanadium (V), arsenic (As), chromium (Cr), rare earth elements (REEs), scandium (Sc)) in the Bayer process. The assessment was designed as a case study in an alumina plant in operation to provide an overview of the trace elements behaviour in an actual industrial setup. A combination of analytical techniques was used, mainly inductively coupled plasma mass spectrometry and optical emission spectroscopy as well as instrumental neutron activation analysis. It was found that Ga, V and As as well as, to a minor extent, Cr are principally accumulated in Bayer process liquors. In addition, Ga is also fractionated to alumina at the end of the Bayer processing cycle. The rest of these elements pass to bauxite residue. REEs and Sc have the tendency to remain practically unaffected in the solid phases of the Bayer process and, therefore, at least 98% of their mass is transferred to bauxite residue. The interest in such a study originates from the fact that many of these trace constituents of bauxite ore could potentially become valuable by-products of the Bayer process; therefore, the understanding of their behaviour needs to be expanded. In fact, Ga and V are already by-products of the Bayer process, but their distribution patterns have not been provided in the existing open literature.

Keywords: Bayer process; trace elements; vanadium; gallium; rare earth elements; lanthanum; yttrium; scandium; karst bauxite; bauxite residue; red mud

1. Introduction

The ever-increasing growth in the electronics industry, the production of light-weight electric vehicles as well as devices for generating renewable energy have imposed an accelerating demand for specific raw materials such as the rare earth elements (REEs) [1–3]. Several economic regions like the European Union, Japan or USA have identified raw materials that are categorised as "critical", relating to their supply risk and economic importance [1,4]. The search for new sources for critical metals has encouraged research work in, amongst others, the alumina industry, in which the caustic process liquor as well as its by-product known as bauxite residue or red mud are prospective sources of certain critical metals [5–8].

More than 50 chemical elements existing typically in bauxites occur in higher concentrations than 1 mg/kg [9] and, therefore, can be considered as trace elements. The alumina industry is particularly interested in the bauxite trace elements for the following reasons: (1) some of them (such as vanadium—V) are undesired impurities that might end up in the product and, therefore, their fractionation through the process must be controlled [10]; (2) some of them (such as gallium—Ga—and in some cases scandium—Sc) can be extracted mainly as a by-product of the primary alumina industry [6,11–13]; (3) some of them (such as beryllium—Be) are considered hazardous from environmental and occupational health points of view [14]. Therefore, keeping track on the fate of trace elements within the Bayer process is very important for the best plant performance and the desired purity of products.

In the present work we set our focus on the Bayer process related trace elements that have a prospect or already possess an existing value of being or becoming a profitable by-product. It is well-known that Ga is worldwide mainly produced as a by-product of Bayer process [8,15] and V can also be recovered as vanadium sludge during the production of alumina [7,16]. There exists a growing interest in the extraction of the relatively valuable REEs and Sc from bauxite residue by exploiting various hydrometallurgical or combined pyro- and hydrometallurgical routes [5–7,11,17,18]. In addition, the demand for these metals has been increasing steadily and further increase is projected in the future [1,3,15].

For a better clarity in further discussion, it should be noted that REEs are a group of chemical elements known as the lanthanides as well as yttrium (Y). Sc is also considered often as a REE, but there is no conclusive consensus in this question. Based on the chemical properties, REEs are usually divided to light REEs (LREE, lanthanum to europium) and heavy REEs (HREE, gadolinium to lutetium and Y) [19], which is also the official IUPAC definition.

To name a few applications where some of the metals discussed in the present paper are used, it can be mentioned that Ga is dominantly used in semi-conductors as well as in light emitting diodes [8,20]. The primary use of V is in the steel industry, where this alloying metal provides grain refinement and hardenability [21]. A wide matrix of applications exists for the REEs, such as strong permanent magnets used in electric motors, catalysts, batteries, phosphors, polishing and many more [19]. The interest is growing in the so far relatively scarcely used Sc, which can be utilised to produce light-weight aluminium (Al) alloys beneficial in the aerospace industry. Another rapidly growing field of Sc applications is in solid oxide fuel cells, which accounts now for about 90% of the use of this metal [1,22].

Regardless of the long history of Ga research in alumina industry [9,12,23], it is not easy to retrieve published data of the distribution of Ga in the Bayer process. It has been emphasised that this missing gap in the accessible information has affected the compilation of worldwide resource estimation exercises, as there is no source available that relates the known Ga concentrations with actual material mass flows in the Bayer process [8,15]. Similar gaps exist for V, chromium (Cr), arsenic (As) and the REEs. For instance, Deady et al. [24] have identified this gap in the available literature and recommend to assess the enrichment of REEs from bauxite to its residues by relating the actual source and resulting materials. Some information can be retrieved about the distribution patterns of lanthanum (La) and Sc in the Bayer process [25], but given the date of this study (1981) it is useful to update and build new knowledge upon that existing study. For the rest of the elements considered in this work, at most the fractionation indexes between bauxite and derived residue or merely concentrations can be found [26–28], and those are also often given based on lab-scale experiments [29,30]. Comprehensive descriptions exist that provide the behaviour and distribution patterns of other bauxite trace elements like molybdenum (Mo), zinc (Zn) [31], Be [14,32], thorium (Th) and uranium (U) [33] as well as mercury (Hg) in the Bayer process [34,35]. It is of high interest to examine in particular the karst/diasporic bauxite trace element distribution in the Bayer process, because these types of bauxites are relatively more enriched in trace elements compared to lateritic/gibbsitic bauxites [36,37].

The aim of this work is therefore to establish the distribution patterns of selected trace elements (Ga, V, As, Cr, REEs, Sc) in the Bayer process and its by-products by formulating mass balance models of the named elements based on a case study. The results of the present work can be further utilised as one of the sources for compiling global resource estimations, Bayer-process-related resource estimations, planning of by-product production from Bayer process materials, etc.

2. Materials and Methods

The Bayer process is a cyclic method that utilises sodium hydroxide leaching of bauxite ore to produce technically pure alumina (>98.3% Al_2O_3) [38–41]. Aluminium of Greece plant (Metallurgy Business Unit, Mytilineos S.A.; hereafter denoted as AoG) uses a set of processing conditions that are known in the industrial sector as high temperature digestion (HTD). These conditions (T > 250 °C, elevated pressure) are dictated by the utilisation of mainly karst bauxite, in which primary alumina-containing minerals are diaspore (α-AlO(OH), digested mostly at >250 °C in the presence of lime) and boehmite (γ-AlO(OH), digested mostly at >240 °C) that dissolve less readily than the more commonly exploited gibbsite mineral ($Al(OH)_3$, digested mostly at 140–150 °C) [9,42,43].

A simplified flow diagram of AoG's process is shown in Figure 1. Because Parnassos-Ghiona bauxite in its natural position is situated between limestones, it is necessary to remove the unwanted limestone from the ore that is inevitably partly mined as a contaminant together with bauxite. Limestone, mineralogically composed mainly of calcite, is removed by heavy media separation (HMS) in ferrosilicon slurry [44,45], also referred as "decalcitation" (sic) process in the literature [44]. This operation unit is shown in Figure 1 as "HMS", marked by a dotted line, because it is not strictly a part of the conventional Bayer process. The primary output of this unit is mixed karst bauxite (also the main input to the Bayer process, 73% of total bauxite mass) and secondary output is "decalcitation residue". Karst bauxite is ground in the presence of concentrated leach liquor to achieve granulometry <315 µm and the resulting suspension is pre-heated to about 180 °C. Digestion of the karst bauxite suspension is performed at about 255 °C and a pressure of about 5.8–6.0 MPa for approximately one hour. To increase the productivity of the Bayer process, AoG also utilises an optimisation step that is termed as the "sweetening" process. In the "sweetening" process, lateritic/gibbsitic bauxite is digested at a lower temperature after the digestion of karst/diasporic bauxite. Lateritic bauxite suspension passes through a pre-desilication step with a residence time of about 24 h, to allow the formation of desilication products (sodalite and cancrinite) and to avoid the problems of reactive silica (i.e., kaolinite) during digestion. In the case of AoG, lateritic bauxite suspension is introduced to the main karst bauxite slurry in the appropriate flash stage after the HTD of karst bauxite suspension. Lime is added to the process during HTD as a reaction catalyst, as well as during causticisation step that reduces soda losses and as a filter aid during the security filtration of pregnant liquor after the settling stage (liquor "polishing") [44,46]. From the leached effluent slurry after digestion, the solid fraction is separated as residue slurry or red mud (bauxite residue in the suspended form) by settling and washing. To obtain de-watered bauxite residue that helps to reduce the losses of soda and eases the stacking as well as further utilisation of residue, AoG makes use of the plate and frame filter pressing of the initial residue slurry after the settling and washing unit [7]. The clear pregnant liquor, rich in sodium aluminate, passes to the next processing step where crystalline aluminium hydroxide ($Al(OH)_3$) is precipitated. Precipitation is initiated by the introduction of aluminium hydroxide seed crystals. The spent liquor after precipitation unit is concentrated in the evaporation unit, to create the necessary sodium hydroxide concentration level for the next processing cycle. Aluminium hydroxide, which is the final product of the Bayer process, is calcined at >1000 °C to produce anhydrous alumina (Al_2O_3) [38,39]. Sometimes, the Bayer process is divided into the "red side" to denote the units where bauxite and its residue are present, and to "white side" to indicate the stages after residue removal (clarification) until precipitation and evaporation stages [40,47].

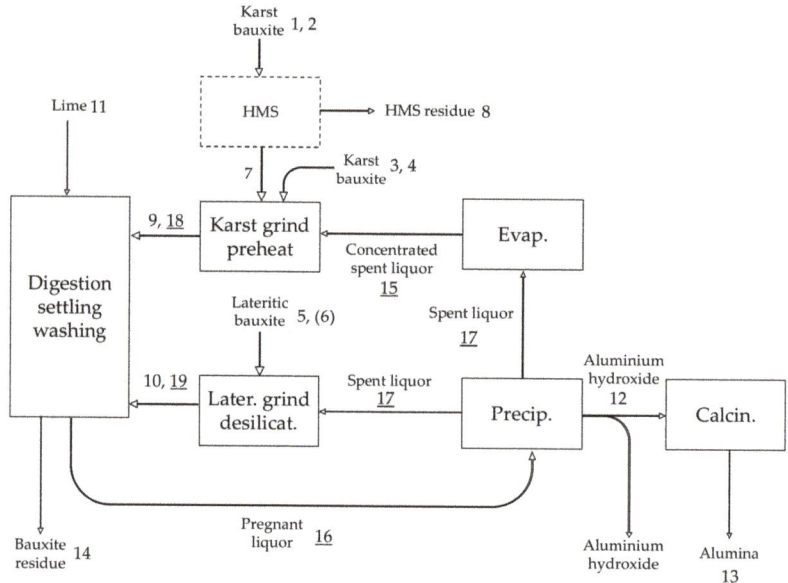

Figure 1. Simplified flowsheet of the Bayer process. Numbers indicate the sampled materials, whilst underlined numbers refer to samples obtained in liquor form. Numbers correspond to Table 1. HMS: Heavy media separation.

2.1. Sampling and Technological Data

Sampling took place over a three-day period and materials were collected from key points in the process flow sheet to provide a snapshot of the whole process. The precondition of such sampling procedure is that all the input constituents should appear in the output materials of the process and sampled output material corresponds to the sampled input material. The sampling points are shown in Figure 1, and sample descriptions are detailed in Table 1. AoG uses largely two types of bauxite feed: locally mined karst (diasporic/boehmitic) bauxite and imported lateritic (gibbsitic) bauxite. The Greek karst bauxite samples originate from the Parnassos-Ghiona B3 stratigraphic horizon, which is the youngest and most exploited horizon of the deposit [48]. A minor amount of B2 stratigraphic horizon Parnassos-Ghiona bauxite was also used at the time of sampling, but the exploitation of this material is currently suspended. More details about the Parnassos-Ghiona deposit can be found for example from Deady et al. [24]. Another minor source of karst bauxite at the time of sampling was diasporic bauxite from Turkey, Milas area. The lateritic bauxites used at AoG originate from Brazil (Porto Trombetas) [49] and Ghana (Awaso), while only Brazilian bauxite was processed in the period of sampling campaign.

Bauxite samples were collected from the one-tonne test batches to provide the best representation of the feed material. Bayer liquors, aluminium hydroxide, alumina and lime samples were collected from the appropriate sampling points according to the internal protocols of AoG. A composite sample of bauxite residue was collected after the filter pressing of the residues. Bauxite residue from AoG is known to have a relatively stable REEs and Sc concentration (8% variation in 15 years), known from the long-term research experience related to this material [6]. Fresh sodium hydroxide addition to the process was negligible during the sampling period and was therefore excluded from the analysis. Process data for both solid and the liquid mass flows were acquired for the same period as the sampling took place.

Table 1. Description of the sampled materials.

Material		No.	Code	Description
Input bauxite	Karst/diasporic	1	DD-BX	Parnassos-Ghiona bauxite from B3 horizon. Extracted by company Delphi-Distomon S.A., a subsidiary of Mytilineos S.A. In the plant jargon it is termed as "Delphi-Distomo" bauxite. This material is subjected to limestone removal before the Bayer process.
		2	ST-BX	Parnassos-Ghiona bauxite from B3 horizon. Extracted by S&B Industrial Minerals S.A. In the plant jargon it is termed as "standard" bauxite. This material is subjected to limestone removal before the Bayer process.
		3	HS-BX	Parnassos-Ghiona bauxite from horizon B2. It represents a minor input (1%) to the process. At present day, it is no longer exploited.
		4	TU-BX	Bauxite from Turkey, western Taurides range and Milas area.
	Lateritic/gibbsitic	5	TR-BX	Bauxite from Brazil, Porto Trombetas deposit. It represents the main input of lateritic bauxite to the process (21% of total bauxite input). Input of lateritic bauxite alters between this (Porto Trombetas) and Ghanaian (Awaso) bauxite.
		6	GH-BX	Bauxite from Ghana, Awaso deposit. At the time of sampling, this material was not processed.
	HMS unit	7	DC-BX	Mixed karst/diasporic bauxite, from which limestone is separated ("decalcitated"). Inputs to this unit are DD-BX and ST-BX. It represents the main bauxite input (73% of total bauxite input) to the Bayer process as a mixture of karst bauxite from two different mining locations that exploit B3 horizon Parnassos-Ghiona bauxite.
		8	DC-RE	HMS residue. Limestone residue that is separated from Parnassos-Ghiona B3 horizon bauxite. Besides limestone, contains also a significant proportion of rejected bauxite material.
Intermediate materials/additives		9	BF-DG	Solid fraction of karst bauxite slurry, from grinding and preheating units of karst bauxite.
		10	SW-DS	Solid fraction of lateritic/gibbsitic bauxite slurry from pre-desilication unit.
		11	CA-OX	Lime, CaO.
Products/by-products		12	HY-AL	Aluminium hydroxide, Al(OH)$_3$, output from precipitation unit.
		13	CA-AL	Calcined alumina, Al$_2$O$_3$, output from calcination unit.
		14	RM-FP	Bauxite residue after the filterpressing of residue slurry.
Liquid Samples				
Liquors		15	CL	Concentrated spent liquor, from the output of evaporation unit, routed to karst bauxite grinding.
		16	PL	Pregnant liquor, from the outlet of settling and security filtration.
		17	SL	Spent liquor, from the outlet of precipitation. Largest proportion is routed to evaporation and a small proportion to lateritic bauxite grinding.
Slurries		18	BF	Liquid phase from grinding and preheating units of karst bauxite, corresponds to sampling point 9 and solid sample "BF-DG".
		19	SW	Liquid phase from grinding and desilication unit of lateritic bauxite, corresponds to sampling point of 10 "SW-DS".

2.2. Analytical Methods

Solid samples were prepared for the analysis using standard techniques (drying, crushing, splitting, grinding, pulverising). Elemental compositions of the samples were determined by a combination of techniques listed in Table 2. Lithium borate fusion was chosen as the appropriate method prior to inductively coupled plasma mass spectrometry (ICP-MS) that ensures with high efficiency the total dissolution of bauxite and bauxite residue mineral matrix [29,50]. Instrumental neutron activation analysis (INAA) has been outlined as a good analytical technique for determining trace element concentrations in bauxite and bauxite residue as it is a non-destructive method and does not require any sample pre-treatment. Also, chemical interferences such as the matrix effect are avoided. The negative property of INAA is, however, that it is a relatively slow technique and not all chemical elements can be measured simultaneously [29,51]. The specifications of applying INAA in analysing geological materials, as practiced by Activation Laboratories Ltd., are given by Hoffman [52]. The quality of the trace element analysis was assessed by measuring certified bauxite reference material BX-N [53,54] with both methods, ICP-MS and INAA. INAA measurements were also verified with certified reference material DMMAS 120.

Bayer liquors were prepared for analyses either by (1) simply dilution, (2) acidification with concentrated HNO_3 [55], or (3) dewatering the liquors to obtain dry pulps of the liquor (Table 2). The latter method also provides a guarantee that trace constituents are not precipitated from the liquid phase during sample preparation. Besides, dewatering enhances the concentration of each component contained in the sample. Analysis of La and Sc in Bayer process solid as well as liquid samples was exercised previously by Derevyankin et al. [25].

2.3. Compiling of the Mass Balance

The results from chemical analysis were used in combination with mass flow data of the plant and normalised to the mass of produced aluminium hydroxide (on dry and calcined basis) according to Equation (1). The mass balance approach to describe trace element distribution was based on the method given by Papp et al. [31]. Original mass flow data was corrected only for the output units of "grinding and preheating of karst bauxite" and "grinding and desilication of lateritic bauxite", assuming a constant Fe_2O_3 total mass in solids [31]:

$$C = \frac{c \times m_1}{m_2}, \quad (1)$$

where:

C: product-normalised concentration of trace element, mg/kg;
c: measured concentration of trace element in solid, mg/kg; or liquid, mg/L;
m_1: mass flow of material on dry basis, kg/d; or liquor flow m^3/d;
m_2: mass flow of aluminium hydroxide on dry calcined basis, kg/d.

Table 2. Analytical methods and preparation techniques used for the determination of trace elements in Bayer process solid and liquid samples.

	Abbreviation	Method	Preparation and Specifications
Solid samples	XRF-st	X-ray fluorescence, standardised (PerformX, Thermo Fisher Scientific™, Waltham, MA, USA)	Fusion of solids with $Li_2B_4O_7/LiBO_2$ (66:33) flux, sample to flux ratio 1:11 [56]. Standardised with appropriate standard materials.
	ICP-MS	Inductively coupled plasma mass spectrometry (Xseries 2, Thermo Fisher Scientific™, Waltham, MA, USA)	Fusion of solids with $Li_2B_4O_7/LiBO_2$ (66:33) flux, sample to flux ratio 1:20, glass bead dissolved in 10% v/v nitric acid.
	INAA	Instrumental neutron activation analysis (Activation Laboratories Ltd., Ancaster, ON, Canada)	About 2 g of sample is inserted in a polyethylene vial [52].
	titr.	Thermometric acid-base titration (855 Robotic Titrosampler, Metrohm, Herisau, Switzerland)	Details given in method description [57].
	AAS	Atomic absorption spectrophotometer (2100, PerkinElmer, Waltham, MA, USA)	Appropriate dilution with deionised water.
	ICP-MS	Inductively coupled plasma mass spectrometer (specified above)	Sample is diluted with deionised water and then acidified with concentrated nitric acid while gently heating the sample in proportions 1:10:1, additional dilutions are made [55].
Liquor samples	ICP-OES	Inductively coupled plasma optical emission spectrometer (Optima 8000, PerkinElmer, Waltham, MA, USA)	Sample is diluted with deionised water and then acidified with concentrated nitric acid while gently heating the sample in proportions 1:10:1, additional dilutions are made [55].
	INAA	Instrumental neutron activation analysis (specified above)	Dewatering of liquor until the creation of dry pulps (Büchi Syncore, vacuum pump Büchi V-700, controller Büchi V-850; Flawil, Switzerland). Then, about 2 g of sample is inserted in a polyethylene vial.
	XRF	X-ray fluorescence, no standardisation, semi-quantitative (Xepos, Spectro, Kleve, Germany)	Dewatering of liquor until the creation of dry pulps (specified above).
	UV	UV Photometer (Cary 60 UV-Vis, Agilent, Santa Clara, CA, United States)	Acidification with conc. HCl and dilution with deionised water in proportions 1:3:20; only for analysing Fe.

3. Results and Discussion

The main and trace element compositions of the analysed solid materials are presented in Supplementary Tables S1 and S2. Analysis of the different bauxites supports the existing knowledge that karst bauxites are more enriched in certain trace elements compared to lateritic bauxites (Supplementary Table S2) [37]. The most prominent trace elements in all analysed bauxites and in derived residue are Cr and V. Among the REEs, Ce is always the most abundant metal in all the materials where the REEs are present. This is in accordance with the review and a case study of REEs in Greek Parnassos-Ghiona diasporic bauxite and derived residues, where the positive anomaly of Ce is always noted [24].

Bayer liquor from various production stages is relatively enriched in the concentration of Ga, V, As and K (Table 3, Supplementary Table S3). Ga, V, As and K concentrations in various Bayer plants are relatively well known, while K is considered highly soluble in the process liquors and Ga, V as well as As medium soluble [10]. Note that the concentration of some analytes, like Ga and V, is higher in the spent liquor compared to pregnant liquor. This is because the total volume of spent liquor is smaller than the total volume of pregnant liquor and therefore the concentrations appear higher. At the same time, the mass balances of these elements are in equilibrium, as explained further (Section 3.1). The same accounts for the concentration of total caustic. Mo was also accumulated to process liquor, but the behaviour and mass balance of this element in Bayer process is already given by Papp et al. [31]. Detectable concentrations of Cr and Ni are also present in Bayer liquor, but these metals are not particularly accumulated into Bayer liquor compared to their concentration in bauxite feed. Other metals, such as Ce, La or Y that were of high interest within the scope of this study, do not occur in dissolved form in Bayer liquor in detectable concentrations (Table 3). This is to be expected as the REEs are not predicted to have soluble species in highly alkaline conditions (pH > 14) [58,59], which is further supported by the mineralogical observations indicating the REEs remain in solid forms during the Bayer digestion [60]. For the case of Sc, INAA found its levels being <0.05 mg/L, which is in accordance with Suss et al. who report that Sc concentration in Bayer liquor remains <1 mg/L [13]. Bayer liquor also contains low concentrations of U (~1 mg/L), which is in accordance with previously known facts [33]. It is interesting to note that 20–30 mg/L concentration of tungsten (W) was also found in process liquors by INAA and XRF. Previous studies that have compared W concentrations in bauxite and derived residue have indicated a depletion of W in bauxite residue compared to bauxite feed, suggesting that current detection of W in Bayer liquor is realistic [29,61].

Table 3. Composition of Bayer process pregnant (PL) and spent (SL) liquors. Extended overview of Bayer process liquors composition is available in Supplementary Table S3.

Sample	Al$_2$O$_3$ *	Na$_2$O **	As	Br	Ca	Ce	Cr	Fe	Ga	Gd	K
	g/L	g/L	mg/L	mg/L	mg/L	mg/L	mg/L	mg/L	mg/L	mg/L	g/L
	titr.	titr.	INAA	INAA	AAS	ICP-MS	ICP-MS	UV	ICP-OES	ICP-MS	AAS
PL	192.2	159.4	110.8	33.6	14.9	<0.04	1.4	9.6	267.2	<0.04	13.7
SL	108.6	171.7	99.6	31.4	16.5	<0.04	1.3	3.4	279.7	<0.04	13.8

	La	Mg	Mo	Ni	Sc	Si	Th	U	V	W	Y
	mg/L	mg/L	mg/L	mg/L	mg/L	mg/L	mg/L	mg/L	mg/L	mg/L	mg/L
	ICP-MS	AAS	INAA	AAS	INAA	AAS	INAA	ICP-MS	ICP-OES	INAA	ICP-MS
PL	<0.04	<0.1	318	4.8	<0.05	544	<0.1	1.28	295.2	27	<0.04
SL	<0.04	<0.1	273	<4	<0.05	520	<0.1	1.27	314.7	21	<0.04

* The aluminate content of the liquor, expressed as Al$_2$O$_3$ [62]. ** Total caustic, the sum of the free Na(OH) and Na bound with sodium aluminate, expressed as Na$_2$O [62].

Based on the preceding information about trace element concentrations in process liquors, we divided the mass distribution description of the trace elements into two main categories. The first one describes the metals (or metalloids) that accumulate to Bayer liquor or dissolve sparingly (V, Ga,

As, Cr) while the second one describes the metals for which the distribution is controlled only by solid materials (REEs and Sc).

Full data describing the distribution and mass balance inventory of all analysed trace elements is given in Supplementary Tables S4–S6. Processing steps are divided into seven principal units: (I) heavy media separation (HMS), (II) grinding and preheating of karst bauxite, (III) grinding and pre-desilication of lateritic bauxite, (IV) digestion, settling and washing, (V) precipitation, (VI) evaporation and (VII) calcination. Overall mass balance is summarised in "internal balance" which includes the process liquors in addition to solids and "external balance" that includes only solid materials input and output. For the metals which do not occur in process liquors, units V–VII are omitted, because the metal concentrations relating to those units were below detection limits (which are specified in Table S3).

3.1. Metals (and Metalloids) that Accumulate to Liquor

There was insignificant difference in the Ga concentrations when comparing lateritic and karst bauxites (Supplementary Table S2). This is in line with the report by U.S. Geological Survey, where they concluded a similar presence of Ga in karstic and lateritic bauxites. They summarise the world's average Ga concentration in all analysed bauxite deposits as being 57 mg/kg [63], which is comparable to the present analysis of 57–66 mg/kg in all currently analysed bauxites. Concentrations are said to be ranging from 12–52 mg/kg Ga with an average of 40 mg/kg in bauxite districts of Greece and Turkey [63].

Gallium possesses a close relation to Al and therefore occurs prevalently in Al-minerals. Similar properties include atomic radius, trivalent oxidation state, tetrahedral or octahedral coordination and amphotericity [20,64]. The mass distribution of Ga is mainly controlled by process liquors (Figure 2). During bauxite digestion, Ga is released from aluminium-bearing minerals like gibbsite, boehmite and diaspore [20]. The Ga digestion reaction is described by Equation (2) [9]:

$$Ga_2O_3 + 6NaOH = 2Na_3GaO_3 + 3H_2O, \qquad (2)$$

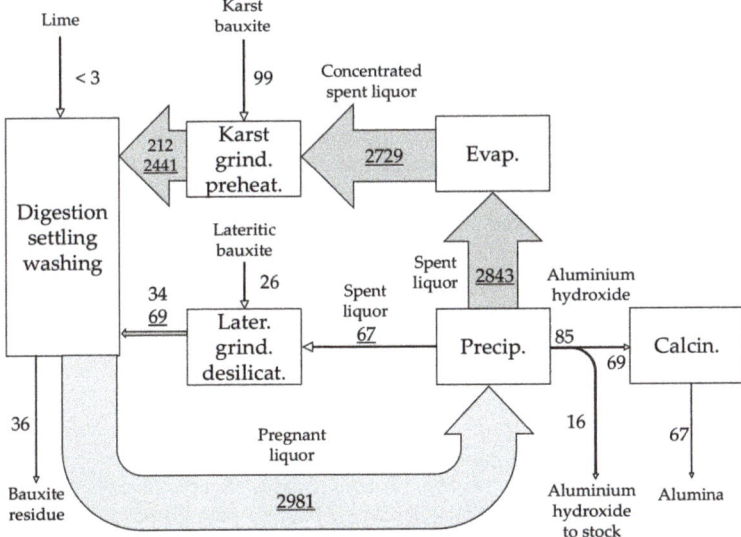

Figure 2. Mass distribution of Ga (mg/kg) normalised to mass of aluminium hydroxide produced, based on ICP-MS (solids) and ICP-OES (liquors) data.

Ga accumulates in process liquors, achieving saturation at levels exceeding 300 mg/L. This is about the average of that reported across earlier publications (60–600 mg/L Ga), yet typically shown values remain between 100–200 mg/L Ga [65–70]. The present Ga saturation levels are prospective for economic extraction given that Frenzel et al. suggest a conservative cut-off concentration for profitable production of Ga from process liquor being 240 mg/L [8]. Ga is about 25 times enriched into pregnant liquor compared to bauxite input. Even though the highest concentration of Ga was detected in concentrated spent liquor, the highest relative amount of Ga (allowing for volumetric changes from gibbsite precipitation and liquor evaporation) was found in pregnant liquor. This is because freshly leached Ga in digestion is present in pregnant liquor, while some Ga is precipitated along with gibbsite during precipitation and so removed from the concentrated spent liquor stream.

From the pregnant liquor, 68% of Ga entering the process is precipitated with aluminium hydroxide, resulting in the concentration of 85 mg/kg. This impurity, however, has no adverse effect on the quality of smelter grade alumina [9]. A smaller proportion of 29% reports to bauxite residue (36 mg/kg on product-normalised basis). The mass difference between the entering and exiting portion of Ga is negligible (3%). Note that the normalised concentration of Ga in liquid fraction decreases from spent liquor to concentrated spent liquor and then to the slurry after preheating stage. In the latter, the decrease of concentration in liquid fraction is accompanied by the simultaneous increase in the solid fraction. Since this is a systematic observation occurring also in the distribution of other trace elements, it will be discussed further in the text.

For the purposes of theoretical modelling of Ga distribution, Hudson [12] has indicated, and Frenzel et al. [8] have applied the partitioning of Ga as 35% going to bauxite residue and 65% to hydrate product [8,12]. From this analysis, the partitioning is more in line with that reported by Figueiredo et al. [65] with 30% of Ga going to bauxite residue, and 70% to hydroxide product, although they do not refer to the source of their data [65]. This case study therefore supports the literature that suggests about 70% of bauxite Ga is digested in the Bayer process, and this part is subsequently precipitated into aluminium hydroxide. About 30% of Ga is separated from the process with bauxite residue [65].

Almost twice as much V is contained in the karst bauxite (336–650 mg/kg) compared to the lateritic bauxite (201–258 mg/kg). Given the different proportions of bauxites in the feed, the major input of V is therefore from karst bauxite (87%). Mass distribution of V is given based on XRF-st data since it provided considerably better fit in the mass balance model compared to ICP-MS data (Figure 3).

The mass distribution of V is again mainly regulated by process liquors, where the concentration of V exceeds 400 mg/L in concentrated spent liquor. This is in accordance with the range of V saturation levels in Bayer liquors reported elsewhere in publications (100–2800 mg/L V) [10,65,69,71]. Authier-Martin et al. refer to earlier studies indicating that V is about 30% soluble during Bayer digestion [9]. Compared to bauxite feed on alumina normalised basis, V is enriched in pregnant liquor up to 4 times in this study. In process liquors, V appears in the form of VO_4^{3-} [71]. This impurity is unwanted in hydroxide and metal production due to its known property of decreasing the electric conductivity of metallic Al, causing a green hue in fused Al, and the scale it can form in the piping of a Bayer refinery when precipitated from the liquor in the cooler parts of the circuit [9,71,72]. The removal of V from process liquors is a side benefit of process lime addition. V precipitates as calcium vanadate, as an impurity in tri-calcium aluminate ($Ca_3Al_2(OH)_{12}$), or as $Na_7(VO_4)_2F \cdot 19H_2O$ [71,73,74]. Our study as well as the regular monitoring in the plant materials did not detect any V in the aluminium hydroxide product (<10 mg/kg). Therefore, lime addition that mainly reduces soda losses among other beneficial effects [46], is simultaneously providing a way to remove excess V from the Bayer cycle and preventing V precipitation to product. In the existing case study, V is separated from the process and is accumulated in the bauxite residue.

The input of As to the system from lateritic bauxite (3%) is negligible compared to karst bauxite (97%). Once again, the accumulation of As to the liquor-based circuit is evident, as seen from the diagram in Figure 4. The saturation of As to process liquor is achieved at about 130 mg/L concentration. In earlier studies, As has been detected in the alkaline liquor of bauxite residue suspension as well as in Bayer liquors [10,75]. Teas and Kotte have classified As as a medium soluble impurity in the Bayer process with a similar behaviour to V [10], which is evident also from this case study.

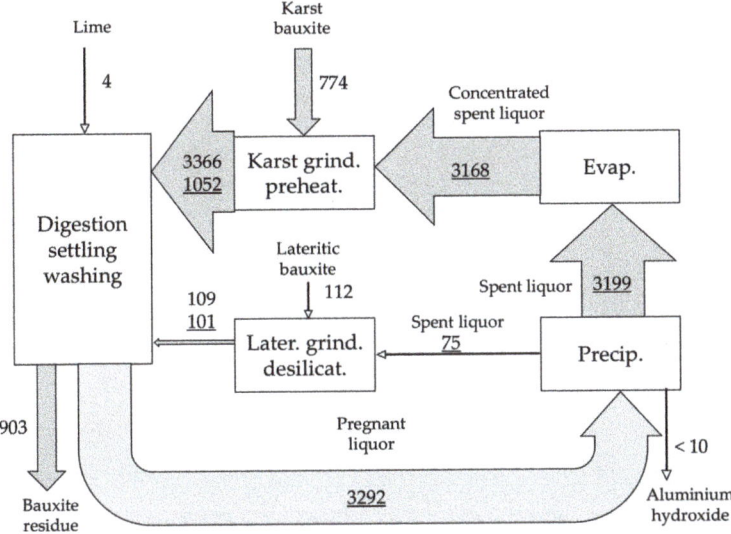

Figure 3. Mass distribution of V (mg/kg) normalised to mass of aluminium hydroxide produced, based on XRF-st and ICP-OES data.

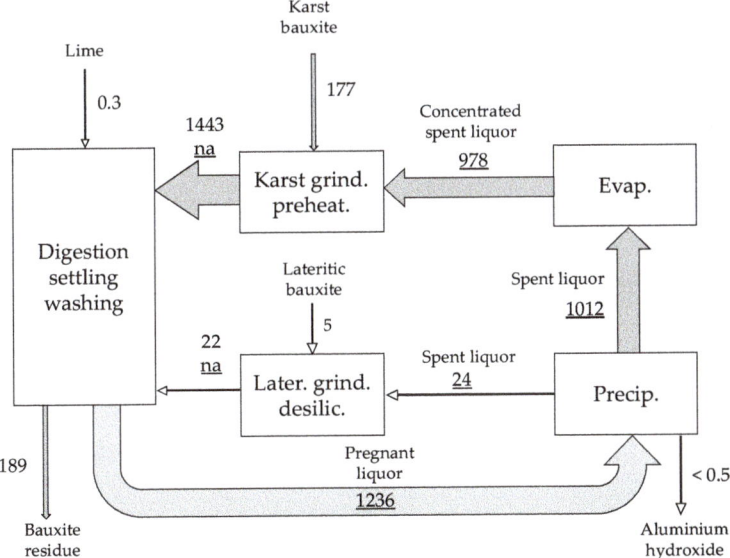

Figure 4. Mass distribution of As (mg/kg) normalised to mass of aluminium hydroxide produced, based on ICP-MS (solids) and INAA (liquors) data.

Higher than usual discrepancies in the input and output masses of As in different units are noted compared to other analysed elements. This could be an indication of the need to further develop analytical techniques relating to As in Bayer process materials. However, it is clear that dissolved As must exist in the system when we observe the "karst bauxite grinding and preheating" unit. It is apparent from there that As concentration increases significantly when comparing the solids entering and exiting the unit. This implies that the entering of As to solid fraction must originate from the concentrated spent liquor. In any case, in the end of the processing cycle, all of the As in found in bauxite residue and in the context of available detection limits is not shown to be present in aluminium hydroxide product (<0.5 mg/kg).

The majority of Cr input (95%) originates from karst bauxite. A minor fraction of Cr, about one percent of input, can be dissolved into process liquor giving rise to a concentration of 1.4 mg/L (Figure 5). During precipitation, Cr was not detected to enter product (<5 mg/kg), or it does in a very small quantity (~2 mg/kg), as could be suggested from difference in the balance of precipitation stage and the small deficiency (one percent) of Cr mass in the output material. An earlier study has pointed out a 5 mg/kg concentration of Cr in hydroxide product [26]. All the quantity of Cr that entered to the process is found in bauxite residue as the sole output carrier of this metal.

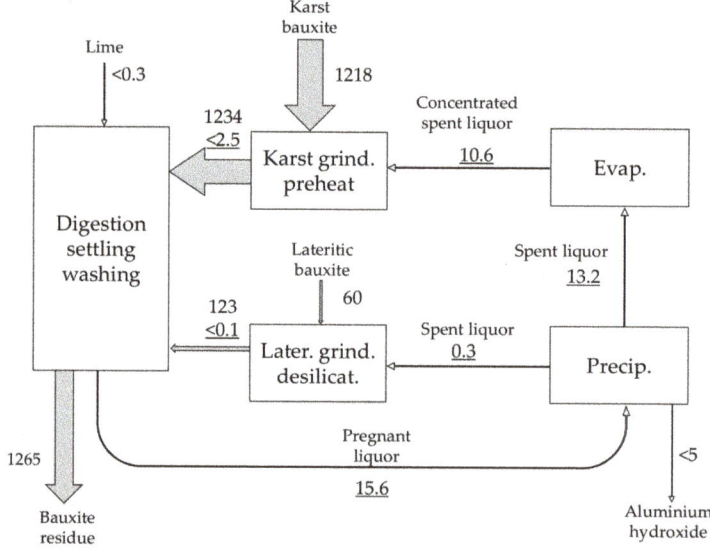

Figure 5. Mass distribution of Cr (mg/kg) normalised to mass of aluminium hydroxide produced, based on ICP-MS (solids and liquors) data.

In the mass balance models of Ga, V, As and Cr it can be noted that during the "karst bauxite grinding and preheating" as well as in smaller scale during "lateritic bauxite desilication", a pronounced increase in trace element concentration is observed in the solid fraction flows. This increase of trace constituents is occurring with the simultaneous decrease in concentrations in liquid flows. Thus, the trace constituents appear to precipitate during these processing phases. Probably, the trace elements precipitate in the composition of Bayer process characteristic solid phases, that are a group of Ca-, Al-, Na- and Si-containing phases, including desilication products (sodalite and cancrinite), hydrogarnet (hydrogrossular) type phases as well as calcium titanate in the form of perovskite ($CaTiO_3$) [42,73]. Calcium vanadate or $Ca_3Al_2(OH)_{12}$ are already known species that contain V in Bayer process-specific solid phases [71,73]. As mentioned before, lime addition position on the schemes is a simplification and it is added in more processing steps, including the preheating, thus the possibility of forming Ca-containing species is not limited to the digestion stage. Therefore, during the preheating stage, the trace elements occurring in the spent liquor (Ga, V, As and Cr) are thought to precipitate in the composition of Bayer process characteristic solid phases. While this effect occurs, the trace element concentration in the liquid fraction decreases and in the solid fraction increases. At the same time, the mass balance equilibrium of the trace elements is maintained. During digestion, the pregnant process liquor becomes saturated again in the trace elements on the account of leaching of the newly added bauxite feed.

Another characteristic that can be observed from the mass distributions of Ga, V, As and Cr is that their product-normalised content shows a decreasing trend from pregnant liquor to spent liquor and then to concentrated spent liquor. Only for Ga it is evident that part of its mass is removed from the liquor during precipitation. Since it is observed in the distribution patterns of all the named trace elements, it can be concluded being a systematic behaviour. The working hypothesis is that minor deposition of the trace constituents occurs throughout the mentioned production steps in the form of secondary precipitates or solid formations like scales in the cooler parts of piping or in the solids of filter cakes (e.g., from security filtration of pregnant liquor, "liquor polishing") [10,44,47]. As already mentioned, some trace elements (Y, Nb, Zr) have been detected in perovskite-based scales in the Bayer circuit [76]. Enhanced concentrations of trace elements like Ni, Cr and V in the range of 700–4900 mg/kg were detected in perovskite-dominated matrix of a scale sample formed in the AoG's digestion autoclave. The cancrinite-dominated matrix of the same sample was, however, relatively depleted in trace elements (e.g., 140–170 mg/kg V) [77]. Sometimes, enhanced concentrations of V (112 mg/kg) and Ga (28 mg/kg) have been identified in the alumina dust from calciner electrostatic filters, making this material an attractive source of V and Ga [78]. The former examples therefore support the hypothesis that a proportion of trace elements is deposited to minor by-products of the Bayer process. Scales, filter cakes and electrostatic filter dust are regularly cleaned during the production. The volumes of these materials being created are not easily quantifiable, but an assumption can be made that the decrease in the trace element concentrations in the liquor stream can account to the passing of the trace elements to the formerly mentioned minor by-products of the Bayer process as result of the described systematic behaviour.

For the previously discussed elements (Ga, V, As and Cr), it can be concluded that they first accumulate to Bayer process liquor (although Cr in very small extent) and once their saturation level in liquor is achieved, their input and output flows equilibrate. Minor output of those trace elements probably occurs into minor Bayer process by-products from the liquor-based circuit as the concentrations in liquors drop systematically in the consecutive production steps.

3.2. Metals that Do Not Accumulate to Liquor

Cerium (Ce) distribution is presented as the representative of LREE elements due to its highest concentration in analysed materials and good analytical stability (Figure 6, Supplementary Table S2). The rest of the mass balances of REEs can be found in Supplementary Table S5 (based on ICP-MS data) and Table S6 (based on INAA data). The input of Ce from lateritic bauxite is practically insignificant and almost the sole source of it is karst bauxite (99%). Cerium distribution is dictated by solid materials only. On the "white side" of the Bayer process, all the analysed concentrations of Ce are below detection limits (aluminium hydroxide < 3 mg/kg, process liquors < 0.04 mg/kg). Ce content remains steady from bauxite to intermediate suspension solid fraction and then to bauxite residue. In the end of the process, there is only one percent difference in the input and output quantities.

The distribution of Y is presented as the representative of HREEs given its highest concentration among this group of elements (Figure 7). Except for the quantities, the distribution of Y is identical to the one of Ce. In all processing stages, its distribution follows the solid materials and Y does not dissolve in the process liquor and thus does not enter into aluminium hydroxide product. The difference in the quantity of input and output of Y (four percent) is higher than for most of the REEs, but still acceptable for presenting its mass balance model.

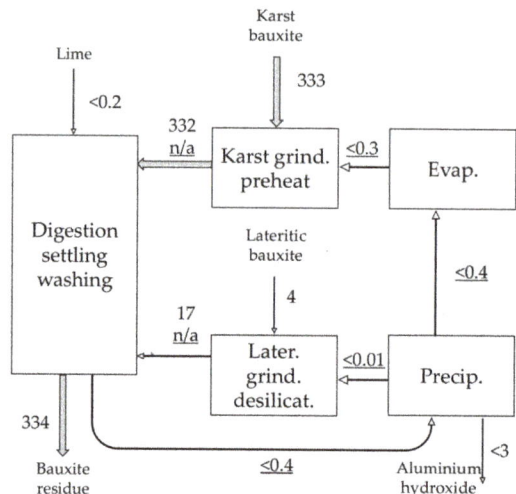

Figure 6. Mass distribution of Ce (mg/kg) normalised to mass of aluminium hydroxide produced, based on INAA (solids) and ICP-MS (liquors) data.

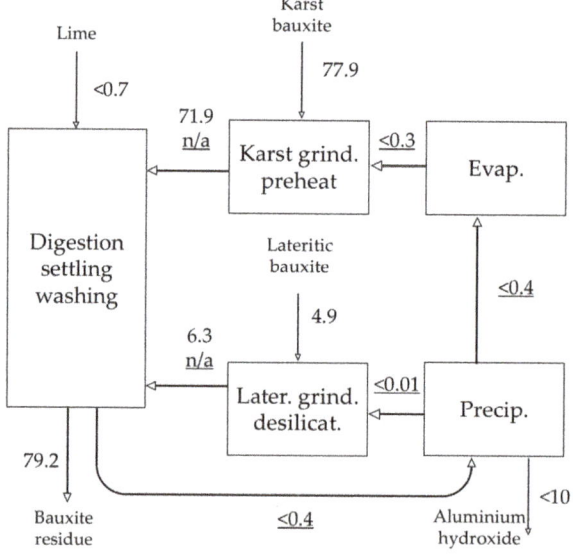

Figure 7. Mass distribution of Y (mg/kg) normalised to mass of aluminium hydroxide produced, based on ICP-MS (solids and liquors).

Finally, the distribution of Sc is presented in Figure 8. Note that the quantities of Sc were analysed by INAA method. The majority of Sc originates from karst bauxite source (96%).

Suss et al. report that Sc is expected to occur in a dissolved form during Bayer digestion, but, it probably precipitates rapidly in an unknown form that might be ScO(OH) or Sc(OH)$_3$ [13]. The progression of Sc through the process is once more regulated by solid material matrix from bauxite to intermediate solids and then to residue. There is no missing quantity of Sc throughout the processing. The present result is slightly different from the previous results of Sc distribution patterns, where

0.6–1.5 mg/kg bauxite-feed-normalised concentration of Sc was detected in aluminium hydroxide products in Alumina Plant of Urals and Bogoslovski Alumina Plant [25]. It can be noted, though, that the processing conditions between the Russian alumina plants and AoG are different, since the former also partly included sintering of the bauxite ore, although it was not performed for the total amount of bauxite feed.

For most the REEs as well as for Sc distribution, a deficiency in the mass balance in the output of limestone separation HMS unit is noted. This can be regarded as a problem of material representativeness. However, since this is a pre-processing step, it does not affect the mass balance models of the Bayer process.

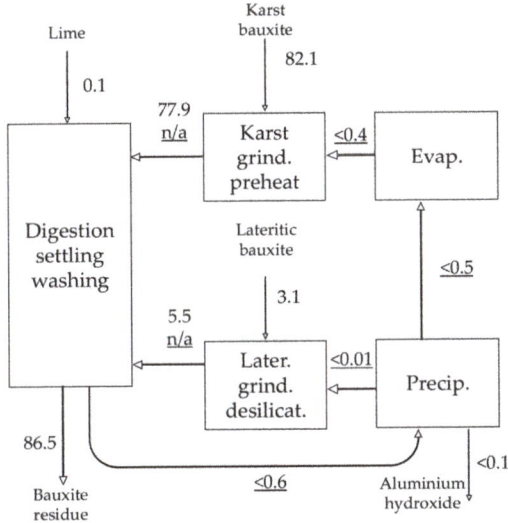

Figure 8. Mass distribution of Sc (mg/kg) normalised to mass of aluminium hydroxide produced, based on INAA (solids and liquors).

Besides the REEs distribution presented graphically (Figures 6–8), all the mass balance inventories of the REEs (except for Tb, Ho and Tm due to low concentrations) are available in the compiled dataset (Supplementary Tables S5 and S6). Even lutetium (Lu), the last chemical element in the lanthanides group, mass balance was possible to be quantified given the very low detection limit (0.05 mg/kg) available in INAA method for this element. The data shows consistently that all the analysed REEs behave similarly during bauxite processing and at least 95% of the REEs entering into Bayer process are transferred in the composition of solid matrix to bauxite residue. Furthermore, in most of cases the transfer rate of REEs to bauxite residue is more than 98%. None of the REEs or Sc enter into aluminium hydroxide, given the available detection limits, e.g., La < 0.5 mg/kg, Sm <0.1 mg/kg or Sc < 0.1 mg/kg (Supplementary Table S2). The fact that REEs and Sc are transferred to bauxite residue only in the composition of solid material is also supported by mineralogical studies [60,77]. One of the investigations has shown that the form of Sc occurrence mainly in the composition of hematite remains the same after bauxite processing [77]. On the other hand, the precursor REE phases found in bauxite are affected by the Bayer process conditions and REE ferrotitanate type compounds are created, but the transformations taking place seem to occur in situ on mineral grain surfaces without the dissolution of the precursor REE phases [60]. Present case study was not able to repeat the result that up to five percent of total La content can be passed to aluminium hydroxide product [25]. This difference can be again explained by the differences between the operational conditions of refineries, as mentioned above. It can be noted, however, that low La presence has been semi-quantitatively found in some

2 *w/v*% aluminium hydroxide suspensions (Chemtrade Rehydragel® LV and SPI Pharma Aluminum hydroxide wetgel VAC 20; 0.14 and 0.72 mg/L La, respectively) that are used as adjuvants in vaccines by applying very sensitive ICP-MS techniques. At the same time, the concentrations of Ce, Nd and Sc were below detection limits, <0.0005, <0.002 and <0.01 mg/L, respectively [79].

Inconsistencies in the mass balance of some REEs such as Ce can be noted in the "lateritic bauxite grinding and desilication" unit (Figure 6). Such situation is best explained with the possibility that during sampling, some contribution of the other lateritic bauxite from Ghana was also present in the lateritic bauxite slurry. Presently sampled Ghanaian bauxite contains higher concentration of REEs and Sc compared to the existing Brazilian bauxite and by hypothetically replacing the two bauxites in the mass balance calculation resolves the inconsistency. However, in a broad sense this discrepancy is not an issue because the total input and output flows are well within acceptable balance and besides, the input of REEs and Sc from lateritic bauxites has a minor magnitude regardless of the two lateritic bauxite types.

3.3. Fractionation Indexes and Systemic Predictions

All the analysed elements except for Ga are enriched in the bauxite residue. This can be emphasised by calculating the fractionation indexes by dividing trace element concentration in bauxite residue with the same parameter in bauxite feed (Figure 9). The fractionation indexes of all the elements except for Ga are like the (1) ratio of bauxite feed mass to bauxite residue mass created during sampling period (2.33), (2) fractionation index of Fe_2O_3 during sampling period (2.34), and (3) fractionation index of Fe_2O_3 during one-year period (2.31). Iron oxide fractionation index is considered for a comparison here because it represents a largely inert oxide in the Bayer process, as well as because it has been used for a similar comparison before [29]. All the indexes of trace elements (except Ga) differ from the three major indexes by a maximum of six percent and for most cases less than two percent. The differences are essentially negligible and probably account for errors in sample representativeness and/or analytical variations. This similarity of indexes is well-reasoned, because raw-material-to-residue ratio or Fe_2O_3 fractionation index set a logical boundary, what can be the maximum possible fractionation index of a constituent in the process. Basically, new material cannot be created during the process and if the constituent does not fraction to aluminium hydroxide product, then the fractionation index must be like the one for Fe_2O_3 or the bauxite-feed-to-residue coefficient. Present result is similar to what was concluded in lab-scale testing of trace element enrichment from bauxite to bauxite residue, although higher variations were noted in lab testing [29].

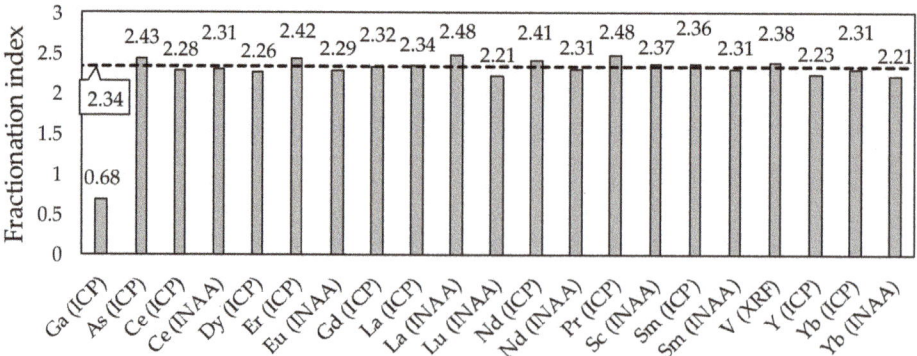

Figure 9. Fractionation indexes of trace elements calculated as trace element concentration in bauxite residue divided by trace element concentration in bauxite. Fractionation indexes are compared to the one of Fe_2O_3 during sampling period (horizontal dashed line).

The former reasoning provides opportunities for predicting the trace element concentrations in bauxite residue based on existing information about bauxite feed. First option is based on Fe_2O_3 concentration in bauxite and bauxite residue, as shown in Equation (3). It is not uncommon that the conditionally inert Fe_2O_3 is used as an aid in mass balance estimations relating to the Bayer process [80].

$$C_{BR} = \frac{C_{Fe\_BR}}{C_{Fe\_BX}} \times C_{BX}, \qquad (3)$$

where:

C_{BR}: Predicted concentration of trace element in bauxite residue, mg/kg;
C_{Fe_BR}: Fe_2O_3 concentration in bauxite residue, %;
C_{Fe_BX}: Fe_2O_3 concentration in bauxite, %;
C_{BX}: Average concentration of trace element in bauxite feed, mg/kg.

Another option is to consider just the mass flows of bauxite feed and resulting residue created, as shown in Equation (4) and combine it with trace element concentration in bauxite feed. In any case, care must be taken on the representativeness of the bauxite trace element concentration values, because considerably high variations can occur in some bauxite deposits [27].

$$C_{BR} = \frac{M_{tot\_BX}}{M_{tot\_BR}} \times C_{BX}, \qquad (4)$$

where:

C_{BR}: Predicted concentration of trace element in bauxite residue, mg/kg;
M_{tot_BX}: Total dry mass of bauxite fed into system, kg;
M_{tot_BR}: Total dry mass of bauxite residue leaving the system, kg;
C_{BX}: Average concentration of trace element in bauxite feed, mg/kg.

Equations (3) and (4) can be used only for the trace elements which do not fraction to aluminium hydroxide product, therefore it is not applicable for such metals as Ga.

4. Conclusions

Bauxite trace elements under the Bayer process extraction conditions and the types of bauxite used in this study are roughly divided in two categories: (1) those that are at least partly soluble in the caustic leaching and accumulate to an extent into process liquor, namely Ga, V, As as well as Cr, and (2) those that are not soluble in the caustic leaching, namely Sc, and the REEs. The trace elements in the first category accumulate in the process liquor until the specific saturation level of each metal, and then the input and output flows equilibrate. In the Bayer process output flows, only Ga possesses the property of entering into the composition of aluminium hydroxide product to the extent of 70% of total Ga output mass. The rest of the Ga is separated from the process with bauxite residue. With respect to the second category, those metals (Sc, REEs) are transferred through the process only in the composition of solid material flows. Sc and REEs are not found in the aluminium hydroxide product and their mass transfer to bauxite residue is mostly at least 98%.

It is evident that Bayer process materials, whether they are bauxite residue or process liquors, are enriched in certain trace elements. Together with the continuous development and improvement of extraction technologies, the trace elements could be recovered as valuable by-products of the Bayer process.

Supplementary Materials: The following are available online at http://www.mdpi.com/2075-4701/8/5/327/s1, Supplementary file containing Table S1: main element composition of sampled solid materials, Table S2: trace element composition of sampled solid materials, Table S3: composition of Bayer liquors, Table S4: mass distribution of metals (and metalloids) that accumulate to process liquor, Table S5: mass distribution of trace elements that do not accumulate to process liquors, based on ICP-MS data, Table S6: mass distribution of trace elements that do not accumulate to process liquors, based on INAA data.

Author Contributions: J.V., V.V. and D.P. conceived and designed the experiments; J.V. prepared the materials and A.A. performed the experiments; J.V. and D.P. analysed the data; V.V. contributed the reagents and materials; J.V. wrote the paper; D.P. and A.A. contributed to the writing of the paper.

Acknowledgments: Athina Filippou and her team of chemical analysts in AoG are thanked for their valuable help in several analytical tasks. The help of Sokratis Tekidis with material mass flow data and sample collection is greatly appreciated. György (George) Bánvölgyi and Ken Evans are kindly acknowledged for their invaluable support throughout the project. The research leading to these results has received funding from the European Community's Horizon 2020 Programme (H2020/2014–2019) under Grant Agreement no. 636876 (MSCA-ETN REDMUD). This publication reflects only the authors' views, exempting the Community from any liability. Project website: http://www.etn.redmud.org.

Conflicts of Interest: The authors declare no conflict of interest.

References

1. Binnemans, K.; Jones, P.T.; Müller, T.; Yurramendi, L. Rare Earths and the Balance Problem: How to Deal with Changing Markets? *J. Sustain. Metall.* **2018**, *4*, 126–146. [CrossRef]
2. Christmann, P.; Arvanitidis, N.; Martins, L.; Recoché, G.; Solar, S. Towards the Sustainable Use of Mineral Resources: A European Geological Surveys Perspective. *Miner. Energy Raw Mater. Rep.* **2007**, *22*, 88–104. [CrossRef]
3. Christmann, P. Towards a More Equitable Use of Mineral Resources. *Nat. Resour. Res.* **2018**, *27*, 159–177. [CrossRef]
4. European Commission; Deloitte Sustainability; TNO; British Geological Survey; Bureau de Recherches Géologiques et Minières. *Study on the Review of the List of Critical Raw Materials*; Publications Office of the European Union: Luxembourg, 2017; pp. 1–93.
5. Balomenos, E.; Davris, P.; Deady, E.; Yang, J.; Panias, D.; Friedrich, B.; Binnemans, K.; Seisenbaeva, G.; Dittrich, C.; Kalvig, P.; et al. The EURARE Project: Development of a Sustainable Exploitation Scheme for Europe's Rare Earth Ore Deposits. *Johns. Matthey Technol. Rev.* **2017**, *61*, 142–153. [CrossRef]
6. Davris, P.; Balomenos, E.; Taxiarchou, M.; Panias, D.; Paspaliaris, I. Current and Alternative Routes in the Production of Rare Earth Elements. *Berg. Huettenmaenn. Monatsh.* **2017**, *162*, 245–251. [CrossRef]
7. Evans, K. The history, challenges, and new developments in the management and use of bauxite residue. *J. Sustain. Metall.* **2016**, *2*, 316–331. [CrossRef]
8. Frenzel, M.; Ketris, M.P.; Seifert, T.; Gutzmer, J. On the current and future availability of gallium. *Resour. Policy* **2016**, *47*, 38–50. [CrossRef]
9. Authier-Martin, M.; Forte, G.; Ostap, S.; See, J. The mineralogy of bauxite for producing smelter-grade alumina. *JOM* **2001**, *53*, 36–40. [CrossRef]
10. Teas, E.B.; Kotte, J.J. The effect of impurities on process efficiency and methods for impurity control and removal. In Proceedings of the JBI-JGS Symposium Titled "Bauxite/Alumina Industry in the Americas", Kingston, Jamaica, 1980; Volume 23, pp. 100–129.
11. Davris, P.; Balomenos, E.; Panias, D.; Paspaliaris, I. Selective leaching of rare earth elements from bauxite residue (red mud), using a functionalized hydrophobic ionic liquid. *Hydrometallurgy* **2016**, *164*, 125–135. [CrossRef]
12. Hudson, L.K. Gallium as a by-product of alumina manufacture. *J. Met.* **1965**, *17*, 948–951. [CrossRef]
13. Suss, A.; Kuznetsova, N.V.; Kozyrev, A.; Panov, A.; Gorbachev, S. Specific features of scandium behavior during sodium bicarbonate digestion of red mud. In Proceedings of the 35th International ICSOBA Conference, Hamburg, Germany, 2–5 October 2017; Volume 42, pp. 491–504.
14. Suss, A.; Paromova, I.; Panov, A.; Shipova, O.V.; Kutkova, N.N. Behaviour of Berylium in Alumina Production from Bauxites by Bayer Process and Development of Reliable Method of Its Determination. In Proceedings of the Light Metals Conference, New Orleans, LA, USA, 9–12 March 2008; pp. 107–112.
15. Løvik, A.N.; Restrepo, E.; Müller, D.B. The global anthropogenic gallium system: Determinants of demand, supply and efficiency improvements. *Environ. Sci. Technol.* **2015**, *49*, 5704–5712. [CrossRef] [PubMed]
16. Pradhan, R.J.; Das, S.N.; Thakur, R.S. Vanadium sludge—An useful byproduct of alumina plant. *J. Sci. Ind. Res.* **1999**, *58*, 948–953.

17. Binnemans, K.; Jones, P.T.; Blanpain, B.; Van Gerven, T.; Pontikes, Y. Towards zero-waste valorisation of rare-earth-containing industrial process residues: A critical review. *J. Clean. Prod.* **2015**, *99*, 17–38. [CrossRef]
18. Borra, C.R.; Blanpain, B.; Pontikes, Y.; Binnemans, K.; Van Gerven, T. Recovery of rare earths and other valuable metals from bauxite residue (red mud): A review. *J. Sustain. Metall.* **2016**, *2*, 365–386. [CrossRef]
19. Atwood, D.A. *The Rare Earth Elements: Fundamentals and Applications*; John Wiley & Sons: Lexington, KY, USA, 2012; ISBN 978-1-119-95097-4.
20. Gray, F.; Kramer, D.A.; Bliss, J.D. Gallium and gallium compounds. In *Kirk-Othmer Encyclopedia of Chemical Technology*; John Wiley & Sons, Inc.: Hoboken, NJ, USA, 2013. [CrossRef]
21. Bauer, G.; Güther, V.; Hess, H.; Otto, A.; Roidl, O.; Roller, H.; Sattelberger, S.; Köther-Becker, S.; Beyer, T.; Bauer, G.; et al. Vanadium and Vanadium Compounds. In *Ullmann's Encyclopedia of Industrial Chemistry*; John Wiley & Sons: Hoboken, NJ, USA, 2017; ISBN 978-3-527-30673-2.
22. Wang, W.; Pranolo, Y.; Cheng, C.Y. Metallurgical processes for scandium recovery from various resources: A review. *Hydrometallurgy* **2011**, *108*, 100–108. [CrossRef]
23. Habashi, F. Gallium update. In Proceedings of the 17th International Symposium ICSOBA, Montréal, QC, Canada, 1–4 October 2006; Volume 33, pp. 141–153.
24. Deady, É.A.; Mouchos, E.; Goodenough, K.; Williamson, B.J.; Wall, F. A review of the potential for rare-earth element resources from European red muds: Examples from Seydişehir, Turkey and Parnassus-Giona, Greece. *Mineral. Mag.* **2016**, *80*, 43–61. [CrossRef]
25. Derevyankin, V.A.; Porotnikova, T.P.; Kocherova, E.K.; Yumasheva, I.V.; Moiseev, V.E. Behaviour of scandium and lanthanum in the production of alumina from bauxite (in Russian). *Izvestiya Vysshikh Uchebnykh Zavedenii Tsvetnaya Metallurgiya* **1981**, *4*, 86–89.
26. Mohapatra, B.K.; Mishra, B.K.; Mishra, C.R. Studies on metal flow from khondalite to bauxite to alumina and rejects from an alumina refinery, India. In *Light Metals*; Suarez, C.E., Ed.; John Wiley & Sons, Inc.: Hoboken, NJ, USA, 2012; pp. 87–91. ISBN 978-1-118-35925-9.
27. Ochsenkühn-Petropulu, M.; Lyberopulu, T.; Parissakis, G. Direct determination of lanthanides, yttrium and scandium in bauxites and red mud from alumina production. *Anal. Chim. Acta* **1994**, *296*, 305–313. [CrossRef]
28. Wagh, A.S.; Pinnock, W.R. Occurrence of scandium and rare earth elements in Jamaican bauxite waste. *Econom. Geol.* **1987**, *82*, 757–761. [CrossRef]
29. Feret, F.R.; See, J. A comparative study of analytical methods of trace elements in bauxite and red mud. In Proceedings of the 18th International Symposium ICSOBA Travaux, Zhengzhou, China, 25–28 November 2010; pp. 68–83.
30. Logomerac, V.G. Distribution of rare-earth and minor elements in some bauxite and red mud produced. In Proceedings of the Second International Symposium of ICSOBA, Budapest, Hungary, 6–10 October 1969; Volume 3, pp. 383–393.
31. Papp, E.; Zsindely, S.; Tomcsanyi, L. Molybdenum and zinc traces in the Bayer process. In Proceedings of the 2nd International Symposium ICSOBA, Budapest, Hungary, 6–10 October 1969; Volume 3, pp. 395–402.
32. Eyer, S.; Nunes, M.; Dobbs, C.; Russo, A.; Burke, K. The analysis of beryllium in Bayer solids and liquids. In Proceedings of the 7th International Alumina Quality Workshop, Perth, Australia, 16–21 October 2005; Volume 1, pp. 254–257.
33. Sato, C.; Kazama, S.; Sakamoto, A.; Hirayanagi, K. Behavior of radioactive elements (uranium and thorium) in Bayer process. In *Reprinted in Essential Readings in Light Metals*; Donaldson, D., Raahauge, B.E., Eds.; John Wiley & Sons, Inc.: Hoboken, NJ, USA, 2013; pp. 191–197. ISBN 978-1-118-64786-8.
34. Bansal, N.; Vaughan, J.; Tam Wai Yin, P.; Leong, T.; Boullemant, A. Chemical thermodynamics of mercury in the Bayer process. In Proceedings of the 7th International Symposium on Hydrometallurgy (Hydro2014), Victoria, BC, Canada, 22–25 June 2014; Volume II, pp. 559–569.
35. Bansal, N.; Vaughan, J.; Boullemant, A.; Leong, T. Determination of total mercury in bauxite and bauxite residue by flow injection cold vapour atomic absorption spectrometry. *Microchem. J.* **2014**, *113*, 36–41. [CrossRef]
36. Bárdossy, G. *Karst Bauxites: Bauxite Deposits on Carbonate Rocks*; Developments in Economic Geology 14; Elsevier Scientific Pub. Co.: Amsterdam, The Netherlands, 1982; ISBN 978-0-444-99727-2.
37. Valeton, I. *Bauxites*; Developments in Soil Science 1; Elsevier Publishing Company: Amsterdam, The Netherlands, 1972; ISBN 978-0-444-40888-4.

38. Adamson, A.N.; Bloore, E.J.; Carr, A.R. Basic Principles of Bayer Process Design. In *Reprinted in Essential Readings in Light Metals*; Donaldson, D., Raahauge, B.E., Eds.; John Wiley & Sons, Inc.: Hoboken, NJ, USA, 2013; pp. 100–117. ISBN 978-1-118-64786-8.
39. Chin, L.A.D. The state-of-the-art in Bayer process technology. *Light Met.* **1988**, *1988*, 49–53.
40. Power, G.; Gräfe, M.; Klauber, C. Bauxite residue issues: I. Current management, disposal and storage practices. *Hydrometallurgy* **2011**, *108*, 33–45. [CrossRef]
41. S&P Global Platts, a Division of S&P Global Inc. Methodology and Specifications Guide, Nonferrous. Available online: https://www.platts.com/methodology-specifications/metals (accessed on 3 March 2018).
42. Gräfe, M.; Power, G.; Klauber, C. Bauxite residue issues: III. Alkalinity and associated chemistry. *Hydrometallurgy* **2011**, *108*, 60–79. [CrossRef]
43. Hudson, L.K.; Misra, C.; Perrotta, A.J.; Wefers, K.; Williams, F.S. Aluminum Oxide. In *Ullmann's Encyclopedia of Industrial Chemistry*; Wiley-VCH Verlag GmbH & Co.: Wienheim, Germany, 2000; Volume 2, pp. 607–645. ISBN 978-3-527-30673-2.
44. Lavalou, E.; Bosca, B.; Keramidas, O. Alumina production from diasporic bauxites. *Light Met.* **1999**, CD-ROM Collection, 55–62.
45. Papanastassiou, D.; Contaroudas, D.; Solymár, K. Processing and marketing of Greek diasporic bauxite for metallurgical and non-metallurgical applications. In Proceedings of the 17th International Symposium of ICSOBA, "Aluminium: From Raw Materials to Applications", Montréal, QC, Canada, 1–4 October 2006.
46. Whittington, B.I. The chemistry of CaO and $Ca(OH)_2$ relating to the Bayer process. *Hydrometallurgy* **1996**, *43*, 13–35. [CrossRef]
47. Bánvölgyi, G. Scale formation in alumina refineries. In Proceedings of the 34th International Conference and Exhibition ICSOBA, Quebec, QC, Canada, 3–6 October 2016; pp. 1–14.
48. Laskou, M.; Economou-Eliopoulos, M. The role of microorganisms on the mineralogical and geochemical characteristics of the Parnassos-Ghiona bauxite deposits, Greece. *J. Geochem. Explor.* **2007**, *93*, 67–77. [CrossRef]
49. Boulangé, B.; Carvalho, A. The bauxite of Porto Trombetas. In *Brazilian Bauxites*; Carvalho, A., Boulangé, B., Melfi, A.J., Lucas, Y., Eds.; NUPEGEL, Departamento de Geologia Geral Universidade de Sao Paulo, Brasil: São Paulo, Brazil, 1997.
50. Adam, C.; Krüger, O. Challenges in analysing rare earth elements in different waste matrices to determine recovery potentials. In *Book of Abstracts, 2nd Conference on European Rare Earth Resources*; Heliotopos Conferences Ltd.: Santorini, Greece, 2017; pp. 237–239.
51. Ochsenkühn-Petropoulou, M.; Ochsenkühn, K.; Luck, J. Comparison of inductively coupled plasma mass spectrometry with inductively coupled plasma atomic emission spectrometry and instrumental neutron activation analysis for the determination of rare earth elements in Greek bauxites. *Spectrochim. Acta Part B At. Spectrosc.* **1991**, *46*, 51–65. [CrossRef]
52. Hoffman, E.L. Instrumental neutron activation in geoanalysis. *J. Geochem. Explor.* **1992**, *44*, 297–319. [CrossRef]
53. Govindaraju, K. Report (1967–1981) on four ANRT rock reference samples: Diorite DR-N, serpentine UB-N, bauxite BX-N and disthene DT-N. *Geostand. Newsl.* **1982**, *6*, 91–159. [CrossRef]
54. Govindaraju, K.; Roelandts, I. 1988 compilation report on trace elements in six ANRT rock reference samples: Diorite DR-N, serpentine UB-N, bauxite BX-N, disthene DT-N, granite GS-N and potash feldspar FK-N. *Geostand. Geoanal. Res.* **1989**, *13*, 5–67. [CrossRef]
55. Singh, U.; Mishra, R.S. Simultanious multielemental analysis of alumina process samples using inductively coupled plasma spectrometry (ICP-AES). *Anal. Chem. Indian J.* **2012**, *11*, 1–5.
56. Yamada, Y. X-ray fluorescence analysis by fusion bead method for ores and rocks. *Rigaku J.* **2010**, *26*, 15–23.
57. Metrohm, Determination of Total Caustic, Total Soda and Alumina in Bayer Process Liquors with 859 Titrotherm. Available online: http://partners.metrohm.com/GetDocument?action=get_dms_document&docid=693348 (accessed on 23 March 2018).
58. Brookins, D.G. *Eh-pH Diagrams for Geochemistry*; Springer Science & Business Media: Berlin/Heidelberg, Germany, 1988; ISBN 978-3-642-73093-1.
59. Brookins, D.G. Eh-pH diagrams for the rare earth elements at 25 C and one bar pressure. *Geochem. J.* **1983**, *17*, 223–229. [CrossRef]

60. Vind, J.; Malfliet, A.; Blanpain, B.; Tsakiridis, P.E.; Tkaczyk, A.H.; Vassiliadou, V.; Panias, D. Rare Earth Element Phases in Bauxite Residue. *Minerals* **2018**, *8*, 77. [CrossRef]
61. Gamaletsos, P.N. Mineralogy and Geochemistry of Bauxites from Parnassos-Ghiona Mines and the Impact on the Origin of the Deposits. Ph.D. Thesis, National and Kapodistrian University of Athens, Athens, Greece, 2014.
62. Wellington, M.; Valcin, F. Impact of Bayer Process Liquor Impurities on Causticization. *Ind. Eng. Chem. Res.* **2007**, *46*, 5094–5099. [CrossRef]
63. Schulte, R.F.; Foley, N.K. *Compilation of Gallium Resource Data for Bauxite Deposits*; U.S. Geological Survey: Reston, VA, USA, 2014. [CrossRef]
64. Shaw, D.M. The geochemistry of gallium, indium, thallium—A review. *Phys. Chem. Earth* **1957**, *2*, 164–211. [CrossRef]
65. Figueiredo, A.M.G.; Avristcher, W.; Masini, E.A.; Diniz, S.C.; Abrão, A. Determination of lanthanides (La, Ce, Nd, Sm) and other elements in metallic gallium by instrumental neutron activation analysis. *J. Alloys Compd.* **2002**, *344*, 36–39. [CrossRef]
66. Riveros, P.A. Recovery of gallium from Bayer liquors with an amidoxime resin. *Hydrometallurgy* **1990**, *25*, 1–18. [CrossRef]
67. Lamerant, J.-M. Process for Extracting and Purifying Gallium from Bayer Liquors. Patent No US5102512 A, 10 April 1991.
68. Lamerant, J.-M. Process for Extracting Gallium from Bayer Liquors Using an Impregnated Absorbent Resin. Patent No. US5424050 A, 13 June 1995.
69. Selvi, P.; Ramasami, M.; Samuel, M.H.P.; Sripriya, R.; Senthilkumar, K.; Adaikkalam, P.; Srinivasan, G.N. Gallium recovery from Bayer's liquor using hydroxamic acid resin. *J. Appl. Polym. Sci.* **2004**, *92*, 847–855. [CrossRef]
70. Ilić, Z.; Mitrović, A. Determination of gallium in Bayer process sodium aluminate solution by inductively coupled plasma atomic emission spectrometry. *Anal. Chim. Acta* **1989**, *221*, 91–97. [CrossRef]
71. Zhao, Z.; Long, H.; Li, X.; Fan, Y.; Han, Z. Precipitation of vanadium from Bayer liquor with lime. *Hydrometallurgy* **2012**, *115–116*, 52–56. [CrossRef]
72. Fenerty, M.J. Production of Fused Alumina. Patent No. US2961296 A, 22 November 1960.
73. Smith, P. Reactions of lime under high temperature Bayer digestion conditions. *Hydrometallurgy* **2017**, *170*, 16–23. [CrossRef]
74. Okudan, M.D.; Akcil, A.; Tuncuk, A.; Deveci, H. Effect of parameters on vanadium recovery from by-products of the Bayer process. *Hydrometallurgy* **2015**, *152*, 76–83. [CrossRef]
75. Burke, I.T.; Mayes, W.M.; Peacock, C.L.; Brown, A.P.; Jarvis, A.P.; Gruiz, K. Speciation of Arsenic, Chromium, and Vanadium in Red Mud Samples from the Ajka Spill Site, Hungary. *Environ. Sci. Technol.* **2012**, *46*, 3085–3092. [CrossRef] [PubMed]
76. Zhong-Lin, Y.; Song-Qing, G. Development of research on the scaling problem in Bayer process. *Light Met.* **1995**, 155–159.
77. Vind, J.; Malfliet, A.; Bonomi, C.; Paiste, P.; Sajó, I.E.; Blanpain, B.; Tkaczyk, A.H.; Vassiliadou, V.; Panias, D. Modes of occurrences of scandium in Greek bauxite and bauxite residue. *Miner. Eng.* **2018**, *123*, 35–48. [CrossRef]
78. Gladyshev, S.V.; Akcil, A.; Abdulvaliyev, R.A.; Tastanov, E.A.; Beisembekova, K.O.; Temirova, S.S.; Deveci, H. Recovery of vanadium and gallium from solid waste by-products of Bayer process. *Miner. Eng.* **2015**, *74*, 91–98. [CrossRef]
79. Schlegl, R.; Weber, M.; Wruss, J.; Low, D.; Queen, K.; Stilwell, S.; Lindblad, E.B.; Möhlen, M. Influence of elemental impurities in aluminum hydroxide adjuvant on the stability of inactivated Japanese Encephalitis vaccine, IXIARO®. *Vaccine* **2015**, *33*, 5989–5996. [CrossRef] [PubMed]
80. Santana, F.; Tartarotti, F. Alumina recovery estimation through material balance in Alumar refinery. In Proceedings of the 19th International Symposium ICSOBA, Belém, Brazil, 29 October–2 November 2012; pp. 1–6.

© 2018 by the authors. Licensee MDPI, Basel, Switzerland. This article is an open access article distributed under the terms and conditions of the Creative Commons Attribution (CC BY) license (http://creativecommons.org/licenses/by/4.0/).

Article

Recovery of Valuable Metals from Lithium-Ion Batteries NMC Cathode Waste Materials by Hydrometallurgical Methods

Wei-Sheng Chen and Hsing-Jung Ho *

Department of Resources Engineering, National Cheng Kung University, Tainan 70101, Taiwan; kenchen@mail.ncku.edu.tw
* Correspondence: e44016013@gmail.com; Tel.: +886-06-275-7575 (ext. 62828)

Received: 30 March 2018; Accepted: 3 May 2018; Published: 6 May 2018

Abstract: The paper focuses on the improved process of metal recovery from lithium-ion batteries (LIBs) lithium nickel manganese cobalt oxide (NMC) cathode waste materials by using hydrometallurgical methods. In the acid leaching step, the essential effects of acidity concentration, H_2O_2 concentration, leaching time, liquid-solid mass ratio, and reaction temperature with the leaching percentage were investigated in detail. The cathode material was leached with 2M H_2SO_4 and 10 vol. % H_2O_2 at 70 °C and 300 rpm using a liquid-solid mass ratio of 30 mL/g. In order to complete the recovery process, this paper designs the proper separation process to recover valuable metals. The leach liquor in the recovery process uses Cyanex 272 to first extract Co and Mn to the organic phase. Secondly, Co and Mn are separated by using D2EHPA, and a high purity of Co is obtained. Thirdly, Ni is selectively precipitated by using DMG, and Ni is completely formed as a solid complex. Finally, in the chemical precipitation process, the remaining Li in the leach liquor is recovered as Li_2CO_3 precipitated by saturated Na_2CO_3, and Co, Mn, and Ni are recovered as hydroxides by NaOH. This hydrometallurgical process may provide an effective separation and recovery of valuable metals from LIBs waste cathode materials.

Keywords: NMC batteries; recycling; leaching; solvent extraction; selective precipitation; hydrometallurgy

1. Introduction

Nowadays, as a result of the rapid development of modern society and technology, the use of lithium-ion batteries (LIBs) has become indispensable. These are commonly applied in our lives and play an important role in power sources and diverse devices such as mobile phones, laptop computers, digital cameras, and even the developing electric vehicles (EVs) and hybrid electric vehicles (HEVs) [1–4]. In the period between 2000 and 2010, the annual production of LIBs increased by 800% worldwide [5]. With the popularity of LIBs' development, the resulting use of LIBs is also growing prominently [6,7]. Consequently, the recycling of spent LIBs by means of the recovery of the valuable metals contained in the cathode material, such as lithium, cobalt, nickel, and manganese, is considered as a progressively more substantial process to prevent environmental problems and meet sustainable and environmentally friendly regulations.

LIBs are frequently classified according to their cathode materials, into lithium cobalt oxide ($LiCoO_2$) batteries, lithium manganese oxide ($LiMn_2O_4$) batteries, and lithium iron phosphate ($LiFePO_4$) batteries, for example [8–12]. However, these types of batteries are gradually being replaced by lithium nickel manganese cobalt oxide ($LiNiMnCoO_2$ or NMC) batteries. Hence, in this study, we focus on dealing with the NMC cathode waste materials.

At present, the recycling process of LIBs has been investigated in several studies [13,14]. In order to recover the valuable metals from several types of LIBs, measures have been reported involving mechanical processes [15], mechano-chemical processes [16], thermal treatment [17,18], and dissolution processes. Subsequently, chemical processes mainly involve the use of hydrometallurgical operations to carry out the recovery of valuable metals; these include acid leaching [19–21], chemical precipitation [22,23], solvent extraction [24,25], ion-exchange [26], and electrochemistry [27]. Thus far, approaches for recycling valuable metals in LIBs mainly use pyrometallurgical and hydrometallurgical processes. In hydrometallurgical processes, the recovery of valuable metals from spent LIB cathode materials is dealt with via acid leaching. According to different leaching ways to cope with spent LIBs, this is mostly carried out using an inorganic acid as an acidic leaching agent, such as H_2SO_4 [28–30], HCl [31], or HNO_3 [32]. Compared with those given in the literature [33], H_2SO_4 has a great effect on lithium cobalt oxide batteries and is cheaper than others. Therefore, we chose H_2SO_4 as the leaching agent to process the NMC cathode materials. The research focused on finding the most suitable acid concentration, reaction temperature, liquid-solid mass ratio, and other parameters.

In the separation process, mainly used are solvent extraction, ion-exchange, and chemical precipitation to cope with the waste materials. Because several extractants, resins, and precipitating agents, such as PC88A [34], D2EHPA [35], Cyanex 272/Cyanex 301/Cyanex 302 [36,37], Mextral272P [38], Dowex M4195, Diaion CR-11, Lewatit TP-272 [26], and dimethylglyoxime (DMG) [39,40], were experimented with, we found that the most common way to separate cobalt, nickel, and manganese from other sources is solvent extraction. In comparison with literature, as an extractant, Cyanex 272 has a great selectivity between nickel and cobalt; D2EHPA has the effect of separating cobalt and manganese. However, when the target materials are no longer only two metals, such as nickel/cobalt or cobalt/manganese, but four metals, such as nickel/cobalt/manganese/lithium, the effect of separation is limited by co-extraction. Furthermore, as a precipitating agent, DMG precipitates nickel very selectively in the absence of cobalt. Hence, we designed an improved recovery process and combined the advantages of the extractants and precipitating agent mentioned above to overcome the inadequate abilities, achieving effective results. In this study, we concentrate on discussing the experimental parameters and design a recycling process for LIB NMC cathode waste materials.

2. Materials and Methods

2.1. Materials and Reagents

The NMC cathode waste materials mainly contain cobalt, nickel, manganese, and lithium. In our experiment, the source of materials was from the LIB industry, and these were produced and acquired during the manufacture of LIBs. The materials were analyzed by scanning electron microscopy (SEM; Hitachi, S-3000N), energy-dispersive X-ray spectroscopy (EDS; Bruker, XFlash6110), X-ray diffraction (XRD; Dandong DX-2700), and inductively coupled plasma optical emission spectrometry (ICP-OES; Varian, Vista-MPX). Figure 1 shows that the valuable metals in the cathode materials were present as $LiCoO_2$ and $LiNi_{1/3}Co_{1/3}Mn_{1/3}O_2$. Figure 2a,b shows the result of the cathode material analysis by SEM and EDS. It was found that the metals were distributed evenly, and we could also determine the presence of $LiCoO_2$ and $LiNi_{1/3}Co_{1/3}Mn_{1/3}O_2$. The chemical composition of the LIB cathode materials was analyzed by ICP-OES and mainly contained 25.83% Co, 26.29% Ni, 14.41% Mn, and 8.31% Li. Compared with previous literature [41], we found that the total amount of Co and Ni in this study was nearly 15% higher than in previous literature. Because of the similar chemical properties of Co and Ni, a large proportion of Co and Ni causes serious co-precipitation and co-extraction, as well as other negative effects. Therefore, an improved separation process is required in subsequent studies.

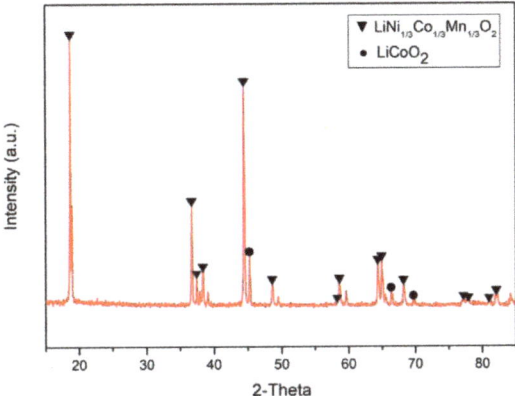

Figure 1. The X-ray diffraction (XRD) pattern of waste lithium nickel manganese cobalt oxide (NMC) cathode materials.

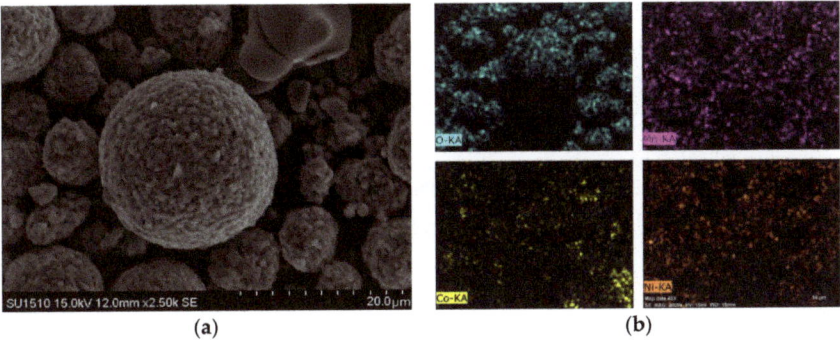

Figure 2. (**a**) The scanning electron microscopy (SEM) analysis of cathode materials; (**b**) the energy-dispersive X-ray spectroscopy (EDS) analysis of cathode materials.

The sulfuric acid used as the leaching agent and stripping agent in the experiment was from Sigma-Aldrich (St. Louis, MO, USA) (H_2SO_4, 98%) and was diluted in deionized water. The commercial extractants Cyanex 272 (CYTEC, 85%) and D2EHPA (Alfa Aesar, 95%) were diluted in kerosene, and both were saponified by the addition of a stoichiometric amount of sodium hydroxide solution. In this study, the preparation of Cyanex 272 and D2EHPA was saponified to 60% Na-Cyanex 272 and 50% Na-D2EHPA. Sodium hydroxide was from Showa (NaOH, 97%), and the selective precipitation reagent in this experiment was dimethylglyoxime (DMG, Sigma-Aldrich, $\geq$99%). All other chemical reagents used in the experiment were of analytical grade and were prepared or diluted with deionized water.

2.2. Leaching

Leaching procedures were carried out using standard laboratory leaching equipment. The cathode material was dissolved in sulfuric acid. The leaching parameters, such as the acid concentration, reaction temperature, reaction time, and liquid-solid mass ratio, were investigated. The acidity was set from 0.25 M to 8 M, the reducing reagent concentration was set from 0.2–21%, and the liquid-solid mass ratio was set from 2.5 mL/g to 50 mL/g. The effect of temperature was tested at different temperatures from 25 °C to 85 °C to achieve a better leaching percentage. The leaching percentage was calculated according to Equation (1):

$$X_B = (m1/m2) \times 100\% \tag{1}$$

where X_B is the leaching percentage, m_1 is the measured quantity of metal leached, and m_2 is the quantity of metal in the raw material.

The following chemical equations of dissolution demonstrated that the cathode waste materials from LIBs were dissolved in the sulfuric acid solution with hydrogen peroxide:

$$2LiCoO_{2(s)} + 3H_2SO_{4(aq)} + H_2O_{2(aq)} \rightarrow 2CoSO_{4(aq)} + Li_2SO_{4(aq)} + 4H_2O_{(g)} + O_{2(g)} \tag{2}$$

$$6LiNi_{1/3}Mn_{1/3}Co_{1/3}O_{2(s)} + 9H_2SO_{4(aq)} + H_2O_{2(aq)} \rightarrow 2MnSO_{4(aq)} + 2NiSO_{4(aq)} + 2CoSO_{4(aq)} + 3Li_2SO_{4(aq)} + 10H_2O_{(g)} + 2O_{2(g)} \tag{3}$$

2.3. Solvent Extraction

In the experiment, Na-Cyanex 272 was used as the extractant to efficiently separate cobalt and nickel in the sulfate solution. Na-D2EHPA was used as the extractant to entirely separate cobalt and manganese in the sulfate solution. The extractant was diluted into kerosene and was partially saponified by NaOH. The saponification reaction of Cyanex 272 and D2EHPA can be written as Equation (4), and the extraction mechanism of Na-Cyanex 272 and Na-D2EHPA can be written as Equation (5) [42–44]:

$$Na^+_{(aq)} + 1/2\,(HA)_{2(org)} \rightarrow NaA_{(org)} + H^+_{(aq)} \tag{4}$$

$$M^{2+}_{(aq)} + NaA_{(org)} + 2(HA)_{2(org)} \Leftrightarrow (MA_2 \bullet 3HA)_{(org)} + Na^+_{(aq)} + H^+_{(aq)} \tag{5}$$

The distribution ratio, D, was calculated as the concentration ratio of the metal present in the organic phase to that in the aqueous phase at equilibrium:

$$D = \frac{C_0 - C}{C} \times \frac{V_{aq}}{V_{org}} \tag{6}$$

where C_0 is an initial total concentration of metal ions in an aqueous phase; C is the equilibrium concentration of metal ions in an aqueous phase; and V_{aq} and V_{org} are the volumes of the aqueous and organic phases, respectively.

From the distribution ratio, D, the extraction percentage, %E, could be calculated by Equation (7):

$$\%E = \frac{D}{D + \frac{V_{aq}}{V_{org}}} \times 100\% \tag{7}$$

where D is the distribution ratio; and V_{aq} and V_{org} are the volumes of the aqueous and organic phases, respectively.

2.4. Stripping Process

The stripping agent in the experiment was H_2SO_4, which was mixed with the organic phases after the solvent extraction step. The metals, such as cobalt ions and manganese ions, were stripped into the aqueous phase owing to their high solubility in H_2SO_4. After the first solvent extraction, the acidity of H_2SO_4 (0.01–0.15 mol/L) and the organic-aqueous ratio (0.5–4) of the stripped cobalt and manganese were investigated. After the second solvent extraction, the acidity of H_2SO_4 (0.005–0.15 mol/L) and the organic-aqueous ratio (1–8) of the stripped manganese were investigated.

2.5. Selective Precipitation and Chemical Precipitation

The selective precipitation process used DMG ($C_4H_8N_2O_2$) as the reagent to separate nickel and lithium. The molar ratio of $C_4H_8N_2O_2$ to nickel (MRDN) and the equilibrium pH value were adjusted respectively from 1 to 3 and 3 to 10 to obtain the best precipitation percentage under optimal parameters. In the chemical precipitation process, sodium hydroxide was dissolved in deionized water and modified the pH value. In addition, the best precipitation percentages of cobalt hydroxide, manganese hydroxide, and nickel hydroxide were investigated. The precipitation percentages were calculated by Equation (8):

$$P = \frac{[M]_0 - [M]}{[M]_0} \times 100\% \qquad (8)$$

where P is the precipitation percentage, is the metal concentration of the leach liquor, and $[M]_0$ is the metal concentration of the leach liquor after precipitation.

3. Results and Discussion

3.1. Leaching Process

3.1.1. Effect of Acid Concentration and H_2O_2 Concentration

Figure 3a shows the leaching behavior of the metals cobalt, nickel, lithium, and manganese from spent LIB cathode materials by sulfuric acid and hydrogen peroxide. The effect of the H_2SO_4 concentration was investigated by varying the H_2SO_4 concentration from 0.25 M to 8.0 M. The results indicated that Co increased steeply from 43.7% to 91.6% as the H_2SO_4 concentration increased up to 2.0 M, while Mn, Ni, and Li respectively increased to 91.8%, 91.4%, and 94.0%. The effect was ascribed to the fact that a higher acid concentration assisted and speeded up the forward reaction, resulting in a higher leaching percentage [45].

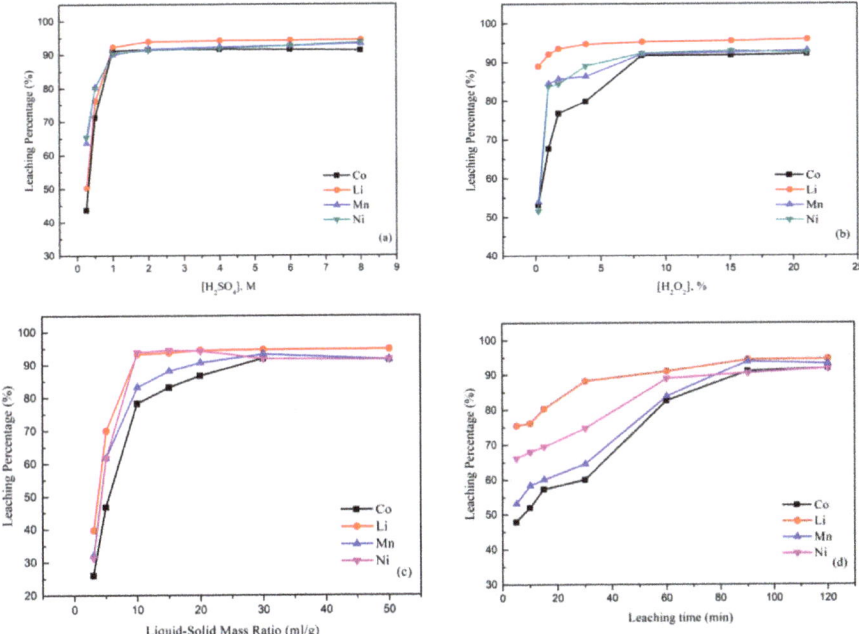

Figure 3. *Cont.*

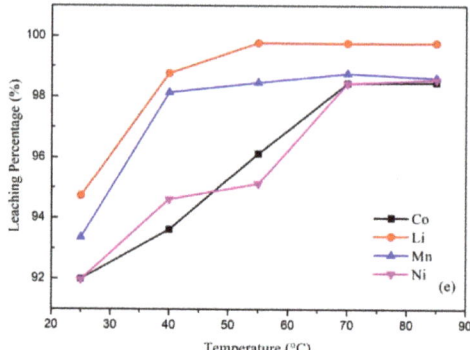

Figure 3. (a) Effect of the leaching percentage on H_2SO_4 concentration (reaction conditions: liquid-solid mass ratio of 50, 10.0% H_2O_2, 120 min, and 25 °C). (b) Effect of the leaching percentage on H_2O_2 concentration (reaction conditions: 2.0 M H_2SO_4, liquid-solid mass ratio of 50, 120 min, and 25 °C). (c) Effect of the leaching percentage on liquid-solid mass ratio (reaction conditions: 2.0 M H_2SO_4, 10.0% H_2O_2, 120 min, and 25 °C). (d) Effect of the leaching percentage on reaction time (reaction conditions: 2.0 M H_2SO_4, 10.0% H_2O_2, liquid-solid mass ratio of 30, and 25 °C). (e) Effect of the leaching percentage on temperature (reaction conditions: 2.0 M H_2SO_4, 10.0% H_2O_2, liquid-solid mass ratio of 30, and 90 min).

In order to examine the effect of the H_2O_2 concentration on the leaching process, the concentration of H_2O_2 was varied in the range from 0.2% to 21%. Figure 3b illustrates that the leaching efficiency of the metal significantly increased when the H_2O_2 concentration was 8.16%. The results indicated that the Co, Ni, Mn, and Li percentages steeply increased up to 91.72%, 92.34%, 92.12%, and 95.27%. This phenomenon was attributed to the fact that the reductions of Co^{3+} to Co^{2+} and Mn^{4+} to Mn^{2+} would help these metals to dissolve more readily. However, there was no apparent influence when H_2O_2 was added with a concentration of more than 10%. Therefore, the concentration of H_2O_2 was chosen as 10% to be optimal in the leaching process.

3.1.2. Effect of Liquid-Solid Mass Ratio

The effect of the liquid-solid mass ratio is shown in Figure 3c. The leaching percentages of all metals investigated were generally increased, while the liquid-solid mass ratio increased from 3/1 to 30/1 and percentages were increased to 91.99% for Co, 94.72% for Li, 93.35% for Mn, and 91.97% for Ni. The reason was that when the liquid-solid mass ratio was low, there was insufficient acid to react in the process. In other words, when the liquid-solid mass ratio was high, there was more acid readily able to react and available to obtain a higher leaching percentage. Hence, the liquid-solid mass ratio was chosen to be 30 mL/g as optimal.

3.1.3. Effect of Reaction Time and Temperature

Figure 3d shows the effect of the leaching percentage with the reaction time. The leaching percentages of Co, Li, Ni, and Mn substantially increased by about 44%, 19.8%, 25.7%, and 40.1% when the reaction time was increased from 5 to 90 min. The reason was that with the increase in the leaching time, a greater and greater surface area of the unreacted particle cores could react with the sulfuric acid.

The effect of the temperature is shown in Figure 3e. The leaching percentage increased with the increasing temperature because the temperature has a great effect on the leaching process. A higher temperature can increase the speed of the molecular motion and increase the energy of the particles' collisions. The optimal leaching parameters from this study are illustrated in Table 1. The leaching percentages under the optimal conditions were Co: 98.46%; Ni: 98.56%; Li: 99.76%; and Mn: 98.62%.

Table 1. The optimal parameters of leaching process.

	[H_2SO_4]	[H_2O_2]	Liquid-Solid Ratio	Temperature	Leaching Time
NMC battery cathode material	2.0 mol/L	10.0%	30 mL/g	70 °C	90 min

3.2. Solvent Extraction with Na-Cyanex 272

Because NMC cathode waste material contains an extraordinarily large proportion of cobalt and nickel, we used Na-Cyanex 272 to first separate the nickel and cobalt. If the nickel-cobalt separation was not treated first, it would have been difficult to achieve a high purity of the product and would have impeded the following processes. Hence in this study, we used Na-Cyanex 272 to handle this problem. The extraction pH value was considered the key variable for separation in the extraction process. The metals' concentration was set as the concentration ratio of Co/Ni/Mn/Li = 2590:2610:1400:800 mg/L; this was set according to the leaching condition and was analyzed by ICP-OES to calculate the extraction percentage.

3.2.1. Effect of Equilibrium pH Value

The equilibrium pH value was set from 1 to 7.5 by using 0.1 M 60% Na-Cyanex 272 with an organic-aqueous ratio of 1:1 over 15 min. Figure 4a shows that the extraction percentages of cobalt and manganese were observed to increase significantly from almost 0% to 98.8% and 98.9% respectively when the pH value was raised from 4 to 6. The extraction was not desirable when the pH value was higher than 6, as the extraction percentage of nickel began to increase rapidly. Moreover, when the pH value was higher than 7, cobalt began to partially precipitate to cobalt hydroxide. Hence, in this step, the equilibrium pH value equal set to 6 was optimal.

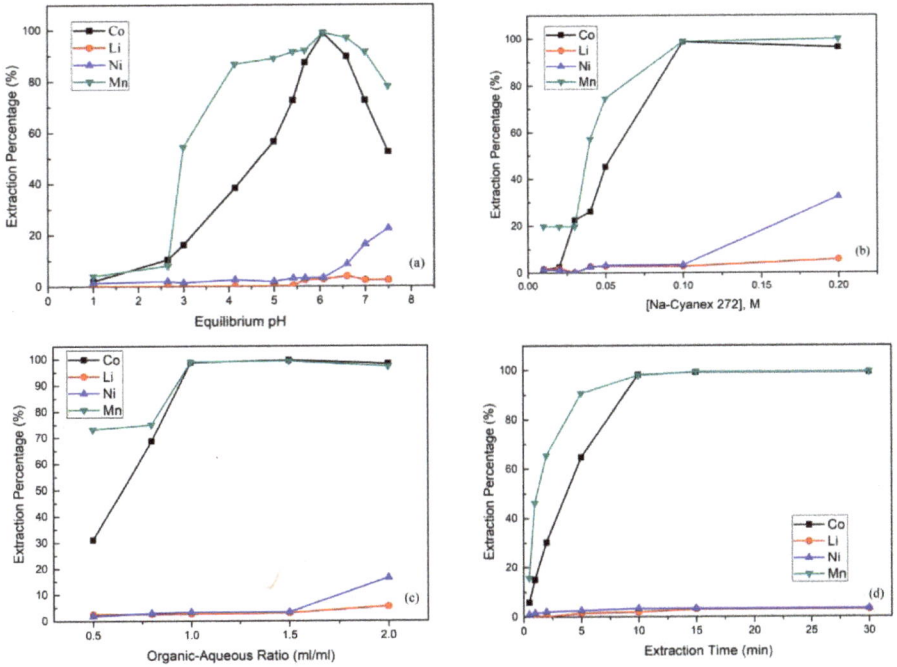

Figure 4. *Cont.*

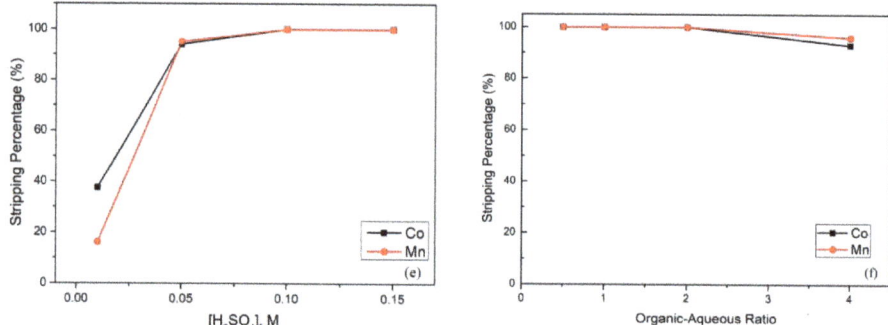

Figure 4. (**a**) Effect of the extraction percentage on equilibrium pH value (reaction conditions: 0.1 M Na-Cyanex 272, organic-aqueous ratio of 1, and 15 min). (**b**) Effect of the extraction percentage on Na-Cyanex 272 concentration (reaction conditions: equilibrium pH value of 6, organic-aqueous ratio of 1, and 15 min). (**c**) Effect of the extraction percentage on organic-aqueous ratio (reaction conditions: equilibrium pH value of 6, 0.1 M Na-Cyanex 272, and 15 min). (**d**) Effect of the extraction percentage on extraction time (reaction conditions: equilibrium pH value of 6, 0.1 M Na-Cyanex 272, and organic-aqueous ratio of 1.5). (**e**) Effect of the stripping percentage on sulfuric acid concentration (reaction conditions: organic-aqueous ratio of 1 and 10 min). (**f**) Effect of the stripping percentage on organic-aqueous ratio (reaction conditions: 0.1 M H_2SO_4 and 10 min).

3.2.2. Effect of Na-Cyanex 272 Concentration

The extraction of Co, Mn, Ni, and Li from the leach liquor of spent NMC batteries was studied with conditions of Na-Cyanex 272 concentration from 0.01 to 0.2 M at pH 6 and an organic-aqueous ratio of 1:1 over 15 min. Figure 4b shows that by increasing the Na-Cyanex 272 concentration from 0.01 M to 0.1 M, the extraction percentages of cobalt and manganese increased strictly. The reason was that a higher concentration of the extractant enabled more Co^{2+} and Mn^{2+} ions to be caught. However, when the Na-Cyanex 272 concentration was higher than 0.1 M, the extraction percentage of nickel started to increase. This was because the excess extractant resulted in an extraction effect that was too strong and that thus had an adverse effect on the separation process.

3.2.3. Effect of Organic-aqueous Ratio

Figure 4c shows that an organic-aqueous ratio from 0.5 to 2.0 was studied using 0.1 M Na-Cyanex 272 at pH 6 over 15 min. The result shows that the extraction percentages of cobalt and manganese increased as the organic-aqueous ratio increased, which means that the cobalt and manganese were not yet extracted completely. However, when the ratio was greater than 1.5, the extraction percentage of nickel increased rapidly. Hence, in order to avoid the extraction problem of nickel, an organic-aqueous ratio of 1.5 was better for the cobalt and nickel separation.

3.2.4. Effect of Extraction Time

In Figure 4d, it is clear that the extraction time was a significant influence in the extraction process. The effect of the extraction time was studied using 0.1 M Na-Cyanex 272 when the pH was 6 and the organic-aqueous ratio was 1.5. The extraction percentage increased substantially from 0.5 to 15 min, and the reaction was balanced after 15 min. In this case, the extraction percentages of lithium and nickel were found to have almost no increases with the increasing extraction time. Finally, the extraction percentages of Co, Mn, Ni, and Li were about 99.2%, 99.3%, 3.3%, and 3.0%, respectively.

3.2.5. Stripping of Co and Mn from the Organic Phase by Sulfuric Acid

After extraction, the cobalt and manganese in the organic phase continued to the stripping process. In this process, we chose H_2SO_4 as the stripping agent, and the effect of the H_2SO_4 concentration is presented in Figure 4e. As the figure shows, when the H_2SO_4 concentration increased, the stripping percentage increased simultaneously. It was clear that the reason was the shortage of H_2SO_4. Hence, we found that when the H_2SO_4 concentration was increased up to 0.1 M, the stripping percentages of cobalt and manganese achieved almost 100%. Figure 4f shows that the stripping percentage declined as the organic-aqueous ratio increased, which means that the cobalt and manganese were not yet stripped completely when the organic phase increased. We found that when the organic–aqueous ratio was 2:1, the stripping percentage started to decrease. Hence, in the stripping process, an organic-aqueous ratio of 2 was optimal, and the stripping percentage of cobalt and manganese was almost 100%.

3.3. Solvent Extraction with Na-D2EHPA

After the separation by using Na-Cyanex 272 as the extractant, the metals were separated into two sides. One side contained cobalt with manganese, and the other side contained nickel with lithium. In the previous step, most of the nickel in the material had been separated. Compared with previous literature [46,47], the biggest problem in the recovery process was the poor separation effect resulting from the co-extraction of cobalt and nickel. Hence, the problem of the high proportion of cobalt and nickel had been already solved. In this step, in order to separate the cobalt and manganese effectively, Na–D2EHPA was used as the extractant.

3.3.1. Effect of Equilibrium pH Value

The effect of the equilibrium pH value in the extraction and separation of cobalt and manganese from the sulfate solution is shown in Figure 5a. The extraction percentage of manganese increased as the equilibrium pH value increased. However, when the equilibrium pH value was greater than 2.95, the extraction percentage of cobalt started to increase rapidly. In order to obtain a good recovery of cobalt, we chose a pH value of 2.95 as the optimal.

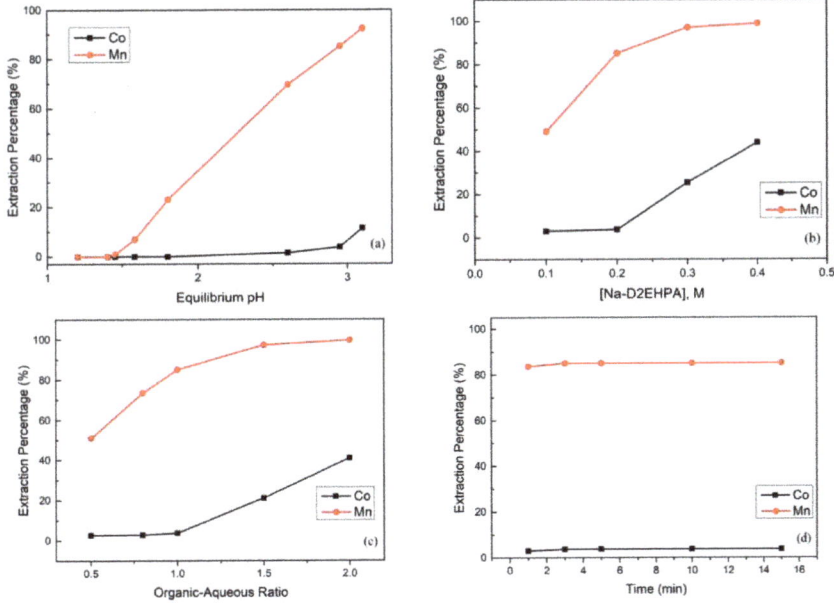

Figure 5. *Cont.*

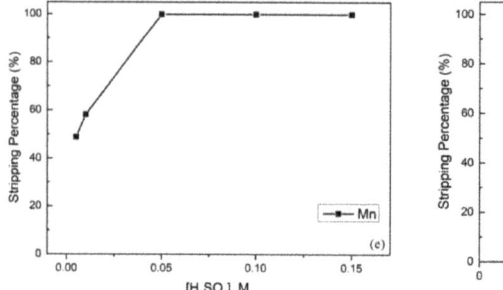
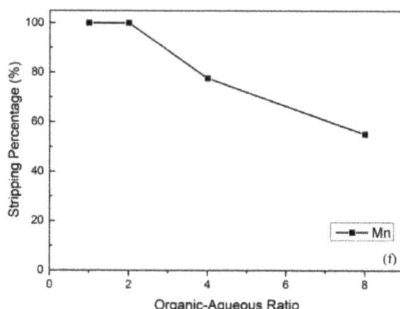

Figure 5. (a) Effect of the extraction percentage on equilibrium pH value (reaction conditions: 0.2 M Na-D2EHPA, organic-aqueous ratio of 1, and 15 min). (b) Effect of the extraction percentage on Na-D2EHPA concertation (reaction conditions: equilibrium pH value of 2.95, organic-aqueous ratio of 1, and 15 min). (c) Effect of the extraction percentage on organic-aqueous ratio (reaction conditions: equilibrium pH value of 2.95, 0.2 M Na-D2EHPA, and 15 min). (d) Effect of the extraction percentage on reaction time (reaction conditions: equilibrium pH value of 2.95, 0.2 M Na-D2EHPA, and organic-aqueous ratio of 1.0). (e) Effect of the stripping percentage on sulfuric acid concentration (reaction conditions: organic-aqueous ratio of 1 and 5 min). (f) Effect of the stripping percentage on organic-aqueous ratio (reaction conditions: 0.05 M H_2SO_4 and 5 min).

3.3.2. Effect of Na-D2EHPA Concentration

The competitive extraction of cobalt and manganese was studied with Na-D2EHPA concentrations from 0.1 M to 0.4 M. Figure 5b shows that the extraction percentage of manganese increased simultaneously with the Na-D2EHPA concentration. However, when the Na-D2EHPA concentration was higher than 0.2 M, owing to the extraction capacity becoming too strong, the extraction percentage of cobalt also started to increase. Therefore, the extraction percentages of manganese and cobalt were 85.1% and 3.7% respectively under 0.2 M Na-D2EHPA at the equilibrium pH value of 2.95 and organic-aqueous ratio of 1.0 over 15 min.

3.3.3. Effect of Organic-Aqueous Ratio and Extraction Time

In the experiment, the organic-aqueous ratio also influenced the extraction percentage. Figure 5c shows that when the organic-aqueous ratio increased from 0.5 to 1.0, the extraction percentage of manganese increased to 85.1% and the extraction percentage of cobalt only slightly increased. However, when the organic-aqueous ratio was increased up to 1.5, the extraction percentage of cobalt increased to 21.1%. The reason was that the organic phase was too great or the aqueous phase was too little to selectively extract quantities of metal ions. Therefore, an organic-aqueous ratio of 1.0 was chosen as optimal.

In order to avoid consuming too much energy, we controlled the reaction time and investigated the interaction between the organic and aqueous phases. The effect of the extraction time was investigated with the optimized parameters. Figure 5d shows that the extraction time needed to approach the reaction balance was very short. Hence, an extraction time of 5 min was chosen as optimal.

According to the above studies, under optimal parameters, the extraction efficiency of manganese was up to 85.14%. On the basis of this distribution ratio, the efficiency could also increase to over 99.6% by three stages of extraction, retaining cobalt in the aqueous phase. The optimal parameters of both extractions are illustrated in Table 2.

Table 2. The optimal parameters of solvent extraction.

	Equilibrium pH Value	Concentration (M)	Organic-Aqueous Ratio	Extraction Time (min)
Na-Cyanex 272	6.0	0.1	1.5	15
Na-D2EHPA	2.95	0.2	1.0	5

3.3.4. Stripping of Mn from the Organic Phase by Sulfuric Acid

After extraction, the manganese in the organic phase continued to the stripping process. In this process, we chose H_2SO_4 as the stripping agent, and the effect of the H_2SO_4 concentration is presented in Figure 5e. To obtain the best stripping percentage, the H_2SO_4 concentration was analyzed from 0.005 M to 0.15 M, and the stripping percentage of manganese achieved almost 100% when the H_2SO_4 concentration increased up to 0.05 M. Furthermore, the effect of the organic-aqueous ratio was also important. The organic-aqueous ratio was analyzed from 1 to 8, and the stripping percentage of manganese started to decline when the organic-aqueous ratio was greater than 2. The reason was that the acidity was insufficient to strip metal ions from the organic phase. Hence, we chose 0.05 M H_2SO_4 and an organic-aqueous ratio of 2 as optimal parameters.

3.4. Selective Precipitation with DMG

After the separation by using Na-Cyanex 272 as the extractant, the cobalt and manganese were extracted from the leach liquor; on the other hand, nickel and lithium were retained in the aqueous phase. The experiment was designed to employ DMG reagent ($C_4H_8N_2O_2$) to separate the nickel and lithium efficiently. Compared with other methods [48,49], using DMG produces excellent selectivity of Ni^{2+}. DMG is often used as an analytical chemistry reagent and reacts with Ni^{2+} to form a nickel DMG chelating precipitate. According to previous literature [39], DMG slightly precipitates cobalt at higher pH value; however, we extracted nearly all of the cobalt in the previous solvent extraction process to prevent this problem. Additionally, lithium cannot react with DMG, and thus the separation could be carried out completely. The equilibrium pH value and the MRDN were investigated under conditions of 25 °C, 300 rpm, and 30 min. Figure 6a shows the effect of the equilibrium pH value on the selective precipitation of nickel and lithium. It is clear from this data that DMG was completely unreactive toward lithium, and thus the highest precipitation percentage of nickel was the optimal parameter. When the equilibrium pH was increased, the precipitation percentage of nickel also increased gradually and reached almost 99.5% when the equilibrium pH value was 9; the precipitation percentage of lithium was almost 0% relatively. However, the precipitation percentage of nickel slightly decreased at a higher equilibrium pH value. The reason was attributed to an inadequate reaction between the nickel and DMG chelating precipitate [39]. Figure 6b shows the effect of MRDN in the selective precipitation process. The results indicate that when the MRDN was lower than 2, Ni^{2+} was not completely reacting with DMG and only formed a small amount of red complex. Hence, the optimal MRDN was 2. This also represented the theoretical ratio of 0.5 for the nickel DMG chelating precipitate in the process. Furthermore, the red complex could be dissolved easily using 4 M HCl solution, and almost 100% of the nickel was dissolved back to the solution. The acid dissolution reaction was the reverse reaction, and DMG is essentially insoluble in strong acid; thus the DMG could be recovered by filtration and reused as a reagent in the selective precipitation process.

3.5. Chemical Precipitation

After the solvent extraction, stripping process, and selective precipitation, four elements had already been separated. Then, chemical precipitation was conducted to obtain the final product under the best operational conditions found in the study. In order to obtain the highest precipitation percentage in the experiment, pH values from 7 to 13 was investigated. In this case, the pH was adjusted to 11 by using the saturated solution of NaOH, and cobalt was precipitated as a red precipitate, cobalt hydroxide. The solution of manganese was precipitated by adding the saturated solution of NaOH at pH 13, and the manganese ions eventually totally transferred to manganese hydroxide. The nickel ion solution could be recovered as nickel hydroxide by using the saturated solution of NaOH when the pH value was increased up to 12. On the other hand, the solution of lithium could be recovered as Li_2CO_3 by adding a saturated solution of Na_2CO_3; moreover, hot water could wash out the remaining sodium ions. Finally, the purity analysis was conducted by using ICP-OES; the purity of cobalt, nickel, and lithium products was over 99.5%, and the manganese product also achieved over 93.3% purity.

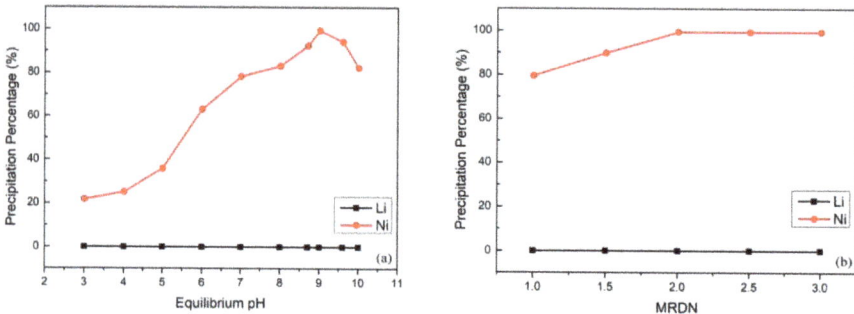

Figure 6. (a) Effect of dimethylglyoxime (DMG) precipitation percentage on equilibrium pH value (reaction conditions: molar ratio of $C_4H_8N_2O_2$ to nickel (MRDN) of 2.5, 300 rpm, and 30 min). (b) Effect of DMG precipitation percentage on MRDN (reaction conditions: equilibrium pH value of 9, 300 rpm, and 30 min).

4. Conclusions

The recovery and separation processes of metals from NMC cathode waste materials has been proven in this work to be successful and effective. The suggested recovery process is shown in Figure 7. The NMC cathode waste materials were treated by leaching, solvent extraction, stripping, selective precipitation, and chemical precipitation processes to recover cobalt, manganese, nickel, and lithium. The optimal parameters obtained in leaching were 2.0 mol/L of H_2SO_4, 30 mL/g, 70 °C, and 90 min. In this study, we used several agents and combined the advantages of each extractant and precipitating agent to improve the recovery process. The results showed that 0.1 M Na-Cyanex 272 should first be used as the extractant to separate cobalt and nickel under the optimal condition of pH 6, with an organic-aqueous ratio of 1.5 and over 15 min. Then, cobalt and manganese should be separated by using 0.2 M Na-D2EHPA at equilibrium pH 2.95, with an organic-aqueous ratio of 1.0 and over 5 min. On the other hand, nickel and lithium can be separated by using DMG at pH 9, with the molar ratio of DMG to Ni^{2+} (MRDN) of 2. Finally, the four elements can be precipitated separately by using a saturated solution of NaOH and Na_2CO_3. By this process, the purity of the cobalt, nickel, and lithium products produced was over 99.5%, and the manganese product also achieved over 90%.

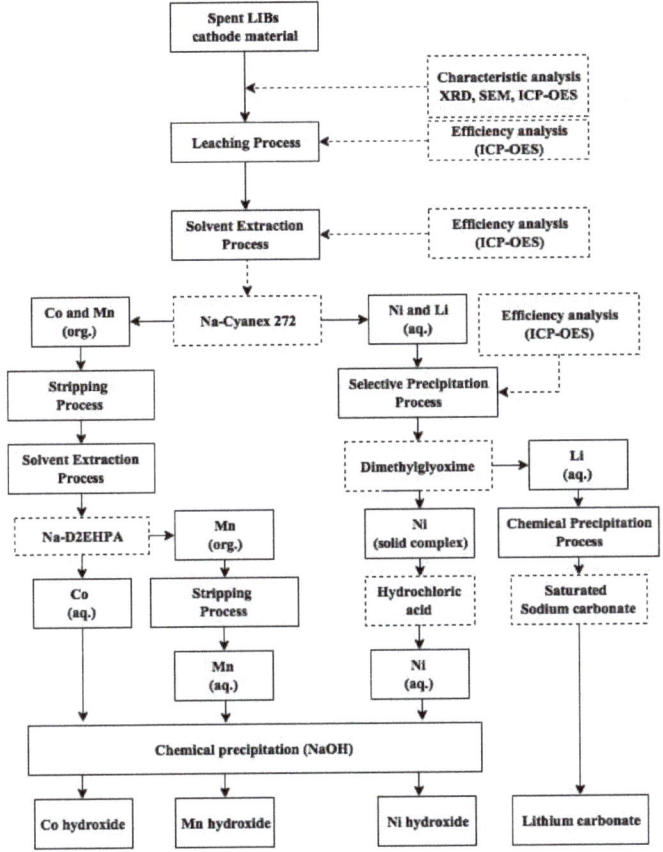

Figure 7. Suggested recovery process to separate the four metals.

Author Contributions: W.-S.C. and H.-J.H. conceived and designed the experiments; H.-J.H. performed the experiments, analyzed the data, and wrote the paper.

Funding: This research was funded by NCKU Research and Development Foundation (106S281).

Acknowledgments: We wish to acknowledge the support of the Laboratory of Resources Circulation (LRC) at National Cheng Kung University.

Conflicts of Interest: The authors declare no conflict of interest.

References

1. Wang, X.; Gaustad, G.; Babbit, C.W.; Richa, K. Economies of scale for future lithium-ion battery recycling infrastructure. *Resour. Conserv. Recycl.* **2014**, *83*, 53–62. [CrossRef]
2. Yamaji, Y.; Dodbiba, G.; Matsuo, S.; Okaya, K.; Shibayama, A.; Fujita, T. A Novel Flow Sheet for Processing of Used Lithium-ion Batteries for Recycling. *Resour. Process* **2011**, *58*, 9–11. [CrossRef]
3. Jha, M.K.; Kumari, A.; Jha, A.K.; Kumar, V.; Hait, J.; Pandey, B.D. Recovery of lithium and cobalt from waste lithium ion batteries of mobile phone. *Waste Manag.* **2013**, *33*, 1890–1897. [CrossRef] [PubMed]
4. Scrosati, B.; Garche, J. Lithium batteries: Status, prospects and future. *J. Power Sources* **2010**, *195*, 2419–2430. [CrossRef]

5. Zeng, X.L.; Li, J.H.; Singh, N. Recycling of spent lithium-ion battery: A critical review. *Crit. Rev. Environ. Sci. Technol.* **2014**, *44*, 1129–1165. [CrossRef]
6. Bernardes, A.M.; Espinosa, D.C.R.; Tenório, J.A.S. Recycling of batteries: A review of current processes and technologies. *J. Power Sources* **2004**, *130*, 288–293. [CrossRef]
7. Shin, S.M.; Kim, N.H.; Sohn, J.S.; Yang, D.H.; Kim, Y.H. Development of a metal recovery process from Li-ion battery wastes. *Hydrometallurgy* **2005**, *79*, 172–181. [CrossRef]
8. Zhao, J.M.; Shen, X.Y.; Deng, F.L.; Wang, F.C.; Wu, Y.; Liu, H.Z. Synergistic extraction and separation of valuable metals from waste cathodic material of lithium ion batteries using Cyanex272 and PC-88A. *Sep. Purif. Technol.* **2011**, *78*, 345–351. [CrossRef]
9. Nayaka, G.P.; Pai, K.V.; Santhosh, G.; Manjanna, J. Recovery of cobalt as cobalt oxalate from spent lithium ion batteries by using glycine as leaching agent. *J. Environ. Chem. Eng.* **2016**, *4*, 2378–2383. [CrossRef]
10. Fouad, O.A.; Farghaly, F.I.; Bahgat, M. A novel approach for synthesis of nanocrystalline γ-LiAlO$_2$ from spent lithium-ion batteries. *J. Anal. Appl. Pyrolysis* **2007**, *78*, 65–69. [CrossRef]
11. Kang, J.G.; Senanayake, G.; Sohn, J.S.; Shin, S.M. Recovery of cobalt sulfate from spent lithium ion batteries by reductive leaching and solvent extraction with Cyanex 272. *Hydrometallurgy* **2010**, *100*, 168–171. [CrossRef]
12. Paulino, J.F.; Busnardo, N.G.; Afonso, J.C. Recovery of valuable elements from spent Li-batteries. *J. Hazard. Mater.* **2008**, *150*, 843–849. [CrossRef] [PubMed]
13. Meshram, P.; Pandey, B.D.; Mankhand, T.R. Extraction of lithium from primary and secondary sources by pre-treatment, leaching and separation: A comprehensive review. *Hydrometallurgy* **2014**, *150*, 192–208. [CrossRef]
14. Xu, J.; Thomas, H.R.; Francis, R.W.; Lum, K.R.; Wang, J.; Liang, B. A review of processes and technologies for the recycling of lithium-ion secondary batteries. *J. Power Sources* **2008**, *177*, 512–527. [CrossRef]
15. Diekmanna, J.; Hanisch, C.; Fröböse, L.; Schälicke, G.; Loellhoeffel, T.; Fölster, A.-S.; Kwade, A. Ecological Recycling of Lithium-Ion Batteries from Electric Vehicles with Focus on Mechanical Processes. *J. Electrochem. Soc.* **2017**, *164*, A6184–A6191. [CrossRef]
16. Pagnanelli, F.; Moscardini, E.; Altimari, P.; Atia, T.A.; Toro, L. Leaching of electrodic powders from lithium ion batteries: Optimization of operating conditions and effect of physical pretreatment for waste fraction retrieval. *Waste Manag.* **2017**, *60*, 706–715. [CrossRef] [PubMed]
17. Barik, S.P.; Prabaharan, G.; Kumar, B. An innovative approach to recover the metal values from spent lithium-ion batteries. *Waste Manag.* **2016**, *51*, 222–226. [CrossRef] [PubMed]
18. Zhang, X.; Xue, Q.; Li, L.; Fan, E.; Wu, F.; Chen, R. Sustainable Recycling and Regeneration of Cathode Scraps from Industrial Production of Lithium-Ion Batteries. *ACS Sustain. Chem. Eng.* **2016**, *4*, 7041–7049. [CrossRef]
19. Nayl, A.A.; Elkhashab, R.A.; Badawy, S.M.; El-Khateeb, M.A. Acid leaching of mixed spent Li-ion batteries. *Arab. J. Chem.* **2014**, *10*, 3632S–3639. [CrossRef]
20. Takacova, Z.; Havlik, T.; Kukurugya, F.; Orac, D. Cobalt and lithium recovery from active mass of spent Li-ion batteries: Theoretical and experimental approach. *Hydrometallurgy* **2016**, *163*, 9–17. [CrossRef]
21. Meshram, P.; Pandey, B.D.; Mankhand, T.R.; Deveci, H. Acid baking of spent lithium ion batteries for selective recovery of major metals: A two-step process. *J. Ind. Eng. Chem.* **2016**, *43*, 117–126. [CrossRef]
22. Zhang, X.; Xie, Y.; Cao, H.; Nawaz, F.; Zhang, Y. A novel process for recycling and resynthesizing LiNi$_{1/3}$Co$_{1/3}$Mn$_{1/3}$O$_2$ from the cathode scraps intended for lithium-ion batteries. *Waste Manag.* **2014**, *34*, 1715–1724. [CrossRef] [PubMed]
23. Pant, D.; Dolker, T. Green and facile method for the recovery of spent Lithium Nickel Manganese Cobalt Oxide (NMC) based Lithium ion batteries. *Waste Manag.* **2017**, *60*, 689–695. [CrossRef] [PubMed]
24. Nayl, A.A.; Hamed, M.M.; Rizk, S.E. Selective extraction and separation of metal values from leach liquor of mixed spent Li-ion batteries. *J. Taiwan Inst. Chem. Eng.* **2015**, *55*, 119–125. [CrossRef]
25. Nguyen, V.T.; Lee, J.C.; Jeong, J.; Kim, B.S.; Pandey, B.D. Selective recovery of cobalt, nickel and lithium from sulfate leachate of cathode scrap of Li-ion batteries using liquid-liquid extraction. *Met. Mater. Int.* **2014**, *20*, 357–365. [CrossRef]
26. Chiu, K.L.; Chen, W.S. Recovery and Separation of Valuable Metals from Cathode Materials of Spent Lithium-Ion Batteries (LIBs) by Ion Exchange. *Sci. Adv. Mater.* **2017**, *9*, 2155–2160. [CrossRef]
27. Song, D.; Wang, X.; Zhou, E.; Hou, P.Y.; Guo, F.X.; Zhang, L.Q. Recovery and heat treatment of the Li(Ni$_{1/3}$Co$_{1/3}$Mn$_{1/3}$)O$_2$ cathode scrap material for lithium ion battery. *J. Power Sources* **2013**, *232*, 348–352. [CrossRef]

28. Meshram, P.; Pandey, B.D.; Mankhand, T.R. Hydrometallurgical processing of spent lithium ion batteries (LIBs) in the presence of a reducing agent with emphasis on kinetics of leaching. *Chem. Eng. J.* **2015**, *281*, 418–427. [CrossRef]
29. Zhang, X.; Cao, H.B.; Xie, Y.B.; Ning, P.G.; An, H.J.; You, H.X.; Nawaz, F. A closed-loop process for recycling $LiNi_{1/3}Co_{1/3}Mn_{1/3}O_2$ from the cathode scraps of lithium-ion batteries: Process optimization and kinetics analysis. *Sep. Purif. Technol.* **2015**, *150*, 186–195. [CrossRef]
30. He, L.P.; Sun, S.Y.; Song, X.F.; Yu, J.G. Leaching process for recovering valuable metals from the $LiNi_{1/3}Co_{1/3}Mn_{1/3}O_2$ cathode of lithium-ion batteries. *Waste Manag.* **2017**, *64*, 171–181. [CrossRef] [PubMed]
31. Wang, X.; Gaustad, G.; Babbitt, C.W.; Bailey, C.; Ganter, M.J.; Landi, B.J. Economic and environmental characterization of an evolving Li-ion battery waste stream. *J. Environ. Manag.* **2014**, *135*, 126–134. [CrossRef] [PubMed]
32. Ferreira, D.A.; Prados, L.M.Z.; Majuste, D.; Mansur, M.B. Hydrometallurgical separation of aluminium, cobalt, copper and lithium from spent Li-ion batteries. *J. Power Sources* **2009**, *187*, 238–246. [CrossRef]
33. Meshram, P.; Pandey, B.D.; Mankhand, T.R. Recovery of valuable metals from cathodic active material of spent lithium ion batteries: Leaching and kinetic aspects. *Waste Manag.* **2015**, *45*, 306–313. [CrossRef] [PubMed]
34. Wang, F.; Sun, R.; Xu, J.; Chen, Z.; Kang, M. Recovery of cobalt from spent lithium ion batteries using sulphuric acid leaching followed by solid-liquid separation and solvent extraction. *RSC Adv.* **2016**, *88*, 85303–85313. [CrossRef]
35. Hung, S.H.; Lin, C.F.; Chiang, P.C.; Tsai, T.H.; Peng, C.Y. Recovery of metal ions from spent Lithium Ion Batteries (LIBs) using sodium salts of D2EHPA or P507: Performance evaluation and life cycle assessment. *Res. J. Chem. Environ.* **2014**, *18*, 39–47. [CrossRef]
36. Swain, B.; Mishra, C.; Jeong, J.; Lee, J.C.; Hong, H.S.; Pandey, B.D. Separation of Co(II) and Li(I) with Cyanex 272 using hollow fiber supported liquid membrane: A comparison with flat sheet supported liquid membrane and dispersive solvent extraction process. *Chem. Eng. J.* **2015**, *271*, 61–70. [CrossRef]
37. Mantuano, D.P.; Dorella, G.; Elias, R.C.A.; Mansur, M.B. Analysis of a hydrometallurgical route to recover base metals from spent rechargeable batteries by liquid-liquid extraction with Cyanex 272. *J. Power Sources* **2006**, *159*, 1510–1518. [CrossRef]
38. Chen, X.P.; Xu, B.; Zhou, T.; Liu, D.; Hu, H.; Fan, S.Y. Separation and recovery of metal values from leaching liquor of mixed-type of spent lithium-ion batteries. *Sep. Purif. Technol.* **2015**, *144*, 197–205. [CrossRef]
39. Wang, R.C.; Lin, Y.C.; Wu, S.H. A novel recovery process of metal values from the cathode active materials of the lithium-ion secondary batteries. *Hydrometallurgy* **2009**, *99*, 194–201. [CrossRef]
40. Chen, X.P.; Zhou, T.; Kong, J.R.; Fang, H.X.; Chen, Y.B. Separation and recovery of metal values from leach liquor of waste lithium nickel cobalt manganese oxide based cathodes. *Sep. Purif. Technol.* **2015**, *141*, 76–83. [CrossRef]
41. Joo, S.H.; Shin, D.J.; Oh, C.H.; Wang, J.P.; Park, J.T.; Shin, S.M. Application of Co and Mn for a Co-Mn-Br or Co-Mn-$C_2H_3O_2$ Petroleum Liquid Catalyst from the Cathode Material of Spent Lithium Ion Batteries by Hydrometallurgical Route. *Metals* **2017**, *7*, 439. [CrossRef]
42. Devi, N.B.; Nathsarma, K.C.; Chakravortty, V. Separation and recovery of cobalt (II) and nickel (II) from sulphate solutions using sodium salts of D2EHPA, PC 88A and Cyanex 272. *Hydrometallurgy* **1998**, *49*, 47–61. [CrossRef]
43. Mohapatra, D.; Kim, H.I.; Nam, C.W.; Park, K.H. Liquid-liquid extraction of aluminium(III) from mixed sulphate solutions using sodium salts of Cyanex 272 and D2EHPA. *Sep. Purif. Technol.* **2007**, *56*, 311–318. [CrossRef]
44. Sarangi, K.; Reddy, B.R.; Das, R.P. Extraction studies of cobalt(II) and nickel(II) from chloride solutions using Na-Cyanex 272. Separation of Co(II)/Ni(II) by the sodium salts of D2EHPA, PC88A and Cyanex 272 and their mixtures. *Hydrometallurgy* **1999**, *52*, 253–265. [CrossRef]
45. Park, K.H.; Kim, H.I.; Parhi, P.K.; Mishra, D.; Nam, C.W.; Park, J.T.; Kim, D.J. Extraction of metals from Mo-Ni/Al_2O_3 spent catalyst using H_2SO_4 baking-leaching-solvent extraction technique. *J. Ind. Eng. Chem.* **2012**, *18*, 2036–2045. [CrossRef]
46. Mubarok, M.Z.; Hanif, L.I. Cobalt and Nickel Separation in Nitric Acid Solution by Solvent Extraction Using Cyanex 272 and Versatic 10. *Procedia Chem.* **2016**, *19*, 743–750. [CrossRef]

47. Flett, D.S. Cobalt-Nickel Separation in Hydrometallurgy: A Review. *Chem. Sustain. Dev.* **2004**, *12*, 81–91.
48. Lewis, A.; Van Hille, R. An exploration into the sulphide precipitation method and its effect on metal sulphide removal. *Hydrometallurgy* **2006**, *81*, 197–204. [CrossRef]
49. Hammack, R.W.; Edenborn, H.M. The removal of nickel from mine waters using bacterial sulfate reduction. *Appl. Microbiol. Biotechnol.* **1992**, *37*, 674–678. [CrossRef]

© 2018 by the authors. Licensee MDPI, Basel, Switzerland. This article is an open access article distributed under the terms and conditions of the Creative Commons Attribution (CC BY) license (http://creativecommons.org/licenses/by/4.0/).

Article

Concentration and Separation of Scandium from Ni Laterite Ore Processing Streams

Şerif Kaya [1,2,*], Carsten Dittrich [2], Srecko Stopic [1] and Bernd Friedrich [1]

1 IME Institute of Process Metallurgy and Metal Recycling, RWTH Aachen University, 52056 Aachen, Germany; sstopic@ime-aachen.de (S.S.); bfriedrich@metallurgie.rwth-aachen.de (B.F.)
2 MEAB Chemie Technik GmbH, 52068 Aachen, Germany; carsten@meab-mx.com
* Correspondence: serifkaya@gmail.com; Tel.: +49-163-460-2568

Received: 7 November 2017; Accepted: 8 December 2017; Published: 12 December 2017

Abstract: The presence of a considerable amount of scandium in lateritic nickel-cobalt ores necessitates the investigation of possible processing alternatives to recover scandium as a byproduct during nickel and cobalt production. Therefore, in this study, rather than interfering with the main nickel-cobalt production circuit, the precipitation-separation behavior of scandium during a pH-controlled precipitation process from a synthetically prepared solution was investigated to adopt the Sc recovery circuit into an already existing hydrometallurgical nickel-cobalt hydroxide processing plant. The composition of the synthetic solution was determined according to the hydrometallurgical nickel laterite ore processing streams obtained from a HPAL (high-pressure sulphuric acid leaching) process. In order to selectively precipitate and concentrate scandium with minimum nickel and cobalt co-precipitation, the pH of the solution was adjusted by $CaCO_3$, MgO, Na_2CO_3, and NaOH. It was found that precipitation with MgO or Na_2CO_3 is more advantageous to obtain a precipitate containing higher amounts of scandium with minimum mass when compared to the $CaCO_3$ route, which makes further processing more viable. As a result of this study, it is proposed that by a simple pH-controlled precipitation process, scandium can be separated from the nickel and cobalt containing process solutions as a byproduct without affecting the conventional nickel-cobalt hydroxide production. By further processing this scandium-enriched residue by means of leaching, SX (solvent extraction), and precipitation, an intermediate $(NH_4)_2NaScF_6$ product can be obtained.

Keywords: laterites; scandium; leaching; precipitation; solvent extraction

1. Introduction

Scandium is classified as a rare earth element, together with yttrium and lanthanides, and it is widely distributed in the Earth's crust without the affinity of forming exploitable, high-grade primary scandium deposits. This geochemical nature hindered its extensive and economical production and, up to now, it is mainly obtained as a byproduct from the hydrometallurgical processing of iron-uranium, titanium, rare earth elements, tungsten, and zirconium ores, tailings, and residues. Depending on the rate of recovery from these sources, its supply was reported to be only 5–12 tons/year with an unsteady price of 2000–4500 \$/kg of 99.9 Sc_2O_3 [1].

Although its superior performance has been reported in several publications, historically it has only been used in applications where the performance is much more important than the cost. Primarily, it was used in aluminum alloys as a minor alloying element (0.2–0.8%) due to its superior contribution to mechanical, corrosion, and welding properties. Because they offer the advantage of high strength-low weight, these alloys are mainly used in military, sporting goods, and aerospace applications. Nowadays, there is a great demand forcing the replacement of heavier structural components with lighter ones in applications such as airplanes and automobiles to decrease fuel consumption and emissions. According to the estimates, aircrafts made from welded aluminum scandium alloys would be 15% lighter and 15% cheaper to build compared to present materials [2]. Therefore, aviation companies have already produced and tested prototypes of Al-Sc alloy components with the aid of 3D printing technology, offering additional benefits during manufacturing.

Besides these alloys, the use of scandium in solid oxide fuel cells (SOFCs) is another important and promising area where an urgent demand for scandium exists. The addition of scandium oxide in the solid electrolyte of fuel cells provides many advantages in terms of cell efficiency and prolonged cell life.

Finally, the outstanding properties of scandium in laser and lighting applications, as well as its potential use in transmission lines and the marine industry, make scandium an indispensable element.

In short, industrial applications are waiting for a sufficient, reliable, and reasonably priced scandium supply. In the light of up-to-date information in the literature; besides the aforementioned sources in which scandium is presently obtained, industrial wastes from aluminum and lateritic nickel-cobalt processing seem to be a new, very abundant and promising source for the huge and urgent scandium need of the industry in the immediate future. For example, in China, ores are generally reported as worthy of exploitation if the Sc content ranges between 20 and 50 g/t and, recently, various lateritic nickel and cobalt deposits were reported to contain from 50 g/t up to 600 g/t of Sc [3–5]. Therefore, this study aimed to investigate the possibility of scandium by-production from lateritic nickel-cobalt deposits without affecting the conventional hydrometallurgical production of the MHP (mixed nickel-cobalt hydroxide) product.

2. Previous Studies and State of the Art

In previous studies, a lateritic ore with a Sc grade of 106 g/t, was digested under high-pressure sulphuric acid leaching and it was reported that 80.6% of scandium within the ore could be extracted into the leach solution together with nickel, cobalt, and impurity elements [6]. After digesting the ore, scandium was aimed to be selectively precipitated, concentrated, separated from the main nickel-cobalt hydroxide processing circuit, and it was used as a secondary scandium source in a two-step pH-controlled solution purification process [7].

In the first step of the solution purification, maximum iron and some of the aluminum and chromium precipitation was obtained with minimum nickel, cobalt and scandium co-precipitation depending on the hydrolysis behavior of ions with increasing the solution pH. For this aim, the precipitation temperature and duration was selected to be 90 °C and 120 min, respectively. The pH of the slurry was adjusted and kept constant at 2.75 by adding $CaCO_3$ slurry. According to the analytical results given in Table 1, iron, aluminum, and chromium impurity levels were reduced in the first step with minimum nickel, cobalt, and scandium loss [7].

Table 1. Chemical composition of the solutions.

Element	Ni	Co	Sc	Fe	Al	Cr	Mn	Mg	Cu	Zn	Ca
Initial Leach Solution (mg/L)	5827	371	30	1814	4317	150	2056	1369	29	74	-
After First Impurity Removal (mg/L)	4650	268	23	227	3599	98	1905	1211	28	67	570
Prepared Synthetic Solution (mg/L)	4919	266	20	102	3279	55	2244	3277	141	76	487

It was reported that only 8% scandium was lost in this step by adjusting the pH to 2.75. Therefore, in order to test and improve the process proposed by previous researchers, a similar synthetic solution was prepared, as given in Table 1. Alternatively, precipitation tests were conducted and compared with different precipitation reagents to make the process more favorable.

3. Materials and Methods

A representative synthetic solution was prepared by using analytical grade chemical reagents and deionized water in order to simulate and test the precipitation behavior of scandium as given in previous study [7]. In addition, the pH of the stock solution was adjusted to 2.75. In the second impurity removal and scandium precipitation tests, a four-necked glass vessel attached to a condenser, contact thermometer, and pH meter was used during the experiments. The pH of the second impurity removal and scandium precipitation experiments was adjusted and kept at 4.75 at 60 °C for 3 h by adding CaCO$_3$, MgO slurry (12.5 g/100 cc water), and Na$_2$CO$_3$ (12.5 g/100 cc water), 1 M NaOH solutions dropwise via a micropipette in order to prevent high local pH changes during the chemical reactions. The amount of reagent added was recorded throughout the experiments to compare the reagent consumption during the reactions. At the end of the experiments, the slurry was filtered via vacuum filtration. The solid remaining after filtration was washed well with pH 4.75 deionized water to eliminate the possibility of precipitation at a higher pH during washing. After washing, the leach residue was dried overnight at 60 °C and ground for chemical analyses. For mass balance calculations, the solid and liquid samples were analyzed by a Spectro Arcos ICP-OES analyzer (SPECTRO Analytical Instruments GmbH, Kleve, Germany) capable of true-axial and true-radial plasma observation. Before analysis, the solid samples were put into solution at 240 °C by a microwave aqua-regia digester with a microwave power of 1300 W. In order to eliminate high salt contents, the samples were diluted where necessary. All of the scandium was aimed to be precipitated and concentrated in this precipitate with minimum nickel and cobalt co-precipitation. Then, preliminary leaching and solvent extraction tests were conducted to obtain an intermediate (NH$_4$)$_2$NaScF$_6$ product. In the solvent extraction test, Baysolvex D2EHPA diluted in Ketrul D85 (kerosene) was mixed with scandium-containing pregnant leach solution at room temperature in a beaker within 10 min and loaded organic separated from the raffinate phase via a separation funnel. No purification steps were conducted on the reagents utilized in SX tests. A scandium stripping test with reagent grade ammonium fluoride solution was accomplished similarly.

4. Results and Discussion

As previously stated, the aim of the second impurity removal and scandium precipitation experiments was to precipitate and separate all of the scandium present in the solution (20 mg/L) from the main circuit together with remaining impurity elements before nickel and cobalt precipitation. Therefore, minimum nickel and cobalt co-precipitation is one of the most important considerations in this step, together with the level of impurity elements present in the solution. When the analytical results given in Table 2 were analyzed in detail, scandium concentration in the solution after all of the precipitation experiments decreased to a level of <1 mg/L, which indicates that the separation of scandium from the main MHP circuit is possible with all of the precipitating reagents used during the experiments.

Table 2. Overall chemical analyses of the precipitation experiments.

Experiment	Analysis of the Initial Solution at pH 2.75										
Precipitation Reagent	Fe	Al	Cr	Ni	Co	Ca	Mg	Na	Sc	Cu	Zn	Mn
	mg/L	mg/L	mg/L	mg/L	mg/L	mg/L	mg/L	mg/L	mg/L	mg/L	mg/L	mg/L
-	102	3279	55	4919	266	487	3277	88	20	141	76	2242
Experiment	Analysis of the Solution after Precipitation at pH 4.75										
Precipitation Reagent	Fe	Al	Cr	Ni	Co	Ca	Mg	Na	Sc	Cu	Zn	Mn
	mg/L	mg/L	mg/L	mg/L	mg/L	mg/L	mg/L	mg/L	mg/L	mg/L	mg/L	mg/L
CaCO$_3$	<1	<1	<1	3863	208	596	2880	78	<1	1	23	1870
Na$_2$CO$_3$	<1	<1	<1	3840	209	420	2800	8602	<1	4	34	1865
MgO	<1	<1	<1	4040	218	486	7642	93	<1	6	34	1952
NaOH	<1	<1	0	3102	166	348	2460	7043	<1	2	20	1635
Experiment	Analysis of the Obtained Precipitate at pH 4.75										
Precipitation Reagent	Fe	Al	Cr	Ni	Co	Ca	Mg	Na	Sc	Cu	Zn	Mn
	wt.%	wt.%	wt.%	wt.%	wt.%	wt.%	wt.%	wt.%	wt.%	wt.%	wt.%	wt.%
CaCO$_3$	0.43	0.71	0.11	1.74	0.06	15.6	0.05	<0.02	0.04	0.22	0.08	0.03
Na$_2$CO$_3$	1.01	2.06	0.29	4.48	0.15	0.04	0.26	0.63	0.10	0.63	0.11	0.20
MgO	0.91	1.85	0.22	3.39	0.12	0.03	1.55	<0.02	0.09	0.57	0.14	0.22
NaOH	0.73	1.78	0.28	6.84	0.28	0.11	0.82	1.80	0.09	0.59	0.20	0.68

Similar to Sc, nearly all of the impurity elements Fe, Al, Cr, Cu and Zn were precipitated and removed at the end of the precipitation reactions. So, in terms of Sc separation and Fe, Al, Cr, Cu and Zn removal, the outcome of all of the precipitation reagents is comparable. The main difference is seen in the increased levels of Mg and Na when MgO and (Na_2CO_3 or NaOH) were used instead of $CaCO_3$, respectively. The similar behavior may also be seen from the chemical analysis of the precipitates obtained, as shown in Table 2. When the analytical results of the solutions were investigated in detail for Ni and Co, it was found that Ni and Co behaved nearly similar during precipitation with Na_2CO_3 and $CaCO_3$. The highest undesired Ni and Co co-precipitation was observed when NaOH was used as a precipitant, probably due to an uncontrollable high local pH rise after the addition of 1 M NaOH during pH adjustment. On the other hand, in terms of the lowest Ni and Co co-precipitation, the best result was obtained when MgO was used as a precipitant. Therefore, when the different precipitation reagents are compared in terms of impurity removal, scandium precipitation-separation, and Ni-Co co-precipitation, the use of MgO offers more advantages during this stage of the process. The only drawback is the increased level of Mg; however since Mg precipitates at a higher pH compared to the co-precipitation of Ni and Co as an intermediate hydroxide product, it is predicted that the increased level of Mg will not be a major problem in obtaining the desired mixed Ni-Co hydroxide product.

Besides obtaining the desired elemental composition in the leach solution after precipitation treatment, it is also very important to compare the amount of reagent consumed during the reaction in terms of operational cost and the amount of Sc in each precipitate for further treatment. For this aim, the amount of reagent used per liter of leach solution was reported together with the percentage of elements precipitated, listed in Table 3 for easy comparison.

Table 3. Percent precipitation of elements and the amount of reagent consumed/precipitate obtained.

Reagent Used	Fe	Al	Cr	Ni	Co	Ca	Mg	Na	Sc	Cu	Zn	Mn	Reagent Consumed	Precipitate Obtained
	%	%	%	%	%	%	%	%	%	%	%	%	g/L soln.	g/L soln.
$CaCO_3$	100	100	100	6.5	6.9	-	-	-	100	99.1	64.3	0.7	23.7	57.7
Na_2CO_3	100	100	100	8.7	8.1	-	-	-	100	97.1	48.3	1.7	21.7	20.4
MgO	100	100	100	2.3	2.5	-	-	-	100	94.8	46.6	1.6	11.6	25.2
NaOH	100	100	100	11.7	12.6	-	-	-	100	98.5	62.6	2.1	17.0	23.4

When the data given in Tables 2 and 3 were analyzed in detail, highest amount of reagent was consumed when $CaCO_3$ is used in order to increase and keep the pH from 2.75 to 4.75. Additionally, when we compare the amount of precipitates obtained after the reaction, there is a drastic difference in the amounts of these precipitates. More specifically, by using 21.7 g Na_2CO_3, 20.4 g precipitate was obtained, containing, 0.10% Sc. On the other hand, by using nearly the same amount of $CaCO_3$ (23.7 g), the amount of precipitate increased to 57.7 g and the amount of Sc in the precipitate decreased to 0.04% due to the increase in the amount of the precipitate. Similarly, lower amounts of MgO and NaOH were consumed to reach the desired pH level and lower amount of precipitates were obtained having a Sc concentration of 0.09%. In short, when MgO or Na_2CO_3 is used instead of $CaCO_3$ during the second impurity removal and scandium precipitation step, a more concentrated precipitate containing 0.09–0.10% Sc can be obtained instead of that containing 0.04% Sc. This difference is due to the formation of gypsum ($CaSO_4 \cdot xH_2O$), which is insoluble during the precipitation reaction and leads to increased residue weight and decreased Sc concentration. Therefore, in the light of experimental data obtained from the precipitation reactions, it is thought that the use of MgO or Na_2CO_3 will be a more suitable alternative compared to the use of $CaCO_3$ in terms of reagent consumption and obtaining a more concentrated precipitate, which decreases the further processing costs of Sc recovery from this precipitate.

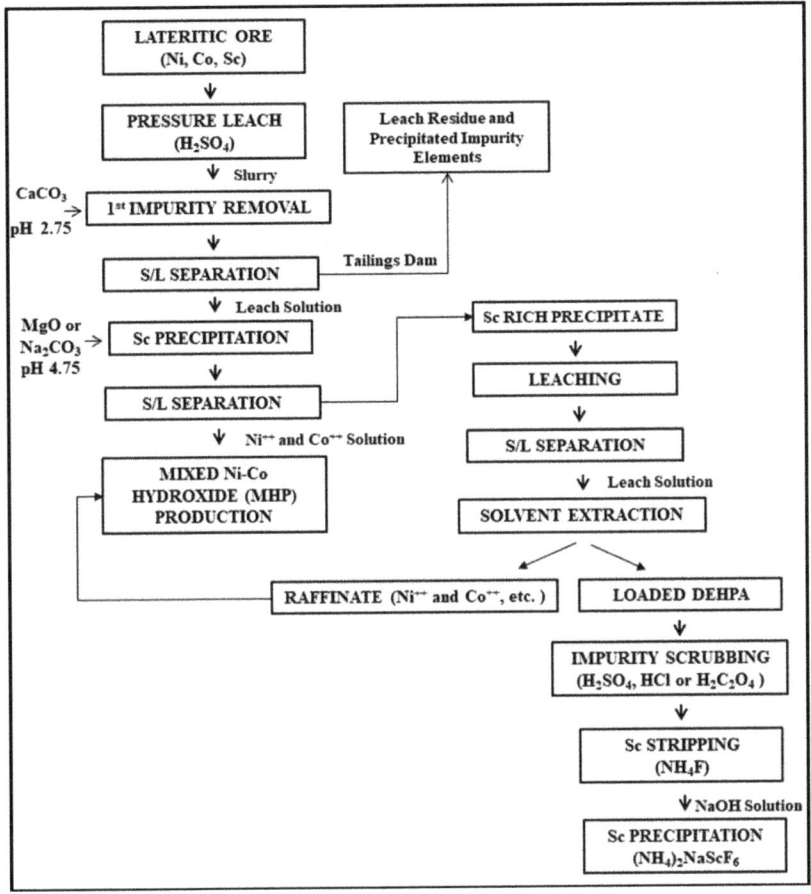

Figure 1. Conceptual process flowsheet for Sc recovery via leaching, SX and precipitation from precipitates obtained via MgO or Na_2CO_3 precipitation.

In order to test the possibility of Sc recovery from the obtained precipitates, only one preliminary leaching test was conducted due to the low amount of residue obtained. During leaching, 20 g of precipitate obtained from MgO precipitation was leached with 100 cc of 100 g/L H_2SO_4 solution at 60 °C for 60 min. According to the leaching results, >98% of Sc dissolved into the pregnant leach solution (PLS). After filtration, 50 cc of the obtained PLS was mixed with a 10% D2EHPA-90% kerosene mixture at room temperature for 10 min with an organic to aqueous ratio of O/A:1/1. The SX parameters were selected according to the previous experience obtained in the course of the European Commission-funded Horizon 2020 SCALE Project. After separating the organic and aqueous phases, the aqueous raffinate phase was analyzed for Sc and it was found that the Sc concentration in the raffinate phase is <1 mg/L, implying, that almost all of the Sc present in the PLS was extracted by D2EHPA. During this single SX test, only Fe and Al were observed to be co-extracted together with D2EHPA, which affects the SX operation adversely. Thus, they should be removed by suitable reagents during the scrubbing stage of the SX operation. However, in this study, due to the limited amount of precipitate and PLS, the scrubbing behavior of impurity elements was not investigated. As a future work, a stock loaded D2EHPA should be obtained by studying the optimum extraction conditions to reach the minimum impurity co-extraction. In removing the co-extracted impurities, the scrubbing behavior of H_2SO_4, HCl, or $H_2C_2O_4$ may be investigated as being possible scrubbing reagents. To finalize the study, the stripping behavior of Sc-loaded D2EHPA was tested in one preliminary test with the loaded D2EHPA obtained in the previous step. To strip and investigate the possibility of recycling the extractant, the loaded D2EHPA was mixed with 3 M NH_4F solution at room temperature for 10 min with an O/A ratio of 1/1. According to the result of the stripping experiment, 80% of Sc was stripped and back-extracted from D2EHPA to an ammonium fluoride solution within one stage. Also, it was previously found in the course of the SCALE project that Sc can be precipitated in the form of $(NH_4)_2NaScF_6$ precipitate by the use of NaOH solution at pH 9.0. So, after calcining this intermediate product, the obtained cryolite type phase ($NaScF_4$-Na_3ScF_6) can be used in aluminum electrolysis instead of Na_3AlF_6 to obtain Al-Sc alloys. Therefore; according to the results of single re-leaching, solvent extraction, and stripping test, it seems that there is a possibility to extract and recover Sc by the solvent extraction method after leaching the Sc-concentrated precipitate obtained via MgO or Na_2CO_3 precipitation. For this aim, a conceptual flowsheet is given in Figure 1; however, it should be kept in mind that the SX behavior (extraction, scrubbing, and stripping) of Sc from the leach liquor was not studied in detail and should be verified in depth as a future work according to the conceptual process flowsheet given in Figure 1.

5. Conclusions

In this study, the precipitation behavior of Sc and other elements originating from pressure acid leaching of a lateritic nickel ore (106 g/t Sc) was investigated by the use of $CaCO_3$, MgO, Na_2CO_3, and NaOH as precipitation reagents. According to the experimental findings, the use of MgO and Na_2CO_3 was found to be more advantageous when compared to $CaCO_3$ to selectively precipitate, separate, and concentrate Sc in a smaller mass up to 900–1000 g/t with a 9–10-fold increase in Sc concentration. After investigating the precipitation behavior, the possibility of Sc recovery was tested with preliminary experiments, and a conceptual process flowsheet is proposed in order to extract and recover Sc independent of the hydrometallurgical nickel-cobalt processing streams. This flowsheet presents the possibility of adopting this approach to already operating hydrometallurgical plants instead of processing a huge volume of main HPAL streams prior to nickel-cobalt recovery, and gives the flexibility of operating nickel-cobalt and scandium streams independently. As a future prospect, the experimental findings of this study indicate that lateritic nickel-cobalt process streams seem to be a valuable resource for the extraction and recovery of scandium in addition to the nickel and cobalt value present in the ore.

Acknowledgments: The research leading to these results has received funding from the European Union's Horizon 2020 research and innovation program under Grant Agreements No. 730105-SCALE. MEAB Chemie Technik GmbH and RWTH Aachen University-IME Department of Process Metallurgy and Metal Recycling are gratefully acknowledged for providing the necessary chemicals and conducting the chemical analyses, respectively.

Author Contributions: Şerif Kaya conceived, designed and performed the experiments; Şerif Kaya, Carsten Dittrich, Srecko Stopic and Bernd Friedrich analyzed the data; Carsten Dittrich contributed reagents/materials; Srecko Stopic and Bernd Friedrich contributed analysis tools; Şerif Kaya wrote the paper.

Conflicts of Interest: The authors declare no conflict of interest.

References

1. Riggall, S. Australian scandium supply—A paradigm shift for a strategic metal. In Proceedings of the Latest Word on Aerospace Materials, Long Beach, CA, USA, 11–14 May 2015.
2. The Critical Metals Report. Richard Karn: Australian Scandium Could Create New Market. Streetwise Reports. Available online: https://www.streetwisereports.com/pub/na/richard-karn-australian-scandium-could-create-new-market (accessed on 8 September 2011).
3. Xu, S.; Li, S. Review of the extractive metallurgy of scandium in China. *Hydrometallurgy* **1996**, *42*, 337–343.
4. Ricketts, N.; Duyvesteyn, W. The current status of scandium supply projects and their technical challenges. In Proceedings of the ALTA'17 Nickel-Cobalt-Copper Conference, Perth, Australia, 20–27 May 2017.

5. Chasse, M.; Griffin, W.L.; O'Reilly, S.Y.; Calas, G. Scandium speciation in a world-class lateritic deposit. *Geochem. Perspect. Lett.* **2017**, 105–114. [CrossRef]
6. Kaya, Ş.; Topkaya, Y.A. Extraction behaviour of scandium from a refractory nickel laterite ore during the pressure acid leaching process. In *Rare Earths Industry: Technological, Economic and Environmental Implications*; de Lima, I.B., Filho, W.I., Eds.; Elsevier: Amsterdam, The Netherlands, 2016; pp. 171–181.
7. Kaya, Ş.; Topkaya, Y.A.; Dittrich, C. Hydrometallurgical Extraction of Scandium from Lateritic Nickel Ores. In Proceedings of the Bauxite Residue Valorisation and Best Practices Conference BR 2015, Leuven, Belgium, 5–7 October 2015; pp. 347–354.

© 2017 by the authors. Licensee MDPI, Basel, Switzerland. This article is an open access article distributed under the terms and conditions of the Creative Commons Attribution (CC BY) license (http://creativecommons.org/licenses/by/4.0/).

Article

Characterization and Feasibility Studies on Complete Recovery of Rare Earths from Glass Polishing Waste

Chenna Rao Borra [1,*], Thijs J. H. Vlugt [2], Jeroen Spooren [3], Peter Nielsen [3], Yongxiang Yang [1] and S. Erik Offerman [1]

1. Department of Materials Science and Engineering, Delft University of Technology, Mekelweg 2, 2628 CD Delft, The Netherlands; y.yang@tudelft.nl (Y.Y.); s.e.offerman@tudelft.nl (S.E.O.)
2. Process & Energy Department, Delft University of Technology, Leeghwaterstraat 39, 2628CB Delft, The Netherlands; t.j.h.vlugt@tudelft.nl
3. VITO–Vlaamse Instelling voor Technologisch Onderzoek (Flemish Institute for Technological Research), Boeretang 200, B-2400 Mol, Belgium; jeroen.spooren@vito.be (J.S.); peter.nielsen@vito.be (P.N.)
* Correspondence: c.r.borra@tudelft.nl; Tel.: +919-603-610-403

Received: 14 January 2019; Accepted: 21 February 2019; Published: 28 February 2019

Abstract: One of the main applications of ceria (CeO_2) is its use in glass polishing. About 16,000 tonnes of rare earth oxides, which is about 10% of total rare earth production, are used for polishing applications. The waste generated in glass polishing contains rare earths, along with other impurities. In this study, two different glass polishing waste samples were characterized and two different processes were proposed for the complete recovery of rare earths from polishing waste, i.e., an acid-based process and an alkali-based process. The polishing waste samples were characterized with inductively coupled plasma optical emission spectrometry (ICP-OES), X-ray fluorescence spectroscopy (XRF), X-ray diffraction (XRD), scanning electron microscopy (SEM), thermo-gravimetric analysis (TGA) and particle size analysis. Chemical analysis showed that sample A (CeO_2-rich waste from plate glass polishing) contained a high amount of impurities compared to sample B (CeO_2-rich waste from mirror polishing). XRD analysis showed that sample B contained CeO_2, $LaO_{0.65}F_{1.7}$ and $LaPO_4$ compounds, whereas sample A contained $CaCO_3$ in addition to rare earth compounds. SEM-EDX analysis showed the presence of alumino-silicates in sample A. Leaching experiments were carried out at 75 °C at different acid concentrations for the recovery of rare earths from polishing waste samples. The leaching results showed that it is difficult to dissolve rare earths completely in acid solutions due to the presence of fluorides and phosphates. Hence, undissolved rare earths in the leach residue were further recovered by an alkali treatment with NaOH. In another approach, polishing waste samples were directly treated with NaOH at 500 °C. After alkali treatment followed by water leaching, rare earths can be completely dissolved during acid leaching. Rare earths from polishing waste can be recovered completely by both the acid-based process and the alkali-based process.

Keywords: polishing waste; rare earths; waste utilization; characterization; leaching

1. Introduction

Cerium is the most abundant rare earth element (REE) [1]. The total estimated global reserves of cerium minerals are about 30 million tonnes [2]. The current production of ceria (CeO_2) is about 54,400 tonnes, which is about 32% of rare earth oxide (REO) production [3]. Cerium is mainly used in catalysts, glass additives, polishing, ceramics, phosphors and LEDs, etc. [1]. The consumption of ceria in glass polishing is about 16,000 tonnes, which is about 10% of total RE oxide production [4].

Ceria is the primary compound in glass polishing powder as it removes the silica from the glass surface efficiently, not only by abrasion but also with chemical action [5]. The material removal rate is relatively higher and the surface is smoother when ceria based polishing powders are used compared to other commercial polishing powders. These ceria based polishing powders generally contain La together with Ce as there is a cost involved with separating La from Ce. Furthermore, La compounds do not affect the polishing quality. Ce-based polishing powders are also used for polishing silicon wafers, gems, and ceramics, etc. [4,6].

The polishing powder is used in the form of a slurry during polishing [7]. With prolonged use of the polishing powder, the particle size distribution (PSD) changes and impurities get accumulated in the powder during polishing and during the powder settling process from the slurry [6]. Hence, the powder can no longer be used in the polishing process due to the decrease in material removal rates and eventually it ends up in landfill [8]. This leads to the wastage of natural resources and causes environmental problems [9]. Therefore, the recovery of REEs is essential for the sustainable usage of glass polishing materials.

REEs can be recovered from polishing waste by physical, physico-chemical and chemical methods [10]. However, it is difficult to remove all the impurities by physical separation due to their fine particle sizes. Silica and alumina can be removed from polishing waste by alkali treatment [11–14]. However, if the waste contains other impurities or the PSD changes, it can no longer be used for polishing purposes even after alkali leaching. Therefore, acid leaching or other chemical treatments are required to recover REEs or remove impurities. Leaching of REEs from polishing waste has been reported by several authors [15–22]. However, there has been a lack of comprehensive characterisation of polishing waste and the residues or intermediate products generated during the recovery process. Furthermore, there has been no study made available on complete recovery of REEs from polishing waste and/or its acid leach residue. In this study, two different polishing waste materials generated from the glass polishing industry were characterised and the feasibility for the complete recovery of REEs from polishing waste was proposed and demonstrated using two different processes. REEs can be recovered completely from polishing waste by both acid-based and alkali-based processes.

2. Materials and Methods

Two different polishing waste samples were obtained from a mirror production industry and a plate glass producer. The sample from glass polishing was denoted sample A and the sample from mirror polishing was denoted sample B. These samples were dried at 105 °C until they reached a constant mass. Next, the material was passed through a 90 μm size mesh to remove any foreign matter before it was used for characterization and recovery studies. Analytical reagent grade HCl (37%) (Sigma-Aldrich, Zwijndrecht, The Netherlands), NaOH (Sigma-Aldrich, Zwijndrecht, The Netherlands), Na_2CO_3 (Sigma-Aldrich, Zwijndrecht, The Netherlands) and sodium tetraborate decahydrate (Sigma-Aldrich, Zwijndrecht, The Netherlands) were used in the present study. Chemical analysis of polishing waste was performed using two methods: (1) wavelength dispersive X-ray fluorescence spectroscopy (WDXRF, Panalytical PW2400, Almelo, The Netherlands) and (2) alkali fusion followed by acid digestion in a 1:1 (v/v) HCl solution, followed by inductively coupled plasma optical emission spectrometry (ICP-OES, Spectro Arcos-OEP, Kleve, Germany) analysis. The alkali fusion was carried out with a mixture of 0.5 g of polishing waste, 1.5 g of sodium carbonate and 1.5 g of sodium tetraborate decahydrate in a platinum crucible at 1000 °C for 60 min. The crystalline phase analysis of the samples was carried out by X-ray diffraction technique (XRD, Bruker D8 Discover, Bruker AXS GmbH, Karlsruhe, Germany). Scanning electron microscopy (SEM, Joel 6500F, Tokyo, Japan) was used for studying the powder morphology. The particle size distribution (PSD) of the samples was measured by laser particle size analysis (Microtrac S3500, Pennsylvania, USA). The leaching experiments were carried out at 75 °C for 4 h and with a liquid to solid ratio of 10:1 in a vibratory shaker (VWR Thermoshake, Zwijndrecht, The Netherlands) at 600 rpm. The acid concentration was varied from 2 to 5 M of HCl. One gram of solid sample was leached with 10 ml of HCl solution. The leach liquor was filtered using a syringe filter (pore size 0.45 μm) and diluted with distilled water for ICP-OES analysis. For alkali roasting studies, the sample was thoroughly mixed with NaOH powder by pestle and mortar. The roasting experiments were carried out at 500 °C for 4 h in a nickel crucible. After roasting, samples were leached with water at 60 °C for 1 h for the removal of sodium fluoride, sodium phosphate, and excess NaOH.

3. Results

3.1. Characterisation

Two different samples were chosen as one sample contained only a small amount of impurities other than RE compounds and the other sample contained a very high amount of impurities. It is difficult to measure the exact composition of La and Ce in a sample using XRF due to the overlapping peaks of Ce and La. Hence, chemical analysis of ICP-OES is more accurate in finding the elemental composition of La and Ce. Chemical analysis based on XRF and ICP-OES is reported in Table 1. The table shows that sample A (from plate glass polishing) contained a high amount of impurities like Ca, Si, Al, F and P. In sharp contrast, sample B (from mirror polishing) contained mainly F and P as the main impurities. Fluoride and phosphate are used for neutralizing the basic La oxide during polishing powder manufacturing [10]. Figure 1 shows the XRD patterns of two polishing waste samples. It shows that REEs come in the form of oxides, oxyfluorides, and phosphates. Sample A contains calcite ($CaCO_3$) together with the Ce and La compounds. However, compounds consisting of Si and Al were not observed in the XRD pattern. Nevertheless, these compounds were found in the SEM-EDX analysis of sample A and they were observed in both fine and coarse sizes. This may be due to their presence as amorphous compounds. The XRD peaks of the RE compounds showed peak shifts with respect to compounds that contained one REE (Ce or La). This is due to the presence of Ce and La together in a compound, which changes the lattice parameter (d-spacing) due to the small difference in ionic radii. For example, the CeO_2 peak was shifted from 3.136 to 3.159 Å. From this peak shift, the concentration of La in CeO_2 can be estimated [23]. It was found from the analysis that about 15–20% of La is present in the CeO_2 phase. Similarly, La-oxyfluoride and La-phosphate phases have also shifted from their original d-spacing. La preferentially forms fluoride and phosphate over Ce as La trioxide is more basic than ceria.

Figure 2 shows the SEM micrographs of the two waste samples. RE particles are finer in sample B compared to sample A. Particle size analysis of the two samples is shown in Figure 3. The particle size analysis (Figure 2) shows that sample B is a fine material (<10 μm) compared to sample A (<100 μm). Both the samples show a multimodal particle size distribution. Impurity particles in sample A are mainly larger than 10 microns. However, these impurities were also found in the fine particle sizes together with RE particles. The high amount of submicron particles was observed in sample B and the particle size was decreased compared to the original polishing powder. As discussed earlier, the change in particle size distribution may be a reason to discard sample B, though the impurity concentration is very low in this sample. An increase in the impurity concentration is a reason to discard sample A. The particles larger than 10 μm in sample A were mainly calcite and alumino-silicates, as shown by SEM-EDX analysis. These compounds were also observed in the finer particle sizes (<10 μm).

The results of the thermogravimetric analyses of sample A and B are given in Figure 4. The high amount of weight loss in sample A is due to the presence of $CaCO_3$, which decomposes within the temperature range 650–850 °C. Two different slopes were observed above 600 °C, which represent two different temperatures of decomposition of $CaCO_3$. This is due to the different sizes of calcite present in the sample as observed from SEM and PSD analyses. The weight loss between 300 and 500 °C was due to organic material present in the sample. This organic material was mainly found in particle sizes above 100 μm.

Table 1. Chemical analysis of polishing waste samples. Sample A (from plate glass polishing) contains a high amount of impurities whereas sample B (from mirror polishing) contains a small amount of impurities. Legend: ICP-OES, inductively coupled plasma optical emission spectrometry; WD-XRF: wavelength dispersive X-ray fluorescence spectroscopy.

Elements	Sample A (wt.%)	Sample B (wt.%)	Analysis
Ce	23.3 ± 0.7	52.2 ± 1.5	ICP-OES
La	9.1 ± 0.3	18.7 ± 0.6	
F	2.3	5.9	WD-XRF
Si	2.6	0.4	
Al	2.2	-	
Ca	20.8	0.2	
Fe	0.4	0.6	
P	0.4	0.9	
Ba	0.2	0.4	
Na	0.2	-	
Mg	0.1	-	
Ti	0.1	-	
Sn	0.1	0.4	
Ag	-	0.2	

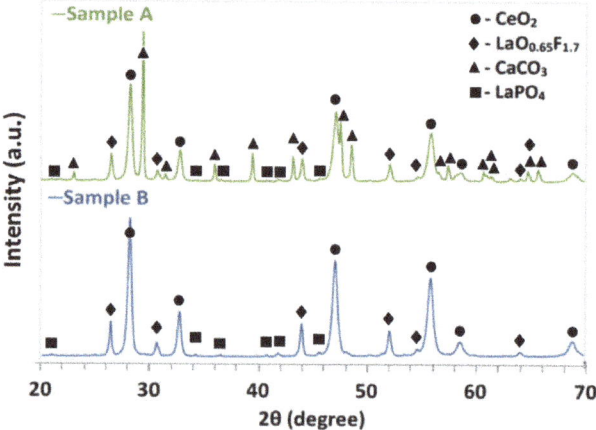

Figure 1. X-ray diffraction (XRD) patterns of polishing waste samples. Sample A is from plate glass polishing and sample B is from mirror polishing.

Figure 2. Scanning electron microscopy (SEM) images of the polishing waste samples: (**a**) sample A shows the coarser particles and (**b**) sample B shows the finer particles. Sample A is from plate glass polishing and sample B is from mirror polishing.

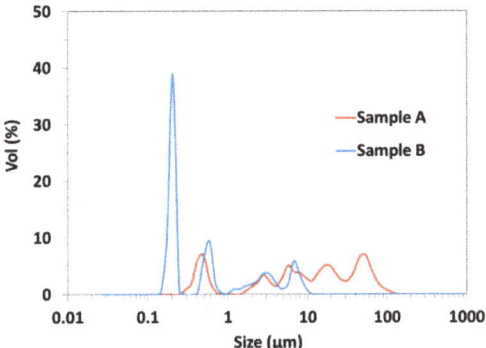

Figure 3. Particle size distribution of polishing waste samples. Sample B shows finer particles, whereas sample A shows coarser particles. Sample A is from plate glass polishing and sample B is from mirror polishing.

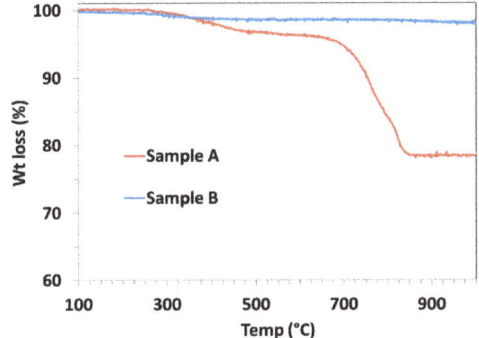

Figure 4. Thermo-gravimetric analysis (TGA) analysis of the two polishing waste samples shows high amount of weight loss in sample A compared to sample B upon heating. Sample A is from plate glass polishing and sample B is from mirror polishing.

3.2. Ca Removal

Ca is the major impurity in sample A. It forms insoluble oxalates during selective precipitation of RE oxalates from the leach solution and contaminates the product [24]. Hence, it needs to be removed before the leaching of REEs from the polishing waste. Although SEM analysis shows that some of the calcite particles are liberated from other particles, physical beneficiation is difficult due to the small particle size of the material. Hence, $CaCO_3$ was removed from polishing waste by HCl leaching. Leaching was carried out with the slow addition of HCl to the slurry (Liquid to solid (L/S) ratio = 5:1) until the pH reached about 6. After leaching the residue was thoroughly washed with DI water. The dried residue after calcium removal and the leach solution were analysed. Chemical analysis of the residue showed that more than 98% of calcium was removed from the polishing waste without any removal of La and Ce from the sample. This residue was used for further studies for recovering REEs.

3.3. Alkali Leaching for Al and Si Removal

Sample A contained alumina and silica as main impurities after calcium removal. If these impurities can be removed from the waste and the material removal rate is similar to the pure polishing powder then the polishing waste can be reused. Hence, alkali leaching was carried out with 2.5 M NaOH aqueous solution and an L/S ratio of 10: 1 at 60 °C for 1 h to dissolve the Al and Si. These conditions were applied based on the literature data [11]. However, there was little to no dissolution of Si and Al in the solution. This was due to the presence of alumina and silica in compound form (confirmed from the SEM), which are difficult to dissolve in alkali solution under ambient conditions.

3.4. Acid-Based Process

High acid concentrations and high temperatures or costly reductants are required to dissolve CeO_2 in nitric and sulfuric acid solutions [25]. Furthermore, sulfuric acid leaching requires additional cold-water leaching treatment to dissolve RE sulphates. HCl was chosen as it can reduce the tetravalent Ce to trivalent Ce (Reaction 1), which is easily soluble in the leach solution. Hence, hydrochloric acid was used in the current study to dissolve REEs from the polishing powder, although it does generate chlorine gas.

$$CeO_2 + 4HCl = CeCl_3 + 2H_2O + \frac{1}{2} Cl_{2(g)} \quad (1)$$

Figure 5 shows the effect of acid concentration (2–5 M) on the recovery of REEs from two different waste samples by HCl acid leaching. The temperature (75 °C) and time (4 h) were chosen for the experiments based on literature data [10] and some preliminary experiments. The recovery of Ce and La increased with the increase in HCl concentration. There was no significant increase in the recovery of Ce and La above 4 M HCl. The recovery of La and Ce were lower in sample A compared to sample B at 2 M acid concentration. This may be due to the presence of finer particles in sample B compared to sample A. However, the highest recovery for Ce was achieved at 3 M for sample A. Due to the fact that the Ce content was low in sample A compared to sample B, less acid was required for its complete dissolution. The maximum recoveries of La and Ce for both samples were between 70 to 80%. La and Ce were not completely leached due to the presence of stable compounds, i.e., fluorides and phosphates of La and Ce. As discussed earlier, polishing waste contains REEs in the form of oxides, oxyfluorides and phosphates. RE oxides readily dissolve during HCl leaching but RE fluoride and phosphates are not dissolved and report to the residue. The XRD patterns of the residues (Figure 6) show the presence of phosphate and fluoride compounds of La and Ce. Thus, according to the following reaction, oxyfluorides reacted to form RE fluorides and partially dissolve REEs.

$$3REOF + 6H^+ = 2RE^{3+} + REF_3 + 3H_2O \quad (2)$$

REE phosphates remained unchanged during leaching. RE phosphate and fluoride can be treated at high temperatures with sodium hydroxide to be converted to RE oxide/hydroxide. as bastnaesite (fluoride) ores are generally treated with alkali to convert stable fluorides to oxides [1]. These RE oxide/hydroxide phases are acid soluble and hence they can be selectively recovered. Hence, alkali treatment was carried out at 500 °C for 4 h with an alkali to residue ratio of 1:2. Figure 7 shows the XRD patterns of alkali treated and water leach residue samples of A and B after water leaching at 60 °C for 1 h. This figure shows that most of the fluoride and phosphate phases disappeared and converted to oxide/hydroxide phases. These samples were subsequently leached at 75 °C with 4 M HCl and an L/S ratio of 5 for 1 h, and more than 95% of the REEs were dissolved into the leach solution. REEs can be selectively recovered from the leach solution by oxalic acid precipitation. The precipitated RE oxalates can be converted to RE oxides by calcination. If the product purity is not sufficient then the REEs can be selectively separated by solvent extraction before the oxalic acid precipitation [1]. The acid regenerated after oxalic acid precipitation can be reused in leaching after adding make up acid. The flowsheet for complete recovery of REEs from polishing waste by acid leaching followed by alkali treatment is shown in Figure 8.

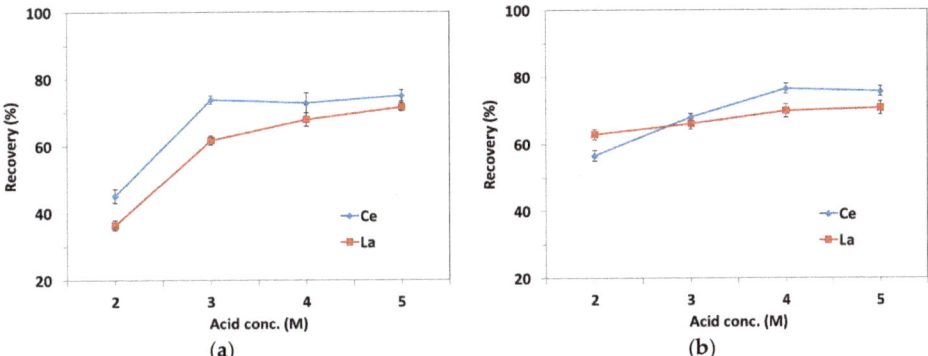

Figure 5. Effect of HCl concentration on leaching of rare earth elements (REEs) from polishing waste samples (T: 75 °C, t: 4 h, L/S: 10). The recovery of La and Ce was lower in (**a**) sample A compared to (**b**) sample B at low acid concentrations. However, the maximum recoveries of La and Ce were similar for both samples at high acid concentrations, which were about 70%. Sample A is from plate glass polishing and sample B is from mirror polishing.

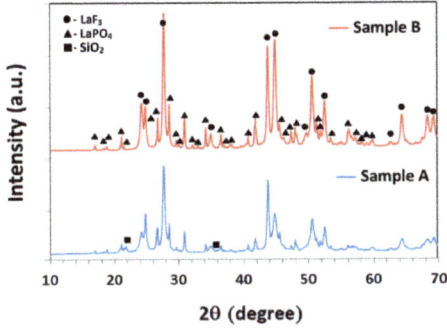

Figure 6. XRD patterns of acid leach residue show mainly the peaks of rare earth (RE) fluorides and phosphates. Sample A is from plate glass polishing and sample B is from mirror polishing.

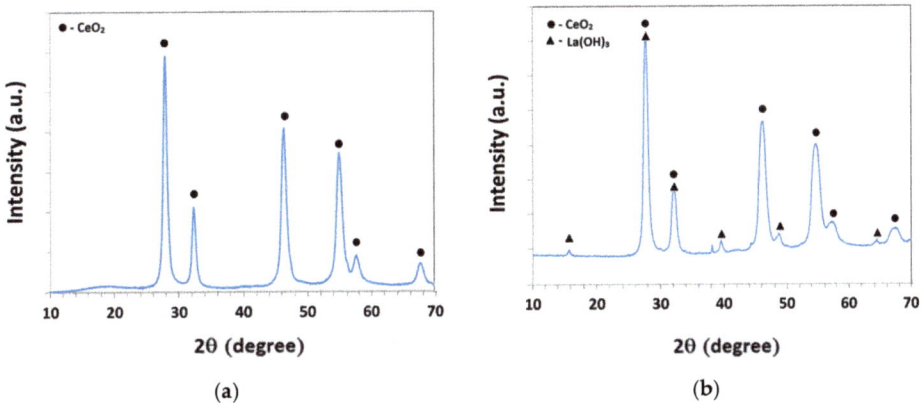

Figure 7. XRD patterns of acid leach residue samples after alkali treatment and water leaching. (**a**) Sample A is from plate glass polishing and (**b**) sample B is from mirror polishing.

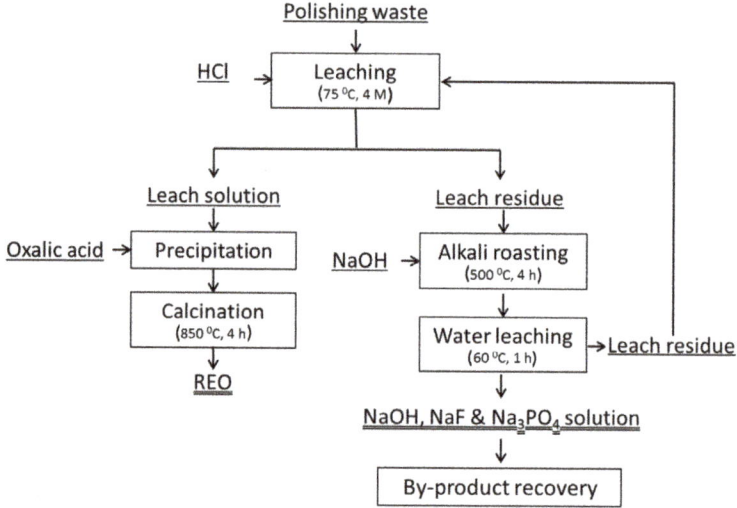

Figure 8. Flowsheet for complete recovery of REEs from polishing waste by acid-based process. This flowsheet shows that polishing waste was treated by acid leaching followed by alkali treatment.

3.5. Alkali-Based Process

As explained earlier, it is difficult to completely recover REEs from polishing waste only by acid leaching. Therefore, in an alternative approach, alkali treatment of the polishing waste was carried out to convert acid insoluble compounds to acid-soluble compounds. Figure 9 shows the flowsheet for an alkali-based process for complete recovery of REEs from glass polishing waste. Alkali treatment of polishing waste converts oxyfluorides and phosphates to oxide/hydroxides of REEs according to the following reactions:

$$REOF + NaOH = NaF + RE_2O_3/REO_2/RE(OH)_3 + H_2O \qquad (3)$$

$$REPO_4 + NaOH = Na_3PO_4 + RE_2O_3/REO_2/RE(OH)_3 + H_2O \qquad (4)$$

The waste samples were roasted with two different ratios of NaOH at 500 °C for 4 h. After alkali treatment, the roasted samples were washed with water at 60 °C for 1 h to dissolve NaF, Na_3PO_4 and other water-soluble compounds. NaOH, NaF and Na_3PO_4 can be recovered from the solution by evaporation/crystallization [1]. The recovered NaOH can be reused in the roasting process. Figure 10 shows the XRD analysis of two samples after alkali treatment followed by water washing. Fluoride was not completely removed from the samples at a NaOH to sample ratio of 0.3. Most of the F and P was removed from the samples at a NaOH to sample ratio of 0.6. The sample after alkali treatment and water washing was dissolved in 6 M acid solution for the recovery of REEs at 75 °C for 4 h. The recovery of La and Ce exceeded 95%. After filtration, the pH of the solution was raised to 5 to remove impurities. Next, the REEs were precipitated from the solution with oxalic acid at a ratio of REEs to oxalic acid of 1:2.

The recovered REO could be used again as a polishing agent or in other direct applications, or, alternatively, the REO could be reduced to metallic Ce or Misch metal and form a new alloy with other metals such as Al through molten slat electrolysis. The further processing of REO for alloy preparation is under way and will be published in the near future.

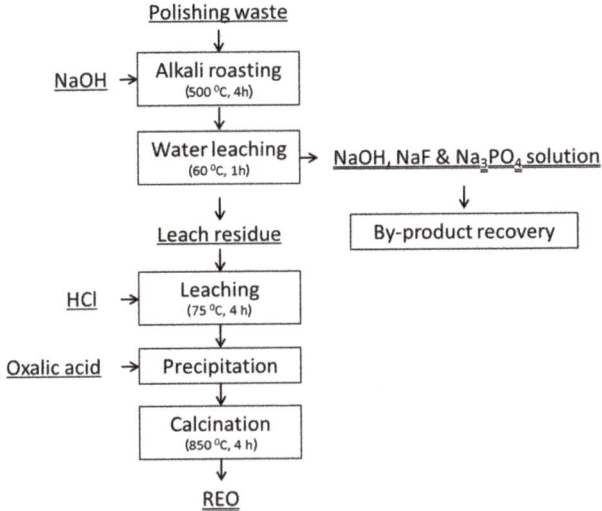

Figure 9. Flowsheet for complete recovery of REEs from polishing waste via an alkali-based process. This flowsheet shows that polishing waste was treated by alkali roasting, followed by acid leaching.

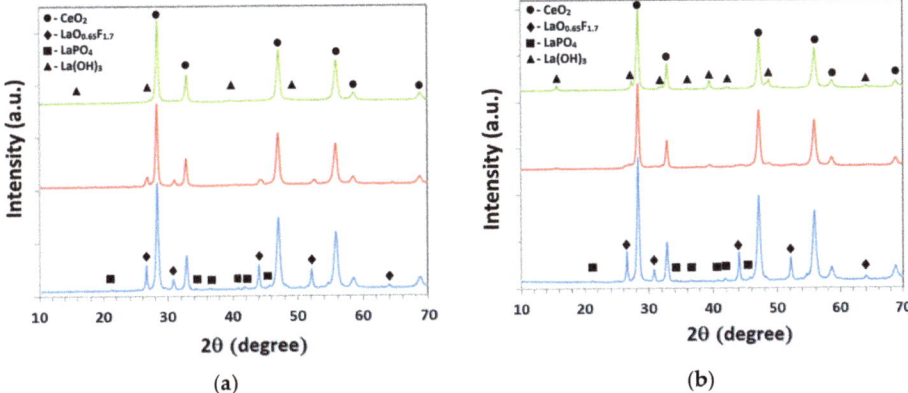

Figure 10. XRD patterns of alkali treated and water leached samples at different alkali to polishing waste samples (blue line: original sample; red line: alkali to sample ratio—0.3; green line: alkali to sample ratio—0.6). Sample A (**a**) is from plate glass polishing and sample B (**b**) is from mirror polishing.

4. Conclusions

In this study, two different polishing waste materials (A and B) were analysed using XRF, XRD, SEM, ICP-OES, PSD and TGA analyses. Sample A (plate glass polishing) contained large amounts of impurities like Ca, Al, Si, F and P, whereas sample B (mirror polishing) contained mainly F and P. Sample B was a finer material compared to sample A. REEs were found in the form of CeO_2, $LaO_{0.65}F_{1.7}$ and $LaPO_4$. Direct acid leaching of the samples with HCl at 75 °C for 4 h with a L/S ratio of 10 dissolved only a maximum of 80% of the REEs. However, an additional alkali treatment at 500 °C for 4 h was able to convert undissolved RE fluoride and phosphates to hydroxides or oxides. These hydroxides or oxides were further purified by acid leaching followed by selective precipitation or solvent extraction, followed by REE precipitation. REEs were also to be recovered by direct alkali treatment of polishing waste at 500 °C for 4 h followed by water leaching at 60 °C for 1 h and subsequent acid leaching at 75 °C for 4 h, which yielded a REE leachability of >95%. This study shows that REEs can be completely recovered from polishing waste via both acid-based and alkali-based processes. However, a detailed study is required to optimise the process conditions and to study the process economics.

Author Contributions: Funding acquisition, T.J.H.V., Y.Y. and S.E.O.; investigation, C.R.B.; methodology, C.R.B., T.J.H.V., Y.Y. and S.E.O.; resources, T.J.H.V., J.S., P.N., Y.Y. and S.E.O.; supervision, T.J.H.V., Y.Y. and S.E.O.; writing—original draft, C.R.B.; writing—review & editing, C.R.B., T.J.H.V., J.S., P.N., Y.Y. and S.E.O.

Acknowledgments: T.J.H.V. acknowledges Nederlandse Organisatie voor Wetenschappelijk Onderzoek-Chemische Wetenschappen (NWO-CW) for a VICI grant. C.R.B. acknowledges TU Delft for funding the 3mE Cohesion project: "Sustainable Rare-earth Cycle". The authors acknowledge Michel van den Brink of TU Delft for ICP-OES and PSD analyses and Ruud Hendrikx of TU Delft for XRD and XRF analyses.

Conflicts of Interest: The authors declare no conflict of interest.

References

1. Krishnamurthy, N.; Gupta, C.K. *Extractive Metallurgy of Rare Earths*; CRC Press: Boca Raton, FL, USA, 2015; ISBN 1466576383.
2. Argus Media Analysing the Changing Global Rare Earths Supply and Demand Outlook. Available online: http://www.argusmedia.jp/~{}/media/files/pdfs/regional-specific/jp/downloads/argus-metal-pages-forum082016-rareearths.pdf/?la=en (accessed on 16 February 2017).
3. Gambogi, J. *USGS 2014 Minerals Yearbook: Rare Earths*; USGS: Reston, VA, USA, 2016.
4. Tercero Espinoza, L.; Hummen, T.; Brunot, A.; Hovestad, A.; Peña Garay, I.; Velte, D.; Smuk, L.; Todorovic, J.; Van Der Eijk, C.; Joce, C. *Critical Raw Materials Substitution Profiles*; Fraunhofer Institute for Systems and Innovation Research: Karlsruhe, Germany, 2015.
5. Lucas, J.; Lucas, P.; Le Mercier, T.; Rollat, A.; Davenport, W. Chapter 12—Polishing with Rare Earth Oxides Mainly Cerium Oxide CeO_2. In *Rare Earths*; Elsevier: Amsterdam, The Netherlands, 2015; pp. 191–212, ISBN 978-0-444-62735-3.
6. Binnemans, K.; Jones, P.T.; Blanpain, B.; Van Gerven, T.; Yang, Y.; Walton, A.; Buchert, M. Recycling of rare earths: A critical review. *J. Clean. Prod.* **2013**, *51*, 1–22. [CrossRef]
7. Kasai, T.; Bhushan, B. Physics and tribology of chemical mechanical planarization. *J. Phys. Condens. Matter* **2008**, *20*, 225011–225023. [CrossRef]
8. Um, N.; Hirato, T. A hydrometallurgical method of energy saving type for separation of rare earth elements from rare earth polishing powder wastes with middle fraction of ceria. *J. Rare Earths* **2016**, *34*, 536–542. [CrossRef]
9. Mishima, F.; Terada, T.; Akiyama, Y.; Nishijima, S. High Gradient Superconducting Magnetic Separation for Iron Removal from the Glass Polishing Waste. *IEEE Trans. Appl. Supercond.* **2011**, *21*, 2059–2062. [CrossRef]
10. Borra, C.R.; Vlugt, T.J.H.; Yang, Y.; Offerman, S.E. Recovery of Cerium from Glass Polishing Waste: A Critical Review. *Metals* **2018**, *8*, 801. [CrossRef]
11. Kato, K.; Yoshioka, T.; Okuwaki, A. Study for recycling of ceria-based glass polishing powder. *Ind. Eng. Chem. Res.* **2000**, *39*, 943–947. [CrossRef]
12. Kato, K.; Yoshioka, T.; Okuwaki, A. Recyle of Ceria-Based Glass Polishing Powder Using NaOH Solution. *Nippon Kagaku Kaishi* **2000**, *10*, 725–732. [CrossRef]
13. Moon, W.; Na, S.; Oh, H. Method for Recycling Cerium Oxide Abrasive. U.S. Patent 20110219704A1, 15 September 2011.
14. Matsui, H.; Harada, D.; Takeuchi, M. Method for Recovery of Cerium Oxide. U.S. Patent 20130152483A1, 20 June 2013.
15. Janoš, P.; Kuráň, P.; Ederer, J.; Šastný, M.; Vrtoch, L.; Pšenička, M.; Henych, J.; Mazanec, K.; Skoumal, M. Recovery of Cerium Dioxide from Spent Glass-Polishing Slurry and Its Utilization as a Reactive Sorbent for Fast Degradation of Toxic Organophosphates. *Adv. Mater. Sci. Eng.* **2015**, *8*. [CrossRef]
16. Poscher, A.; Luidold, S.; Schnideritsch, H.; Antrekowitsch, H. Extraction of Lanthanides from Spent Polishing Agent. In *Rare Earths Industry: Technological, Economic, and Environmental Implications*; Elsevier Inc.: Amsterdam, The Netherlands, 2015; pp. 209–222, ISBN 9780128023280.
17. Poscher, A.; Luidold, S.; Antrekowitsch, H. Extraction of cerium and lanthanum from spent glass polishing agent. In *Materials Science & Technology 2013*; London, I.M., Goode, J.R., Moldoveanu, G., Rayat, M.S., Eds.; Canadian Institute of Mining, Metallurgy and Petroleum: Montréal, QC, Canada, 2013; pp. 543–552.

18. Henry, P.; Lamotte, S.; Bier, J. Recycling of rare earth materials at Hydrometal (Belgium). In *52nd Conference of Metallurgists (COM), Hosting by Materials Science Technology Conference (MS&T)*; London, I.M., Goode, J.R., Moldoveanu, G., Rayat, M.S., Eds.; Canadian Institute of Mining, Metallurgy and Petroleum: Montréal, QC, Canada, 2013; pp. 537–542.
19. Byeon, M.S.; Kim, J.Y.; Hwang, K.T.; Kim, U.; Cho, W.S.; Kang, W.K. Recovery and purification of cerium from glass polishing slurry. In Proceedings of the 18th International Conference on Composite Materials, Jeju, Korea, 21–26 August 2011.
20. Kim, J.Y.; Kim, U.S.; Byeon, M.S.; Kang, W.K.; Hwang, K.T.; Cho, W.S. Recovery of cerium from glass polishing slurry. *J. Rare Earths* **2011**, *29*, 1075–1078. [CrossRef]
21. Janoš, P.; Novak, J.; Broul, M. A Procedure for Obtaining Salts of Rare Earth Elements. U.S. Patent 21,039,151, 31 October 1988.
22. Terziev, A.L.; Minkova, N.L.; Todorovsky, D.S. Regeneration of waste rare earth oxides based polishing materials. *Bulg. Chem. Commun.* **1996**, *29*, 274–284.
23. Ryan, K.M.; McGrath, J.P.; Farrell, R.A.; O Neill, W.M.; Barnes, C.J.; Morris, M.A. Measurements of the lattice constant of ceria when doped with lanthana and praseodymia—The possibility of local defect ordering and the observation of extensive phase separation. *J. Phys. Condens. Matter* **2003**, *15*, L49–L58. [CrossRef]
24. Chi, R.; Zhang, X.; Zhu, G.; Zhou, Z.A.; Wu, Y.; Wang, C.; Yu, F. Recovery of rare earth from bastnasite by ammonium chloride roasting with fluorine deactivation. *Miner. Eng.* **2004**, *17*, 1037–1043. [CrossRef]
25. Um, N.; Hirato, T. Dissolution Behavior of La_2O_3, Pr_2O_3, Nd_2O_3, CaO and Al_2O_3 in Sulfuric Acid Solutions and Study of Cerium Recovery from Rare Earth Polishing Powder Waste via Two-Stage Sulfuric. *Acid Leach. Mater. Trans.* **2013**, *54*, 713–719. [CrossRef]

© 2019 by the authors. Licensee MDPI, Basel, Switzerland. This article is an open access article distributed under the terms and conditions of the Creative Commons Attribution (CC BY) license (http://creativecommons.org/licenses/by/4.0/).

Article

Study of Nd Electrodeposition from the Aprotic Organic Solvent Dimethyl Sulfoxide

Evangelos Bourbos [1,*], Antonis Karantonis [2], Labrini Sygellou [3], Ioannis Paspaliaris [1] and Dimitrios Panias [1]

1. School of Mining & Metallurgical Engineering, National Technical University of Athens, 9 Iroon Polytechniou, 15780 Zografou, Athens, Greece; paspali@metal.ntua.gr (I.P.); panias@metal.ntua.gr (D.P.)
2. School of Chemical Engineering, National Technical University of Athens, 9 Iroon Polytechniou, 15780 Zografou, Athens, Greece; antkar@central.ntua.gr
3. Institute of Chemical Engineering Sciences, Foundation of Research and Technology Hellas, Platani, 26504 Patras, Greece; sygellou@iceht.forth.gr
* Correspondence: ebourbos@metal.ntua.gr; Tel.: +30-210-772-2123

Received: 22 September 2018; Accepted: 5 October 2018; Published: 8 October 2018

Abstract: The use of organic solvents in an electrolytic system for neodymium electrorecovery by electrolysis at low temperatures is studied in the current work. More specifically, an alternative route, that of the system of DMSO (Dimethyl sulfoxide) with dissolved $NdCl_3$ has been researched and has given promising results. The study of this electrolytic system has been divided into two stages. Firstly, the characteristics of the electrolyte, the dissolution of $NdCl_3$ in DMSO, the conductivity and the viscosity of $NdCl_3$ solutions in DMSO at various temperatures, and the Nd complexation in the solution were studied and secondly, the electrolysis parameters and their impact on the Nd electrodeposition process were evaluated. Finally, the deposits were submitted to SEM-EDS (Scanning Electron Microscopy-Energy Dispersive X-Ray Spectroscopy) analysis and metallic Nd was confirmed to be electrodeposited by X-ray Photoelectron Spectroscopy (XPS) spectroscopy.

Keywords: neodymium; dimethyl sulfoxide; electrodeposition

1. Introduction

In June 2010, the European Commission published a list of 14 raw materials that are critical for many important emerging technologies. The list was revised in 2014 and 2017, including 20 critical raw materials [1]. In all three lists published, rare earths elements were identified as critical raw materials, while in the last two, greater detail was provided for rare earth elements by splitting them into heavy and light rare earth elements, and scandium. It is evident that rare earth elements are of great interest for the European Union (EU) due to the EU's high import dependency rate, low substitution, and low recycling rate. Among rare earth elements, neodymium gains significant attention, since its production is considered to be a critical technology metal mostly used for permanent magnets in wind turbines, electric vehicles, hard-disc drives, mobile phones, and more [1–3]. The first neodymium and neodymium alloys in industrial scale were produced by calciothermic reduction of neodymium fluoride and chloride electrolysis. However, due to the high demand of pure neodymium for $Nd_2Fe_{14}B$ permanent magnets, the calciothermic reduction as a costly batch process was found to be economically inviable for industrial production [4]. On the other hand, neodymium chloride electrolysis could only produce mischmetal or Nd-Fe alloy [5]. It was created with a voltage 10–14 V and a current between 1 and 25 kA, while the electrolysis bath consisted of a mixture of $NdCl_3$-KCl-NaCl and also an addition of LiCl [6] and $CaCl_2$ at 1050 °C. The serious drawbacks of this technology are the fact that above 1000 °C, large quantities of electrolytes evaporate and at the same time a high amount of chlorine off-gas is released, causing environmental and health issues [7]. The fluoride electrolysis can be realized at the required temperatures, but presents similar drawbacks, such as, environmental issues along with the constraint to use neodymium salts as raw material [5,7].

The neodymium oxide electrolysis in a fluoride bath conducted by the United States (US) Bureau of Mines in the 1960s had a lot of problems to overcome, such as the dissolution of neodymium oxide in the salt. The best results were presented by the system composed of NdF_3 and LiF [8]. It was suggested that higher than 87% of NdF_3 resulted in higher current efficiency and better metal quality. However, the use of LiF is needed in order to decrease the melting point and boost the electrical conductivity. In 1984 an industrial 3 kA electrolysis cell was developed in Baotou China; it had a cylindrical shape with an inner crucible [9]. A tungsten rod was used as a cathode in the center of the cell surrounded by a one-piece tube-like graphite cylinder acting as a consumable anode. Neodymium was electrodeposited on the inert cathode and was dropped as liquid metal into a collecting molybdenum crucible. The substitution of the one-piece anode to divided four-arc shaped anode blocks made it possible to perform electrolysis continuously. This electrolysis cell type is the most wide spread in neodymium production, and in Baotou, the design of four cathode rods inserted vertically and surrounded by block anodes has been implemented [8,9].

During electrolysis for neodymium production, the main off-gas products are CO and CO_2, along with CF_4 emissions released from the anode. The emission of CF_4 is strongly enhanced when the anode effect takes place. The emission of carbon fluorides from neodymium electrolysis can have a significant impact on global warming [10].

The scope of the current work is the study of an alternative to molten salts electrolysis technology for Nd reduction by using a common organic aprotic solvent, namely dimethyl sulfoxide (DMSO), as an electrolyte and to design a potential process for the electrorecovery of metallic neodymium at ambient temperature based on such an electrolytic system.

2. Materials and Methods

The anhydrous organic solvent DMSO was supplied by Sigma Aldrich (Saint Louis, MO, USA) and neodymium chloride by Johnson Matthey (London, UK). The moisture content was less than 50 ppm for the anhydrous DMSO solvent according to the data provided by the company. The organic solvent was placed over 3 A molecular sieves under vacuum for 24 h in order to eliminate the presence of residual water. Cyclic voltammetry and electrolysis tests were performed in a three electrodes cell connected to a VersaSTAT 3 potentiostat by Princeton Applied Research (AMETEK SI, Berwyn, PA, USA); the obtained experimental data were analyzed with the VersaStudio software by Princeton Applied Research (AMETEK SI, Berwyn, PA, USA). In cyclic voltammetry experiments, the working electrode was a platinum disk of 1 mm diameter. The working electrode was polished with 1 μm alumina paste on a velvet pad and by performing voltammetric cycles in 1 M sulfuric acid. As a counter electrode, a Pt wire was used, immersed directly into the solution, whereas, a Pt wire was used as a pseudoreference electrode, calibrated against the reversible couple Fc/Fc^+. The ferrocene/ferrocenium (Fc/Fc^+) redox potential was recorded vs. the Pt pseudoreference electrode after the direct dissolution of 10 mM of ferrocene in the organic solvent under study and was used as the reference potential. In electrolysis tests, the set up used was identical, with the only difference being a copper sheet as a working electrode, of dimensions 10×10 mm^2, with a total reactive surface equal to 20 mm^2. Before the experiments, the Cu working electrode was polished with a series of abrasive papers.

Although DMSO is stable under normal atmospheric conditions, all electrochemical measurements and tests were performed under inert conditions, to minimize oxygen and moisture contamination, in an Ar atmosphere inside a Pure Lab glove box supplied by INERT (Amesbury, MA, USA) where oxygen and moisture were kept below 20 and 50 ppm, respectively. The viscosity of the solutions was measured with a Brookfield DV-I+LV viscometer (AMETEK SI, Berwyn, PA, USA) supplied with an electric thermomantle. The conductivity was measured by a 4-Pt rings electrode conductometer Si-Analytics HandyLab 200 (Xylem, NY, USA). Chemical analysis, in order to determine neodymium concentration in the solutions prepared, was performed by the use of an inductively coupled plasma optical emission spectrometer (ICP-OES) Perkin Elmer 8000 Optimal (Perkin Elmer, Waltham, MA, USA). The infra-red spectrum was collected from 650 to 4000 cm^{-1} on an attenuated total reflection (ATR) module with a Perkin Elmer model FTIR Spectrum 100 (Perkin Elmer, Waltham, MA, USA). To perform the measurement, a droplet of the sample was placed on the ATR crystal. The morphology of electrodeposits was examined by Scanning Electron Microscope (JEOL model 6380LV, JEOL, Tokyo, Japan), provided with an Energy Dispersive Spectrometer (JEOL, Tokyo, Japan). The photoemission experiments were carried out in an ultra-high vacuum system (UHV) consisting of a fast entry specimen assembly, a sample preparation chamber, and an analysis chamber. The base pressure in both chambers was 1×10^{-9} mbar. The analysis chamber was equipped with a hemispherical electron energy analyzer (SPECS LH-10, Scanwel, Gwynedd, UK) and a twin anode X-ray gun for X-ray Photoelectron Spectroscopy (XPS) (Scanwel, Gwynedd, UK) measurements. The preparation chamber consisted of an ion gun for Ar$^+$ sputtering. The unmonochromatized Mg-Kα line at 1253.6 eV and an analyzer pass energy of 97 eV, giving a full width at half maximum (FWHM) of 1.7 eV for the Au 4f7/2 peak, were used in all XPS measurements. The XPS core level spectra were analyzed using a fitting routine, which can decompose each spectrum into individual mixed Gaussian-Lorentzian peaks after a Shirley background subtraction. The samples were in vials in an inert atmosphere and inserted in the UHV system through a glove bag attached to the fast entry specimen assembly of the UHV system. The glove bag was kept under continuous He flow in order to prevent further surface oxidation. Survey scans were recorded for all samples, while the core level peaks that were recorded in detail were: Nd3d, S2p, F1s, O1s, and C1s.

3. Results and Discussion

3.1. Study of the Electrolyte

3.1.1. Dissolution Tests of NdCl$_3$ in DMSO at Different Temperatures

Rare earth metals are known to form strong complexes with halide anions, rendering their dissolution a challenging task [11,12]. The halide salts in non-aqueous electrolytes present great interest for electrorecovery applications, because they are oxidized at mild anodic potentials prior to the oxidation of the electrolyte, thus preventing its anodic decomposition. The first point that had to be elucidated in the system under consideration was the dissolution of the halide salt in the organic solvent chosen as the electrolytic medium. It was decided to investigate the dissolution of NdCl$_3$ in DMSO. The dissolution experiments were performed in a mini-reactor of working volume V = 60 mL, under continuous argon purging, stirring set at 300 rpm for a total duration of 24 h. The amount of NdCl$_3$ added to the mini-reactor was determined to be equal to the quantity needed to form a solution of 1 M Nd concentration, if total dissolution occurred. The experiments were performed for three selected temperatures (30 °C, 60 °C, and 90 °C) and samples were collected after the 1st, 3rd, and 6th hours and at the end of the experiment in order to be analyzed by ICP-OES and to measure Nd concentration in the solution. The chemical analyses performed showed that by the first hour almost the total amount of Nd added in DMSO has been dissolved (Figure 1). The concentration of Nd was stable after 3 h. Nd concentration in the final solution produced at 90 °C and 60 °C was 1 M, while it was slightly lower for the solution produced at 30 °C (0.96 M).

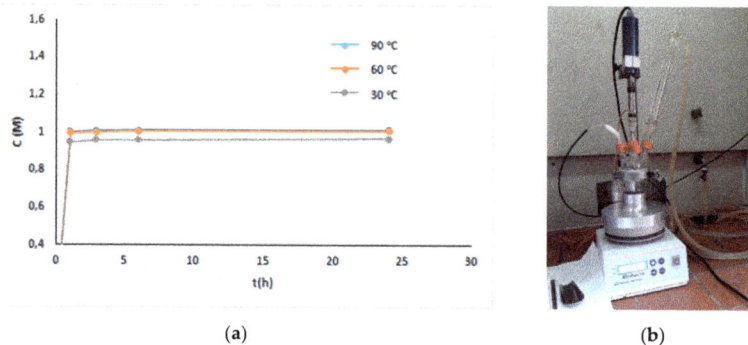

Figure 1. (a) Nd concentration measured by inductively coupled plasma optical emission spectrometer (ICP-OES) for the solutions prepared and (b) the mini reactor used for the dissolution experiments.

3.1.2. Viscosity and Conductivity Study of NdCl$_3$ and DMSO Solutions

The conductivity and the viscosity of the solutions produced after the dissolution experiments were studied at various temperatures. Figure 2a presents the results for the viscosity measurements performed for the solutions that were previously produced and for a dilute solution of 0.1 M Nd. It was revealed that the more concentrated solutions present higher viscosity. The solution with 1 M Nd, as it was anticipated, presented significant viscosity, while the dilute solution of 0.1 M Nd at room temperature was almost 16 times less viscous than the concentrated one. It is estimated that the higher Nd concentration leads to the complexation of Nd with the solvent's molecules and the subsequent formation of chemical species that contribute to the increase of the viscosity of the system under study. As the temperature is increased, however, the viscosity of 1 M Nd solution is drastically decreased, and at temperatures higher than 80 °C it is almost comparable with the viscosity of the dilute solution of 0.1 M Nd. The increase of temperature affects, mainly, the concentrated solution and not the dilute one, since for the latter the viscosity at room temperature is already substantially low and it cannot be drastically further decreased. On the other hand, the conductivity measurements that were realized for the concentrated solution (1 M) and the dilute one (0.1 M) and are presented in Figure 2b, demonstrate that at room temperature the conductivity of the two solutions is comparable. Nevertheless, as the temperature is increased above 30 °C, the more concentrated solutions exhibit higher conductivity in comparison to the dilute solution. Apparently, higher conductivity is due to the higher amount of charged species present in the solution. The increase of temperature has, as a result, the consequent decrease of system's viscosity, as it was reported previously, and, therefore, the charged species present higher mobility. Since the concentrated solutions are present in higher amounts, they lead to the increase of the system's conductivity.

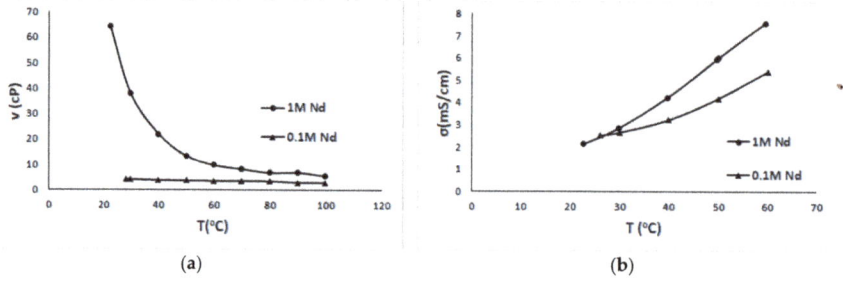

Figure 2. (a) Viscosity measurements vs. temperature (T); (b) Conductivity measurements vs. temperature (T).

3.1.3. FTIR Study of NdCl$_3$ and DMSO Solutions

It is strongly recognized that the complexation of cations in a solution plays an important role in the properties and characteristics of a potential electrolyte [12,13]. The next step of the study was to determine the state of Nd in the prepared solutions by Infra-Red spectroscopy. The Infra-Red spectra were collected in the range of 4000 to 650 cm^{-1}. The spectra are presented in Figure 3 for the pure DMSO, 0.1 M NdCl$_3$ in DMSO, and 1 M NdCl$_3$ in DMSO. Dimethyl sulfoxide has a sulfonyl group and the normal absorption of the S=O bond occurs at 1050 cm^{-1} [13,14], as confirmed in Figure 3. Metals can bond to DMSO either through its oxygen or its sulfur [13,15]. If the bonding is to the sulfur, the metal donates electrons from its π orbitals into an empty π orbital on the DMSO ligand, thereby increasing the S–O bond order. Thus, if the metal is bonded to the DMSO at the sulfur, the frequency of the S=O absorption increases. If the bonding is to the oxygen of the DMSO, the metal forms a bond with one of the lone pairs on the oxygen and thereby withdraws electron density from the oxygen [13]. Therefore, the S=O bond order declines and the S=O absorption appears at a lower frequency.

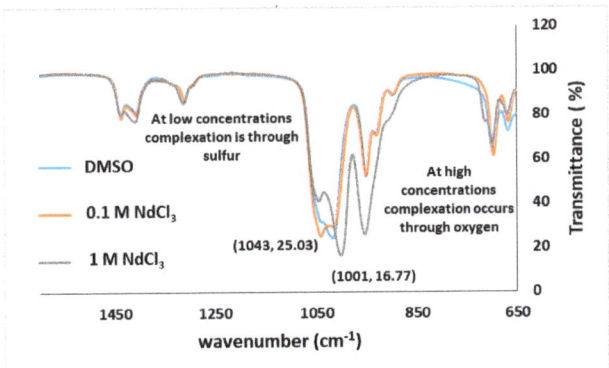

Figure 3. Infra-Red spectra for pure dimethyl sulfoxide (DMSO) (in blue line), DMSO_0.1 M NdCl$_3$ (in orange line), and DMSO_1 M NdCl$_3$ (grey line).

Taking that into account and the FTIR spectra presented in Figure 3 for the different concentrations of Nd, it is implied that at low Nd concentrations, Nd cations are complexed through sulfur since a shift to increased frequency wavenumbers is observed in the spectrum. However, at higher Nd concentrations (1 M) a difference in the complexation trend is noticed and the transmittance peak appears at a lower frequency wavenumber, suggesting that the complexation, in this case, occurs through oxygen.

3.1.4. Cyclic Voltammetry of NdCl$_3$ and DMSO Solutions

The extremely negative reduction potential, E_{red} of Nd^{3+} imposes severe constraints on the electrodeposition process. It restricts the electrolyte to aprotic materials, because the applied potential, necessary for deposition of this element, will vigorously reduce water and other protic solvents. DMSO has been used as a solvent in non-aqueous polarography and voltammetry and various cathodic processes have been reported to occur [16–19]. Therefore, DMSO was selected as the electrolyte medium because it is stable at high reducing potentials and is a good, polar solvent [19–23]. In Figure 4 the cyclic voltammograms recorded with a scan rate of 20 mV/s at room temperature for the pure DMSO and 0.1 M Nd in DMSO are reported. The cyclic voltammogram of the solution 0.1 M NdCl$_3$ in DMSO presents a reductive loop that begins when scanning the potential to values more cathodic than −1.8 V and shapes a peak at −2.45 V vs. the reference, which is attributed to the reduction of Nd trivalent cations to the zerovalent state. Moreover, an oxidative peak is observed in the reverse scan at

−1.1 V that is ascribed to the oxidation of the deposited metallic Nd, implying that the overall reaction is irreversible.

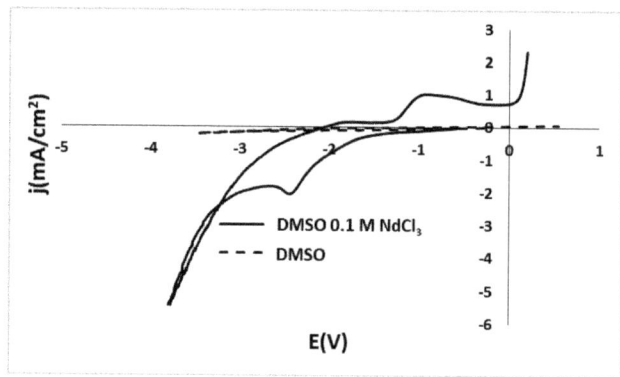

Figure 4. Cyclic voltammogramm of DMSO_0.1 M NdCl$_3$ (solid line) and DMSO (dash line).

3.2. Study of Electrolysis Parameters

In the second phase, the parameters affecting electrolysis of Nd from solutions of NdCl$_3$ in DMSO were studied with the aim of concluding to the optimum conditions for Nd electrorecovery. The parameters studied were the mode of electrolysis, the use of supporting electrolyte, concentration, temperature, and stirring. The criteria used to determine the optimum conditions were the presence of impurities in the electrodeposit and the mass of the electrodeposit produced after electrolysis is performed.

3.2.1. Electrolysis Mode

Electrolysis was performed either galvanostatically or potentiostatically and by imposing potential or galvanic pulses. The experiments were performed with dilute solutions, i.e., 0.1 M NdCl$_3$ in DMSO. Four experiments were performed to determine the most efficient electrolysis mode; all experiments were performed at room temperature for a total duration of 24 h. Galvanostatic electrolysis was performed at −0.8 µA. Potentiostatic electrolysis was performed at −2 V. For pulsed current electrolysis, current pulses at −2 mA for t = 1 s and −0.1 µA for t = 3 s were used. Finally, for pulsed potential electrolysis, potential pulses at −2.3 V and −1.5 V were used for the same time intervals as for pulsed current electrolysis. After the end of each electrolysis test, the cathode was thoroughly rinsed with acetone in order to remove the electrolyte, and the cathode was weighed. The difference between the mass of the cathode prior to and after the end of electrolysis was considered the mass of the deposit and is stated in Table 1.

Table 1. The mass of the deposit (in mg) for each electrochemical test performed.

Electrochemical Mode	Galvanostatic Polarization	Potentiostatic Polarization	Potential Pulsed Electrolysis	Current Pulsed Electrolysis
Mass	3	3	8	12

The electrodeposits were evaluated by SEM-EDS analysis in order to be identified. To define the preferable conditions for electrolysis, it was decided to take into consideration both the mass and the quality of the deposit (presence of impurities in the electrodeposited metal). Consequently, after the end of each electrolysis test, rinsing, and weighing, the cathode was removed from the glove box to proceed with SEM-EDS analysis. Nevertheless, the contact with ambient conditions caused the immediate oxidation of the electrodeposited metal forming a greyish film. The oxidized film was identified by SEM-EDS images and the results of the EDS analysis for the electrodeposits are reported in Figure 5.

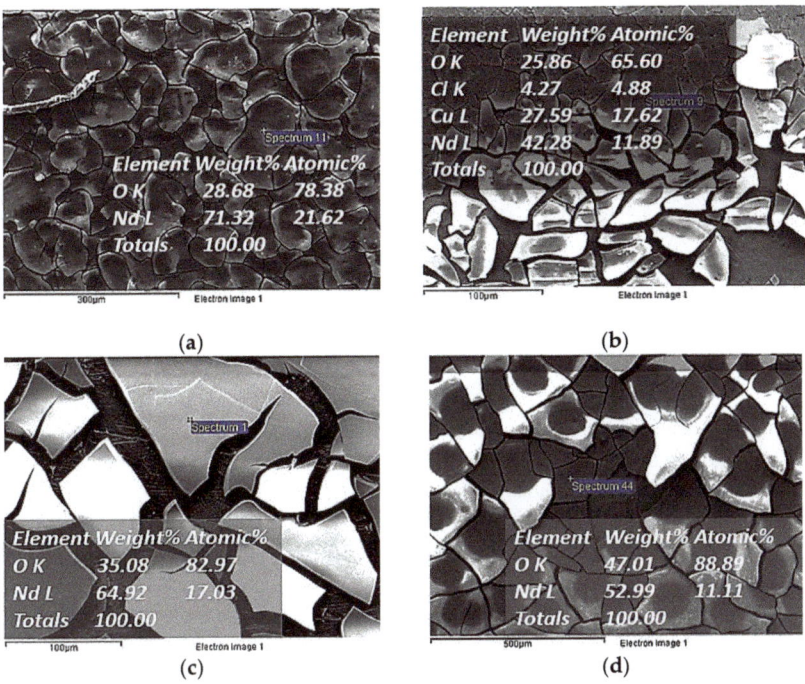

Figure 5. Scanning Electron Microscopy-Energy Dispersive X-Ray Spectroscopy (SEM-EDS) analysis of the electrodeposits produced by (**a**) Galvanostatic polarization, (**b**) Potentiostatic polarization, (**c**) Current pulsed electrolysis, and (**d**) Potential Pulsed electrolysis.

The EDS analysis confirmed the presence of Nd and O, thus permitting the assumption that neodymium metal was electrodeposited and oxidized after the removal of the electrolyte. It was determined from the SEM-EDS analysis and the mass of the deposit that the optimum electrochemical mode is pulsed electrolysis since the deposits were of higher mass and of higher purity.

3.2.2. Supporting Electrolyte

As was previously stated, DMSO is a polar organic solvent, and solutions of salts can exhibit significant conductivity. However, the use of a supporting electrolyte was tested in order to examine if their addition can boost the electrodeposition of Nd. Aliquat 336 was used as a supporting electrolyte. Aliquat 336 is a common ionic liquid that presents electrochemical stability, which is a prerequisite in this application [24,25]. In addition, Aliquat 336 shares the same anion with the Nd precursors in our system, which is the chloride anion, thus eliminating the introduction of new charged species. In the system, DMSO 0.1 M $NdCl_3$ Aliquat 336 was added as a supporting electrolyte in a volumetric ratio 5:1 ($V_{DMSO}/V_{Aliquat} = 5/1$), and electrolysis was performed by imposing current pulses at −1.8 mA for 1 s and −0.1 µA for 3 s. The total duration of the experiment was 24 h. After the end of electrolysis, the electrodeposit was thoroughly rinsed, its mass was measured, and subsequently, it was submitted to SEM-EDS analysis (Figure 6).

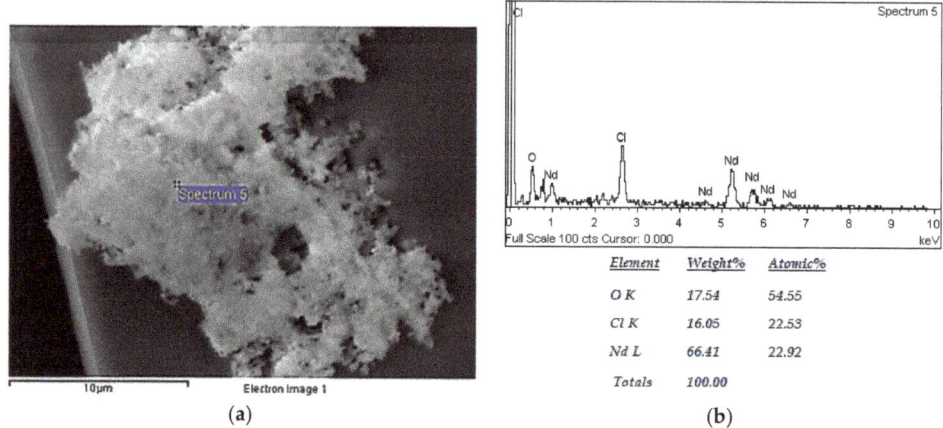

Figure 6. (a) SEM image and (b) EDS analysis of the electrodeposit.

The EDS analysis showed the presence of high amounts of chloride present in the deposit, suggesting its contamination by the electrolyte. It was concluded that the use of this specific supporting electrolyte was detrimental for the process.

3.2.3. Concentration

The electrolysis was performed by imposing current pulses −2 mA for 1 s and −0.1 µA for 3 s at room temperature for a total electrolysis duration of 24 h. The deposits were weighed and examined by SEM-EDS analysis after the end of electrolysis. The neodymium electrodeposited in each electrolysis test is reported in Table 2.

Table 2. The mass of the deposit (in mg) for the three different concentration solutions (in M).

Concentration	0.1	0.5	1
Mass	12	14	19

Larger electrodeposits were found for the more concentrated solutions as is presented in Table 2. The SEM-EDS analysis confirmed the presence of Nd and O due to the oxidation of the metal when in contact with the air (Figure 7).

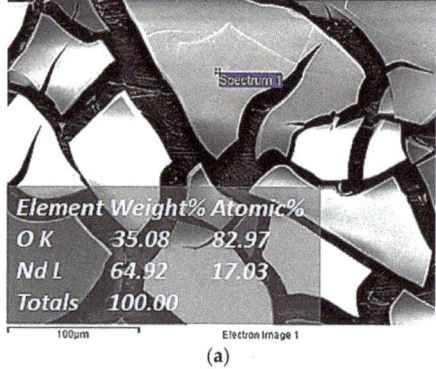
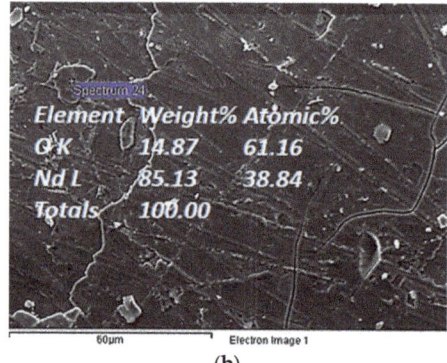

Figure 7. Cont.

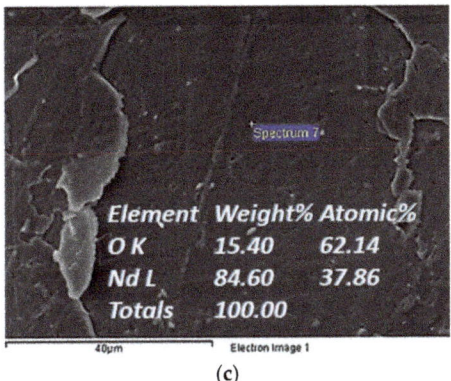

(c)

Figure 7. SEM-EDS analysis of the electrodeposits for the three different concentration solutions (**a**) 0.1 M; (**b**) 0.5 M; (**c**) 1 M.

3.2.4. Temperature

Electrolysis was performed at 30 °C, 50 °C, and 70 °C. The electrolysis was realized by imposing current galvanic pulses at −2 mA for 1 s and −0.1 µA for 3 s for a total duration of 5 h, and the solution used for electrolysis was 1 M $NdCl_3$ in DMSO. DMSO has a boiling point at 189 °C that permits the increase of temperature during electrolysis. Nevertheless, there is an unavoidable evaporation at lower temperatures that could alter the volume and by consequence the concentration of the solution. This is the reason that electrolysis was performed for 5 h when studying the effect of the temperature, instead of 24 h. A 24 h test could have been performed by the use of a condenser, but that would render the set up rather difficult to install inside the glove box. The masses of the electrodeposits are presented on Table 3.

Table 3. The mass of the deposit (in mg) produced at the three tested temperatures (in °C).

Temperature	30	50	70
Mass	4	4	3

As was mentioned, DMSO has a boiling point at 189 °C that permits the increase of temperature during electrolysis. Furthermore, the increase of temperature is considered a simple way to boost the rate of a reaction. Yet, in the system under study, it is observed that at temperatures above 50 °C the final mass of the electrodeposit is decreased, implying that either cracking phenomena take place or a competitive reaction, such as the decomposition of the electrolyte, happens [22]. Among the three electrodeposits, the higher mass was measured for the electrodeposit produced at 50 °C.

Nonetheless, the EDS analysis performed on all three electrodeposits revealed that the increase of temperature to 50 °C and 70 °C increased the impurities found in the electrodeposit, which is detrimental for the process (Figure 8). Hence, it was derived by SEM-EDS analysis that the higher temperatures enhance the incorporation of impurities from the electrolyte, and the preferable temperature to perform electrolysis is T = 30 °C.

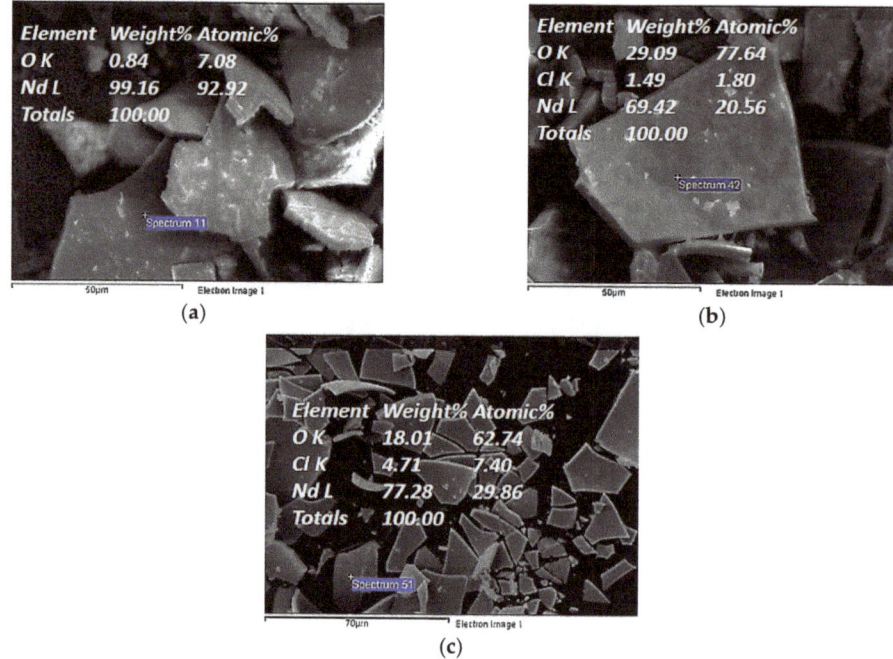

Figure 8. SEM-EDS analysis of the electrodeposits produced at (**a**) 30 °C, (**b**) 50 °C, and (**c**) 70 °C.

3.2.5. Stirring

Current pulsed electrolysis was performed in 1 M NdCl$_3$ in DMSO at −2 mA for 1 s and −0.1 µA for 3 s for a total duration of 24 h at T = 30 °C. Under mild stirring, 21 mg of Nd were electrodeposited, while without stirring that value was 19 mg. As it was anticipated, stirring improved the electrodeposition rate, whereas the deposit in both cases did not present contaminations from the electrolyte (Figure 9).

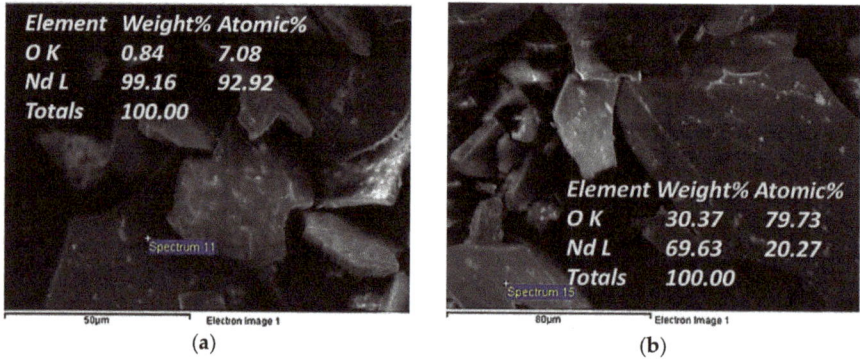

Figure 9. SEM-EDS analysis of the electrodeposits produced (**a**) without and (**b**) with stirring.

3.3. Optimum Conditions

Taking into consideration the results from the study of electrochemical parameters, the optimum conditions for electrolysis were defined. The pulsed electrolysis provided the most satisfying results, while the use of supporting electrolyte did not improve the behavior of the system. The more concentrated solutions produced larger deposits. The higher temperatures enhanced the incorporation of impurities from the electrolyte, thus, the preferable temperature was T = 30 °C, while stirring enhanced the electrodeposition of Nd. Electrolysis was performed in the optimum conditions and the electrodeposit was examined by X-ray Photoelectron Spectroscopy in order to confirm metallic Nd electrodeposition. The deposits, after the end of electrolysis, were rinsed and stored inside the glovebox under inert conditions in sealed serum bottles, to perform XPS measurements. The photoemission survey showed the presence of the atoms Nd, O, C, and S. For that reason, Ar^+ sputtering cycles (2 kV, 1.2×10^{-6} mbar Ar) were performed, in order to remove the first atomic layers, and after the first sputtering cycle, no sulfur or carbon was detected, thus revealing that their presence was due to inadequate electrolytes removal from the surface of the deposit and not due to a deposit's contamination. In Figure 10, the spectrum shown with a black line, which is collected before sputtering cycles, shows the Nd 3d peak of the sample, and the binding energy of $Nd3d_{5/2}$ is at 983.3 eV and is attributed to the Nd_2O_3 chemical state. It is obvious that the sealing of the samples after preparation in the serum bottles as well as the introduction to the UHV chamber using a glove bag is not enough to avoid the oxidation of the surface. Sputtering caused the appearance of a second peak component assigned to Nd^0 as evidenced by the XPS measurements. In Figure 10, the spectra of Nd3d collected after different periods of sputtering cycles of the sample are presented and show that a second peak at ~3 eV lower binding energy appeared, which is assigned to the Nd^0 chemical state [26], thus confirming that metallic Nd is electrodeposited.

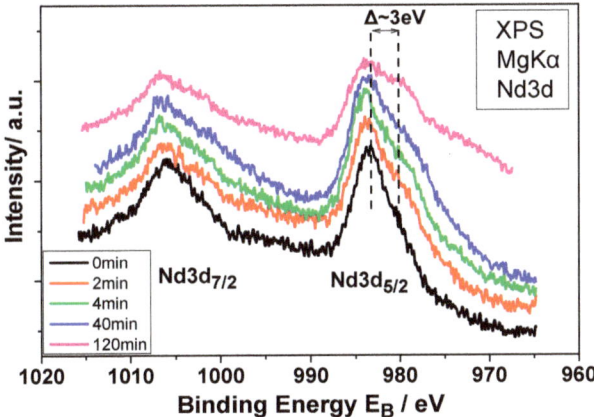

Figure 10. X-ray Photoelectron Spectroscopy (XPS) core level spectra of Nd3d of the electrodeposit after introduction in UHV before and after Ar^+ sputtering for 2 min, 4 min, 40 min, and 120 min.

4. Conclusions

In the present study, the use of an organic solvent was investigated as a potential electrolytic medium for the reduction of drastic metals of high technological and economic importance, such as Nd. The high-temperature molten salts electrolysis that is the current technology used for the production of Nd is an energy-intensive process with a severe environmental impact. The experimental results showed that $NdCl_3$ can be dissolved rapidly in the aprotic organic solvent DMSO and produce concentrated solutions. In the first part of the study, the viscosity, the conductivity, the way Nd is complexated with the electrolyte, and the electrochemical behavior of the electrolytic system were presented, while in the second part, the research was focused on defining the optimum conditions for electrolysis. The study of the electrolysis parameters revealed that pulsed electrolysis, at low temperatures, under stirring exhibited the best results, and the deposit produced was identified by XPS analysis, and it was confirmed that metallic Nd was electrodeposited indicating that the metallurgical process under development is promising.

Author Contributions: Conceptualization, E.B. and D.P.; Methodology, E.B. and D.P.; Validation, E.B., D.P., and A.K.; Formal Analysis, E.B. and D.P.; Investigation, E.B. and L.S.; Resources, E.B., D.P. and I.P.; Data Curation, E.B.; Writing-Original Draft Preparation, E.B.; Writing-Review and Editing, D.P. and A.K.; Supervision, D.P.; Project Administration, D.P. and I.P.; Funding Acquisition, D.P. and I.P.

Funding: This research was funded by European Community's Seventh Framework Programme ([FP7/2007-2013]) under grant agreement n°309373. This publication reflects only the author's view, exempting the Community from any liability. Project web site: www.eurare.eu.

Conflicts of Interest: The authors declare no conflict of interest.

References

1. European Union. *European Commission, Study on the Review of the List of Critical Raw Materials. Critical Raw Materials Factsheets*; Publications Office of the European Union: Luxembourg, 2017; pp. 403–411, ISBN 978-92-79-72119-9.
2. Rahman, M.A.; Slemon, G.R. Promising Applications of Neodymium Boron Iron Magnets in Electrical Machines. *IEEE Trans. Magn.* **1985**, *21*, 1712–1716. [CrossRef]
3. Le Mercier, T. Applications of Rare Earth Luminescent Materials. In *Rare Earths, Science, Technology, Production and Use*, 1st ed.; Lucas, J., Lucas, P., Le Mercier, T., Rollat, A., Davenport, W., Eds.; Elsevier: Amsterdam, The Netherlands, 2015; pp. 281–317, ISBN 978-0-444-62735-3.
4. Davensport, W. Metallothermic Rare Earth Metal Reduction. In *Rare Earths, Science, Technology, Production and Use*, 1st ed.; Lucas, J., Lucas, P., Le Mercier, T., Rollat, A., Davenport, W., Eds.; Elsevier: Amsterdam, The Netherlands, 2015; pp. 109–122, ISBN 978-0-444-62735-3.
5. Firdaus, M.; Rhamdhani, M.A.; Durandet, Y.; Rankin, W.J.; McGregor, K. Review of high-temperature recovery of Rare Earth (Nd/Dy) from magnet waste. *J. Sustain. Metall.* **2016**, *2*, 276–295. [CrossRef]
6. Miao, Y.W.; Horng, J.S.; Hoh, Y.C. The Preparation of Nd Metal from Taiwan Black Monazite. In Proceedings of the International Symposium on Production and Electrolysis of Light Metals, Halifax, NS, Canada, 20–24 August 1989.
7. Vogel, H.; Friedrich, B. An estimation of PFC emission by rare earth electrolysis. In *Light Metals 2018*; Martin, O., Ed.; The Minerals, Metals & Materials Series; Springer: Cham, Switzerland, 2018; pp. 1507–1516, ISBN 978-3-319-72283-2.
8. Krishnamurthy, N.; Gupta, C.K. Rare earth metals and alloys by electrolytic methods. *Min. Proc. Ext. Met. Rev.* **2001**, *22*, 477–507. [CrossRef]
9. Krishnamurthy, N.; Gupta, C.K. *Extractive Metallurgy of Rare Earths*, 2nd ed.; CRC Press: Boca Raton, FL, USA, 2005; pp. 333–442, ISBN 978-1-4665-7638-4.
10. Keller, R. Electrolytic production of neodymium with and without emission of greenhouse gases. *Electrochem. Soc. Proc.* **1998**, *97*, 143–145. [CrossRef]
11. Konings, R.J.M.; Kovács, A. Thermodynamic Properties of the Lanthanide(III) Halides. In *Handbook on the Physics and Chemistry of Rare Earths*, 1st ed.; Gschneidner, K.A., Jr., Bünzli, J.-C.G., Pecharsky, V.K., Eds.; Elsevier Science B.V.: Amsterdam, The Netherlands, 2003; Volume 33, pp. 147–248, ISBN 0-444-51323-33.
12. Reynolds, W.L.; Silesky, H.S. Solubilities of potassium chloride and sodium iodide in dimethylsulfoxide—Water mixtures. *J. Chem. Eng. Data* **1960**, *5*, 250. [CrossRef]
13. Reynolds, W.R. Dimethyl sulfoxide in inorganic chemistry. In *Progress in Inorganic Chemistry*, 1st ed.; Lippard, S.J., Ed.; Interscience: New York, NY, USA, 1970; Volume 12, pp. 3–21, ISBN 978-0-471-540823.
14. Mercer, A.; Trotter, J. Crystal and molecular structure of dichlorotetrakis(dimethyl sulphoxide)ruthenium(II). *J. Chem. Soc. Dalton Trans.* **1975**, *23*, 2480–2483. [CrossRef]
15. Meek, D.W.; Straub, D.K.; Drago, R.S. Transition metal ion complexes of dimethyl sulfoxide. *J. Am. Chem. Soc.* **1960**, *82*, 6013–6016. [CrossRef]
16. Maxfield, M.; Eckhardt, H.; Iqbal, Z.; Reidinger, F.; Baughman, R.M. Bi-Sr-Ca-Cu-O and Pb-Bi-Sr-Ca-Cu-O superconductor films via an electrodeposition process. *Appl. Phys. Lett.* **1989**, *54*, 1932–1933. [CrossRef]
17. Kolthoff, I.M.; Reddy, T.B. Polarography and Voltammetry in Dimethylsulfoxide. *J. Electrochem. Soc.* **1961**, *108*, 980–985. [CrossRef]

18. Simka, W.; Puszczyk, D.; Nawrat, G. Electrodeposition of metals from non-aqueous solutions. *Electrochim. Acta* **2009**, *54*, 5307–5319. [CrossRef]
19. Handley, T.H.; Cooper, J.H. Quantitative electrodeposition of actinides from dimethylsulfoxide. *Anal. Chem.* **1969**, *41*, 381–382. [CrossRef]
20. Li, G.; Tong, Y.; Liu, G. Preparation of Lu–Bi–Ni thin films in dimethylsulfoxide by cylic electrodeposition method. *Mater. Lett.* **2004**, *58*, 3839–3843. [CrossRef]
21. Yuan, D.; Liu, G.; Tong, Y. Electrochemical behavior of Tm^{3+} ion and cyclic electrodeposition of a Tm–Co alloy film in dimethylsulfoxide. *J. Electroanal. Chem.* **2002**, *536*, 123–127. [CrossRef]
22. Li, G.; Tong, Y.; Wang, Y.; Liu, G. Electrodeposition of Lu–Ni alloy thin films. *Electrochim. Acta* **2003**, *48*, 4061–4067. [CrossRef]
23. Giordano, M.C.; Bazan, J.C.; Arvia, A.J. The electrolysis of dimethylsulphoxide solutions of sodium chloride and sodium iodide. *Electrochim. Acta* **1966**, *11*, 741–747. [CrossRef]
24. Giridhar, P.; Venkatesan, K.A.; Subramaniam, S.; Srinivasan, T.G.; Vasudeva Rao, P.R. Electrochemical behavior of uranium (VI) in 1-butyl-3-methylimidazolium chloride and in 0.05 M aliquat-336/chloroform. *Radiochim. Acta* **2006**, *94*, 415–420. [CrossRef]
25. Litaiem, Y.; Dhahbi, M. Measurements and correlations of viscosity, conductivity and density of a hydrophobic ionic liquid (Aliquat 336) mixtures with a non-associated dipolar aprotic solvent (DMC). *J. Mol. Liq.* **2012**, *169*, 54–62. [CrossRef]
26. Otaa, H.; Matsumiya, M.; Sasayaa, N.; Nishihatab, K. Investigation of electrodeposition behavior for Nd(III) in [P2225][TFSA] ionic liquid by EQCM methods with elevated temperatures. *Electrochim. Acta* **2016**, *222*, 20–26. [CrossRef]

© 2018 by the authors. Licensee MDPI, Basel, Switzerland. This article is an open access article distributed under the terms and conditions of the Creative Commons Attribution (CC BY) license (http://creativecommons.org/licenses/by/4.0/).

Article

Reduction Characteristics of Carbon-Containing REE–Nb–Fe Ore Pellets

Bo Zhang [1,2,*], Yong Fan [3,4,*], Chengjun Liu [1,2], Yun Ye [1,2] and Maofa Jiang [1,2]

1. School of Metallurgy, Northeastern University, Shenyang 110819, China; liucj@smm.neu.edu.cn (C.L.); 20152529@stu.neu.edu.cn (Y.Y.); jiangmf@smm.neu.edu.cn (M.J.)
2. Key Laboratory for Ecological Metallurgy of Multimetallic Ores (Ministry of Education), Northeastern University, Shenyang 110819, China
3. Institute of Multidisciplinary Research for Advanced Materials, Tohoku University, Sendai 980-8577, Japan
4. Institute of Iron and Steel Technology, TU Bergakademie Freiberg, 09599 Freiberg, Germany
* Correspondence: zhangbo@smm.neu.edu.cn (B.Z.); Yong.Fan@extern.tu-freiberg.de (Y.F.)

Received: 29 December 2017; Accepted: 19 March 2018; Published: 23 March 2018

Abstract: To separate and recover the valuable metals from low-grade REE (rare earth elements)–Nb–Fe ore in China, the reduction characteristics of carbon-containing REE–Nb–Fe ore pellets, including mineral phase variation, reduction degree, and reaction kinetics, were observed based on thermogravimetry experiments. The results showed that the reduction and separation efficiency of valuable metals in the carbon-containing pellets were superior to the ones in the previous non-compact mixture. After the reduction roasting of the pellets at 1100 °C and a subsequent magnetic separation, the iron powder with a grade of 91.7 wt % was separated, and in magnetic separation tailings the grades of Nb_2O_5 and (REE)O were beneficiated to approximately twice the grades in the REE–Nb–Fe ore. The reaction rate of the reduction of the carbon-containing pellets was jointly controlled by the carbon gasification reaction and the diffusion of CO in the product layer with an activation energy of 139.26–152.40 kJ·mol^{-1}. Corresponding measures were proposed to further improve the kinetics condition.

Keywords: Bayan Obo; REE–Nb–Fe ore; carbothermal reduction; kinetics

1. Introduction

Bayan Obo ore, Inner Mongolia in China is a well-known multimetallic iron ore deposit, which accounts for 35% of the world's REE reserves and 5.5% of the world's Nb reserves, in addition to abundant iron ore [1]. After a dressing process, most of the Fe (~70%) and a part of REE (<10%) could be recovered [2]. Meanwhile, tailings containing iron oxides, REE oxides ((REE)O) and Nb_2O_5 become a precious secondary resource (~190 million tons). To recover the valuable metals from the tailings, some particular dressing processes have been employed to improve the grade of valuable metals. For example, the grade of Nb_2O_5 was enhanced from 0.14 wt % to ≥3 wt % using a combined dressing process including forth flotation and magnetic separation. However, the grade of REE–Nb–Fe ore beneficiated from the tailings could still not meet the requirements of the Nb industry and REE industry [3]. To separate and recover valuable metals from the low-grade REE–Nb–Fe ore, a process comprising of reduction roasting, magnetic separation, and sulfuric acid (H_2SO_4) leaching was proposed in our previous investigation [4]. In this process, a large quantity of iron was separated from the ore as direct-reduction iron (DRI) via reduction roasting and magnetic separation. The mass fractions of Nb_2O_5 and REE oxides in the magnetic separation tailings were also significantly increased by these treatments. Subsequently, Nb and REE were recovered in the leaching of the magnetic separation tailings with a H_2SO_4 solution.

In our previous investigation, the REE–Nb–Fe fine ore and pulverized coal were mixed and then roasted in a graphite crucible at an optimal temperature. Although a desirable reduction was obtained, the loading pattern of the non-compact mixture is adverse to industrial production for the following reasons. Firstly, the loosely mixed powders are found to be inconvenient for transporting and feeding. Secondly, the gap between fine ore and pulverized coal enhances the diffusion resistance of the reducing agent, which limits the reduction rate. Thirdly, excess carbon is added to ensure the sufficient reduction reaction of iron oxides. Lastly, the generated dust easily blocks the equipment and contaminates the environment. Overmatching the non-compact mixture, the composite pellets of ore and carbon can be reduced at moderately high temperatures to produce DRI within a rotary hearth furnace (RHF), which has been proven to be environment-friendly and economically viable for the recycling of the industrial wastes which are rich in iron [5,6]. To improve the reduction process, the carbon-containing pellets were prepared through a pressure molding with the mixture of the REE–Nb–Fe fine ore and the pulverized carbon. The reduction and separation efficiency of valuable metals was also evaluated. Following this, the reduction kinetics of the carbon-containing REE–Nb–Fe pellets was studied using thermogravimetry to improve the reduction efficiency in this work.

2. Experimental Section

Table 1 shows the composition of the low-grade REE–Nb–Fe ore which was obtained from Bayan Obo tailings. The pressure-forming pellets were prepared by adding 0.35–0.50 mL water into the uniform mixed powders, which are comprised of REE–Fe–Nb ore powders (6.13 g) with a particle size of <0.096 mm, and graphite powders (0.87 g, $w(C) \geq 99.85\%$) with a particle size of <0.074 mm. A pressure of 30 MPa was applied to the mold via the pressure-forming device, as shown in Figure 1. Finally, the pellets were heated at 110 °C for four hours. Considering the reduction reaction of the iron oxides and the carburization of the iron product, the molar ratio of carbon and oxygen (only in iron oxides) in the pellets is 1.2.

Table 1. Chemical composition of the low-grade REE–Nb–Fe ore (wt %).

T.Fe	FeO	P	F	SiO_2	CaO	MnO	MgO
44.36	1.35	0.13	2.04	7.17	2.79	0.38	0.6
Na_2O	K_2O	TiO_2	Al_2O_3	S	Sc_2O_3	Nb_2O_5	(REE)O
0.85	0.55	5.82	0.21	1.48	0.036	3.04	2.91

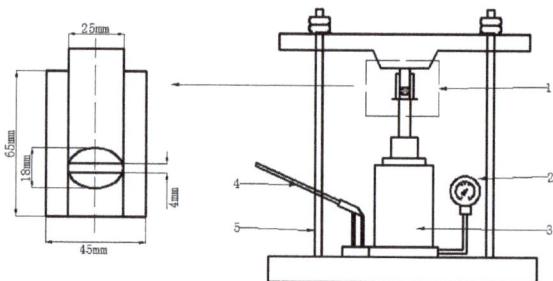

Figure 1. Pressure-forming pellet device (1—Pellet mold; 2—Piezometer; 3—Oil jack; 4—Pressure bar; 5—Pressure support).

The thermogravimetry experiments were carried out in a vertical tube furnace shown in Figure 2, where silicon-molybdenum resistances were used as heating elements. The inner diameter of the furnace tube is Φ50 mm. The accuracy of the thermo-gravimetric electronic balance is ±0.1 mg. When the temperature of the heating zone in the furnace reached the desired temperature (950 °C, 1000 °C, 1050 °C, 1100 °C, 1150 °C), the furnace was flushed with nitrogen gas at 4 L·min^{-1}. Afterwards, the corundum crucible with 3 pellets (21 g) was hung underneath an electronic balance with molybdenum wire. The weight of the samples was recorded at a time interval of 30 s via a computer. It took approximately 300 s for the samples to reach the desired temperature according to the temperature variation measured with Thermocouple II. After the weight data was stabilized for 300 s, the corundum crucible was taken out and cooled via argon gas.

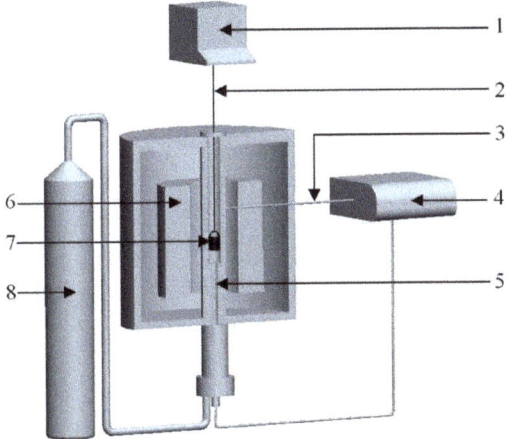

Figure 2. Experimental apparatus of thermogravimetry used in the roasting process (1—Electronic balance; 2—Molybdenum wire; 3—Thermocouple I; 4—Control device; 5—Thermocouple II; 6—Heating element; 7—Graphite basket; 8—Atmosphere control device).

To observe the micrograph of the roasted pellets via scanning electron microscope (SEM), the pellet specimens were mounted in epoxy resin, ground by silicon carbide papers and polished with diamond paste to expose the core cross section. The polished samples were then coated with gold for compositional and microstructural analysis. To separate the Fe from the reduction products, a magnetic separation was carried out in a magnetic tube with a magnetic flux density of 50 mT after milling. The content of Fe in the iron powder magnetically separated from the pellets was determined using the potassium dichromate volumetric method. To measure the content of Nb and REE, the magnetic separation tailing samples were digested with HNO_3–HF–$HClO_4$ and analyzed by inductively coupled plasma-optical emission spectrometry (ICP-OES). The detection limit of ICP-OES is 10 µg·L^{-1}.

3. Results and Discussion

3.1. Mineral Phase Variation and Separation Efficiency of the Valuable Metals

Figure 3 shows the SEM images of the core cross-section of the roasted pellets at various temperatures. The chemical composition of different positions obtained by energy dispersive spectroscopy (EDS) analysis is shown in Table 2. There was no apparent metallic iron at 950 °C as the temperature was assumed to be not high enough for thorough carbothermal reduction. The residual carbon content in the pellets was relatively high (as shown at point A), and the greyish white color appeared to be silicate mineral phases (as shown at point B). At 1000 °C, metallic iron was discovered (as shown at point C) which formed a local iron crystal, and pyrochlore $((Ca,Na,Ce)_2(Nb,Ti)_2O_6(F,OH))$ (as shown at point D) inlaid in the silicate phase. At 1050 °C and 1100 °C, the reduced iron covered the surface of the ore phase and formed a large area of an iron crystal (as shown at point E and point F). At 1150 °C, the low-melting ore phase with high fluorine content started to melt, slag formed locally (as shown at point G), accompanied by crystallization (as shown at point H), and a large area of iron crystal (as shown at point I) was trapped inside the slag.

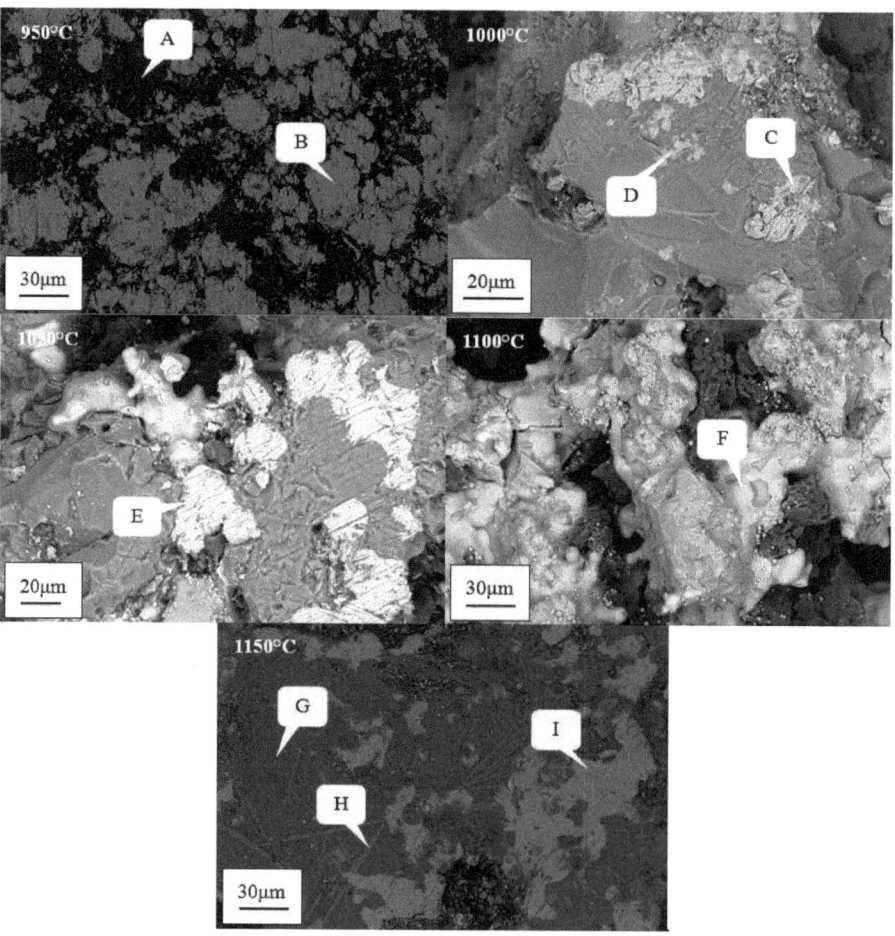

Figure 3. SEM images of the core cross-section of the roasted pellets.

Table 2. EDS results of the characteristic points in the roasted pellets (mass fraction, %).

Analysis Point	C	O	F	Si	Ca	Ti	Fe	Nb	Ce	Nd
A	98.68	-	-	-	-	-	1.32	-	-	-
B	-	27.03	3.23	18.60	10.19	12.89	4.98	8.87	6.44	-
C	11.54	-	-	-	-	-	88.46	-	-	-
D	-	19.29	2.50	0.72	13.61	16.85	4.21	31.85	6.14	4.63
E	7.61	-	-	-	-	-	92.39	-	-	-
F	7.12	-	-	-	-	-	92.88	-	-	-
G	4.45	20.37	22.32	10.52	22.45	7.88	1.88	1.99	3.49	-
H	5.03	31.82	-	9.79	9.47	22.41	1.47	2.41	8.98	4.37
I	7.29	-	-	-	-	-	92.71	-	-	-

After reduction at 1150 °C, a fragmented Nb-enriched region with a particle size smaller than 2 μm can be found around the metallic iron phase. Figure 4 shows the EDS results of the local surface scanning for the reduced pellet at 1150 °C. The green area is metallic iron and Nb enriched in the cavity of the iron phase. The EDS analysis of point J shown in Table 3 shows that the fragments are comprised of niobium and carbon with a small amount of titanium and iron. It was further deduced that the fragments are NbC through a comparison between the XRD patterns of magnetic separation tailings after reduction at 1150 °C and pure NbC, using the data of PDF-65-7964., as shown in Figure 5.

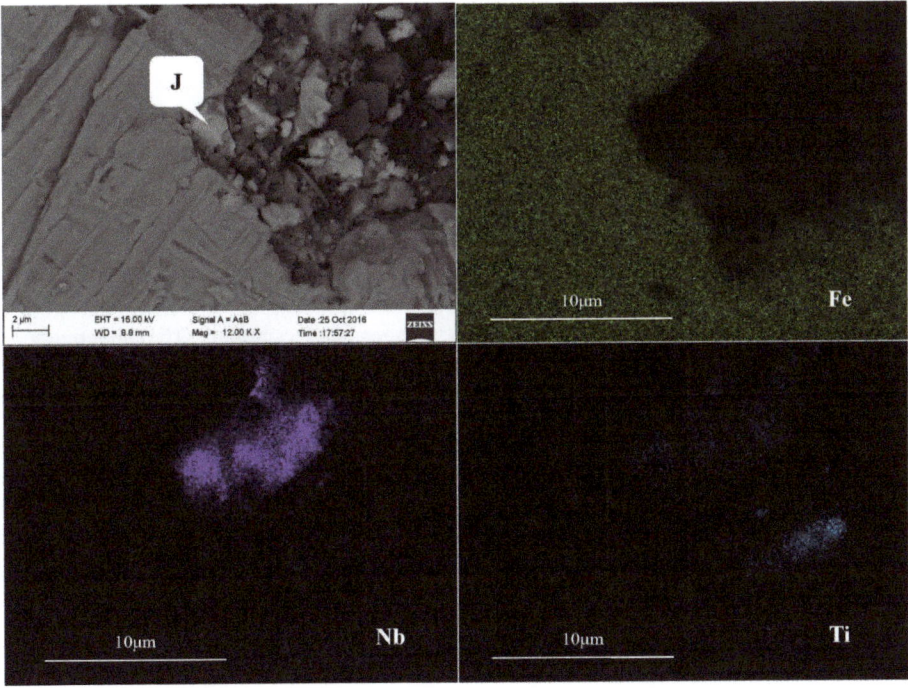

Figure 4. EDS results of the local surface scanning for the reduced pellet at 1150 °C.

Table 3. EDS results of point "J" in the reduced pellet at 1150 °C.

Analysis Point	C	Ti	Fe	Nb
J	18.11	2.63	2.64	76.62

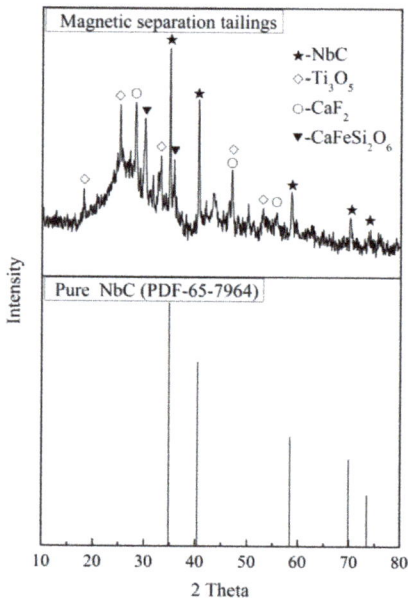

Figure 5. A comparison between the XRD patterns of the magnetic separation tailings and pure NbC.

Figure 6 shows the relationship between the temperature and the Gibbs free energy change of the carbothermal reduction reaction of niobium oxides under the standard conditions. The Gibbs free energy changes of the reactions in Figure 6 were obtained by calculation in virtue of the thermodynamic data in reference [7]. According to Figure 6, the reduction process of niobium oxides follows the sequence of $Nb_2O_5 \rightarrow NbO_2 \rightarrow NbC$, which has been confirmed by Shimada et al. through their carbothermal reduction experiments [8]. Firstly, Nb_2O_5 was reduced to NbO_2, and the skip-level reduction reaction from Nb_2O_5 to NbO, with a higher Gibbs free energy change, is demonstrated to be inexistent. Then, NbO_2 was reduced to NbC instead of NbO, because the Gibbs free energy change of NbC formed by the carbothermal reduction of NbO_2 is apparently lower than that of the NbO formation from NbO_2. According to the experimental results, NbC formed through the reduction of niobium oxide in the reduction of the carbon pellets in the REE–Nb–Fe ore at 1150 °C. Since NbC is insoluble in inorganic acids, NbC generated during the reductive roasting process could not be leached efficiently during the subsequent sulfuric acid leaching process, which affects the recovery efficiency of Nb. Therefore, the reduction temperature of carbon pellets in the REE–Nb–Fe ore should be less than 1150 °C in order to avoid the formation of NbC.

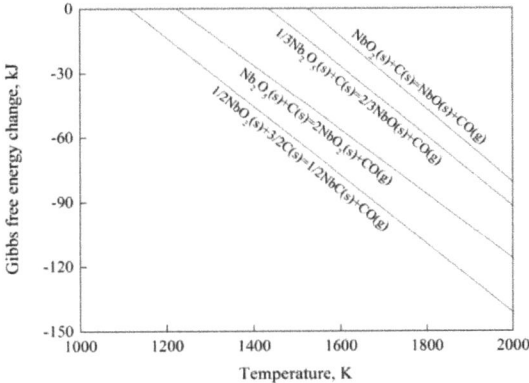

Figure 6. The relationship between the temperature and Gibbs free energy change of the carbothermal reduction reaction of niobium oxides.

Figure 7 shows the grade and the recovery of Fe in the iron powder magnetically separated from the carbon-containing pellets and in the non-compact mixture after reduction roasting, respectively. Compared to the non-compact mixture, the loading pattern of pressure-forming pellets increases the separation efficiency of Fe, including the grade and the recovery, naturally. The grade and recovery both increase with the increasing temperature. Meanwhile, niobium and rare earth elements were better beneficiated in the magnetic separation tailings. As shown in Figure 8, the grades of Nb_2O_5 and (REE)O in the magnetic separation tailings from pressure-forming pellets are higher than the ones from the non-compact mixture due to the higher separation efficiency of Fe and the lesser amount of unreacted carbon. In the carbon-containing pellets, the lower carbon addition leads to less unreacted carbon after reduction roasting. After the reduction roasting of the pellets at 1100 °C and the magnetic separation, the iron powder with a grade of 91.7 wt % was obtained, and the grades of Nb_2O_5 and (REE)O in the magnetic separation tailings were beneficiated to 6.06 wt % and 5.87 wt %, approximately twice the grades in REE–Nb–Fe ore, respectively.

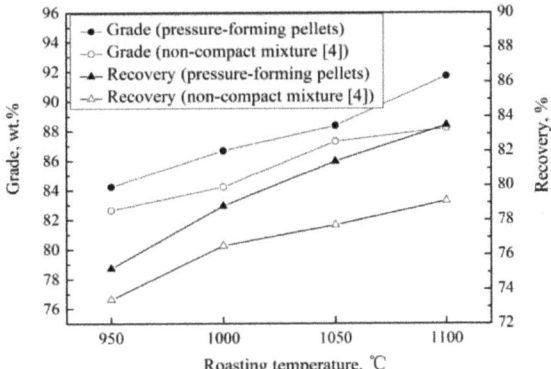

Figure 7. Grade and recovery of Fe in the iron powder.

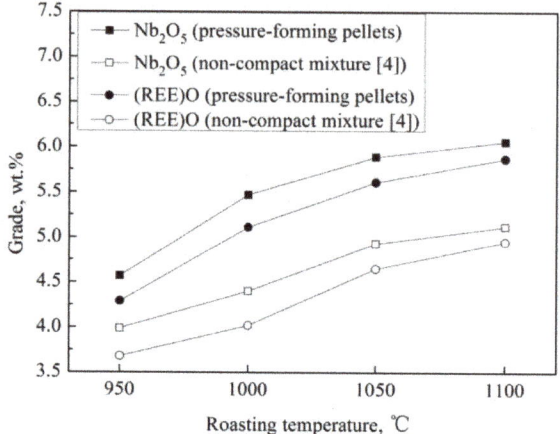

Figure 8. Grades of Nb_2O_5 and (REE)O in the magnetic separation tailings.

3.2. Reduction Degree at Different Temperature

It was shown in our previous work [4] that, at ≤1100 °C, iron oxide is the sole metallic oxide in REE–Nb–Fe ore which can be reduced by carbon. At the initial stage of the roasting process, REE–Fe–Nb ore particles contacted with graphite particles directly. So iron oxides were reduced by carbon, shown as Equations (1)–(3). With the generation of CO, iron oxide can be further reduced by CO which is shown as Equations (4)–(6). Following this, the gasification reaction of carbon (Boudouard reaction) happens, shown as Equation (7) [9]. Man et al. investigated the Boudouard reaction kinetics using the thermogravimetric method at a constant heating rate; the rate of the Boudouard reaction increases significantly at >920 °C [10]. Meanwhile, the generation of metallic iron on the surface of the ore particles isolates the iron oxides and graphite particles. Therefore, CO replaces C as the actual reductant by diffusion through the iron product layer at high temperature [11]. We can obtain Equation (8) by summing Equations (4)–(7). Therefore, the weight loss of the carbon-containing pellets becomes the weight loss of the carbothermal reduction of iron oxides. Based on the weight-loss method, the reaction fraction, *f*, which represents the reduction degree of iron oxides can be calculated using Equation (9) [11].

$$3Fe_2O_3 + C = 2Fe_3O_4 + CO \tag{1}$$

$$Fe_3O_4 + C = 3FeO + CO \tag{2}$$

$$FeO + C = Fe + CO \tag{3}$$

$$3Fe_2O_3 + CO = 2Fe_3O_4 + CO_2 \tag{4}$$

$$Fe_3O_4 + CO = 3FeO + CO_2 \tag{5}$$

$$FeO + CO = Fe + CO_2 \tag{6}$$

$$C + CO_2 = 2CO \tag{7}$$

$$Fe_xO_y + C = Fe_xO_{y-1} + CO \tag{8}$$

$$f = \frac{\Delta W_t}{\Delta W_{max}} = \frac{(W_0 - W_t)}{W_O \cdot \frac{M_{CO}}{M_O}} \times 100\% \qquad (9)$$

where, f is the reaction fraction, ΔW_t is the weight loss at the time of t (g), ΔW_{max} is the maximum theoretical reduction weight loss (g), W_0 is the initial weight of the pellets (g), W_t is the weight of the pellets at the time of t (g), W_O is the oxygen weight of the iron oxides in the pellets (g), M_{CO} is the molar mass of the carbon monoxide (g·mol^{-1}), and M_O is the molar mass of the oxygen (g·mol^{-1}).

The changes in the reaction fraction in the roasting process are shown in Figure 9. If all of the iron oxides were reduced to metallic iron, the weight loss of the pellets should reach the expected maximum weight loss ($\Delta W_t = \Delta W_{max}$ = 6.09 g), and f should be 100%. However, the maximum value of f in the roasting process is 92.6% at 1100 °C, as shown in Figure 9, which means the iron oxides cannot been thoroughly reduced to metallic iron. Moreover, the reduction reaction rate increases significantly with the increasing temperature. At 1100 °C, the reduction degree reaches >90% within 30 min, which is far less than the reaction time (180 min) required in our previous work with the loading pattern of the non-compact mixture. Hence, the reaction efficiency vastly improves using carbon-containing pellets.

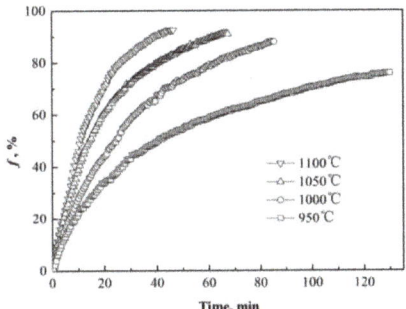

Figure 9. Variation of reaction fraction at different times in the roasting process.

3.3. Kinetic Analysis of Reduction Reaction

As stated in Section 3.1, the reduced iron covered the surface of the unreacted ore and formed a large area of iron crystal. Thus the reduction reaction process of the particles in the carbon-containing pellet can be described via the shrinking unreacted core model, as shown in Figure 10. The carbothermal reduction process includes the following:

(a) The gasification reaction of carbon;
(b) The external diffusion of CO/CO$_2$ through the gas phase boundary layer;
(c) The internal diffusion of CO/CO$_2$ through the iron product layer;
(d) The interfacial reduction reaction of iron oxide particles.

It is generally recognized that the diffusion of CO/CO$_2$ through the gas phase boundary layer is too fast to become the restrictive link of the reduction rate at a high temperature [12]. To discuss the kinetics of the reduction reaction, we assumed that (1) the pellet was isotropic, and that the ore particles and the carbon particles uniformly distributed; (2) the temperature gradient and gas phase concentration gradient were negligible; and (3) the temperature change of the pellets at the initial stage (~300 s) was insignificant. According to the above analysis and assumption, the reduction reaction kinetics can be described by three models according to the restrictive links of rate, as shown in Table 4, where, f is the reaction fraction until time t, k is the apparent reaction rate constant (s^{-1}), and t is the reaction time (s).

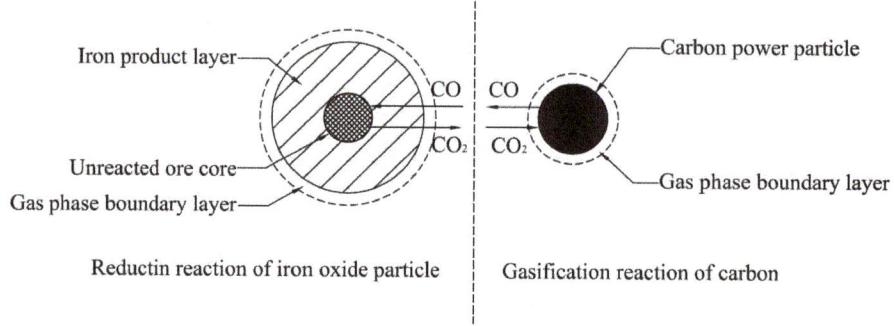

Figure 10. Schematic diagram of the reduction reaction process.

Table 4. Kinetics models of the reduction reaction.

Model	Restrictive Link of Reaction Rate	Equation	No.	References
I	Gasification reaction of carbon	$-\ln(1-f) = kt$	(10)	[13]
II	Interface reaction between CO and iron oxides	$1 - (1-f)^{1/3} = kt$	(11)	[14]
III	Diffusion of CO in product layer	$1 - 2/3f - (1-f)^{2/3} = kt$	(12)	[15]

As shown in Figure 11, the data of $-\ln(1-f)$, $1 - (1-f)^{1/3}$ and $1 - 2/3f - (1-f)^{2/3}$ at different times in the roasting process was calculated according to Equations (10)–(12) in Table 4 and then linearly fitted, and the rate constant k, equal to the slope of the fitted line, was obtained. The rate constant k and the correlation coefficient R^2 of the linear fitting are listed in Table 5. As can be seen, Model I and Model III have good linear relations with time t, whose correlation coefficients are higher than those of Model II. Hence, it can be predicted that the reaction kinetics may be governed by the gasification reaction of carbon or by the diffusion of CO/CO$_2$ in the product layer.

To determine the restrictive link of the reaction rate further, the apparent energy was calculated based on the Arrhenius equation, Equation (13).

$$\ln k = \ln A - \frac{E}{RT} \qquad (13)$$

where, A is the frequency factor, E is the apparent activation energy (J·mol^{-1}), and R is the gas constant, 8.314 (J·mol^{-1}·K^{-1}).

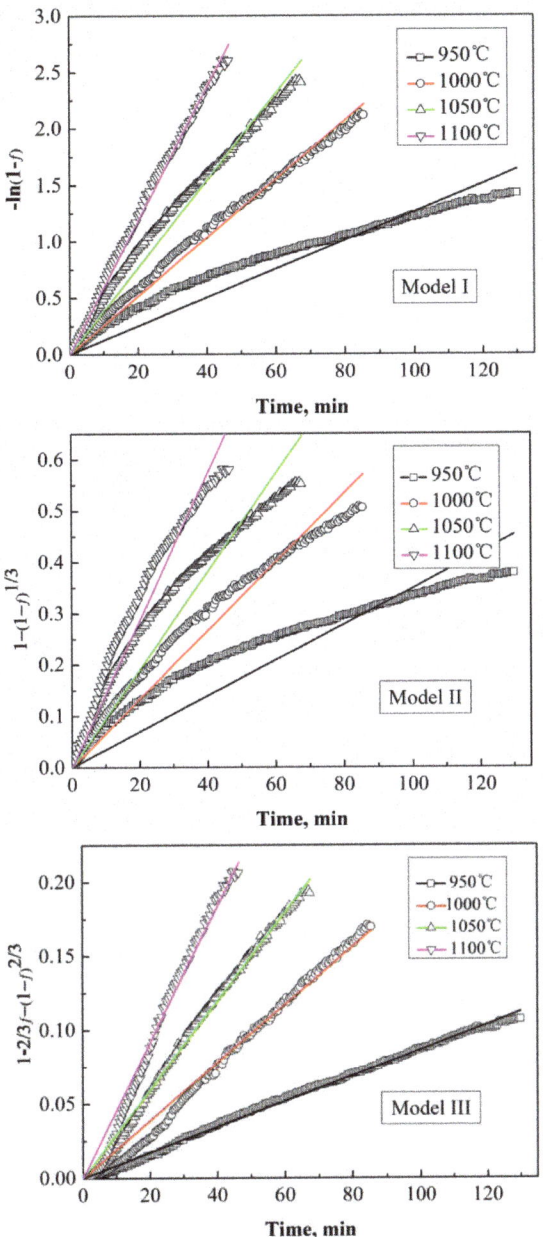

Figure 11. Linear fitting of different kinetics models.

Table 5. Calculated parameters of the linear fitting.

Model	Temperature, °C	Time, min	k, s^{-1}	R^2
I	950	0~129	1.26×10^{-2}	0.9822
	1000	0~85	2.59×10^{-2}	0.9977
	1050	0~67	3.86×10^{-2}	0.9952
	1100	0~46	5.91×10^{-2}	0.9992
II	950	0~129	3.48×10^{-3}	0.9700
	1000	0~85	6.68×10^{-3}	0.9880
	1050	0~67	9.56×10^{-3}	0.9807
	1100	0~46	1.43×10^{-2}	0.9894
III	950	0~129	8.64×10^{-4}	0.9991
	1000	0~85	1.96×10^{-3}	0.9961
	1050	0~67	2.99×10^{-3}	0.9985
	1100	0~46	4.59×10^{-3}	0.9952

In theory, there is a linear relation between ln$k$ and T^{-1}. Therefore E and A can be obtained from the slope and intercept of the line fitted with ln$k$ and T^{-1}, as shown in Figure 12. The activation energy E is found, for Model I and Model III, to be 139.26 kJ·mol^{-1} and 152.40 kJ·mol^{-1}, while the frequency factor A is 12,027.01 and 3064.19 for Model I and Model III respectively. Walker et al. [16] recommended 360 kJ·mol^{-1} as the true activation energy for the reaction of carbon with dioxide, and the approximate values of the activation energy were obtained by Rao [14] (301 kJ·mol^{-1}) and Fruhen [17] (293–335 kJ·mol^{-1}). It has also been said that the apparent activation energy is about half of the true value when the reaction is controlled by both pore diffusion and chemical reaction [16]. Consequently, the reaction kinetics of the carbon-containing pellets in our work are co-controlled by the carbon gasification reaction and the diffusion of CO in the product layer, according to the activation energy at 139.26–152.40 kJ·mol^{-1}.

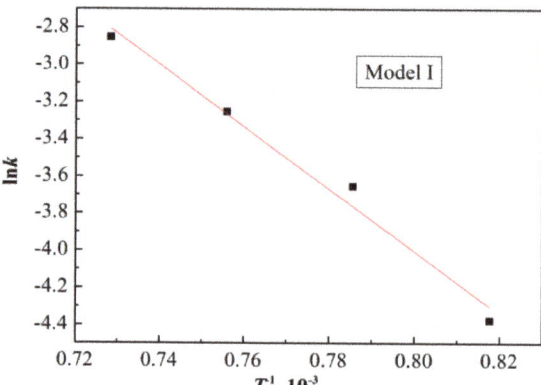

Figure 12. Cont.

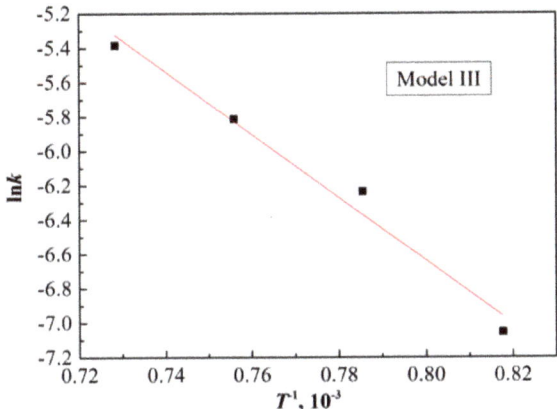

Figure 12. Relationship between ln$k$ and T^{-1} for Model I and Model III.

To improve the reduction kinetics of carbon-containing pellets, two methods can be applied: (1) instead of graphite powders, choosing carbonaceous powders with higher activity such as coal, coke, semi-coke, etc. as the reductant; (2) grinding the ore powders to a finer granularity to restrict the thickness of the product layer and to reduce the diffusion resistance.

3.4. Process Flow

Based on the above observations, a modified process can be proposed, as presented in Figure 13. Firstly, the low grade REE–Nb–Fe ore and carbonaceous powders, such as coal, coke, semi-coke, etc., were finely grinded, homogeneously mixed and pressure-forming briquetted into the carbon-containing pellets. Secondly, the reduction roasting of the pellets was carried out within a rotary hearth furnace at 1100 °C, and the metallized pellets were obtained. Then, the metallized pellets were smashed and magnetically separated. Finally, the iron powders with a high grade (>90%) and recovery rate (>80%) were obtained, and Nb_2O_5 and (REE)O were beneficiated into the magnetic separation tailings with approximatively twice the grades of REE–Nb–Fe ore. Subsequently, Nb and REE can be extracted from the magnetic separation tailings using a hydro-metallurgical method, such as the aforementioned H_2SO_4 leaching process, or another pyro-metallurgical method. Compared to the non-compact mixture, the loading pattern of the pressure-forming pellets is convenient for transporting and feeding, and the roasting reduction of the carbon-containing pellets in RHF is highly efficient, economical, and environment-friendly.

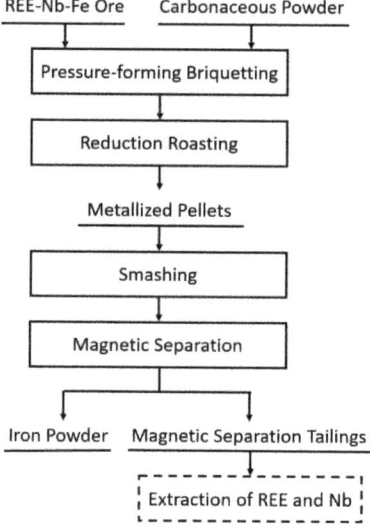

Figure 13. Flow sheet of the modified process.

4. Conclusions

To adapt to the requirement of industrial production and improve the reduction efficiency, the carbon-containing REE–Nb–Fe ore pellets were prepared through a pressure molding to replace the non-compact mixture used in our previous work. The mineral phase variation and separation efficiency of valuable metals was evaluated. The carbothermal reduction kinetics of the pellets was studied using thermogravimetry.

The separation efficiency of valuable metals in the carbon-containing pellets was superior to that of the ones in the previous non-compact mixture. After the reduction roasting of the pellets at 1100 °C and after a magnetic separation, the iron powder with a grade of 91.7 wt % was separated. The grades of Nb_2O_5 and (REE)O in magnetic separation tailings were beneficiated to 6.06 wt % and 5.87 wt %, respectively, approximately twice the grades in REE–Nb–Fe ore. The reduction kinetics of the carbon-containing pellets were jointly controlled by the carbon gasification reaction and the diffusion of CO in the product layer, with an activation energy of 139.26–152.40 kJ·mol^{-1}. The carbonaceous powders with a higher activity and the ore powders with a finer granularity were proposed to improve the reduction kinetics of carbon-containing REE–Nb–Fe ore.

Acknowledgments: Financial support to this project is provided by the National Natural Science Foundation of China (Grant No. 51774087), the National Key R & D Program of China (Grant No. 2017YFC0805100), the Fundamental Research Funds for the Central Universities (Grant No. N162504015), the China Scholarship Council (Grant No. 201706085022).

Author Contributions: Bo Zhang and Chengjun Liu conceived and designed the experiments; Bo Zhang and Yun Ye performed the experiments; Bo Zhang analyzed the data; Maofa Jiang contributed reagents/materials/analysis tools; Bo Zhang and Yong Fan made the figures and tables and wrote the paper.

Conflicts of Interest: The authors declare no conflict of interest.

References

1. Ding, Y.G.; Xue, Q.G.; Wang, G.; Wang, J.S. Recovery behavior of rare earth from Bayan Obo complex iron ore. *Metall. Mater. Trans. B* **2013**, *44*, 28–36. [CrossRef]
2. Li, J.C.; Guo, Z.C. Innovative methodology to enrich britholite ($Ca_3Ce_2[(Si,P)O_4]_3F$) phase from rare-earth-rich slag by super gravity. *Metall. Mater. Trans. B* **2014**, *45*, 1272–1280. [CrossRef]
3. Gibson, C.E.; Kelebek, S.; Aghamirian, M. Niobium oxide mineral flotation: A review of relevant literature and the current state of industrial operations. *Int. J. Miner. Process.* **2015**, *137*, 82–97. [CrossRef]
4. Zhang, B.; Liu, C.J.; Li, C.L.; Jiang, M.F. Separation and recovery of valuable metals from low-grade REE–Nb–Fe ore. *Int. J. Miner. Process.* **2016**, *150*, 16–23. [CrossRef]
5. Lee, Y.S.; Ri, D.W.; Yi, S.H.; Sohn, I. Relationship between the reduction degree and strength of DRI pellets produced from iron and carbon bearing wastes using an RHF simulator. *ISIJ Int.* **2012**, *52*, 1454–1462. [CrossRef]
6. Lu, W.K.; Huang, D.F. The evolution of ironmaking process based on coal-containing iron ore agglomerates. *ISIJ Int.* **2001**, *41*, 807–812. [CrossRef]
7. Barin, I.; Knacke, O.; Kubaschewski, O. *Thermochemical Properties of Inorganic Substances*; Springer-Verlag: Berlin, Germany, 1977.
8. Shimada, S.; Koyama, T.; Kodaira, K.; Mastushita, T. Formation of NbC and TaC by solid-state reation. *J. Mater. Sci.* **1983**, *18*, 1291–1296. [CrossRef]
9. Tiwari, P.; Bandyopadhyay, D.; Ghosh, A. Kinetics of gasification of carbon and carbothermal reduction of iron oxide. *Ironmak. Steelmak.* **1992**, *19*, 464–468.
10. Man, Y. A Study on the Direct Reduction Characteristics of Iron Ore-Coal Composite Pellets under Gas-Solid Base. Ph.D. Thesis, University of Science and Technology Beijing, Beijing, China, 30 October 2014. (In Chinese)
11. Wang, Q.; Yang, Z.; Tian, J.; Li, W.; Sun, J. Mechanisms of reduction in iron ore coal composite pellet. *Steel Res.* **1997**, *24*, 457–460.
12. Man, Y.; Feng, J.X.; Li, F.J.; Ge, Q.; Chen, Y.M.; Zhou, J.Z. Influence of temperature and time on reduction behavior in iron ore-coal composite pellets. *Powder Technol.* **2014**, *256*, 361–366. [CrossRef]
13. Seaton, C.E.; Foster, J.S.; Velasco, J. Reduction kinetics of hematite and magnetite pellets containing coal char. *Trans. Iron Steel Inst. Jpn* **1983**, *23*, 490–496. [CrossRef]
14. Rao, Y.K. The kinetics of reduction of hematite by carbon. *Metall. Trans.* **1971**, *2*, 1439–1447.
15. Ginstling, A.M.; Brownstein, V.I. Concerning the diffusion kinetics of reaction in spherical particles. *J. Appl. Chem. USSR* **1950**, *23*, 1327–1338.
16. Eley, D.D.; Selwood, P.W.; Weisz, P.B. *Advances in Catalysis, Volume XI*; Academic Press: New York, NY, USA, 1959; pp. 134–217.
17. Fruhen, R.J. The rate of reduction of iron oxides by carbon. *Metall. Mater. Trans. B* **1977**, *8*, 175–178. [CrossRef]

© 2018 by the authors. Licensee MDPI, Basel, Switzerland. This article is an open access article distributed under the terms and conditions of the Creative Commons Attribution (CC BY) license (http://creativecommons.org/licenses/by/4.0/).

Review

Recovery of Cerium from Glass Polishing Waste: A Critical Review

Chenna Rao Borra [1,*], Thijs J. H. Vlugt [2], Yongxiang Yang [1] and S. Erik Offerman [1]

1. Department of Materials Science and Engineering, Delft University of Technology, Mekelweg 2, 2628CD Delft, The Netherlands; y.yang@tudelft.nl (Y.Y.); s.e.offerman@tudelft.nl (S.E.O.)
2. Process & Energy Department, Delft University of Technology, Leeghwaterstraat 39, 2628CB Delft, The Netherlands; t.j.h.vlugt@tudelft.nl
* Correspondence: c.r.borra@tudelft.nl; Tel.: +31-620230363

Received: 30 August 2018; Accepted: 30 September 2018; Published: 6 October 2018

Abstract: Ceria is the main component in glass polishing powders due to its special physico-chemical properties. Glass polishing powder loses its polishing ability gradually during usage due to the accumulation of other compounds on the polishing powder or due to changes in the particle size distribution. The recovery of cerium from the glass polishing waste results in the efficient utilization of natural resources. This paper reviews processes for the recovery of rare earths from polishing waste. Glass polishing powder waste can be reused via physical, physico-chemical or chemical processes by removing silica and/or alumina. The removal of silica and/or alumina only improves the life span up to some extent. Therefore, removal of other elements by chemical processes is required to recover a cerium or cerium-rich product. However, cerium leaching from the polishing waste is challenging due to the difficulties associated with the dissolution of ceria. Therefore, high acid concentrations, high temperatures or costly reducing agents are required for cerium dissolution. After leaching, cerium can be extracted from the leach solution by solvent extraction or selective precipitation. The product can be used either in glass polishing again or other high value added applications.

Keywords: cerium; flotation; glass polishing waste; gravity separation; leaching; precipitation; rare-earths; recycling; reuse; solvent extraction

1. Introduction

1.1. Rare-Earth Elements

Rare-earth elements (REEs) are a group of 17 chemically similar elements that include the lanthanides, yttrium and scandium. Rare-earth elements are divided into light rare-earth elements (LREEs), ranging from lanthanum to europium, and the heavy rare-earth elements (HREEs), ranging from gadolinium to lutetium, and yttrium [1]. Total worldwide production of REEs is about 130,000 t out of which 80% of REEs are produced in China [2]. Primary production of REEs generates large amount of pollution such as radiation, acid effluents generation, fluoride contamination, etc. [3]. Some of the REEs (Pr, Nd, Eu, Tb, Dy and Y) are important for the low carbon economy, as they are used in wind turbines, electric vehicles, and energy efficient lighting. These elements comes under the list of critical elements due to the supply risk [4]. However, cerium and lanthanum are not on the critical elements list as they are over produced during critical REEs production [5].

Cerium

Cerium is the most abundant element in all REEs [1]. Principal ores of cerium are the minerals bastnasite ((REE)CO_3F), monazite ((Ce,La,Nd,Th)PO_4), loparite ((Ce,Na,Ca)(Ti,Nb)O_3) and lateritic ion-adsorption clays [6]. The total global cerium containing mineral reserves are estimated to be

30 million tons [7] and the current production of ceria (CeO$_2$) is about 54,400 t (32% of RE oxides). Current price of CeO$_2$ is about 3 USD/kg [6]. Important applications of cerium includes catalysts, glass additives, polishing, ceramics, phosphors, LEDs etc. [1]. About 40,000 t of RE oxides are consumed by glass industry, out of which about 16,000 tons are being used for polishing applications [8].

1.2. Glass Polishing Powder and Its Applications

The hardness of the polishing powder plays very important role in glass polishing [9]. The hardness of polishing powder should be similar to the glass to avoid deep penetration of the polishing grains and formation of large grooves. The average hardness for some glasses and polishing powders are listed in Table 1. In addition to hardness, the polishing rate is also an important factor for the selection of the polishing powder. The polishing rate of glass depends on the isoelectric pH (pH level where the zeta potential is zero) of the polishing powder. Silvernail found a relation between isoelectric pH of different oxides with respect to their polishing rates [10], which is shown in the Figure 1. Being an amphoteric oxide, ceria can exchange both cations (Na$^+$, Ca^{2+}, etc.) and anions (silicate, phosphate, etc.). Based on the two criteria (isoelectric pH and hardness), ceria is the best polishing agent compared to other oxides such as zirconia and alumina.

Table 1. Hardness values of different polishing powders and different type of glasses [9].

Polishing Powder	Hardness (Mohs Scale)	Type of Glass	Hardness (Mohs Scale)
Diamond	10	Silica	7
Alumina	9	Soda lime	5.3
Zirconia	8	Borosilicate	5.8
Ceria	7–8	Lead	4.8

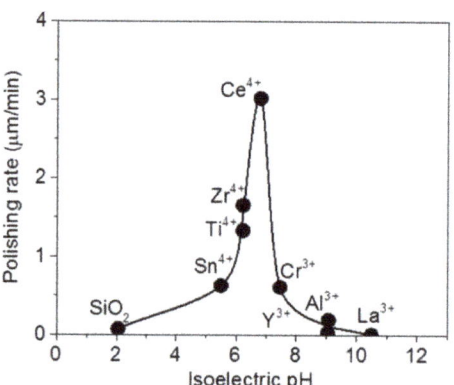

Figure 1. Relation between isoelectric pH (where the concentrations of positively and negatively charged species at the surface are equal) of different polishing compounds with respect to glass polishing rate data. Reproduced from Cook [10], with permission from Elsevier, 2018.

The composition of different glass polishing powders available from the literature is provided in Table 2. The name glass polishing powder is a misnomer as this polishing powder is also used in polishing the silicon wafers, gems and ceramics. The use of cerium compounds as a polishing agent started in the 1930's [8,11]. Prior to 1990's, the growth was mainly due to the demand from the production of CRT screens. Since the 1990's, the demand is rapidly increased due to the rapid growth of electronic products, which require high quality polishing. Seventy five percent (75%) of glass polishing powder is consumed in traditional glass applications (including display panels, flat glass and optical glass) and around 25% in electronic components. The specific application of glass polishing powders include glass, precision glass lenses, glass display panels, liquid crystal displays, glass magnetic memory disks, silicon wafers etc. [9]. A detailed list of applications can be found elsewhere [9].

Table 2. Chemical composition of different glass polishing powders.

Oxide	La_2O_3	CeO_2	Pr_6O_{11}	Nd_2O_3	Reference
wt%	31.5	65	3.5	-	[5]
	34.2	43.8	3.4	10.9	[12]
	-	62.1	-	-	[13]
	0–35	50–99	0–5	0–15	[14]
	-	30–99.9	-	-	[9]

1.3. Production of Glass Polishing Powder

Ceria-based polishing powders are usually prepared by thermal decomposition of cerium oxalates, hydroxides, acetates, or carbonates [15]. A fluorination step (with HF or NH_4F) is typically performed to control the particle size range of the polishing powder [9], or to fix the basic La_2O_3 [16]. Next, sintering (annealing or calcination) is carried out at about 1000 °C to promote the final crystallization and required hardness. Sintering can increase the particle size, therefore, one or several milling steps (dry or wet) are applied to get suitable mean particle size (<5 µm), and a relatively sharp particle size distribution [9,11]. The presence of other rare earths hardly influences the polishing efficiency of the powder [15]. However, the polishing efficiency is mainly affected by calcination time and temperature, particle size, CeO_2 reactivity, additives (l-proline, l-glutamic acid etc.) and particle impurities [11].

The presence of strongly basic lanthanum oxide is likely to cause clogging of a polishing pad during polishing [16]. Therefore, the basicity is decreased by the addition of fluorine to form LaOF. The fluorine content in the polishing powder is preferred within a range from about 3 to 9 mass %. If the fluorine content is low, it is not possible to sufficiently change lanthanum oxide to lanthanum oxyfluoride (LaOF) [16]. If the fluorine content is too high, the excess rare earth fluoride is likely to undergo sintering during firing, which is undesirable [16].

1.4. Polishing Process

A schematic representation of the polishing process with a polishing slurry is shown in Figure 2. In this process, a wafer is polished on a polishing pad using a ceria based slurry. The abrasive particles in the slurry (5–10% of solids) [17] polishes the wafer surface with chemical-mechanical action (tribo-chemical polishing or chemical mechanical polishing (CMP) or planarization). During polishing the grinding process is enhanced by chemical interactions between polishing grains (Ce^{4+} and Ce^{3+}) and the glass substrate (Si-O-H) [18]. These interactions lead to bridge formation (Si–O–Ce) between the glass and the abrasive particles, allowing the detachment of the glass material [19]. Water plays an important role in polishing by providing the hydroxyl ions [10].

1.5. Glass Polishing Powder Waste

The waste slurry generated after glass polishing contains cerium and other REEs (lanthanum, neodymium, and praseodymium), iron, aluminium, zinc, sodium, silicon, etc. [12,13,20–30]. The concentration of the elements other than REEs depends on the chemical composition of the glass, additives (pH modifiers, dispersing agents, etc.) and flocculants [9,24,26]. The abrasive properties of the polishing powder gradually diminishes due to the enrichment of the slurry with other elements and affects the quality of the product adversely. The waste slurry generated after polishing is sent for settling the solids by flocculation, followed by filter press separation. The solid cake formed after the filter press ends up in landfills [26]. This leads to loss of valuable REEs. Furthermore, disposal of glass polishing waste incur costs.

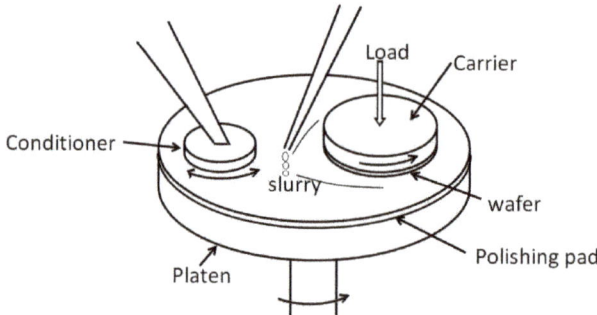

Figure 2. Schematic representation of a wafer polishing process using glass polishing powder slurry as a polishing medium.

Recycling of the polishing waste not only aids to the enhanced utilization of natural resources, but also makes recycling of valuable REEs possible. Recovery and recycling of metals from waste generally consumes less energy and results in less environmental impact when compared with primary production from an ore [31]. Furthermore, it also saves the disposal cost. The recovered cerium can be converted to cerium metal and can be used in high value aluminum and magnesium alloys and steels [32–34]. This generates highly added value from the waste and closes metal loops.

Due to the high prices and limited supply of cerium oxide since 2010, highly efficient polishing technologies that consume less amount of cerium oxide are being developed [17]. These new processes have shorter polishing times, smaller amounts of slurry and slurry re-use where possible. This decreased the polishing powder consumption by 30–50% [35]. Different types of medium hard alumina can also be replaced with ceria for some low-end purposes [8].

Recycling of polishing waste is limited in Europe and USA, however in Japan and China this is significant [36]. Nevertheless, most of the recycling operations are currently stopped due to large price fluctuations of REEs [36]. Therefore, the new processes developed for recycling should be easily managed with small waste streams and with maximum recovery [26]. However, it is difficult to recover REEs from glass polishing waste as it is not easy to treat the waste chemically or physically [37]. The particle size distribution plays an important role in polishing [5]. Hence, the morphology of the regenerated powder from glass polishing waste should be according to the specifications, if it is reused in polishing process.

2. Characterization of Glass Polishing Waste

The chemical composition of different glass polishing waste samples are shown in Table 3. The waste mainly contains cerium together with other rare earths and impurities. Some of the waste contains fluorine and phosphorus which is due to the presence of fluoride and phosphate compounds of rare earths, which are very stable compounds and are not soluble in acid solutions at ambient conditions [25]. As mentioned earlier, iron and aluminum are present in the polishing waste due to the hydrolysis of the flocculants like aluminum chlorides and iron chlorides that are used in the settling of solids from slurry [12]. The presence of other rare earths in the polishing powder may be due to the costs associated with the complete separation of cerium from other rare earths. Neodymium and praseodymium content in the polishing waste is lower in the recent studies when compared to older studies [12,13,20–30]. This may be due to the new process development or process improvement in recent years for separating the adjacent rare earths. This is also due to the high value of neodymium and praseodymium compared to cerium. The presence of fluorine is due to the fluorine treatment (with ammonium fluoride or HF) of the powder during processing in order to increase the CeO_2 concentration in the solid solution [38], or to control the particle size range [9], or to neutralize the basic lanthanum oxide [16].

Polishing waste is a very fine material with d50 around 2 μm [29] and 100% of the particles are less than 9 μm. A typical SEM image of a polishing waste sample is shown in Figure 3. Kato et al. found that the cerium is in the form of CeO_2 and $La_2O_3 \cdot CeF_3$ [12]. However, the peaks are shifted due the substitution of cerium with other rare earths. Poscher et al. [24] found that 21% of cerium is in the form of fluorobritholite and monazite. In a recent study Borra et al. [39] found CeO_2, $LaO_{0.65}F_{1.7}$ and $LaPO_4$ phases in a polishing waste sample.

Table 3. Chemical composition of different glass polishing waste samples available in the literature.

	CeO_2	La_2O_3	Pr_6O_{11}	Nd_2O	F (elem.)	SiO_2	Al_2O_3	Fe_2O_3	Na_2O	P_2O_5	ZnO	K_2O	CaO	PbO	Reference
1	8.8	1.5	-	-	-	57.6	1	-	13.3	-	2.0	6.9	2.9	-	[23]
2	69	8	-	1	-	11	1.3	-	1	2	2	-	1	-	[24–26]
3	49–80	1–10	-	0.1–0.8	-	5–10	-	1.4–2.9	-	-	1.2–2.5	-	-	1–2	[27]
4	54	-	-	-	-	12	-	-	-	-	-	-	-	-	[13]
5	22.3	17.7	1.41	0.21	1.8	0.1	11.5	0.2	-	-	-	-	9.3	-	[28]
6	48.4	23.5	-	-	3.7	0.6	8.1	0.9	0.4	0.7	0.1	-	0.3	-	[29,30]
7	60	25.8		13.6 (elem.)						15.3 (elem.)					[20]
8	22.1	17.8	2.30	5.1	-	12.6	24.8	-	-	-	-	-	-	-	[12]
9	50	28	7.4	4	9.5	0.6	-	0.3	-	0.1	-	-	0.7	-	[21]
10	38	28	3.8	10	12	1.7	-	0.2	-	1.7	-	-	1.3	-	[22]

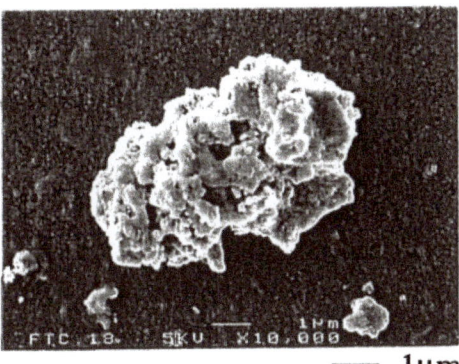

— 1μm

Figure 3. SEM image of a glass polishing waste sample shows fine particles. Reproduced from Kato et al. [12], with permission from American Chemical Society, 2018.

3. Recovery of Polishing Powder (Cerium Compounds)

As explained earlier, the polishing ability of the powder decreases after several uses. This is shown in the Figure 4. This figure also shows the effect of different kinds of polishing powders on surface roughness of the glass. It can be observed from the figure that after the removal of silica and alumina (by alkali treatment), polishing powder can be reused in polishing as the surface roughness is similar to fresh polishing powder.

Cerium can be reused or recycled from glass polishing waste by physical, physico-chemical and/or chemical methods [26,40]. Figure 5 shows an overview of different processes used for reusing and recycling of rare earths from polishing waste. Table 4 summarizes the different studies carried out on polishing waste to recover and recycle rare earths. It also gives details of reagents used and the conditions applied.

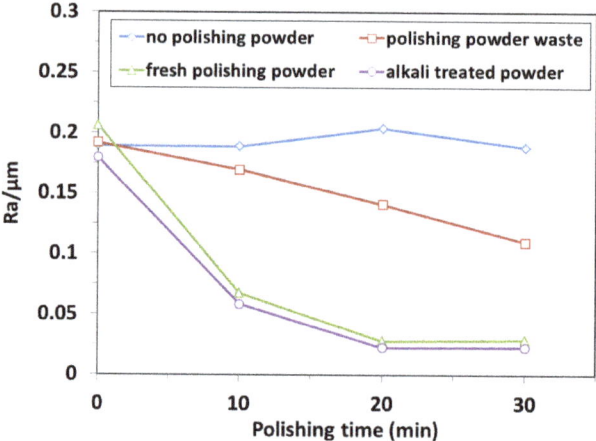

Figure 4. Effect of different polishing agents on glass polishing: the surface roughness (Ra/μm) as a function of polishing time. Adapted from Kato et al. [12], with permission from American Chemical Society, 2018.

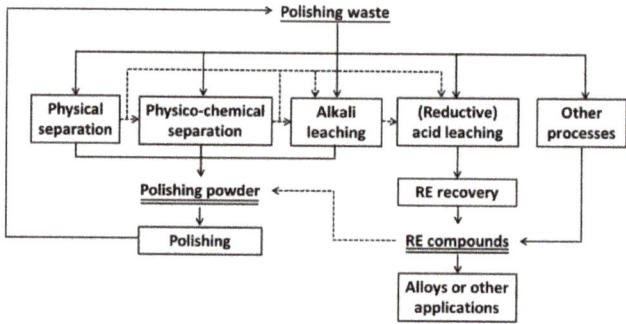

Figure 5. Different processes used for recovery of rare earths from glass polishing waste for reusing and recycling applications.

3.1. Physical Separation

Silica can be removed from polishing waste by physical or physicochemical separation processes like sieving, gravity separation, flotation, selective flocculation, etc. Kato et al. and Sato and Kato developed an elutriation method for removing silica [40]. In this process, the glass particles sank rapidly from polishing slurry in a first tank followed by slow separating (three days) of the supernatant solution in a second tank for settling the polishing powder particles.

Kim et al. partially removed the glass particles from waste by flotation with citric acid aqueous solution at pH ~1.5 [29]. These authors found that application of ultrasound enhances the removal of silica during flotation. Liang et al. applied flotation using sodium dodecylsulfate as a collector [40]. The authors were able to recover over 99% of REEs from a synthetic mixture of REOs and glass. However, the recovery was only 40% with a purity of 44% (35% in feed sample) when they used real slurry. Japan Oil, Gas, and Metals National Corporation (JOEGMEC) studied the removal of flocculants by homogenization followed by removing coarse particles with sieving [40]. In this study fine particles were dispersed with shearing dispersion followed by filtration using a vibrating membrane. It was proposed that the fines generated after the filtration can be used in glass polishing again.

Polishing waste can be reused after silica removal up to some extent, if the particle size distribution is similar to the original polishing powder. However, it is difficult to remove all the silica particles as the particle size is very small (<5 μm). Due to van der Waals forces, it is difficult to separate silica from polishing waste selectively. Therefore, chemical separation processes are required for further purification (for silica and/or other elements removal).

3.2. Chemical Separation

3.2.1. Alkali Leaching—for Reuse of Polishing Powder

Alkali leaching is mainly used for the removal of silica and alumina from the polishing waste. The following reactions take place during the leaching [12].

$$SiO_2 + OH^- \rightleftharpoons HSiO_3^- \quad (1)$$

$$Al(OH)_3 + OH^- \rightleftharpoons Al(OH)_4^- \quad (2)$$

Kato et al. removed most of Al_2O_3 and SiO_2 from glass polishing waste by treating it with 4 mol/kg of NaOH at 50–60 °C for 1 hour [12]. These authors removed the residue (product) from the leach solution by centrifugation and also produced zeolites from the leach solution by heating it to 100 °C [12,41]. The flowsheet proposed by Kato et al. is shown in the Figure 6.

Kim et al. [29] removed the remaining glass in the waste after a flotation treatment by NaOH leaching with a pH level of 11.5. These authors proposed to reuse the slurry in glass polishing after silica removal as the particle size distribution matches with the original glass polishing slurry. However, the alumina content is still high (about 9%) compared to silica (0.5%). Low removal rate of alumina was observed in this study compared to the previous study by Kato et al. [12]. This may be due to the type of alumina compound or the leaching conditions.

Moon et al. [20] proposed to use sodium fluoride during NaOH leaching to enhance the leaching by forming Na-F-Si-O structure. These authors also claim that the addition of NaF and Na_2CO_3 during leaching helps in solid liquid separation as the particles of Na-F-Si-O are easy to separate from the ceria particles. Matsui et al. added 0.5% of aluminum sulfate to promote solid-liquid separation (flocculation) after leaching [13]. These authors also proposed to treat the leach residue with nitric acid (pH 5.8) for complete removal of alkali compounds.

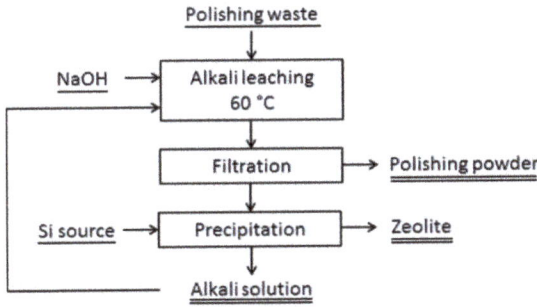

Figure 6. Alkali leaching of polishing powder for the recovery of polishing powder followed by recovery of silica and alumina as zeolite from leach solution. Adapted from Kato et al. [12], with permission from American Chemical Society, 2018.

Table 4. Different studies on recovery and recycling of rare earths from glass polishing waste by using physical, chemical and/or physico-chemical processes.

Physical or Physio-Chemical Separation	Chemical Separation	Recovery of REEs/Reuse of Polishing Powder	Reference
Coarse glass particles: sieving Filtration with vibrating screen		Product: for reuse in polishing	[40]
Elutriation Rapid settling: glass Slow settling (3 days): rare earths		Product: for reuse in polishing	[40]
Flotation: dodecylesulfate (collector)		40% with a purity of 44% (35% in feed sample)	[40]
Flotation	HCl (1.4 kmol/m^3) 55 °C, 60 min Selective Ce leaching	D2EHPA and PC-88ACe recovered preferentially over La	[40]
Flocculent removal: oxalic or citric acid leaching (pH-1.5) Silica removal: flotation aided by sonication	Silica: alkali leaching (pH-11.5) Roasting: 600 °C, 2 h Rare earths: Sulfuric acid (3 mol/L), S/L—1:10, 60 °C, 3 h	Other rare earths: Double sulfate precipitation, Na$_2$SO$_4$/REO—0.5 to 1 Ce Yield: 60%	[29,30]
	NaOH: 4 mol/kg, 1 h 50-60 °C	Residue: for reuse in polishing Aluminum and silicon in the leach solution were recovered as zeolites	[12,41]
	Alkali Leaching NaOH, NaF, Na$_2$CO$_3$ 90 °C, 2 h	Residue: for reuse in polishing	[20]
	NaOH (2 mol/L) KOH (3.5 mol/L)	Residue: for reuse in polishing	[13]
	HCl: 15-35 wt% (10-20% excess). Iodide: 10-20% excess 15-60 min		[42]
	HNO$_3$ and H$_2$O$_2$ H$_2$O$_2$/HNO$_3$ = 0.15 80 °C, S/L: 1.5	RE recovery: 80% Precipitation with ammonium carbonate	[43]

Table 4. Cont.

Physical or Physio-Chemical Separation	Chemical Separation	Recovery of REEs/Reuse of Polishing Powder	Reference
	Step 1: 0.4 mol/L Na_2SO_4 and 8 mol/L H_2SO_4 for double salt precipitation Step 2: RE hydroxide precipitation from double salt with NaOH Step 3: Oxidation of cerium Step 4: HCl (selective) leaching of rare earths except cerium Step 5: H_2SO_4 (selective) leaching of cerium	Pure oxides	[44–46]
	H_2SO_4 (1–8 mol/dm^3), 30–90 °C	Pure oxides	[37]
	HCl: 32 wt%, H_2O_2: 30 wt% 80 °C, 4 h Ce recovery (>90%), La recovery (>60%)	Oxalic acid (27% excess) precipitation	[24–26]
	HNO_3: 47%, H_2O_2: 30 wt% 70 °C, 1 h Ce recovery (>90%), La recovery (>60%)		
	H_2SO_4: 98%, 100–200 °C followed by water leaching Ce, La recovery (>90%)		
	HNO_3 and H_2O_2: 20% excess (with respect to the total REE content) 70 °C, 3 h	Carbonate precipitation by CO_2 and NH_3 followed by annealing in rotary kiln	[23]
	HNO_3 and H_2O_2 (stoichiometric) 80 °C	Oxalate precipitation	[27]
	Carbochlorination: Active carbon, N_2-Cl_2 gas, 1000 °C,	Rare earth chloride deposition: <1000 °C to 450 °C.	[21,22]
	Sulfation: H_2SO_4, 250 °C, 2 h Water leaching 50 °C, 2 h, L/S: 10	Double sulfate precipitation	[47]

155

Removal of glass particles by sieving, flotation, and/or alkali leaching can increase the lifespan of polishing powder to some extent [23]. However, the removal of silica and/or alumina alone is insufficient as the increase in concentration of other elements like zinc, calcium, sodium, etc. adversely affects the product quality after several cycles [25]. The main problem with alkali leaching is that it needs large scale facilities and generate high volumes of alkaline waste water [40]. Therefore, complete removal of impurities or dissolution of cerium into the leach solution followed by cerium recovery is required. This will be described and discussed in Section 3.2.2 (acid leaching) and Section 3.2.3 (REEs recovery).

3.2.2. Acid Leaching—for REE Recovery

Direct Leaching

Cerium dioxide is sparingly soluble in most of the acids at ambient conditions [48]. The solubility of CeO_2 in acids solutions is further reduced due to the high temperature sintering (annealing) at 600 to 1000 °C of the powder before its application in polishing [23]. Therefore, leaching needs to be carried out in concentrated acid solutions and/or at elevated temperatures. Different acids such as HCl, HNO_3, H_2SO_4 and HF can be used for leaching of glass polishing waste. HF acid can be used for removal of silica. However, it has some disadvantages like residual HF, which reacts with the glass surface and roughen the surface during polishing, and HF is also hazardous [13]. Sulfuric acid is a low cost reagent and less corrosive. However, the solubility of rare-earth sulfates is very low in leach solutions and the solubility decreases with increasing the temperature. The decrease in solubility is due to the exothermic nature of the rare-earth sulfate dissolution [29,37]. In the following, four different direct leaching methods are briefly discussed.

- Concentrated H_2SO_4 leaching: After removal of silica from glass polishing waste by flotation and after alkali leaching, Kim et al. performed oxidative roasting on dried samples at 600 °C to convert cerium compounds to CeO_2 [29]. Cerium was leached from the roasted mass by concentrated sulfuric acid solution at 60 °C. These authors claimed that high amount of oxidation of cerium oxide dissolves more cerium in to the leach solution. However, for a small increase in cerium recovery (about 10%) the roasting of the powder is difficult to be justified.
- Concentrated H_2SO_4 digestion—water leaching: Poscher et al. [25] digested polishing powder with conc. sulfuric acid (>96 wt%) at temperatures above 100 °C. During digestion the slurry was solidified. The digested material was subsequently leached with water to dissolve rare earth sulfates.
- Two stage H_2SO_4 leaching: Um et al. leached a synthetic mixture with sulfuric acid in two stages for selective recovery of cerium [37]. In the first stage, the mixture was treated with 2 mol/dm^3 sulfuric acid at 90 °C for leaching La, Nd, Pr and Ca. In the second stage, the leach residue from first stage was treated with 12 mol/dm^3 and 120 °C. After leaching, the leach solution was diluted with water to dissolve $Ce(SO_4)_2$.
- HCl leaching: Yamada et al. [40] were able to recover cerium selectively by using 1.4 $kmol/m^3$ HCl at 55 °C. The recovery of cerium is about 65% and that of lanthanum is very low. From this study it looks that cerium can be selectively dissolved. However, the recovery of cerium or lanthanum mainly depends on the presence of different compounds [25]. For example, if the lanthanum is present in a fluoride phase then it is difficult to dissolve. However, part of lanthanum in oxyfluoride may dissolve during leaching according to the following reaction.

$$LaO_{0.65}F_{1.7} + 1.3HCl \rightleftharpoons 0.433LaCl_3 + 0.566LaF_3 + 0.65\ H_2O \qquad (3)$$

The recovery of cerium and/or lanthanum in some studies is low due to the fact that part of the cerium and lanthanum are present in stable compounds like fluorides or phosphates.

Acid Leaching Together with a Reductant

As discussed earlier, leaching of CeO_2 directly is difficult. Therefore, the use of a reductant can help for easier dissolution and in decreasing the acid concentration and leaching temperature. Potassium iodide [42] and H_2O_2 [23–25,27] are used as reductants during acid leaching of glass polishing waste so far. The reaction involved during the reductive leaching of CeO_2 with H_2O_2 is

$$2Ce^{4+} + H_2O_2 \rightleftharpoons 2Ce^{3+} + 2H^+ + O_2 \qquad (4)$$

Poscher et al. leached the polishing waste with HCl and H_2O_2. More than 97% of cerium was recovered by using concentrated acid [24]. These authors proposed that the lanthanum leaching is low due to the presence of fluorine compounds. The optimized conditions are: liquid-to-solid ratio of 5.5, acid concentration of 32% at 80 °C for 4 h duration.

An industrial process was developed by Hydrometal, Belgium for the recovery of cerium from glass polishing waste by reductive leaching of glass polishing powder with nitric acid together with H_2O_2 [27]. The flowsheet developed by Hydrometal is shown in Figure 7. The Ce^{4+} concentration in the leach solution was monitored during leaching using ORP (oxidation/reduction potential) detector. Nitric acid was used as a lixiviant as it is selective towards silica and alumina. The authors used a higher temperature (~80 °C) for the diffusion of the acid in to silica matrix and also for enhancing filtration.

Janoš et al. [49] were able to leach the glass polishing waste (~60% CeO_2) in nitric acid and H_2O_2 solution without any external heating. However, with low cerium content (~9%) the external heating is necessary to maintain the temperature (65–70 °C) [23]. The leaching recovery of cerium was around 70%. In an another study, Janoš et al. [42] studied the leaching of polishing waste with hydrochloric acid and potassium iodide. 15–35% HCl was added together with 10–20% excess of potassium iodide during leaching. These authors claimed that the leaching was complete in 15–60 min.

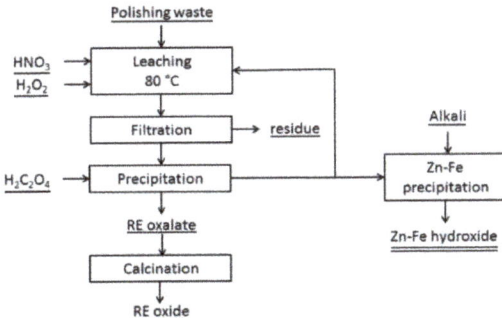

Figure 7. Flowsheet developed by Hydrometal for recovery of rare earth oxides from polishing waste by leaching followed by selective precipitation. Adapted from Henry et al. [27].

3.2.3. Recovery of REEs from Polishing Waste Leach Solutions

Different methods that are used for the recovery of cerium from a leach solution include: double salt precipitation, oxalate precipitation, carbonate precipitation, hydroxide precipitation and solvent extraction followed by precipitation. Separation of individual REEs from a mixture is difficult due to the similar physical and chemical properties. Separation of rare earths are generally carried out by solvent extraction using following steps: (1) rare earth separation in a trivalent state, (2) separation of rare earths into three or four groups, (3) selective recovery of cerium using its tetravalent state, (4) separation of the desired individual rare earth of required purity [1,50].

Lanthanum and cerium are next to each other in the periodic table and their stability constants are very close, therefore their complexation behavior is also very similar [51]. Hence, it is difficult to separate them completely in trivalent state as hundreds of stages of mixers and settlers are required in solvent extraction. Nevertheless, cerium separation is the easiest as it can be oxidized to tetravalent state, which has different chemical properties compared to trivalent rare-earth ions. Figure 8 shows the individual and overlapped cerium and lanthanum Pourbaix diagrams. The overlapped Pourbaix diagram shows a window where cerium can be selectively precipitated by keeping lanthanum in the solution [52]. However, in a recent study, Marsac et al. found more accurate hydrolysis constants for cerium system [53]. This may slightly change the stability regions of different species in cerium Pourbaix diagram.

Cerium can be oxidized and precipitated as tetravalent state in alkaline solution even with air [29]. Other oxidants that may convert Ce(III) to Ce(IV) are persulfate, permanganate, lead oxide, silver oxide or by electrochemical or photochemical oxidation [29].

Poscher et al. precipitated cerium and lanthanum from the solution by 27% excess oxalic acid addition [24]. The precipitated oxalates are calcined at 650 °C to convert oxalates to respective oxides. The solution obtained after oxalic acid precipitation is reused in leaching, after impurity removal by precipitation [27]. Poscher et al. proposed to selectively precipitate cerium from solution with an oxidizing treatment and pH adjustment [24]. This oxidizing agents include H_2O_2 or potassium permanganate. However, oxidants other than H_2O_2 may contaminate the product with the reaction product of the oxidizing agents [24,29]. After cerium removal by oxidation, lanthanum can be removed by oxalic acid treatment. Cerium was also recovered from the solution by a mixture of CO_2 and NH_3 [23]. The obtained carbonate can be converted to oxide by calcination from 200 to 900 °C. The carbonate can also be prepared by precipitation of an aqueous solution of cerous nitrate (0.2 mol/L) with an excess of ammonium bicarbonate (0.5 mol NH_4HCO_3/L) [54].

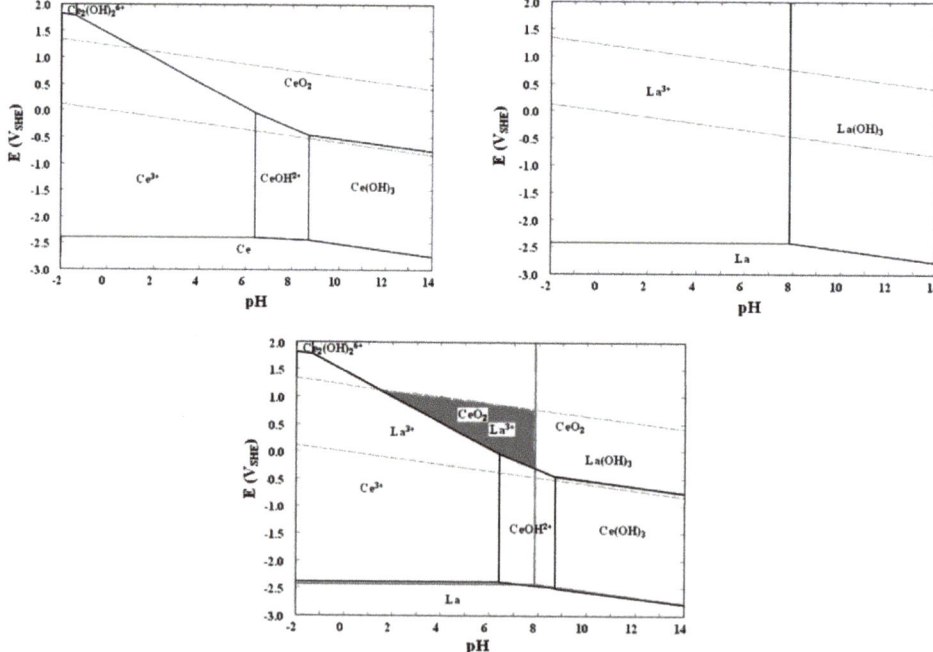

Figure 8. Individual and overlapped Pourbaix diagrams of cerium and lanthanum. The overlapped Pourbaix diagram of lanthanum (gray lines) and cerium (black lines) show the region where cerium can be selectively precipitated in aqueous solutions. La and Ce concentrations are 10^{-3} mol/L each. Reproduced and adapted from Kim and Osseo-Asare [52], with permission from Elsevier, 2018.

Kim et al. [29] removed other REEs from cerium by double sulfate precipitation with sodium sulfate at 50 °C and 90 min. The authors were able to recover over 62% cerium in solution at Na_2SO_4/RE ratio of 0.5 with La/Ce ratio of 0.0307. Yoon et al. separated rare earths from aluminum by double salt precipitation with Na_2SO_4 at 50 °C for 1 h [47]. Um and Hirato precipitated $NaCe(SO_4)_2 \cdot xH_2O$ from CeO_2 by sulfuric acid leaching in presence of Na_2SO_4 at about 125 °C [46,48]. In another study, Um and Hirato proposed to produce rare earth hydroxide from $NaCe(SO_4)_2 \cdot xH_2O$ with the help of NaOH followed by oxidation of $Ce(OH)_3$ to $Ce(OH)_4$. In the next step, the authors proposed to leach rare earths other than Ce at higher pH values (2.5 to 3.5). Cerium with high selectivity was leached in the following step with sulfuric acid [44].

Yamada et al. [40] extracted cerium preferentially over lanthanum using D2EHPA and PC-88A reagents. The authors were able to extract cerium selectively with 50% PC-88A at pH 1.24 with a selectivity of about 2.5.

3.2.4. Other Processes for REE Recovery

Ozaki et al. separated cerium and other REEs from glass polishing waste by a chemical transport method [21,22]. A Cl_2-N_2 gaseous mixture was used for carbo-chlorination together with active carbon. The following reactions of cerium and other lanthanides (Ln) takes place during the carbochlorination:

$$Ln_2O_3 + 2C + 3Cl_2 \rightleftharpoons 2LnCl_3 + 3CO \qquad (5)$$

$$2CeO_2 + 4C + 3Cl_2 \rightleftharpoons 2CeCl_3 + 4CO \qquad (6)$$

After chloride conversion, rare-earth chlorides were exposed to Al_2Cl_6 gas to form low volatile ($LnAl_3Cl_{12}$) complexes. At low temperatures (457–947 °C) these complexes dissociates and forms rare-earth chlorides. By this process RE chlorides were successfully separated from CO gas and other chlorides. However, it is difficult to separate individual REEs, because of their similar chemical properties due to same oxidation state and similar ionic radii.

Yoon et al. developed a sulfation process to convert (decompose) rare earth compounds in the polishing waste to their respective sulfates at 250 °C in presence of sulfuric acid [47]. The sulfated roast was leached in water at 50 °C and able to recover rare earths completely. Wang et al. studied the sulfation process using TGA and FTIR [28]. The recovery yield in their process was about 95%. These authors found that rare earth oxides decompose at 250 °C and rare earth oxyfluorides decompose at 300 °C. The temperature in TGA is higher for complete decomposition when compared to the study of Yoon et al. This difference may be due to non-isothermal heating used in TGA.

4. Applications of Recovered Cerium

Large quantities of cerium can be generated by treating the current glass polishing waste and waste from landfills. Currently, the recovered cerium is reused in polishing process [36]. However, the recovered cerium can also be used in other applications such as redox flow batteries, oxidimetric agent and as a catalyzer [37]. Janoš et al. [54] transformed polishing powder waste into phosphor elimination adsorption material, which could effectively eliminate the phosphate in sewage. The recovered cerium can also be used for preparing valuable Al-Ce alloys [32]. Cerium as an alloying element in aluminum increases the strength at room temperature as well as at high temperature. Hence, these alloys can be used for IC engines. These engines with Al-Ce alloys can be operated at higher temperatures that increase the fuel efficiency. Application of cerium in aluminum alloys can also address the so-called balance problem [32].

5. Conclusions

Large quantities of cerium in glass polishing waste are being lost in landfills. The life of glass polishing waste can be increased by removing silica with physical beneficiation and/or alkali leaching processes. However, it is difficult to remove silica completely by physical beneficiation processes alone. Silica and alumina can be removed from polishing waste by alkali leaching. Nevertheless, an increase in the concentration of impurities other than alumina and silica affects the polishing process, and thus complete removal of impurities is required. Another alternative is to recover all REEs as oxides or RE metals/alloys from the polishing waste. Cerium and other REEs can be recovered from glass polishing waste by acid leaching processes followed by extraction from leach solutions. However, high acid concentration and/or temperatures or costly reagents (reductants) are required during leaching. Furthermore, part of cerium and/or lanthanum are present in the form of fluoride and phosphates and they report to residue during leaching. Hence, an extra processing step is required for complete recovery of rare earths. Cerium can be extracted from solution by precipitation or separated by solvent extraction. The extracted cerium can be used in glass polishing or other high value added applications.

Author Contributions: C.R.B. prepared the manuscript. All authors discussed and reviewed the manuscript.

Funding: This research was funded TU Delft (3mE Cohesie project: "Sustainable Rare-earth Cycle") and NWO-CW (VICI grant).

Conflicts of Interest: The authors declare no conflict of interest.

References

1. Krishnamurthy, N.; Gupta, C.K. *Extractive Metallurgy of Rare Earths*; CRC Press: Boca Raton, FL, USA, 2015; ISBN 1466576383.
2. Gambogi, J. Rare earths. In *Mineral Commodity Summeries*; USGS: Reston, VA, USA, 2018; pp. 132–133.
3. Jha, M.K.; Kumari, A.; Panda, R.; Kumar, J.R.; Yoo, K.; Lee, J.Y. Review on hydrometallurgical recovery of rare earth metals. *Hydrometallurgy* **2016**, *165*, 2–26. [CrossRef]
4. Guyonnet, D.; Planchon, M.; Rollat, A.; Escalon, V.; Tuduri, J.; Charles, N.; Vaxelaire, S.; Dubois, D.; Fargier, H. Material flow analysis applied to rare earth elements in Europe. *J. Clean. Prod.* **2015**, *107*, 215–228. [CrossRef]
5. Binnemans, K.; Jones, P.T.; Blanpain, B.; Van Gerven, T.; Yang, Y.; Walton, A.; Buchert, M. Recycling of rare earths: A critical review. *J. Clean. Prod.* **2013**, *51*, 1–22. [CrossRef]
6. Gambogi, J. *USGS 2014 Minerals Yearbook: Rare Earths*; USGS: Reston, VA, USA, 2016.
7. Argus Media Analysing the Changing Global Rare Earths Supply and Demand Outlook. Available online: http://www.argusmedia.jp/~/media/files/pdfs/regional-specific/jp/downloads/argus-metal-pages-forum082016-rareearths.pdf/?la=en (accessed on 16 February 2017).
8. Tercero Espinoza, L.; Hummen, T.; Brunot, A.; Hovestad, A.; Peña Garay, I.; Velte, D.; Smuk, L.; Todorovic, J.; Van Der Eijk, C.; Joce, C. *Critical Raw Materials Substitution Profiles*; CRM InnoNet: Karlsruhe, Germany, 2015.
9. Lucas, J.; Lucas, P.; Le Mercier, T.; Rollat, A.; Davenport, W.; Le Mercier, T.; Rollat, A.; Davenport, W. Chapter 12—Polishing with Rare Earth Oxides Mainly Cerium Oxide CeO_2. In *Rare Earths*; Elsevier: Amsterdam, The Netherlands, 2015; pp. 191–212. ISBN 978-0-444-62735-3.
10. Cook, L.M. Chemical processes in glass polishing. *J. Non-Cryst. Solids* **1990**, *120*, 152–171. [CrossRef]
11. Hedrick, J.B.; Sinha, S.P. Cerium-based polishing compounds: Discovery to manufacture. *J. Alloys Compd.* **1994**, *207–208*, 377–382. [CrossRef]
12. Kato, K.; Yoshioka, T.; Okuwaki, A. Study for recycling of ceria-based glass polishing powder. *Ind. Eng. Chem. Res.* **2000**, *39*, 943–947. [CrossRef]
13. Matsui, H.; Harada, D.; Takeuchi, M. Method for Recovery of Cerium Oxide. U.S. Patent 20130152483A1, 20 June 2013.
14. Xu, T.; Peng, H. Formation cause, composition analysis and comprehensive utilization of rare earth solid wastes. *J. Rare Earths* **2009**, *27*, 1096–1102. [CrossRef]
15. Janoš, P.; Ederer, J.; Pilařová, V.; Henych, J.; Tolasz, J.; Milde, D.; Opletal, T. Chemical mechanical glass polishing with cerium oxide: Effect of selected physico-chemical characteristics on polishing efficiency. *Wear* **2016**, *362–363*, 114–120. [CrossRef]
16. Komiya, H.; Yamaguchi, S.; Hisatsune, T.; Takenaka, A.; Yonemori, S. Method for Evaluating the Quality of Abrasive Grains, Polishing Method and Abrasive for Polishing Glass. U.S. Patent 7025796B2, 11 April 2006.
17. Ikeda, H.; Akagami, Y. Highly efficient polishing technology for glass substrates using tribo-chemical polishing with electrically controlled slurry. *J. Manuf. Process.* **2013**, *15*, 102–107. [CrossRef]

18. Kaller, A. The basic mechanism of glass polishing. *Naturwissenschaften* **2000**, *87*, 45–47. [CrossRef] [PubMed]
19. Kasai, T.; Bhushan, B. Physics and tribology of chemical mechanical planarization. *J. Phys. Condens. Matter* **2008**, *20*, 225011–225023. [CrossRef]
20. Moon, W.-J.; Na, S.-O.; Oh, H.-Y. Method for Recycling Cerium Oxide Abrasive. U.S. Patent 20110219704A1, 15 September 2011.
21. Ozaki, T.; Machida, K.; Adachi, G. Recovery of rare earths from used polishes by chemical vapor transport process. *Mater. Sci. Forum* **1999**, *315–317*, 297–305. [CrossRef]
22. Ozaki, T.; Machida, K.; Adachi, G. Extraction and mutual separation of rare earths from used polishes by chemical vapor transport. *Metall. Mater. Trans. B* **1999**, *30*, 45–51. [CrossRef]
23. Janoš, P.; Kuráň, P.; Ederer, J.; Šastný, M.; Vrtoch, L.; Pšenička, M.; Henych, J.; Mazanec, K.; Skoumal, M. Recovery of Cerium Dioxide from Spent Glass-Polishing Slurry and Its Utilization as a Reactive Sorbent for Fast Degradation of Toxic Organophosphates. *Adv. Mater. Sci. Eng.* **2015**, *2015*, 241421. [CrossRef]
24. Poscher, A.; Luidold, S.; Antrekowitsch, H. Extraction of cerium and lanthanum from spent glass polishing agent. In *Materials Science & Technology 2013*; London, I.M., Goode, J.R., Moldoveanu, G., Rayat, M.S., Eds.; Canadian Institute of Mining, Metallurgy and Petroleum: Montréal, QC, Canada, 2013; pp. 543–552.
25. Poscher, A.; Luidold, S.; Schnideritsch, H.; Antrekowitsch, H. Extraction of Lanthanides from Spent Polishing Agent. In Proceedings of the ERES2014-1st European Rare Earth Resources Conference, Milos, Greece, 4–7 September 2014; pp. 209–222.
26. Poscher, A.; Luidold, S.; Schnideritsch, H.; Antrekowitsch, H. *Extraction of Lanthanides from Spent Polishing Agent*; Elsevier Inc.: San Diego, CA, USA, 2015; ISBN 9780128023280.
27. Henry, P.; Lamotte, S.; Bier, J. Recycling of rare earth materials at Hydrometal (Belgium). In *52nd Conference of Metallurgists (COM), Hosting by Materials Science Technology Conference (MS&T)*; London, I.M., Goode, J.R., Moldoveanu, G., Rayat, M.S., Eds.; Canadian Institute of Mining, Metallurgy and Petroleum: Montréal, QC, Canada, 2013; pp. 537–542.
28. Wang, X.; Liu, J.; Yang, Q.; Du, J.; Wang, F.; Tao, W. Decomposition process and kinetics of waste rare earth polishing powder TG-DTA-FTIR studies. *J. Therm. Anal. Calorim.* **2012**, *109*, 419–424. [CrossRef]
29. Kim, J.Y.; Kim, U.S.; Byeon, M.S.; Kang, W.K.; Hwang, K.T.; Cho, W.S. Recovery of cerium from glass polishing slurry. *J. Rare Earths* **2011**, *29*, 1075–1078. [CrossRef]
30. Byeon, M.S.; Kim, J.Y.; Hwang, K.T.; Kim, U.; Cho, W.S.; Kang, W.K. Recovery and purification of cerium from glass polishing slurry. In Proceedings of the 18th International Conference on Composite Materials, Jeju, Korea, 21–26 August 2011.
31. Kumar, V.; Jha, M.K.; Kumari, A.; Panda, R.; Kumar, J.R.; Lee, J.Y. Recovery of Rare Earth Metals (REMs) from Primary and Secondary Resources: A Review. In *Rare Metal Technology 2014*; Neale, R., Neelamegghham, N.R., Alam, S., Oosterhof, H., Jha, A., Wang, S., Eds.; Wiley: San Diego, CA, USA, 2014; pp. 81–88.
32. Sims, Z.C.; Weiss, D.; McCall, S.K.; McGuire, M.A.; Ott, R.T.; Geer, T.; Rios, O.; Turchi, P.A.E. Cerium-Based, Intermetallic-Strengthened Aluminum Casting Alloy: High-Volume Co-product Development. *JOM* **2016**, *68*, 1940–1947. [CrossRef]
33. Willbold, E.; Gu, X.; Albert, D.; Kalla, K.; Bobe, K.; Brauneis, M.; Janning, C.; Nellesen, J.; Czayka, W.; Tillmann, W. Effect of the addition of low rare earth elements (lanthanum, neodymium, cerium) on the biodegradation and biocompatibility of magnesium. *Acta Biomater.* **2015**, *11*, 554–562. [CrossRef] [PubMed]
34. Pan, F.; Zhang, J.; Chen, H.-L.; Su, Y.-H.; Kuo, C.-L.; Su, Y.-H.; Chen, S.-H.; Lin, K.-J.; Hsieh, P.-H.; Hwang, W.-S. Effects of rare earth metals on steel microstructures. *Materials* **2016**, *9*, 1–19. [CrossRef] [PubMed]
35. Lucas, J.; Lucas, P.; Le Mercier, T.; Rollat, A.; Davenport, W. Epilogue. In *Rare Earths*; Elsevier: Amsterdam, The Netherlands, 2015; pp. 351–362. ISBN 9780444627353.
36. Lucas, J.; Lucas, P.; Le Mercier, T.; Rollat, A.; Davenport, W. *Chapter 18—Rare Earth Recycle*; Elsevier: Amsterdam, The Netherlands, 2015; pp. 333–350. [CrossRef]
37. Um, N.; Hirato, T. Dissolution Behavior of La_2O_3, Pr_2O_3, Nd_2O_3, CaO and Al_2O_3 in Sulfuric Acid Solutions and Study of Cerium Recovery from Rare Earth Polishing Powder Waste via Two-Stage Sulfuric Acid Leaching. *Mater. Trans.* **2013**, *54*, 713–719. [CrossRef]
38. Lebedeva, M.I.; Dzidziguri, E.L.; Argatkina, L.A. Research of structure and polishing properties of nanopowders based on cerium dioxide. In *Nanostructures, Nanomaterials, and Nanotechnologies to Nanoindustry*; Apple Academic Press: Waretown, NJ, USA, 2014; pp. 215–225.

39. Borra, C.R.; Vlugt, T.J.; Yang, Y.; Offerman, S.E. Characterisation of glass polishing waste samples. In Proceedings of the ERES 2017: The Second Conference on European Rare Earth Resources, Santorini, Greece, 28–31 May 2017; pp. 215–216.
40. Tanaka, M.; Oki, T.; Koyama, K.; Narita, H.; Oishi, T. *Recycling of Rare Earths from Scrap*, 1st ed.; Elsevie: Amsterdam, The Netherlands, 2013; Volume 43, ISBN 9780444595362.
41. Kato, K.; Yoshioka, T.; Okuwaki, A. Recyle of Ceria-Based Glass Polishing Powder Using NaOH Solution. *Nippon Kagaku Kaishi* **2000**, *10*, 725–732. [CrossRef]
42. Janoš, P.; Novak, J.; Broul, M. A Procedure for Obtaining Salts of Rare Earth Elements. U.S. Patent 21039151, 31 October 1988.
43. Terziev, A.L.; Minkova, N.L.; Todorovsky, D.S. Regeneration of waste rare earth oxides based polishing materials. *Bulg. Chem. Commun.* **1996**, *29*, 274–284.
44. Um, N.; Hirato, T. A hydrometallurgical method of energy saving type for separation of rare earth elements from rare earth polishing powder wastes with middle fraction of ceria. *J. Rare Earths* **2016**, *34*, 536–542. [CrossRef]
45. Um, N.; Hirato, T. Conversion kinetics of cerium oxide into sodium cerium sulfate in Na_2SO_4-H_2SO_4-H_2O solutions. *Mater. Trans.* **2012**, *53*, 1992–1996. [CrossRef]
46. Um, N.; Hirato, T. Synthesis of Sodium Cerium Sulfate (NaCe(SO4)2 ·H2O) from Cerium Oxide in Sulfuric Acid Solutions. In *Zero-Carbon Energy Kyoto 2011: Special Edition of Jointed Symposium of Kyoto University Global COE "Energy Science in the Age of Global Warming"*; Yao, T., Ed.; Springer: Tokyo, Japan, 2012; pp. 171–176. ISBN 978-4-431-54067-0.
47. Yoon, H.; Kim, C.; Kim, S.; Lee, J.; Cho, S.; Kim, J. Separation of Rare Earth and Aluminium from the Dried Powder of Waste Cerium Polishing Slurry. *J. Korean Inst. Resour. Recycl.* **2003**, *12*, 10–15.
48. Um, N.; Hirato, T. Precipitation of cerium sulfate converted from cerium oxide in sulfuric acid solutions and the conversion kinetics. *Mater. Trans.* **2012**, *53*, 1986–1991. [CrossRef]
49. Janoš, P.; Novák, J.; Broul, M.; Loučka, T. Regeneration of polishing powders based on REE oxides from glass polishing sludges. *Chem. Prum.* **1987**, *37–62*, 189–194.
50. Xie, F.; Zhang, T.A.; Dreisinger, D.; Doyle, F. A critical review on solvent extraction of rare earths from aqueous solutions. *Miner. Eng.* **2014**, *56*, 10–28. [CrossRef]
51. Byrne, R.H.; Li, B. Comparative complexation behavior of the rare earths. *Geochim. Cosmochim. Acta* **1995**, *59*, 4575–4589. [CrossRef]
52. Kim, E.; Osseo-Asare, K. Aqueous stability of thorium and rare earth metals in monazite hydrometallurgy: Eh–pH diagrams for the systems Th–, Ce–, La–, Nd–(PO_4)–(SO_4)–H_2O at 25 °C. *Hydrometallurgy* **2012**, *113–114*, 67–78. [CrossRef]
53. Marsac, R.; Réal, F.; Banik, N.L.; Pédrot, M.; Pourret, O.; Vallet, V. Aqueous chemistry of Ce(iv): Estimations using actinide analogues. *Dalt. Trans.* **2017**, *46*, 13553–13561. [CrossRef] [PubMed]
54. Janoš, P.; Kuran, P.; Kormunda, M.; Stengl, V.; Grygar, T.M.; Dosek, M.; Stastny, M.; Ederer, J.; Pilarova, V.; Vrtoch, L. Cerium dioxide as a new reactive sorbent for fast degradation of parathion methyl and some other organophosphates. *J. Rare Earths* **2014**, *32*, 360–370. [CrossRef]

© 2018 by the authors. Licensee MDPI, Basel, Switzerland. This article is an open access article distributed under the terms and conditions of the Creative Commons Attribution (CC BY) license (http://creativecommons.org/licenses/by/4.0/).

Article

Cavitation-Dispersion Method for Copper Cementation from Wastewater by Iron Powder

Andrei Shishkin [1], Viktors Mironovs [1], Hong Vu [2,*], Pavel Novak [2], Janis Baronins [3], Alexandr Polyakov [1] and Jurijs Ozolins [4]

1. Scientific Laboratory of Powder Materials, Faculty of Civil Engineering, Riga Technical University, 6B Kipsalas Street, Room 331a, LV-1048 Riga, Latvia; andrejs.siskins@rtu.lv (A.S.); viktors.mironovs@gmail.com (V.M.); 29433183a@gmail.com (A.P.)
2. Department of Metals and Corrosion Engineering, University of Chemistry and Technology Prague, Technická 5, 166 28 Prague 6-Dejvice, Czech Republic; panovak@vscht.cz
3. Department of Mechanical and Industrial Engineering, School of Engineering, Tallinn University of Technology, Ehitajate tee 5, 19086 Tallinn, Estonia; jbaronins@gmail.com
4. Rudolfs Cimdins Riga Biomaterials Innovations and Development Centre of RTU, Institute of General Chemical Engineering, Faculty of Materials Science and Applied Chemistry, Riga Technical University, Pulka 3, LV-1007 Riga, Latvia; jurijs.ozolins@rtu.lv
* Correspondence: vun@vscht.cz; Tel.: +420-220-445-025

Received: 30 September 2018; Accepted: 29 October 2018; Published: 8 November 2018

Abstract: The circular economy for sustainable economic deployment is strongly based on the re-use of secondary products and waste utilization. In the present study, a new effective cementation method for recovering valuable metallic copper from industrial wastewater using Fe^0 powders is reported. A high-speed mixer-disperser (HSMD) capable of providing a cavitation effect was used for the rapid intake, dispersion, and mixing of Fe^0 powder in an acidic wastewater solution (pH ≈ 2.9) containing copper ions mainly in the form of $CuSO_4$. Three iron powders/particles were tested as the cementation agent: particles collected from industrial dust filters (CMS), water-atomized iron-based powder AHC100.29, and sponge-iron powder NC100.24. The effects of mixing regimes and related mixing conditions on the effectiveness of the Cu cementation process were evaluated by comparison between the HSMD and a laboratory paddle mixer. It was observed that the use of cavitation provided more efficient copper removal during the copper cementation process in comparison to the standard experiments with the propeller mixer. Under the cavitation regime, about 90% of copper was cemented in the first five minutes and the final copper removal of 95% was achieved using all three Fe^0 powders after seven minutes of cementation. In comparison, only around 55% of copper was cemented in the first seven minutes of cementation using the traditional mixing method.

Keywords: cementation; copper removal; cavitation

1. Introduction

Copper is one of the most important non-ferrous metals due to its high thermal and electrical conductivity. It can be applied as an electrical and heat conductor, a building material, an alloying component, and in metal finishing, electroplating, plastics, fungicides, etching, etc. [1–4]. Increases in the rate of copper production and widespread demand for applications in many fields have led to a considerable increase of copper ions leaking into the ecosystem, particularly in industrial wastewater [1,4–6]. As a heavy metal, the presence of copper may cause hazardous effects on human health, even at low concentrations [7–9].

The United States Environmental Protection Agency (USEPA) allows a maximum contaminant level (MCL) of only 1.3 mg/L for the concentration of copper ions in industrial effluents.

The standardized MCL in drinking water for copper is 0.25 mg/L [10–13]. Effective copper removal from wastewater has always been of great interest to researchers around the world. Cementation [14–17], adsorption [5,9,10,18,19], membrane filtration [20–22], electrodialysis [20,23], and photocatalysis [19,24] are the most common methods and techniques for the extraction of copper from wastewater. Each method and/or technology has its own advantages and disadvantages [25–29]. Nevertheless, the cementation process is considered to be the most economical and effective method to remove valuable and/or toxic metals from industrial wastewater [1]. This method reduces copper ions in the copper's salt-containing aqueous solution to its zero valent state at the interface of a more positive metal in the electromotive force series [14]. Low cost, low energy consumption, recovery of metals in pure metallic form, ease of control, simplicity of operation, and high efficiency make the cementation process most advantageous [1,30,31]. The most common cementation agents are iron and zinc [32]. The main disadvantage of the cementation technique is its excess sacrificial metal consumption [33].

Intensification of the cementation process has interested researchers in recent years. Nosier et al. has reported the use of ceramic for the enhancement of copper cementation under single-phase flow [17]. A bubble column reactor fitted with horizontal screens was used for copper removal from aqueous solutions [34]. A modified stirred tank reactor in combination with alcoholic additives increased the cementation rate of copper powder from wastewater containing $CuSO_4$ [35]. In his paper, Konsowa used a special jet reactor to cement copper on a zinc disc from a dilute copper sulfate solution [14].

Ultrasonic cavitation is the well-known phenomenon in the ultrasonic cleaning method. In a liquid medium, the ultrasonic waves, generated by an electronic ultrasound generator and a special transducer suitably mounted under the bottom of a stainless-steel tank or directly to a metal cylinder (called a "horn"), produce compression and vacuum waves at a very high speed. Speed depends on the working frequency of the ultrasound generator. Normally, such apparatus operates at a frequency between 28 and 50 kHz. The pressure and vacuum waves in the liquid cause the phenomenon known as "ultrasonic cavitation". However, this method is complicated for implementation on a commercial scale due to the low productivity and high cost of commercial ultrasonic devices. Another important factor is the risk to human health caused by long-term exposure to ultrasonic noise generated by ultrasound generators. In the present work, a rotational cavitation-generated device was employed to study the effect of cavitation on copper cementation from industrial wastewater.

2. Theory

Cementation is an electrochemical process that leverages the difference between the standard electrode potentials of ions of the extracted metal and the cementing metal. The cementing metal should have a more negative standard electrode potential than the extracted metal. This process in acid media can be described by following Equations (1)–(3).

$$CuSO_4 + Fe \rightarrow FeSO_4 + Cu, \tag{1}$$

$$H_2SO_4 + Fe \rightarrow H_2 + FeSO_4, \tag{2}$$

$$Fe_2(SO_4)_3 + Fe \rightarrow 3FeSO_4. \tag{3}$$

However, the following reactions occur simultaneously ((4)–(6)):

$$Fe_2(SO_4)_3 + Cu = CuSO_4 + 2FeSO_4, \tag{4}$$

$$Cu + 1/2 O_2 + H_2SO_4 = CuSO_4 + H_2O, \tag{5}$$

$$2FeSO_4 + 1/2 O_2 + H_2SO_4 = Fe_2(SO_4)_3 + H_2O. \tag{6}$$

Standard electrode potentials of main species are given as follows [6]:

$$Cu \rightarrow Cu^{2+} + 2e, E^0 = +0.34 \text{ V},$$

$$Fe \rightarrow Fe^{2+} + 2e, E^0 = -0.44 \text{ V},$$

$$Fe^{3+} + e \rightarrow Fe^{2+}, E^0 = +0.77 \text{ V},$$

$$O_2 + 4e + 4H^+ \leftrightarrow 2H_2O, E^0 = +1.23 \text{ V}.$$

Reactions (4) and (5) reduce the amount of copper extraction in the cementation process. Also, in kinetic equilibrium, three following reactions occur simultaneously:

$$Cu \leftrightarrow Cu^+ + e, E^0 = +0.52 \text{ V},$$

$$Cu \leftrightarrow Cu^{2+} + 2e, E^0 = +0.34 \text{ V},$$

$$Cu^+ \leftrightarrow Cu^{2+} + e, E^0 = +0.15 \text{ V}.$$

The speed and limit of copper dissolution is determined by steady equilibrium potential. Cuprous ions are unstable in sulfate solutions and are spontaneously transformed into divalent copper to form a powder according to disproportionation (Equation (7)):

$$Cu_2SO_4 \leftrightarrow Cu + CuSO_4. \tag{7}$$

In neutral solutions, copper sulfate is hydrolyzed to form dark red crystals of copper oxide (Equation (8)):

$$Cu_2SO_4 + H_2O \rightarrow Cu_2O + H_2SO_4. \tag{8}$$

In acidic solutions, oxidation with oxygen present in the solution is the main reaction (Equation (9)):

$$Cu_2SO_4 + H_2SO_4 + 1/2 O_2 \rightarrow 2CuSO_4 + H_2O. \tag{9}$$

An insufficient amount of free sulfuric acid in solutions creates favorable conditions for the hydrolysis of iron ions.

$$2Fe_2(SO_4)_3 + 3H_2O = Fe_2(SO_4)_3 \cdot Fe_2O_3 + 3H_2SO_4, \tag{10}$$

$$Fe_2(SO_4)3Fe_2O_3 + 9H_2O = 4Fe(OH)_3 + 3H_2SO_4. \tag{11}$$

Theoretically, the copper cementation process terminates when the concentration of copper ions in the solution decreases to a certain value, at which the electrode potential for copper is equal the electrode potential for iron. The system achieves the equilibrium state. The concentration of copper ions in the solution at equilibrium can be calculated from the Nernst Equations (12) and (13) [6]:

$$E_{Fe^{2+}/Fe} = E^0_{Fe^{2+}/Fe} + \frac{RT}{zF} \ln a_{Fe^{2+}}, \tag{12}$$

$$E_{Cu^{2+}/Cu} = E^0_{Cu^{2+}/Cu} + \frac{RT}{zF} \ln a_{Cu^{2+}}, \tag{13}$$

If

$$E_{Fe^{2+}/Fe} = E_{Cu^{2+}/Cu'} \tag{14}$$

then

$$E^0_{Fe^{2+}/Fe} + \frac{RT}{zF} \ln a_{Fe^{2+}} = E^0_{Cu^{2+}/Cu} + \frac{RT}{zF} \ln a_{Cu^{2+}}, \tag{15}$$

where: E^0 is the standard electrode potential, a is ion activity, R is the gas constant, F is the Faraday constant, and z is the number of electrons participating in reaction.

The solution of this equation at 25 °C gives the following expression:

$$-0.44 + \frac{0.059}{2}\ln a_{Fe^{2+}} = 0.34 + \frac{0.059}{2}\ln a_{Cu^{2+}}, \qquad (16)$$

$$\frac{aCu^{2+}}{aFe^{2+}} = 1.3 \times 10^{-27}. \qquad (17)$$

At equilibrium, the concentration of copper ions in the solution is low ($C_{Cu}{}^{2+} = 1.3 \times 10^{-27}$), that is, cementation reaction can be considered as extending to the end. However, the thermodynamic equilibrium is not achieved due to kinetic difficulties. The mechanism of copper cementation can be represented by successive stages: delivery of ions to the cathode surface and removal of ions from the anode surface through a diffusion layer, discharge of copper ions to the cathode sections, and ionization of iron ions on the anodic sites (electrochemical conversion).

3. Materials and Methods

A high-speed cavitation-dispersion device (HSCD) (CORVUS Ltd., Riga, Latvia) and a paddle laboratory mixing device (LMD) EUROSTAR 40 digital (IKA, Staufen, Germany) were compared to determine the influence of mixing regimes and related parameters on the effectiveness of the copper extraction process. The duration of each test was 10 min. Samples were collected before and during the test after 1, 2, 3, 5, and 10 min. A total of 1500 mL of wastewater was used for every test.

Wastewater for treatment experiments was obtained from a European mining company. According to the non-disclosure agreement between the mining company and researchers, it is prohibited to reveal the source of wastewater. The chemical composition was provided by the supplier as listed in Table 1. The wastewater contained mainly copper ions in the form of $CuSO_4$ with pH = 2.95. Three different iron powders from Höganäs AB, Sweden were provided for cementation experiments: (1) iron particles collected from industrial dust filter (CMS) (Höganäs AB, Höganäs, Sweden); (2) water-atomized iron-based powder AHC100.29 (AHC); and (3) sponge-iron powder NC100.24 (NC). The particle size distribution and surface area of the powders are shown in Table 2.

The HSCD was used for Cu extraction experiments with the setup and principles demonstrated in authors' previous works as shown in Figure 1 [36–38]. As determined in the previous works [36–38], a linear speed over 24.13 m/s, which corresponds to 3847 rpm, is needed in order for the cavitation effect to appear in water media when HSCD is used. To ensure stable cavitation in all treated volumes, a rotor rotational frequency of 6000 rpm was applied. A reference laboratory mixer with an electric drive and a paddle-type sitter was used (Figure 2). Nylon syringe filters (pore size of 0.45 μm, diameter of 25 mm, supplied by Cole-Parmer, Hamburg, Germany) were used with the aim of avoiding the presence of Fe^0, Cu^0, and other solid particles, and to terminate cementation reactions in collected solutions. Each experiment was repeated four times with the deviation in measurement results less than 5%. Surface area was determined by the Brunauer–Emmett–Teller (BET) analysis, using the surface area analyzer Quadrasorb SI-KR/MP (Anton Paar GmbH, Berlin, Germany) with nitrogen adsorption/desorption at 77K.

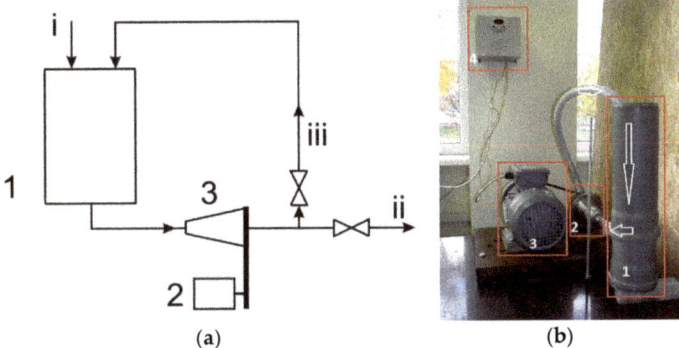

Figure 1. Scheme of the experimental setup for Cu^{2+}-containing wastewater treatment. Principal scheme (**a**) 1—container for suspension; 2—engine; 3—mixer-disperser; i—supply components to be mixed; ii—suspension output; iii—recycle stream. General view (**b**) 1—container for suspension; 2—mixer-disperser; 3—engine; 4—engine rpm control unit Arrows indicate suspension circulation directions.

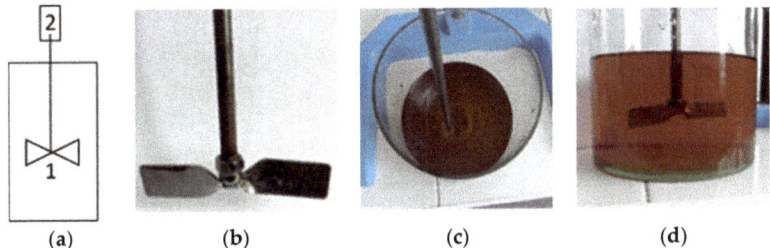

Figure 2. Experimental setup for control experiment of a propeller-type sitter. (**a**) Scheme of process volume with a propeller-type sitter (1), and a laboratory mixer with electric drive (2). (**b**) Propeller-type sitter view. (**c**) Sitter axis located asymmetrically to the vessel. (**d**) Sitter location in processed wastewater volume.

Table 1. Physicochemical parameters of the wastewater sample.

Parameter	Value	Unit
pH *	2.95	-
Copper (Cu) *	750 ± 10	mg/L
Lead (Pb)	0.250	mg/L
Sulfate ions (SO_4^{2-})	20.0	g/L
Cyanide ions (CN^-)	<0.050	mg/L
Chromium ions (Cr^{6+})	≥0.005	mg/L
Nickel (Ni)	≥4.000	mg/L
Mercury ions (Hg^+)	≥0.005	mg/L
Nitrate ions (NO^{3-})	<0.010	mg/L
Fe (II+III) *	6600 ± 100	mg/L
Concentration of other cations: Al^{3+}, Mg^{2+}, Zn^{2+}	<300	mg/L
* determinate by RTU laboratory.	-	-

Table 2. Physical properties of iron powders—CMS: iron particles collected from industrial dust filter; AHC100.29: water-atomized iron-based powder; NC100.24: sponge-iron powder.

Parameter	CMS	AHC 100.29	NC 100.24
Particle size >150 µm	0%	6%	1%
Particle size 150–45 µm	47%	70%	80%
Particle size <45 µm	53%	24%	19%
Specific surface area, m^2/g	0.67 ± 0.07	0.78 ± 0.1	0.73 ± 0.1

The stoichiometric amount of Fe^0 powder needed for cementation tests was calculated according to reaction (1) and the initial concentration of Cu^{2+}. It was found that 0.441 g/L of iron powder was necessary. On the other hand, taking into account simultaneous chemical reactions (2)–(4), four times the excess of the stoichiometric amount of iron powder was proposed and tested, corresponding to the ratio between the amount of iron powder and the solution volume of 1.764 g/L. Figure 3 shows a photograph of iron powder AHC 100.29 with sponge-like morphology. For microstructural characterization, optical imaging was carried out by the optical digital microscope VHX-2000 (Keyence Corporation, Osaka, Japan) equipped with lenses VH-Z20R/W and VH-Z500R/W. Metal concentrations in solutions were analyzed by the atomic absorption spectrometer GBC 932plus (GBC Scientific Equipment Ltd., Dandenong, Australia).

Figure 3. Picture of sponge-like surface structure of AHC 100.29.

4. Results and Discussion

As shown in Figures 4 and 5, cavitation dispersion provided more effective copper cementation from wastewater by Fe^0 powders than the traditional mixing method using propellers. Using the high-speed mixer-disperser (HSMD), the copper concentration in wastewater decreased rapidly in first five minutes and achieved the steady state in only seven minutes, when about 95.9% of copper was cemented. Increasing the cementation duration to ten minutes did not result in a significant increase in cementation efficiency. Although the copper concentrations remaining in wastewater when three types of Fe^0 powders were used varied slightly in the first 5 min of cementation, the final copper concentration after 10 min cementation with all three powders was almost identical, at about 31 mg/L. Therefore, it is more economical and ecological to use the powder from dust filters in practice. In contrast, the copper cementation using the conventional propeller sitter proceeded more slowly than the cementation using the HSMD and achieved only 53%–55% efficiency in the first five minutes of cementation for all three types of Fe^0 powder. After that, the cementation process slowed down, with cementation efficiency varying from 61.5% for AHC powders to 69.2% for NC powders after ten minutes. It was also observed that the temperature of wastewater during cementation using the HSMD increased to about 35 °C after 7 min, while the temperature of wastewater during cementation using the propeller sitter remained unchanged, at 18 °C. It was probable that the increase in cementation temperature also contributed to some increase in the copper cementation efficiency by the cavitation-dispersion method.

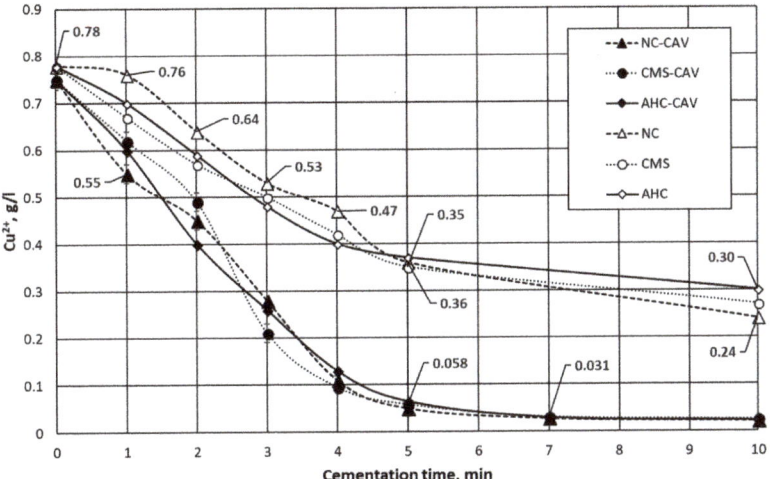

Figure 4. Dependence of Cu^{2+} concentration in the waste water, depending on the treatment time and used setup. NC-CAV: powder NC 100.24 with cavitation; CMS-CAV: powder CMS with cavitation; AHC-CAV: powder AHC 100.29 with cavitation; NC: powder NC 100.24 without cavitation; CMS: powder CMS without cavitation; AHC: powder AHC 100.29 without cavitation.

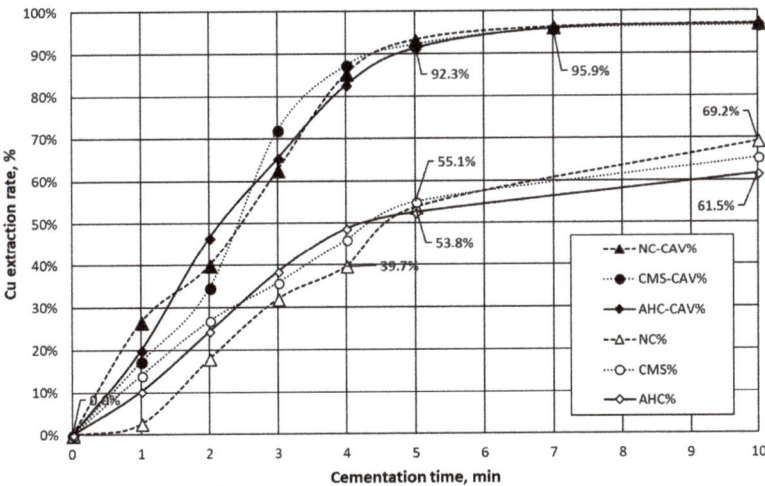

Figure 5. Dependence of the copper extraction rate, depending on the treatment time and setup used.

As described in Section 2, the reduction of Cu^{2+} to Cu^0 occurs on Fe^0 surface, which will be gradually covered by a layer of cemented Cu^0. During the first five minutes of cementation, the opened Fe^0 surface was quite large, so cementation proceeded quickly. After that, cementation slowed down due to decreased accessible Fe^0 surface on which Cu^{2+} ions can be cemented. This behavior was observed during cementation using all three types of Fe^0 powders under both non-cavitated and cavitated cementation conditions. As shown in Table 2, all three Fe^0 powders had similar specific surfaces, so in general, copper cementation courses using these Fe⁰ powders followed similar trends (Figures 4 and 5).

Figure 6 shows the original NC 100.24 Fe⁰ powders and their cemented particles obtained under the traditional mixing regime and the cavitation regime. Under the traditional mixing regime, Fe⁰ powders were covered by cemented Cu⁰. Some of them are agglomerated, probably by bridging of cemented Cu⁰, making the resulting Fe⁰ particles bigger than the original ones (Figure 6b). The cavitated samples consisted of some small particles with particle size ranging from 10 to 30 μm (Figure 6c). The majority of particles had particle size ranging from 50 to 100 μm, much smaller than the original particle size of 150 μm (Table 2). The formation of smaller particles and the reduction in particle size of the majority of original Fe⁰ particles were probably the results of the cavitation effects, which caused newly cemented Cu⁰ to rip off the Fe⁰ surface, creating small Cu⁰ particles and allowing additional cementation on the newly opened Fe⁰ surface. Together with increased diffusion of Cu^{2+} ions to Fe⁰ surfaces and/or through the pores of cemented Cu⁰ on Fe⁰ surfaces, the copper cementation process quickened during the cavitation regime. The other explanation for the formation of smaller particles in cavitated samples is that some of the Fe⁰ particles broke down during cavitation, leading to increased specific surface and increased copper cementation efficiency.

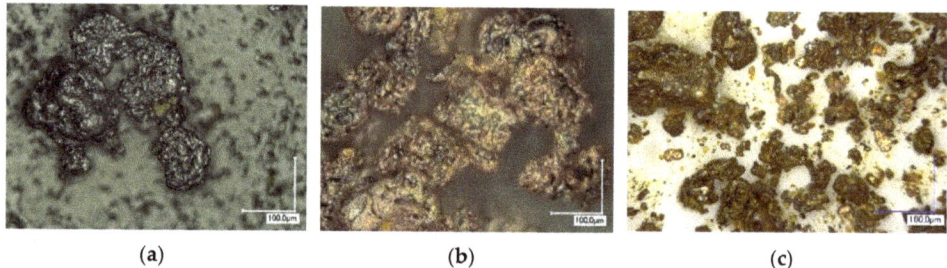

(a) (b) (c)

Figure 6. (**a**) NC 100.24 Fe powder particles before cementation; (**b**) NC 100.24 Fe powder particles after cementation at 10th minute, without cavitation treatment; (**c**) NC 100.24 Fe powder particles after cementation at 10th minute, with cavitation treatment.

The commercial cementation process usually operates the flows at 100–300 m³/h. In order to implement the studied cavitation cementation method, repetition of the cavitation treatment using tanks can be envisaged. Figure 7 shows the following proposed commercial application scheme: Cu^{2+} contaminated waste water and Fe⁰ powder are continuously fed in to a tank (1) with a paddle mixer (2), the mixture flows through the HSMD (3), during cavitation treatment, Fe⁰ particles are cemented with Cu⁰ (i), and then separated and de-agglomerated (ii). The set-up has a multiple number (n) of set tanks, mixers, and HSMD (4). At the end of the process, the suspension with Fe⁰ and Cu⁰ flows through a magnetic separator (5) for unreacted Fe⁰ separation, which is returned to the extraction process via the reactor (1) for mechanical separation by filtration or sedimentation (6) of Cu⁰. According to a common electrochemical Equation (18), it is also possible to extract Cd, In, Tl, Co, Ni, Sn, Pb, Bi, Cu, Hg, Ag, and Au metals.

$$N^{n+} + M^0 \rightarrow N^0 + M^{m+}, \tag{18}$$

where N^{n+} is the reduced metal ion (metal with a higher standard potential value), and M^0 is metal that carries cementation (metal with a lower standard potential value).

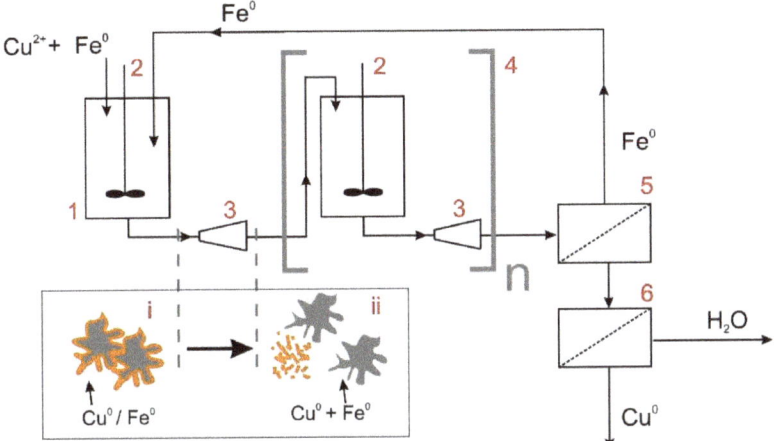

Figure 7. Principal scheme for commercial implementation for Cu extraction from mining waste water: 1—container for suspension; 2—paddle mixer; 3—mixer-disperser; 4—a set of tanks, mixers, and HSMD; 5—magnetic separator; 6—mechanical filter; i—Cu cemented on Fe particles before the cavitation; ii— cemented Cu and Fe particles after the cavitation; Arrows indicating flow directions.

5. Conclusions

Using a high-speed mixer-disperser (HSMD) as the mechanical cavitation source, it is possible to achieve copper extraction from mining wastewater with around 40% higher cementation efficiency, in a significantly faster time than by using the traditional mixing method. After seven minutes, over 95% of copper was cemented on Fe^0 powders by applying the HSMD. Morphology and iron powder origins do not play a significant role in metal extraction by the cementation process. The investigated cementation process improvement could be foreseen as a route for obtaining fine powders of copper and other metals (e.g., Cd, In, Tl, Co, Ni, Sn, Pb, Bi, Cu, Hg, Ag, and Au). It could also be used at the industrial scale for mining and metallurgical wastewater purification.

Author Contributions: A.S., J.B., V.M., A.P. and J.B. conceptualized and designed the experiments; A.S., J.B., J.O. and A.P. performed the experiments; A.S., J.B., V.M. and J.O. analyzed the data; A.P. contributed reagents, materials, analysis, and necessary tools; H.V., A.S. and J.O. prepared the original draft; H.V. and P.N. reviewed and completed the final manuscript.

Funding: This research was partly supported by InterOcean Metals Joint Organization, Szczecin, Poland under international grant No. 106 19 0063.

Acknowledgments: This research was supported by COST Action CA15102.

Conflicts of Interest: The authors declare no conflict of interest.

References

1. Al-Saydeh, S.A.; El-Naasa, M.H.; Zaidi, S.J. Copper removal from industrial wastewater: A comprehensive review. *J. Ind. Eng. Chem.* **2017**, *56*, 35–44. [CrossRef]
2. Trakal, L.; Šigut, R.; Šillerová, H.; Faturíková, D.; Komárek, M. Copper removal from aqueous solutions using biochar: Effect of chemical activation. *Arab. J. Chem.* **2014**, *7*, 43–52. [CrossRef]
3. Ruyters, S.; Salaets, P.; Oorts, K.; Smolders, E. Copper toxicity in soils under established vineyards in Europe: A survey. *Sci. Total Environ.* **2013**, *443*, 470–477. [CrossRef] [PubMed]
4. Zou, Y.; Wang, X.; Khan, A.; Wang, P.; Liu, Y.; Alsaedi, A.; Hayat, T.; Wang, X. Environmental Remediation and Application of Nanoscale Zero-Valent Iron and Its Composites for the Removal of Heavy Metal Ions: A Review. *Environ. Sci. Technol.* **2016**, *50*, 7290–7304. [CrossRef] [PubMed]

5. Kong, Z.; Li, X.; Tian, J.; Yang, J.; Sun, S. Comparative study on the adsorption capacity of raw and modified litchi pericarp for removing Cu(II) from solutions. *J. Environ. Manag.* **2014**, *134*, 109–116. [CrossRef] [PubMed]
6. Barakat, M.A. New trends in removing heavy metals from industrial wastewater. *Arab. J. Chem.* **2011**, *4*, 361–377. [CrossRef]
7. Babel, S.; Kurniawan, T.A. Various treatment technologies to remove arsenic and mercury from contaminated groundwater: An overview. In Proceedings of the First International Symposium on Southeast Asian Water Environment, Bangkok, Thailand, 24–25 October 2003; pp. 433–440.
8. Jaishankar, M.; Tsenten, T.; Anbalagan, N.; Mathew, B.B.; Beegegowda, K.N. Toxicity, mechanism and health effects of some heavy metals. *Interdiscip. Toxicol.* **2014**, *7*, 60–72. [CrossRef] [PubMed]
9. Gong, J.L.; Wang, X.Y.; Zeng, G.M.; Chen, L.; Deng, J.H.; Zhang, X.R.; Niu, Q.Y. Copper (II) removal by pectin–iron oxide magnetic nanocomposite adsorbent. *Chem. Eng. J.* **2012**, *185–186*, 100–107. [CrossRef]
10. Burakov, A.E.; Galunin, E.V.; Burakova, I.V.; Kucherova, A.; Agarwal, S.; Tkachev, A.G.; Gupta, V.K. Adsorption of heavy metals on conventional and nanostructured materials for wastewater treatment purposes: A review. *Ecotoxicol. Environ. Saf.* **2018**, *148*, 702–712. [CrossRef] [PubMed]
11. Aydın, H.; Bulut, Y.; Yerlikaya, Ç. Removal of copper (II) from aqueous solution by adsorption onto low-cost adsorbents. *J. Environ. Manag.* **2008**, *87*, 37–45. [CrossRef] [PubMed]
12. El-Ashtoukhy, E.S.; Amin, N.K.; Abdelwahab, O. Removal of lead (II) and copper (II) from aqueous solution using pomegranate peel as a new adsorbent. *Desalination* **2008**, *223*, 162–173. [CrossRef]
13. Femina Carolin, C.; Senthil Kumar, P.; Saravanan, A.; Janet Joshiba, G.; Naushad, M. Efficient techniques for the removal of toxic heavy metals from aquatic environment: A review. *J. Environ. Chem. Eng.* **2017**, *5*, 2782–2799. [CrossRef]
14. Konsowa, A.H. Intensification of the rate of heavy metal removal from wastewater by cementation. *Desalination* **2010**, *254*, 29–34. [CrossRef]
15. Ahmed, I.M.; El-Nadi, Y.A.; Daoud, J.A. Cementation of copper from spent copper-pickle sulfate solution by zinc ash. *Hydrometallurgy* **2011**, *110*, 62–66. [CrossRef]
16. Gross, F.; Baup, S.; Aurousseau, M. Copper cementation on zinc and iron mixtures: Part 1: Results on rotating disc electrode. *Hydrometallurgy* **2011**, *106*, 121–133. [CrossRef]
17. Nosier, S.A.; Alhamed, Y.A.; Alturaif, H.A. Enhancement of copper cementation using ceramic suspended solids under single phase flow. *Sep. Purif. Technol.* **2007**, *52*, 454–460. [CrossRef]
18. Dil, E.A.; Ghaedi, M.; Asfaram, A. The performance of nanorods materials as adsorbent for removal of azodyes and heavy ions: Application of ultrasound wave, optimization and modeling. *Ultrason. Sonochem.* **2017**, *34*, 792–802. [CrossRef] [PubMed]
19. Kanakaraju, D.; Ravichandar, S.; Chin Lim, Y. Combined effects of adsorption and photocatalysis by hybrid TiO_2/ZnO-calcium alginate beads for the removal of copper. *J. Environ. Sci.* **2017**, *55*, 214–223. [CrossRef] [PubMed]
20. Su, Y.N.; Lin, W.S.; Hou, C.H.; Den, W. Performance of integrated membrane filtration and electrodialysis processes for copper recovery from wafer polishing wastewater. *J. Water Proc. Eng.* **2014**, *4*, 149–158. [CrossRef]
21. Ferrer, O.; Gibert, O.; Cortina, J.L. Reverse osmosis membrane composition, structure and performance. *Water Res.* **2016**, *103*, 256–263. [CrossRef] [PubMed]
22. Xu, Y.C.; Wang, Z.X.; Cheng, X.Q.; Xiao, Y.C.; Shao, L. Positively charged nanofiltration membranes via economically mussel-substance-simulated co-deposition for textile wastewater treatment. *Chem. Eng. J.* **2016**, *303*, 555–564. [CrossRef]
23. Dong, Y.; Liu, J.; Sui, M.; Qu, Y.; Ambuchi, J.J.; Wang, H.; Feng, Y. A combined microbial desalination cell and electrodialysis system for copper-containing wastewater treatment and high-salinity-water desalination. *J. Hazard. Mater.* **2017**, *321*, 307–315. [CrossRef] [PubMed]
24. Satyro, S.; Marotta, R.; Clarizia, L.; Di Somma, I.; Vitiello, G.; Dezotti, M.; Pinto, G.; Dantas, R.F.; Andreozzi, R. Removal of EDDS and copper from waters by TiO_2 photocatalysis under simulated UV–solar conditions. *Chem. Eng. J.* **2014**, *251*, 257–268. [CrossRef]
25. Ahmed, M.J.K.; Ahmaruzzaman, M. A review on potential usage of industrial waste materials for binding heavy metal ions from aqueous solutions. *J. Water Process Eng.* **2016**, *10*, 39–47. [CrossRef]

26. Ruihua, L.; Lin, Z.; Tao, T.; Bo, L. Phosphorus removal performance of acid mine drainage from wastewater. *J. Hazard. Mater.* **2011**, *190*, 669–676. [CrossRef] [PubMed]
27. Nguyen, T.A.H.; Ngo, H.H.; Guo, W.S.; Zhang, J.; Liang, S.; Yue, Q.Y.; Li, Q.; Nguyen, T.V. Applicability of agricultural waste and by-products for adsorptive removal of heavy metals from wastewater. *Bioresour. Technol.* **2013**, *148*, 574–585. [CrossRef] [PubMed]
28. Farooq, U.; Kozinski, J.A.; Khan, M.A.; Athar, M. Biosorption of heavy metal ions using wheat based biosorbents—A review of the recent literature. *Bioresour. Technol.* **2010**, *101*, 5043–5053. [CrossRef] [PubMed]
29. Khan, I.; Abbas, A.; Al-Amer, A.M.; Laoui, T.; Al-Marri, M.J.; Nasser, M.S.; Khraisheh, M.; Atieh, M.A. Heavy metal removal from aqueous solution by advanced carbon nanotubes: Critical review of adsorption applications. *Sep. Purif. Technol.* **2016**, *157*, 141–161. [CrossRef]
30. Aktas, S. Cementation of rhodium from waste chloride solutions using copper powder. *Int. J. Miner. Process* **2012**, *114–117*, 100–105. [CrossRef]
31. Aktas, S. Rhodium recovery from rhodium-containing waste rinsing water via cementation using zinc powder. *Hydrometallurgy* **2001**, *106*, 71–75. [CrossRef]
32. Wu, L.K.; Xia, J.; Zhang, Y.F.; Li, Y.Y.; Cao, Z.H.; Zheng, G.Q. Effective cementation and removal of arsenic with copper powder in a hydrochloric acid system. *RSC Adv.* **2016**, *6*, 70832–70841. [CrossRef]
33. Demirkiran, N.; Kunkul, A. Recovering of copper with metallic aluminum. *Trans. Nonferrous Met. Soc. China* **2001**, *21*, 2778–2782. [CrossRef]
34. El-Ashtoukhy, E.-S.Z.; Abdel-Aziz, M.H. Removal of copper from aqueous solutions by cementation in a bubble column reactor fitted with horizontal screens. *Int. J. Min. Process.* **2013**, *121*, 65–69. [CrossRef]
35. Abdel-Aziz, M.H. Production of copper powder from wastewater containing $CuSO_4$ and alcoholic additives in a modified stirred tank reactor by cementation. *Hydrometallurgy* **2011**, *1091*, 61–167. [CrossRef]
36. Polyakov, A.; Polyakova, E.; Lazarenko, L. Mixer-Homogenizer. Latvia Patent LV13592 (B), 20 September 2007.
37. Polyakov, A.; Polyakova, E. Hydrodynamic cavitation homogenizer. Latvia Patent LV15143 (A), 20 July 2016.
38. Polyakov, A.; Mironovs, V.; Shishkin, A.; Baronins, J. Preparation of Coal-Water Slurries Using a High—Speed Mixer—Disperser. In Proceedings of the 4th International Scientific Conference Civil Engineering' 13, Jeglava, Latvia, 16–17 May 2013; pp. 77–81.

© 2018 by the authors. Licensee MDPI, Basel, Switzerland. This article is an open access article distributed under the terms and conditions of the Creative Commons Attribution (CC BY) license (http://creativecommons.org/licenses/by/4.0/).

Article

Recycling Decisions in 2020, 2030, and 2040—When Can Substantial NdFeB Extraction be Expected in the EU?

Maximilian V. Reimer [1,*], Heike Y. Schenk-Mathes [1], Matthias F. Hoffmann [2] and Tobias Elwert [2]

1. Department of Business Administration and Environmental Management, Institute of Management and Economics, Clausthal University of Technology, Julius-Albert-Straße 2, 38678 Clausthal-Zellerfeld, Germany; heike.schenk-mathes@tu-clausthal.de
2. Department of Mineral and Waste Processing, Institute of Mineral and Waste Processing, Waste Disposal and Geomechanics, Clausthal University of Technology, Walther-Nernst-Straße 9, 38678 Clausthal-Zellerfeld, Germany; matthias.hoffmann@tu-clausthal.de (M.F.H.); tobias.elwert@tu-clausthal.de (T.E.)
* Correspondence: maximilian.reimer@tu-clausthal.de; Tel.: +49-5323-72-7618

Received: 30 September 2018; Accepted: 17 October 2018; Published: 24 October 2018

Abstract: In recent years, China's dominant role in the rare earth market and the associated impacts have strengthened the interest in the recovery of rare earth elements (REE) from secondary resources. Therefore, numerous research activities have been initiated aiming at the recovery of REEs from different types of waste streams, which includes, inter alia, neodymium-iron-boron (NdFeB) magnets. Although several research projects have successfully been completed, most experts do not expect an industrial implementation in Europe within the next years. This article analyses the reasons for this situation, addressing the availability of sufficient amounts of NdFeB wastes, the technology readiness level of the developed processes in Europe, as well as the economic aspects. Based on these analyses, an estimation of a realistic timeframe for the industrial implementation of NdFeB recycling in Europe is deduced and critically discussed.

Keywords: NdFeB magnets; rare earth elements; recycling; recycling potential; neodymium; dysprosium

1. Introduction

Since the 1980s, China has become the dominant producer of rare earth elements (REE), reaching market shares of up to 97% in 2010. In 2017, China mined about 81% of the world production (approx. 130,000 t/year), and is the only country that operates the full production chain from ores to REE metals at the industrial scale. Despite its monopolistic position, the REE prices remained stable on relatively low levels before 2010 due to the uncontrolled and unregulated growth of the Chinese REE industry, which resulted in low production costs but caused tremendous environmental damage. Therefore, the security of the supply was not questioned and interest in alternative REE sources was generally low. This situation changed drastically when China introduced export restrictions and duties for REEs in 2010, leading to massive price increases around 2011 and serious concerns about the security of the supply outside China. Since then, prices for rare earth compounds and metals declined significantly due to excess supply. Furthermore, China abandoned the export restrictions and duties in 2015 after the World Trade Organization upheld a ruling in favor of the United States, the European Union, and Japan's claims that China violated trade rules. Nevertheless, there are still mid- and long-term supply risks, as China will presumably stay the dominant REE producer in the next decade [1–3].

In answer to China's policy, many projects regarding primary production, recycling, substitution, as well as associated topics, such as the material flow analysis of REEs, have been initiated in recent years [4–7]. Regarding primary production, 442 projects were listed in December 2012 [4]. However, less than five have reached industrial scale, the most notable being Lynas Corporation Ltd., which has mining operations in Australia and refining operations in Malaysia [1]. As there are no competitive REE deposits in Europe [4], the European Union (EU) focuses on the recovery of REEs from secondary resources and substitution. Although first research and development projects have already been completed, only very few entered, or will enter, industrial scale [5–7].

This article analyses the reasons for this development in the case of neodymium-iron-boron (NdFeB) magnets, which are the main application for the REEs neodymium (Nd) and dysprosium (Dy) and represent a major share of the REE market by volume (22%) as well as by value (37%) [8]. NdFeB magnets are currently the strongest permanent magnets and are used in various applications (see Figure 1). Due to the high number of applications and expected high growth rates, mainly driven by electromobility and wind turbines, NdFeB magnets presumably bear a high recycling potential [9,10].

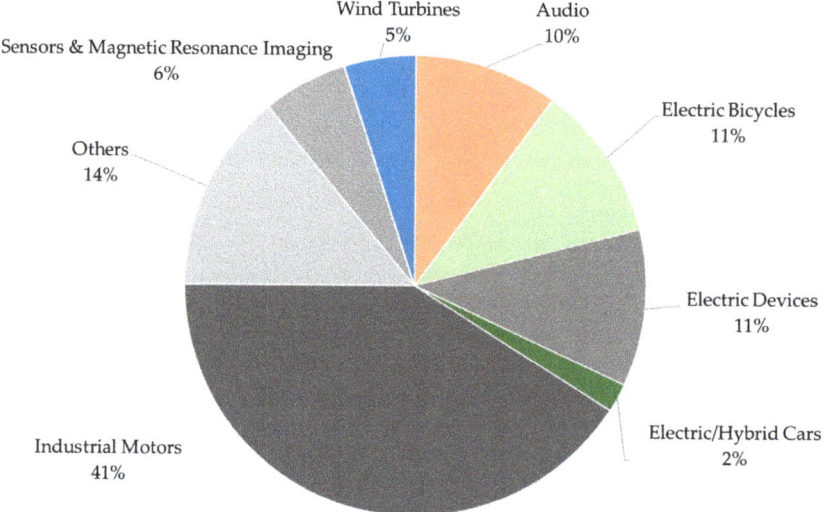

Figure 1. Shares of the different applications in the global neodymium-iron-boron (NdFeB) magnet market in the year 2012 [11].

After providing background knowledge about NdFeB magnets (Section 2), in this study, an attempt was made to enlighten this contradictory situation by using a holistic approach, starting with an estimation of the theoretical European recycling potential until 2040 (Section 3). Based on this, a more realistic scenario for the availability of NdFeB scrap is deduced, considering technological as well as economic aspects (Section 4). In Section 5, the implications are critically discussed.

2. NdFeB Magnets

Sintered NdFeB magnets are the strongest permanent magnets available today. Global production in 2014 was 112,000 t, of which about 88% were produced in China, followed by Japan. Europe and other regions have only minor market shares [12].

The excellent magnetic properties can be traced back to the strongly magnetic matrix phase $Nd_2Fe_{14}B$, featuring high saturation polarization and high magnetic anisotropy. Because of the low Curie temperature and low corrosion resistance of pure $Nd_2Fe_{14}B$, the properties are usually enhanced by alloying other REEs and cobalt. Typically, the alloys contain between 60–70 wt% iron (Fe), 28–35 wt% REEs (Pr, Nd, Tb, and Dy), 1–2 wt% boron (B), and 0–4 wt% cobalt (Co). The benefit of adding Dy in place of Nd is that it improves coercivity and therefore temperature tolerance. At the highest end of possible operating temperatures (approx. 200 °C), NdFeB magnets contain up to 10 wt% Dy. Terbium (Tb) can perform in a similar function but is rarely used due to its high price. Praseodymium (Pr) can directly substitute Nd to some extent without a severe impact on the magnetic properties. Co is commonly added to improve the corrosion resistance. For further improvement of the corrosion resistance, the magnets are phosphated or coated with organic or metallic coatings [6].

The most important trend in research, development, and production of sintered NdFeB magnets aims at an increase in coercivity using less Dy for economic reasons, as well as minimization of the related remanence losses. The most promising approaches to achieve this target are a further reduction of the grain size and the grain boundary diffusion process. Though the first approach has not been industrially implemented so far, the grain boundary diffusion process was introduced industrially some years ago [13].

3. Theoretical Recycling Potential of NdFeB Magnets in the European Union until 2040

In this section, the theoretical recycling potential of NdFeB magnets from the main applications, being electric vehicles (propulsion motors), auxiliary motors in passenger cars (electric and with internal combustion engine), electric two-wheeled vehicles, industrial motors, wind turbines, magnetic resonance imaging (MRI), hard disk drives (HDD), and audio devices, will be estimated for the years 2018–2040. As the relative share of other applications (currently 14%) is supposed to decrease due to expected high growth rates of electromobility and wind power [10], they are excluded.

The theoretical recycling potential was defined as the total amount of NdFeB magnets from end-of-life applications excluding post-production wastes. Post-production wastes were not considered as their amount of approx. 60 t/year [14] was low in comparison to the potential from post-consumer wastes (see Section 3.4) and was not expected to increase in the EU. To establish a benchmark, in this section, it was assumed that there were no losses through export and that 100% of the products were recycled after their life cycle, with a technical recycling efficiency of 100%. In this way, the theoretical recycling potential of NdFeB magnets could be assessed. More realistic recycling rates were later assumed in Section 4 for a direct comparison. The period under review was chosen up to the year 2040 to take into account, on the one hand, the long life cycle of major applications and, on the other hand, to have an acceptable forecast accuracy. A possible substitution of NdFeB magnets was not considered as alternative technologies are often less efficient (e.g., permanent magnet-free propulsion motors for electric vehicles), and extreme price increases of NdFeB magnets, which would favor substitution, are unlikely. Furthermore, a replacement of NdFeB by new materials is not expected within the next years. Therefore, the introduction of a new magnet material would only significantly affect the market after our considered time horizon due to the long lifespan of most considered products. Regarding the magnet composition, a simplified REE composition of 32 wt% Nd and Dy, 67 wt% Fe, and 1 wt% B was assumed. As Pr and Tb are minor constituents and substitutes for the aforementioned elements, they were not considered.

The most important assumptions for the calculations for each application are explained in Section 3.1. In Section 3.2, the sales data estimation is presented. These results are used to calculate the return flows in Section 3.3.

3.1. Key Assumptions for the Considered Applications

In Table 1, assumptions for average magnet content, percentage of application containing NdFeB magnets, Nd/Dy ratio, average lifespan, and related standard deviation are summarized. Further details are given in the following.

Table 1. Literature based assumptions.

Application	Average Magnet Content		Nd/Dy [%] [3]	μ_L; σ_L [a] [4]	Sales Data Sources
Electric Vehicles	2500 g	[15]	27/5	15; 3.75	[16–19]
Hybrid Electric Vehicles	1500 g	[15]	27/5	15; 3.75	[16–19]
Auxiliary Vehicle Motors	175 g	[9]	29/3	15; 3.75	[20]
Electric Bikes	270 g	[21]	29/3	5; 1.25	[22–24]
Industrial Motors [1]	-		28/4	-	[25,26]
Wind Turbines, Direct Drive [2]	650 kg/MW	[11]	28/4	22; 5.50	[27–32]
Wind Turbines, Hybrid Drive [2]	160 kg/MW	[11]	28/4	22; 5.50	[27–32]
Magnetic Resonance Imaging	2.5 t	[33]	31/1	12; 3.00	[11,34]
Hard Disk Drives	5.67 g	[21]	32/0	6; 1.50	[26,35,36]
Audio Devices [1]	-		32/0	-	[34]

[1] Available data was already given in tons of NdFeB. [2] Only wind turbines with integrated NdFeB magnets were considered. [3] Adopted from [34]. [4] Adopted from [37].

The product segment of electric vehicles is grouped into hybrid electric vehicles (HEV) and electric vehicles (EV). HEVs include mild, full, and plug-in hybrid electric vehicles. Whereas HEVs are typically equipped with permanent magnet motors, due to the limited space, some car manufacturers prefer magnet-free motors for EVs despite lower power densities and efficiencies. As propulsion motors are operated at elevated temperatures, comparatively high average Dy contents were assumed [15]. At present, even higher Dy concentrations up to 10% are observed, but these concentrations are expected to decrease (see Section 2).

All passenger vehicles contain auxiliary motors, which are partly equipped with NdFeB magnets. The magnet content varies. For this study, an average magnet content of 175 g/vehicle was assumed based on [9].

In the market of electric two-wheeled vehicles, a distinction is made between pedelecs, electric bicycles (e-bikes), electric scooters, and electric motor bicycles. Pedelecs require pedalling to activate the motor. In contrast, e-bikes can drive solely electric without pedalling. It is important to note that the term e-bike is often synonymously used for pedelecs, e-bikes, and e-scooters in literature [38]. In this paper, only pedelecs and e-bikes were considered and summarized in the following under the term e-bikes. Electric scooters and electric motor bicycles represent niche markets in the EU and were therefore neglected. According to [39], since 2015 all electric bicycles are equipped with permanent magnet motors. In 2007, the share was 47.1%. For the calculations, a linear increase between 2007 and 2015 was assumed.

The main application of NdFeB magnets is in industrial motors (see Figure 1). They exist in a wide range of sizes and power classes, and are used, for example, in conveyers, pumps, and robots. The data and assumptions were mainly based on a study conducted in Germany by the Institute for Applied Ecology [25].

Currently, three types of wind turbines are produced: Traditional drive train, direct drive, and hybrid systems. Only the last two designs contain NdFeB magnets. Direct drive systems require less maintenance than the other systems as they operate without gears. Therefore, they are the preferred technology for the offshore sector [40]. In contrast to most other applications, a detailed database of installed on- and offshore wind turbines in Europe is available and enabled a direct assignment of the respective magnet content.

According to field strength, magnetic resonance imaging (MRI) devices can be classified into low-, mid-, and high-field systems. Only low-field systems contain NdFeB magnets, whereas the other systems use electromagnets. The global compound annual growth rate is estimated at 3% including the high-growth markets, India and China [41]. As this study focusses on the EU, a lower annual growth rate of 2% was assumed.

While e-bikes, electric vehicles, and wind turbines represent fast growing markets, the market share of hard disk drives (HDD) is declining [36]. They are continuously replaced by faster solid-state-drives

(SSD), which do not contain magnets. The HDDs are divided into 2.5″ and 3.5″ drives with global market shares of 40% and 60%, respectively [35].

3.2. Methodology for the Prediction of Sales Data

The general approach to predict the future sales in the different industrial branches is to extrapolate the sales trends from a combination of past sales data and available forecasts.

As an example, the prediction of future HDD sales is outlined. Figure 2 shows exemplarily our predicted sales data for HDDs for the EU market. The data was based on past sales and a prediction from a worldwide study published by Statista [36]. We assumed a proportional relationship between sales and gross domestic product (GDP) in a region, allowing the estimate of sales on the EU market from the global market. Utilizing annual GDP data [26], we derived the annual GDP_{EU}/GDP_{World}-ratio to receive the portion of HDD sales in the EU market. For example, in 2016 the EU accounted for approx. 21% of the global GDP, leading to the assumption that 90.4 million out of the 424.1 million HDDs were sold in this market. In Figure 2, the resulting sales data is depicted (blue diamonds).

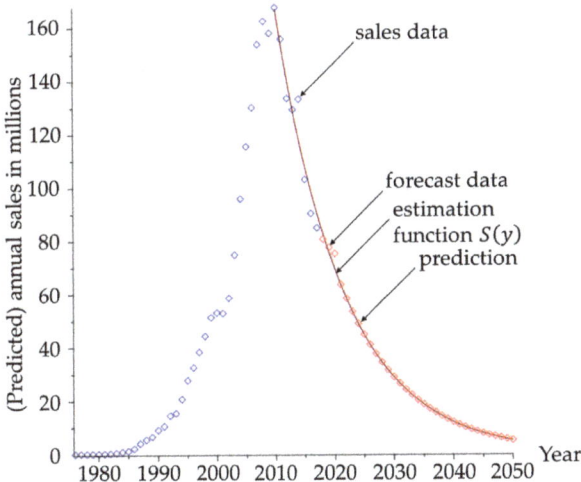

Figure 2. Sales data (blue diamonds), best-fit function (red curve), and forecast (red diamonds) for hard disk drives (HDDs).

It is clear that due to the technical advantages of SSDs over HDDs, the latter experienced a declining market share starting from around the year 2010. To capture this trend, we used a multinomial least-square fit with an exponential estimation function to predict the sales (S) in the year (y):

$$S(y) = e^{a+b \cdot y}, y > 2010 \qquad (1)$$

It can be expected [42] that HDD sales will decline and slowly approach zero as customers switch to the new SSD technology. The function given in Formula (1) was therefore chosen for its asymptotical properties. It is well known that an exponential function of this type can be linearized to find the unknown parameters a and b by ordinary least squares. The best-fit parameters were $\hat{a} = 179.588$ (t-Test, $p < 0.000$) and $\hat{b} = -0.087$ (t-Test, $p < 0.000$), and resulted in an adjusted R-squared value of 0.9998. In Figure 2, the consequential estimation function (red curve) and the corresponding sales values (red diamonds) are depicted.

If the trend for other applications was not clearly asymptotical, polynomial estimation functions were more suitable for complex trends and were selected. A similar approach was used for other applications in the case of incomplete sales data to close gaps via interpolation. Whenever forecasts on sales were available, the data points were incorporated into the predictions.

By applying this method, we were able to estimate annual sales for the time span until 2040, which was relevant to our return flow calculations. It should be noted that a possible increase in uncertainty for sales estimates in the distant future is unproblematic because the corresponding return flows will not be relevant for the considered time span.

3.3. Methodology for the Calculation of Return Flows

To illustrate the calculation of return flows, we utilized the following example. Sales data for the years 2015–2018 were considered: In the year 2015, sales amounted to 1000 units, in the year 2016, to 3000 and, after a year without sales, 5000 units were sold in the year 2018.

We assumed the lifetime to be normally distributed with mean $\mu_L = 5$ and standard deviation $\sigma_L = 1.25$. Therefore, the return year X_i of unit i was a normally distributed random variable $X_i \sim \mathcal{N}(y_{S,i} + 5, 1.25)$, where $y_{S,i}$ was the sales year of that unit i. In general, the probability density function was given by Formula (2):

$$f(y|y_{S,i} + \mu_L, \sigma_L) = \frac{1}{\sqrt{2\pi\sigma_L^2}} \cdot e^{-\frac{(y-y_{S,i}-\mu_L)^2}{2\sigma_L^2}} \qquad (2)$$

The three distinct probability density functions for the units 1 to 1000 with sales year $y_{S,i} = 2015$, the units 1001 to 4000 with $y_{S,i} = 2016$, and the units 4001 to 9000 with $y_{S,i} = 2018$ are shown in Figure 3. As we can see from Figure 3, eventual inaccuracies, caused by a negative lifetime, have a probability close to 0%. With the expected lifetimes, considered in the real data set, being larger than $\mu_L = 5$, the probability was even lower than in this example.

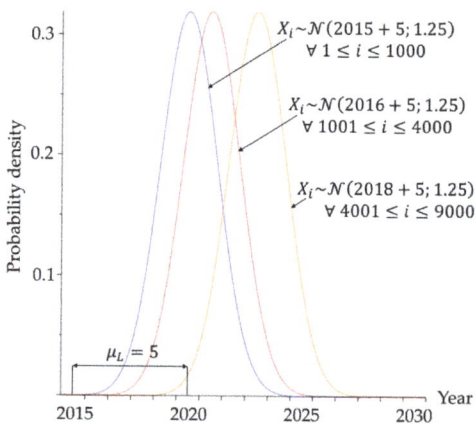

Figure 3. Probability density functions for the return year of units, which were sold in the years 2015, 2016, and 2018, respectively.

The expected returns $R(y_R)$ in a specific year y_R (i.e., returns between y_R and $y_R + 1$) were then given by Formula (3):

$$\mathbb{E}[R(y_R)] = \int_{y_R}^{y_R+1} \sum_i f(y|y_{S,i} + \mu_L, \sigma_L)\, dy \qquad (3)$$

Following the example, we expect $R(2022) = 1929.38$ units to return in the year 2022. A simulation of the exemplary sales data can illustrate the result. In Figure 4, we see the absolute returns obtained from the above probability densities. The blue bars show the returns of the 1000 units that where produced in 2015, the red bars the returns of the 3000 units produced in 2016, and the orange bars the returns of the 5000 units that were produced in the year 2018. In 2022, the returns amounted to 83 + 704 + 1145 units, which were produced in the years 2015, 2016, and 2018, respectively. The resulting 1932 units were close to the theoretical value 1929.38 and we could expect a convergence to the latter value when averaging multiple simulations.

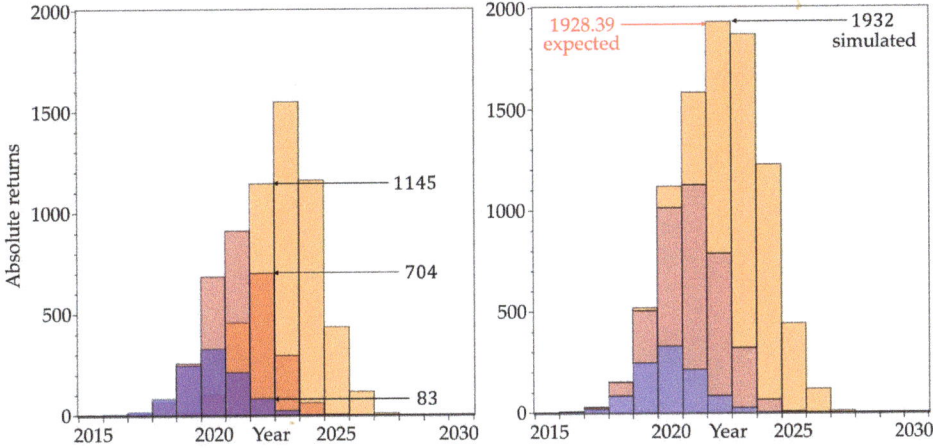

Figure 4. Simulated absolute return flows superimposed (**left**) and stacked (**right**). For the year 2022, the simulation (one run) provides a value of 1932 units. The exact expected return derived from the calculation for this given year was 1928.39 units.

This methodology was applied to the real sales data we derived in the previous section for all sources of NdFeB magnets. The usual timeframe of input data was 1995–2040 and the returns were calculated for the years 2018–2040.

3.4. Results

By applying this stochastic approach to the sales data from Section 3.2, we obtained the expected unit returns for the desired years. Both are shown in Figure 5. Depending on the average lifespan, the sales trends were only depicted for their relevant time spans. For example, in case of wind turbines, the average lifespan was $\mu_L = 22$ years. Consequently, only sales data until around 2025 will be impactful for the return flow estimates.

Considering the magnet content per unit (see Table 1), we obtained the resulting amounts of magnets per application in tons per year (Figure 6). It can be seen that the annual theoretical recycling potential increases only slightly in the next 10 years. In this first period, auxiliary vehicle motors, magnet resonance imaging, and electronic devices dominate. From 2028 on, a strong increase can be expected, which is mainly caused by the expected market penetration of (H)EVs and e-bikes. The share of wind turbines will increase from 2030, but remains comparatively low until 2040. The total theoretical recycling potential from 2016–2040 is about 233,000 t of NdFeB, corresponding to 66,600 t Nd and 7900 t Dy.

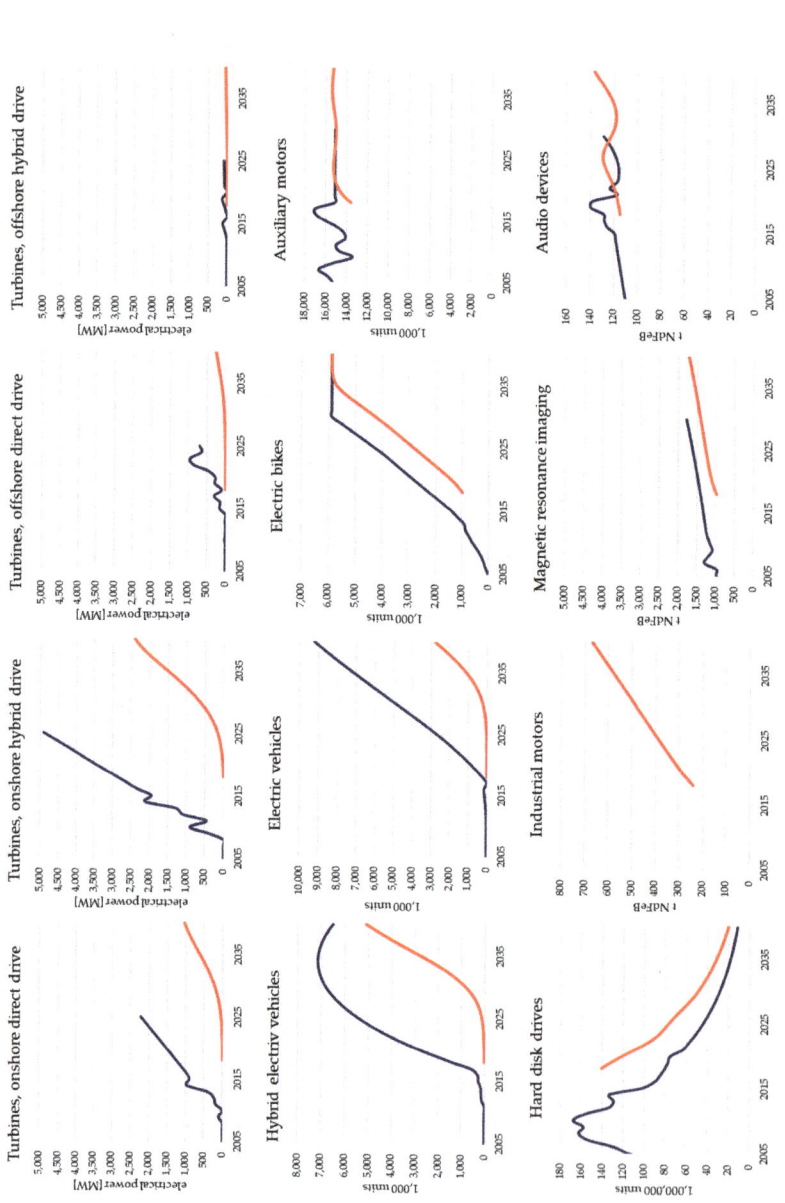

Figure 5. Sales (blue) and returns (red) by application. Y-axes units vary whilst x-axes depict year dates. The relevant range of sales trends is shorter, if the associated average lifespan is high (e.g., wind turbines).

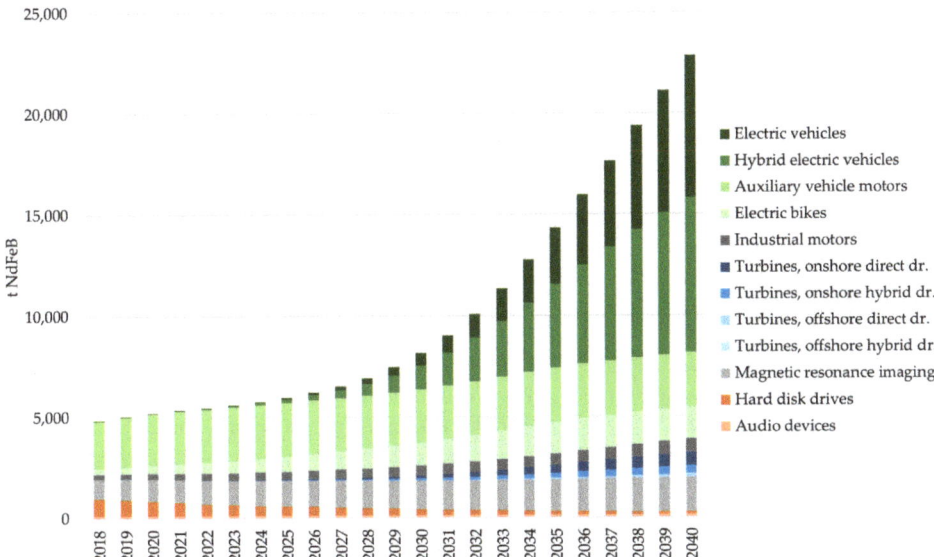

Figure 6. Potential return flows by application in tons of NdFeB.

4. Estimation of the Practical Recycling Potential in the European Union until 2040

The recycling chain of NdFeB magnets consists of the steps: Collection, magnet extraction, and recycling of the extracted magnets, and was recently reviewed by [8]. According to this review, currently, no recycling of NdFeB wastes takes place. However, our own market studies, in collaboration with scrap dealers and electronic scrap recycling companies, revealed that in today's industrial practice, collection of post-production wastes and limited extraction of NdFeB magnets from different applications takes place in the EU [14,43,44]. The magnet scraps are exported to China and Japan for metallurgical recycling. At present, approximately 150 t are exported annually, whereof about 40% originate from magnet production. Whereas the amount of production scrap has been relatively constant in recent years, increasing amounts of scrap from other applications, such as HDDs, MRI devices, and wind turbines, are observed [14,43].

The low traded amount of NdFeB wastes in comparison to the theoretical potential indicates that an extraction is not economic in most cases. The economic feasibility of the magnet extraction depends on the market prices for NdFeB scraps, which are strongly connected to the REE prices, and the extraction costs. The scrap prices fluctuated between 4 and 15 US$/kg in the last years [43].

For the recycling of extracted magnets, different approaches are currently under development in Europe, which can be classified into direct reuse, reprocessing of the alloys, and raw material recovery, all bearing a series of advantages and disadvantages. At present, none of the approaches is developed beyond pilot scale [5,8,15].

Theoretically, a direct reuse of the magnets would be economically and environmentally the most favourable way of recycling, due to the low energy demand as well as consumption of auxiliary and operating materials. However, in practice, reuse cannot be expected due to progress of development regarding NdFeB magnets and applications, difficulties in the non-destructive extraction of the brittle magnets, and cleaning without compromising the dimensions [5,8,15].

Prerequisites for reprocessing of the alloy are that magnet scraps are of known and homogeneous composition and a thorough removal of any impurities, such as glue residues and coatings, is undertaken, as the magnetic properties deteriorate even at low impurity concentrations (ppm range). For the actual reprocessing of the magnetic alloys, several processes are under investigation, such as re-melting of the alloy, purely mechanical comminution, and comminution after hydrogen decrepitation. All these options allow feeding of the cleaned magnet scrap into the established production chain for NdFeB magnets. The main drawback of these routes is that contamination with trace impurities cannot be prevented, which causes a deterioration of the magnetic properties. Therefore, reprocessing of alloys cannot be considered a practical way for the production of high-end magnets, a market segment in which European producers are active [5,8,15].

For raw material recovery, various approaches and processes have been developed in recent years, which can be classified into gas-phase extraction, pyrometallurgical methods, and hydrometallurgical methods. Despite disadvantages such as high consumption of chemicals and the production of wastewater, hydrometallurgical processes can be regarded as the most promising way. Reasons for this are their flexibility, with respect to the chemical composition of the wastes, impurity removal, and their ability to treat metallic and oxidized NdFeB wastes (e.g., grinding sludges from magnet production) [5,8,15].

Within the German research project "Recycling of components and strategic metals of electric drive motors", a life cycle assessment (LCA) was conducted for reuse, reprocessing of the alloy by comminution after hydrogen decrepitation, and a hydrometallurgical process route. The LCA was carried out according to ISO 14040/44 based on small pilot-scale data and reviewed by an independent external expert. The LCA revealed that all these recycling options show clear environmental advantages in comparison to the primary production of neodymium and dysprosium [45].

4.1. Assumptions for the Practical Recycling Scenario

In the following, the assumptions for the realistic recycling scenario are explained for each application based on literature review, interviews with market participants, and our own investigations.

For electric motors, semi-automated extraction technologies for magnets from rotors were developed within a German research project, which can, in principal, be transferred to motors from other applications [15]. However, the scale-up to industrial scale will require at least several years. In case of electric motors, it was assumed that (partly) automated extraction technologies will be available from 2026 onwards at an industrial scale. Therefore, no recycling will take place before 2026 from electric motors. As a European-wide implementation of the extraction technology will require at least several years, a linearly increasing magnet extraction rate from 0% in 2025 to 50% in 2040 was assumed.

For (H)EVs a European recycling rate of 50% was adopted from [46] for all end-of-life vehicles. It was assumed that only traction motors will be extracted from vehicles and dismantled for specific treatment, whereas magnets from auxiliary motors will be lost in the shredding process. An economic assessment of the recycling of propulsion motors concluded that a magnet extraction is economically feasible [15].

For e-bikes, a significantly lower recycling rate of only 5% was estimated. Due to the lack of data on the market for second hand bikes, unknown disposal routes, and the young e-bike market, this rate is a noteworthy uncertainty factor in the calculations. From an economic point of view, magnet extraction is currently uneconomic considering German labour costs, but might be feasible in Eastern and Southern Europe [21].

In case of industrial motors, a recycling rate of 10% was assumed due to high export rate of second-hand motors to non-European countries [25]. Furthermore, the profitability strongly depends on the motor size. Therefore, only mid- and large-size motors were of interest.

In case of wind turbines, it was assumed that 90% of all wind turbines are recycled after their use in the EU, with a magnet extraction efficiency of 90% resulting in an overall rate of 81% from the theoretical potential. The high rates seem to be realistic as the owners are responsible for the dismantling of the turbines, reuse of the turbines in other wind parks after more than 20 years cannot be expected, and the high amounts of large magnets represent a significant monetary value.

Little information is available on the recycling of magnetic resonance imaging devices. As they occasionally appear on the scrap market [43], a recycling rate of 5% was assumed. Like industrial motors, many second-hand devices are probably exported to developing countries. The extraction of magnets from MRI devices requires demagnetization of the magnets to remove them from the steel frame, but has economic potential due to the high amount of magnets (2–3 t) per unit.

For electronic products, it was assumed that only magnets from hard disk drives are extracted to a certain extent, which already takes place today in sheltered workshops and by prison inmates in Germany, due to the lower labor costs in comparison to the regular labor market [44]. An extraction using regular workers is uneconomic [21]. An extraction of magnets from audio devices was neglected, as they do not appear on the scrap market due to the prohibitive extraction costs [21]. The extraction rate for HDDs was estimated to be 2% of the theoretical potential in the EU. According to our market research, the current extraction rate in Germany is approx. 5%, but this rate cannot be projected to the EU due to the lower recycling standards, especially in Eastern and Southern Europe.

4.2. Possible Scenario for the Availability of NdFeB Scrap in the EU

Based on the theoretical recycling potential and the assumptions in Section 4.1, we obtained the return flows depicted in Figure 7. The general trend was comparable with Figure 6, but on a much lower level. The scenario predicted an annual recycling rate increase from 1% to 21% of the theoretical potential. The overall recycling potential from 2018–2040 was about 25,700 t of NdFeB (see Table 2), corresponding to 7100 t Nd and 1100 t Dy. In comparison to the global consumption of these elements for NdFeB magnets, the impact on the REE market will be low. According to [10], the global demand for Nd/Pr and Dy/Tb was 28,900 t and 2000 t, respectively, in 2013, and is expected to increase to an annual demand of 62,400 t Nd/Pr and 7200 t Dy/Tb until 2035.

The highest recycling potential will stem from mobility and wind turbines, which shows strong growth after 2028. Before 2028, the amounts do not exceed 250 t/year, 1,000 t will be reached around 2033, which is considered the minimum amount for an industrial recycling plant [47]. The predicted return flows in 2018 are in accordance with the information we received from interviews with scrap dealers, who estimated the market volume to be 60–80 t without production wastes from magnet producers.

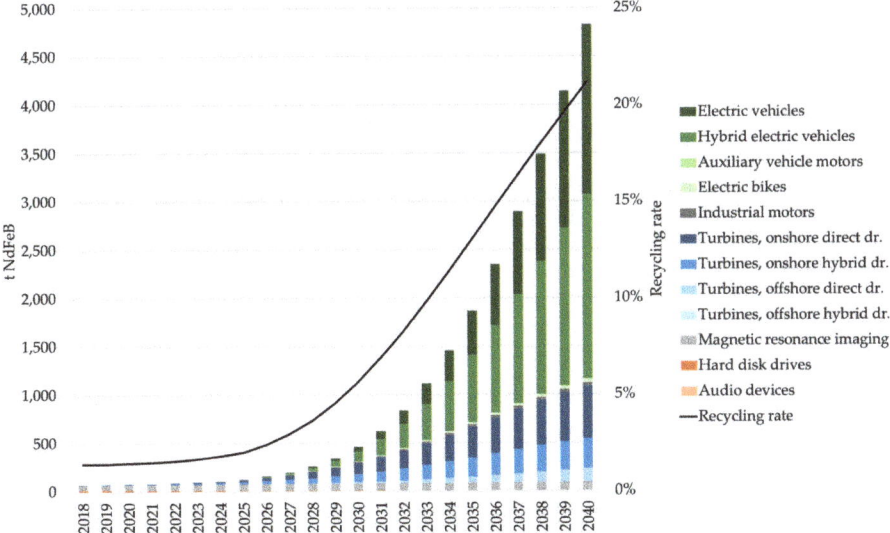

Figure 7. Estimation of the realistic return flows by application in tons of NdFeB and overall recycling rate.

Table 2. Comparison of the theoretical potential and realistic returns of NdFeB in tons accumulated from 2018 up to 2040.

Application	Theoretical Potential Returns [t NdFeB]	Estimated Realistic Returns [t NdFeB]	
Electric Vehicles	36,336	7097	(28%)
Hybrid Electric Vehicles	49,320	9140	(36%)
Auxiliary Vehicle Motors	60,861	0	(0%)
Electric Bikes	23,832	292	(1%)
Industrial Motors	10,575	232	(1%)
Wind Turbines, Direct Drive	5776	4678	(18%)
Wind Turbines, Hybrid Drive	3066	2483	(10%)
Magnetic Resonance Imaging	31,585	1579	(6%)
Hard Disk Drives	8704	174	(1%)
Audio Devices	2836	0	(0%)
Total	232,891	25,675	(100%)

5. Discussion and Conclusions

Since the REE crisis, a lot of research regarding the recycling of NdFeB magnets has been conducted and several companies have discussed the development of NdFeB magnet recycling as a business segment. Despite all these activities, industrial recycling of NdFeB wastes does not take place in the EU. Instead, collected scrap is exported to China and Japan for recycling. This is desirable from an environmental point of view [15], but does not reduce the import dependencies of European companies, a proclaimed objective of many political initiatives.

This study shows that the main reason for this situation is the insufficient amount of NdFeB scrap to feed an industrial plant. According to our appraisal, the necessary volumes above 1000 t/year cannot be expected before 2033. Furthermore, at present, the necessary recycling technology is not available at an industrial scale in Europe. However, considering the predicted slow increase of available NdFeB waste in the upcoming years, there is enough time to scale-up the developed approaches.

Although NdFeB magnet recycling is currently not a profitable business segment for recycling companies, our forecast shows that a strong increase of the available scrap amounts can be expected in the future. Therefore, companies should regularly review industrial trends and scrap market developments to enter the market at an optimal time.

Regarding future research and development, our study shows that an emphasis should be laid on an efficient extraction of NdFeB magnets from mobility applications and wind turbines. In comparison, most other applications are of minor importance for economic reasons, due to expected technological changes, and export to non-EU countries for second use.

Furthermore, a critical discussion regarding the funding of research and development in the field of metallurgical recycling of NdFeB magnets is necessary. Considering the small market share of European magnet producers (approx. 1%), the demand for neodymium and dysprosium metal, which are both primarily used for the production of NdFeB magnets, is low. Both metals are mainly imported to the EU in the form of magnets or magnet containing products. Therefore, the future markets for recycled neodymium and dysprosium are presumably in Asia.

However, a funding stop would inevitably lead to a major loss of metallurgical and material expertise in Europe, which has been (re)built in recent years, and to a further consolidation of Europe's dependency on Asia. Considering the importance of NdFeB magnets for the European automotive and other industries, this development bears major risks. However, to avert these risks a fundamental reorientation in the European raw material policy would be required.

Author Contributions: H.Y.S.-M. and T.E. conceived the paper. T.E. wrote Sections 1, 2 and 5 and interviewed the industry experts. M.F.H. and M.V.R. conducted the literature review. M.V.R. performed the calculations in Sections 3 and 4, which were supervised by H.Y.S.-M. and wrote Sections 3.2 and 3.3. T.E. and M.V.R. wrote all other sections.

Funding: This research received no external funding.

Acknowledgments: We acknowledge support by Open Access Publishing Fund of Clausthal University of Technology.

Conflicts of Interest: The authors declare no conflict of interest.

References

1. Ober, J.A. Mineral Commodity Summaries. 2018. Available online: https://pubs.er.usgs.gov/publication/70194932 (accessed on 18 October 2018).
2. Schüler-Zhou, Y. Chinas Rohstoffpolitik Für Seltene Erden. 2018. Available online: https://www.giga-hamburg.de/en/publication/chinas-rohstoffpolitik-f%C3%BCr-seltene-erden (accessed on 18 October 2018).
3. Rare Earth Elements: A Review of Production, Processing, Recycling, and Associated Environmental Issues. 2012. Available online: https://nepis.epa.gov/Exe/ZyPURL.cgi?Dockey=P100EUBC.txt (accessed on 26 September 2018).
4. Elsner, H. Seltene Erden–Stand heute. In Proceedings of the GDMB Special Metals Expert Committee, Schwäbisch Gmünd, Germany, 8 October 2015.
5. Binnemans, K.; Jones, P.T.; Blanpain, B.; van Gerven, T.; Yang, Y.; Walton, A.; Buchert, M. Recycling of rare earths: A critical review. *J. Clean. Prod.* **2013**, *51*, 1–22. [CrossRef]
6. Lucas, J.; Lucas, P.; Le Mercier, T.; Rollat, A.; Davenport, W. *Rare Earth Recycle*; Elsevier: Amsterdam, The Netherlands, 2015; pp. 333–350.
7. Jowitt, S.M.; Werner, T.T.; Weng, Z.; Mudd, G.M. Recycling of the rare earth elements. *Curr. Opin. Green Sustain. Chem.* **2018**, *13*, 1–7. [CrossRef]
8. Yang, Y.; Walton, A.; Sheridan, R.; Güth, K.; Gauß, R.; Gutfleisch, O.; Buchert, M.; Steenari, B.-M.; van Gerven, T.; Jones, P.T.; et al. REE recovery from end-of-life NdFeB permanent magnet scrap: A critical review. *J. Sustain. Metall.* **2017**, *3*, 122–149. [CrossRef]
9. Glöser-Chahoud, S.; Pfaff, M.; Tercero Espinoza, L.A.; Faulstich, M. Dynamische materialfluss-analyse der magnetwerkstoffe neodym und dysprosium in Deutschland. In Proceedings of the 4th Symposium Rohstoffeffizienz und Rohstoffinnovationen, Tutzing, Germany, 17–18 February 2016; Teipel, U., Reller, A., Eds.; Fraunhofer Verlag: Stuttgart, Germany, 2016; pp. 258–288.
10. Marscheider-Weidemann, F.; Langkau, S.; Hummen, T.; Erdmann, L.; Tercero Espinoza, L.A.; Angerer, G.; Marwede, M.; Benecke, S. *Rohstoffe Für Zukunftstechnologien 2016*; Deutsche Rohstoffagentur (DERA) in der Bundesanstalt für Geowissenschaften und Rohstoffe (BGR): Berlin, Germany, 2016.
11. Globale Verwendungsstrukturen der Magnetwerkstoffe Neodym und Dysprosium: Eine szenariobasierte Analyse der Auswirkung der Diffusion der Elektromobilität auf den Bedarf an Seltenen Erden. Available online: http://hdl.handle.net/10419/142767 (accessed on 18 October 2018).
12. Global and China NdFeB Industry Report, 2015–2018. Available online: https://www.giiresearch.com/report/rinc312750-global-china-ndfeb-industry-report.html (accessed on 18 October 2018).
13. Katter, M. Entwicklungstrends bei pulvermetallurgisch hergestellten Seltenerd-Dauermagneten. Fraunhofer Workshop Magnetwerkstoffe–vom Design bis zum Recycling. Available online: http://www.pulvermetallurgie.com/cms/File/Anmeldung_interaktiv.pdf (accessed on 23 October 2018).
14. Elwert, T.; Clausthal University of Technology, Clausthal-Zellerfeld, Germany. Anon. Magnet Industry Expert. Personal communication, 2015.
15. Elwert, T.; Goldmann, D.; Roemer, F.; Schwarz, S. Recycling of NdFeB magnets from electric drive motors of (hybrid) electric vehicles. *J. Sustain. Metall.* **2017**, *3*, 108–121. [CrossRef]
16. Global EV Outlook 2017. 2017. Available online: https://www.iea.org/publications/freepublications/publication/GlobalEVOutlook2017.pdf (accessed on 18 October 2018).
17. Prognose der Anteile verschiedener Automobil-Antriebsarten an den Kfz-Neuzulassungen in der Europäischen Union in den Jahren 2016 bis 2030. Available online: https://de.statista.com/statistik/daten/studie/666782/umfrage/automobil-antriebsarten-in-der-eu-prognose/ (accessed on 11 September 2018).
18. Number of New Car Registrations in Germany from 1955 to 2019 (In Millions). Available online: https://www.statista.com/statistics/587730/new-car-registrations-germany/ (accessed on 11 September 2018).
19. EU Passenger Car Production. Available online: https://www.acea.be/statistics/article/eu-passenger-car-production (accessed on 11 September 2018).

20. The Automobile Industry Pocket Guide. Available online: https://www.acea.be/uploads/publications/ACEA_Pocket_Guide_2018-2019.pdf (accessed on 13 September 2018).
21. Elwert, T.; Schwarz, S.; Bergamos, M.; Kammer, U. Entwicklung einer industriell umsetzbaren recyclingtechnologiekette für NdFeB-magnete—Semarec. In *Recycling und Rohstoffe*; Thiel, S., Thomé-Kozmiensky, E., Goldmann, D., Eds.; Thomé-Kozmiensky Verlag GmbH: Nietwerder, Germany, 2018; pp. 253–271.
22. Number of electric bicycles sold in the European Union (EU) from 2006 to 2016, (in 1000 units). Available online: https://www.statista.com/statistics/397765/electric-bicycle-sales-in-the-european-union-eu/ (accessed on 11 September 2018).
23. Bicycle Market 2017 Edition: Industry and Market Profile. Available online: http://www.conebi.eu/wp-content/uploads/2018/09/European-Bicycle-Industry-and-Market-Profile-2017-with-2016-data-update-September-2018.pdf (accessed on 11 September 2018).
24. Zweirad-Industrie-Verband. Zahlen—Daten—Fakten zum Deutschen E-Bike-Markt 2017: E-Bikes mit Rekordzuwächsen. Available online: https://www.ziv-zweirad.de/fileadmin/redakteure/Downloads/Marktdaten/PM_2018_13.03._E-Bike-Markt_2017.pdf (accessed on 11 September 2018).
25. Buchert, M.; Manhart, A.; Sutter, J. Untersuchung zu Seltenen Erden: Permanentmagnete im industriellen Einsatz in Baden-Württemberg: Freiburg. 2014. Available online: https://www.oeko.de//oekodoc/2053/2014-630-de.pdf (accessed on 18 October 2018).
26. United Nations Conference on Trade and Development. Gross Domestic Product: Total and per Capita, Current and Constant (2010) Prices, Annual. Available online: http://unctadstat.unctad.org/wds/TableViewer/tableView.aspx?ReportId=96 (accessed on 11 September 2018).
27. Wind in Power 2017: Annual Combined Onshore and Offshore Wind Energy Statistics. Available online: https://windeurope.org/wp-content/uploads/files/about-wind/statistics/WindEurope-Annual-Statistics-2017.pdf (accessed on 11 September 2018).
28. Global Offshore Wind Farms Database. Available online: https://www.4coffshore.com/windfarms/ (accessed on 11 September 2018).
29. Serrano-González, J.; Lacal-Arántegui, R. Technological evolution of onshore wind turbines-a market-based analysis. *Wind Energy* **2016**, *19*, 2171–2187. [CrossRef]
30. Renewables 2017—Analysis and Forecasts to 2022. 2017. Available online: https://www.iea.org/publications/renewables2017/ (accessed on 18 October 2018).
31. World Energy Outlook 2017. 2017. Available online: https://www.iea.org/weo2017/ (accessed on 18 October 2018).
32. Wind Energy in Europe: Scenarios for 2030. Available online: https://windeurope.org/wp-content/uploads/files/about-wind/reports/Wind-energy-in-Europe-Scenarios-for-2030.pdf (accessed on 11 September 2018).
33. Zepf, V. Das verkannte recyclingpotential der seltenen erden—quantitative ergebnisse für neodym in Deutschland. In *Recycling und Rohstoffe*; Thomé-Kozmiensky, K.J., Goldmann, D., Eds.; Thomé-Kozmiensky Verlag GmbH: Nietwerder, Germany, 2015; pp. 463–476.
34. Elwert, T.; Hoffmann, M.; Schwarz, S. Can recycling of NdFeB magnets be expected in Europe before 2030? In *EMC 2017 Proceedings*; Waschki, U., Ed.; GDMB Verlag GmbH: Clausthal-Zellerfeld, Germany, 2017; Volume 3, pp. 1263–1278.
35. WDC Unit Shipments of Hard Disk Drives (HDD) Worldwide by Size from 2013 to 2014 (In Millions). Available online: https://www.statista.com/statistics/407893/wdc-global-shipment-figures-for-hard-disk-drives/ (accessed on 11 September 2018).
36. Worldwide Unit Shipments of Hard Disk drives (HDD) from 1976 to 2020 (In Millions). Available online: https://www.statista.com/statistics/398951/global-shipment-figures-for-hard-disk-drives/ (accessed on 11 September 2018).
37. Schulze, R.; Buchert, M. Estimates of global REE recycling potentials from NdFeB magnet material. *Resour. Conserv. Recycl.* **2016**, *113*, 12–27. [CrossRef]
38. Weinert, J.; VanGelder, E. *Encyclopedia of Electrochemical Power Sources*; Elsevier: Amsterdam, The Netherlands, 2009; pp. 292–301.
39. Hoenderdaal, S.; Tercero Espinoza, L.; Marscheider-Weidemann, F.; Graus, W. Can a dysprosium shortage threaten green energy technologies? *Energy* **2013**, *49*, 344–355. [CrossRef]

40. Lacal Arántegui, R.; Serrano González, J. The Technology, Market and Economic Aspects of Wind Energy in Europe. 2015. Available online: https://www.evwind.es/2015/06/26/the-technology-market-and-economic-aspects-of-wind-energy/52993 (accessed on 18 October 2018).
41. Sriram Radhakrishnan. MRI System Market by Architecture Type (Open MRI and Close MRI), by Field Strength (High field system, Medium field system and Low Field System)—Global Opportunity Analysis and Industry Forecast 2014–2022. 2016. Available online: https://www.giiresearch.com/report/amr676714-mri-system-market-by-architecture-type-open-mri.html (accessed on 18 October 2018).
42. Shipments of Hard and Solid State Disk (HDD/SSD) Drives Worldwide from 2015 to 2021 (In Millions). Available online: https://www.statista.com/statistics/285474/hdds-and-ssds-in-pcs-global-shipments-2012-2017/ (accessed on 29 September 2018).
43. Elwert, T.; Clausthal University of Technology, Clausthal-Zellerfeld, Germany. Anon. Scrap Industry Expert. (Innova Recycling GmbH, Goslar, Germany). Personal communication, 2016.
44. Elwert, T.; Clausthal University of Technology, Clausthal-Zellerfeld, Germany. Anon. WEEE Recycling Expert. (ELPRO Elektronik-Produkt Recycling GmbH, Braunschweig, Germany). Personal communication, 2016.
45. Walachowicz, F.; March, A.; Fiedler, S.; Buchert, M.; Sutter, J.; Merz, C. Recycling von Elektromotoren—MORE: Ökobilanz der Recyclingverfahren. 2014. Available online: https://www.researchgate.net/publication/272809889_Recycling_von_Elektromotoren_MOtor_REcycling_-_MORE (accessed on 18 October 2018).
46. Kohlmeyer, R.; Sander, K.; Jung, M.; Wagner, L. Klärung des verbleibs von außer betrieb gesetzten fahrzeugen. In *Recycling und Rohstoffe*; Thomé-Kozmiensky, K.J., Thiel, S., Thomé-Kozmiensky, E., Goldmann, D., Eds.; Thomé-Kozmiensky Verlag GmbH: Nietwerder, Germany, 2017; pp. 285–304.
47. Elwert, T.; Clausthal University of Technology, Clausthal-Zellerfeld, Germany. Anon. Recycling Industry Expert. (Siemens AG, München, Germany). Personal communication, 2018.

© 2018 by the authors. Licensee MDPI, Basel, Switzerland. This article is an open access article distributed under the terms and conditions of the Creative Commons Attribution (CC BY) license (http://creativecommons.org/licenses/by/4.0/).

Article

Purification of Aluminium Cast Alloy Melts through Precipitation of Fe-Containing Intermetallic Compounds

Marina Gnatko *, Cong Li, Alexander Arnold and Bernd Friedrich

IME Institute of Process Metallurgy and Metal Recycling, RWTH Aachen University, 52056 Aachen, Germany; CLi@ime-aachen.de (C.L.); aarnold@ime-aachen.de (A.A.); bfriedrich@ime-aachen.de (B.F.)
* Correspondence: mgnatko@gmx.net; Tel.: +49-(0)241-80-95751

Received: 19 August 2018; Accepted: 1 October 2018; Published: 4 October 2018

Abstract: Aluminium secondary materials are often contaminated by impurities such as iron. As the alloy properties are affected by impurities, it is necessary to refine aluminium melts. The formation of Fe intermetallics in aluminium melts can be used to develop a purification technology based on the removal of intermetallic compounds. In this study, the temperature range for effective separation of intermetallics was determined in an industrial-relevant Al–Si–Fe–Mn system with 6 to 10 Si wt. %, 0.5 to 2.0 Fe wt. %, and 0 to 2.0 Mn wt. %. Based on DTA (Differential Thermal Analysis) and SEM (scanning electron microscope) results and following the rules of phase boundary drawing, isopleths were drawn. This method allows to derive the temperature ranges of intermetallic phase stability and can be applied for the assessment of melt-refining parameters.

Keywords: aluminium purification; iron removal; intermetallic formation; polythermal section

1. Introduction

In order to achieve legal recycling rate requirements (e.g., regarding end-of-life vehicles, 95% of materials must be recycled), material cycles must be almost completely closed. The recovery of all metals in their pure form, however, is not possible. Secondary recovered materials are often contaminated. The complexity of such materials leads to difficulties in sorting, as well as to impurity pickup during the mechanical treatment processes. As property formation is affected by impurities, aluminium end-of-life scrap is normally used for the production of cast alloys. Since impurities such as iron accumulate in aluminium secondary alloys at values of up to 2 wt. %, it is difficult to produce Al recycling alloys which conform to standards (Table 1). Therefore, it is necessary to refine aluminium melts, as the current practice of diluting primary aluminium is becoming uneconomical. Among all the impurities that need to be removed, iron is a serious challenge.

Table 1. Composition of some Al cast alloys, data from [1,2].

Alloy Identification		Alloy Composition Limits, wt. %					
Numerical	Chemical	Si	Fe	Cu	Mn	Mg	Other
		cast alloys for pressure casting					
EN AC-44300	EN AC-AlSi12(Fe)	10.5–13.5	1.0	0.10	0.55	-	0.55
EN AC-46000	EN AC-AlSi9Cu3(Fe)	8.0–11.0	1.30	2.0–4.0	0.55	0.05–0.55	2.75
		cast alloys for common application					
EN AC-44200	EN AC-AlSi12(a)	10.5–13.5	0.55	0.05	0.35	-	0.40
EN AC-46200	EN AC-AlSi8Cu3	7.5–9.5	0.8	2.0–3.5	0.15–0.65	0.05–0.55	2.45

While many efforts have been made for the removal of iron from primary aluminium and high-purity aluminium [3–6], only limited attention has been paid to that of secondary aluminium, which contains usually more than 2 wt. % Fe. Conventional ways of iron removal from high iron-containing aluminium melts include filtration, centrifugal separation, and electromagnetic (EM) separation [7–9]. All these methods are based on the principle of precipitation of Fe-enriched phases. It is a well-known fact that in the Al–Si–Fe system, a variety of binary and ternary compounds with Al exist, including Al_3Fe, Al_5FeSi, Al_8Fe_2Si, Al_3FeSi, and Al_4FeSi_2 [10,11]. On the one hand, the precipitation of these phases impacts the quality of the end products. On the other hand, it can provide a basis for the development of a refining technology with the help of physical separation process, e.g., filtration. Thus, it was the aim of a six-year project at IME (Institute IME Process Metallurgically and Metal Recycling) to find elements that influence the residue–melt composition in order to reduce the concentration of impurities, above all iron. Even if intermetallic compounds are formed, the conditions and separation technique considered are very important for reaching the highest grade of purity. The aim of this work was to determine suitable temperature ranges in the Al–Si–Fe–Mn system in the industrially relevant concentration areas of 6 to 10 Si wt. %, 0.5 to 2.0 Fe wt. %, and 0 to 2.0 Mn wt. %, in which the separation of intermetallics becomes effective.

The eutectic iron content in a pure binary Al–Fe melt is 1.8 wt. % at 655 °C [10]. Therefore, in the case of hypereutectic alloys (over 1.8 wt. % Fe), the iron content cannot be reduced by segregation below this value. Iron precipitates in the form of the intermetallic compound Al3Fe, if the temperature falls below the liquidus line (Figure 1). Since this intermetallic phase has a melting point of 1060 °C and is insoluble in molten aluminium, it can be mechanically removed from molten aluminium, e.g., by filtration. Nevertheless, this system has no industrial significance.

Industrial cast alloy compositions are based on the binary system Al–Si, where the ternary eutectic iron content is reduced to 0.7 wt. % at 577 °C [10,11]. In the Al corner of this system, iron is present in the phases Al_3Fe, Al_8Fe_2Si, Al_5FeSi, and Al_4FeSi_2 (Figure 2).

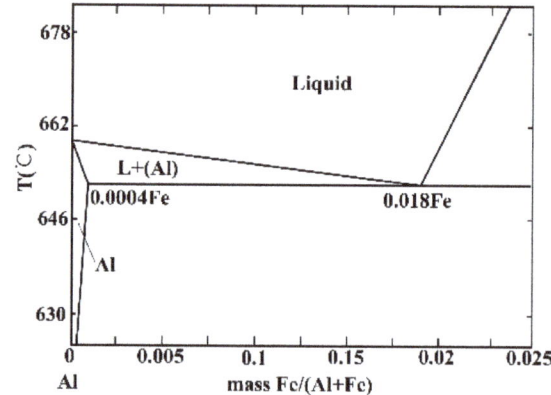

Figure 1. Al–Fe phase diagram calculated with FactSage™.

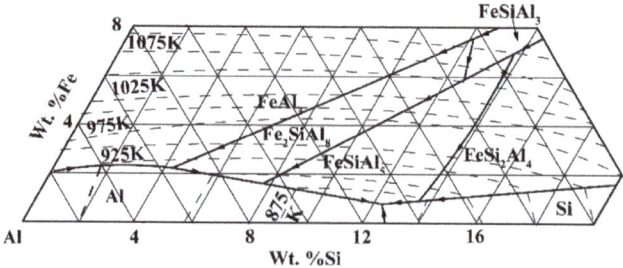

Figure 2. Liquidus surface in the Al corner of the Al–Si–Fe system [10].

The addition of further alloying elements results in the formation of quaternary or higher alloy systems with complex phase relations. Ternary and quaternary intermetallic compounds with iron are formed in the Al–Si–Fe–Mn system, and iron solubility decreases to 0.29 wt. % at the eutectic point [12]. The current German standards regarding maximum Fe content in cast Al–Si–alloys range between 0.2 and 0.9 wt. % (depending on alloying class) [13]. For the current investigation, the Al–Si–Fe–Mn system was applied because numerous intermetallics are formed in this system, and the residue melt composition can be influenced depending on the Mn/Fe ratio [10–12,14,15]. Table 2 summarizes the phases to be expected in the Al–Si–Fe–Mn system.

Table 2. Published data on the expected phases in the Al corner of the Al–Si–Fe–Mn system, data from [10–12,14,15].

Phases	Components, wt. %			
	Al	Mn	Fe	Si
Al_8Fe_2Si	56.0–62.6	–	30.0	7.4–11.0
Al_5FeSi	59.4–60.9	<0,8	25.5–26.5	12.8–13.3
$Al_{16}(FeMn)_4Si_3$	53.0–64.6	14.6–19.7	10.4–15.3	10.4–12.0
$Al_{15}Mn_3Si_2$	58.0–60.3	27.7–29.5	<1.8	10.2–10.7
Al_4FeSi_2	46.9–48.0	<0.8	25.9	25.3–26.4

Until now, no quaternary phase has been clearly identified in this system [10,12,15]. Initially, it was believed that an area of solid solutions existed between Al_8Fe_2Si and $Al_{15}Mn_3Si_2$. Later, this assumption was rejected on the basis of the fact that these compounds had different crystal structures (hexagonal and cubic). The currently accepted version of the phase diagram illustrates a broad range of solid solutions based on the compound $Al_{15}Mn_3Si_2$ extending towards the Al–Si–Fe surface [10]. In this variant, manganese is replaced with iron to form the compound with the composition 31 wt. % Fe, 1.5 wt. % Mn, 8 wt. % Si. This broad range of homogeneity is considered as quaternary phase $Al_{15}(FeMn)_3Si_2$ [10]. On the other hand, Zakharov A. et al. studied alloys containing 10–14 wt. % Si, 0–3 wt. % Fe, 0–4 wt. % Mn, and proposed the existence of the quaternary compound $Al_{16}(FeMn)_4Si_3$ [12]. The formation of this phase would allow a quasi-ternary section $Al-Al_{16}(FeMn)_4Si_3-Si$ and the formation of two secondary systems on both sides of this section: $Al-Al_{16}(FeMn)_4Si_3-Si-Al_5FeSi$ and $Al-Al_{16}(FeMn)_4Si_3-Si-Al_{15}Mn_3Si_2$.

According to reference [10], the solid solution of iron in the $Al_{15}Mn_3Si_2$ phase has a cubic structure with a lattice parameter which decreases because of an increase of Fe content from 1.265 nm (0 wt. % Fe) to 1.25 nm (31.1 wt. % Fe). The quaternary phase found in reference [12] has a face-centered cubic structure with a lattice parameter of $a = 1.252 \pm 0.04$ nm. The similar lattice parameters mean that it cannot be determined which version of the Al–Si–Fe–Mn phase diagram is correct.

In references [11,15], it was proposed that non-equilibrium crystallization had a significant effect on phase composition, especially in Al–Si–Fe alloys. This is because of the inhibition of peritectic reactions, which take a long time to be completed. However, due to numerous intermetallics, this system shows a potential for removing iron and manganese from Al–Si melts. Phase diagrams are a useful tool for presenting the required relations in a metal system.

In comparison with binary systems (only two dimensions), ternary and multi-phase phase diagrams (here and after in this article, "Multi-" refers specially to more than three) are rather complicated. A ternary phase diagram is shown in Figure 3a, where the composition plane forms the base triangle, and phase variations caused by temperature change are illustrated vertically (Figure 3a). Vertical sections (Figure 3b) of a ternary phase diagram—also known as isopleths—have been widely used because of their similarities to binary diagrams. Such sections are two-dimensional planes constructed by cutting the three-dimensional diagrams with a slice which is vertical to the base composition triangle. Once phase areas in an isopleth are clearly clarified, the liquidus and solidus temperatures for certain alloy compositions can be readily read from it.

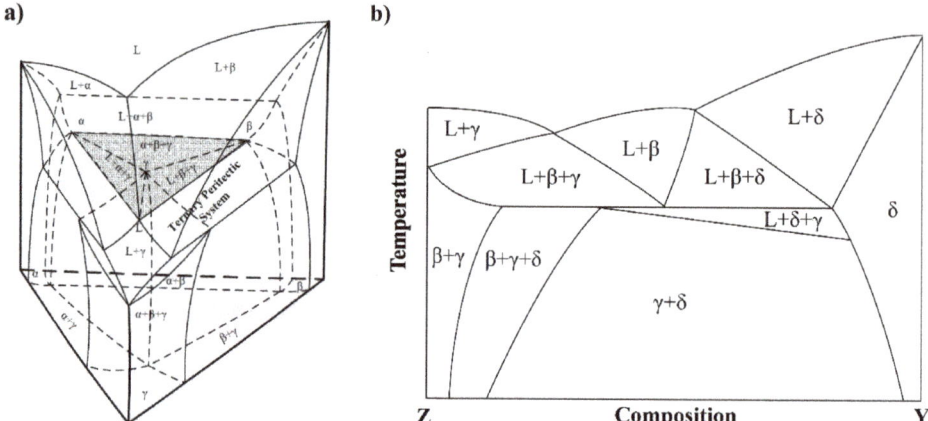

Figure 3. (a) Temperature–composition space diagram of a ternary system (b) Isopleth through a ternary system [16].

From the metallurgical practice point of view, multi-phase alloy diagrams involving four or more elements are needed more than binary or ternary diagrams. This is because most commercial alloys contain more than three alloying elements, even without taking impurity into consideration. However, temperature–composition phase diagrams of multi components are extremely inconvenient and highly complicated.

In order to determine the phase variation caused by temperature changes, as well as the composition difference in complex multi-components system, a feasible way is to draw the corresponding three- or two-dimensional sections, in which temperature and concentration of certain component(s) are represented as variables.

For the construction of a two-dimensional isopleth, i.e., temperature–composition diagrams, the following information is usually needed: (1) the general diagram including the number, disposition, and identity of the phases and the respective invariant reaction, and (2) the temperature and compositions along all boundary lines (and surfaces).

The most widely used method of constitutional investigation is Differential Thermal Analysis (DTA). It is capable of locating the liquidus lines and at the same time indicating the general disposition of phases and invariant reactions in the system. Its principle is extremely simple: every occurrence of phase change is accompanied by exothermic and endothermic effects such as heat from the melt crystallization. The delay and acceleration of the cooling speed compared to a reference material is monitored.

2. Materials and Methods

In this research work, approximately 60 alloy compositions were prepared by induction melting within the following concentration ranges: 6 to 10 wt. % Si, 0 to 2 wt. % Fe, and 0 to 2 wt. % Mn. ICP (Spectro ICP-OES Spectro Ciros Vision, Kleve, Germany) analysis was applied to determine the composition of the samples. Differential Thermal Analysis (DTA) (IME, Aachen, Germany) and Scanning Electron Microscopy (SEM) (JEOL JSM-7000F, Tokyo, Japan) with integrated EDX (Energy Dispersive X-ray analysis) (Oxford Instruments, Oxford, UK) were applied to determine phase precipitations and the temperatures of phase transformations.

In order to allow an evaluation in the form of isopleths, three of four element concentrations were kept constant. The groups of investigated alloys and isopleths are shown in Table 3. The manganese content changed from 0 to 2 wt. % by representation on the isopleths in steps of 0.5 wt. %.

Table 3. Groups of investigated alloys leading to the individual isopleth.

Group	Iron/Manganese Content, wt. %			
	Fe/Mn Step 0.5	Fe/Mn Step 0.5	Fe/Mn Step 0.5	Fe/Mn Step 0.5
AlSi6FeMn	0.5/0–2	1.0/0–2	1.5/0–2	2.0/0–2
AlSi8FeMn	0.5/0–2	1.0/0–2	1.5/0–2	2.0/0–2
AlSi10FeMn	0.5/0–2	1.0/0–2	1.5/0–2	2.0/0–2

Extended experimental equipment for the Differential Thermal Analysis (DTA) (IME, Aachen, Germany) was built, containing a resistance furnace and a differential thermocouple (Figure 4). The differential thermocouple consists of two connected thermocouples. The first one, the working thermocouple, measured the temperature in the sample. The second one, the reference thermocouple, measured the temperature difference which existed during cooling between the samples and the reference substance. Two crucibles, one with the reference substance (Al_2O_3) and the other with the sample, were placed in a steel block to ensure the same external heat conditions for both crucibles during cooling. As steel has a lower thermal conductivity than Al, this block protected the crucibles from temperature changes in the furnace space. Such changes could influence the temperature data and distort the results.

In order to determine an isopleth with sufficient accuracy, a minimum of five alloys must be investigated. After melting the alloy, the differential thermal analysis commenced. The sample, weighing approximately 20 g, was placed in the crucible (Figure 4) and heated to 750 °C–760 °C. This temperature value was chosen to allow a sufficient superheat. As according to literature data, the melting point of the alloys studied was below or near 700 °C. Subsequently, the furnace was switched off, and the cooling curve with a rate of approx. 4.5 °C/min was recorded.

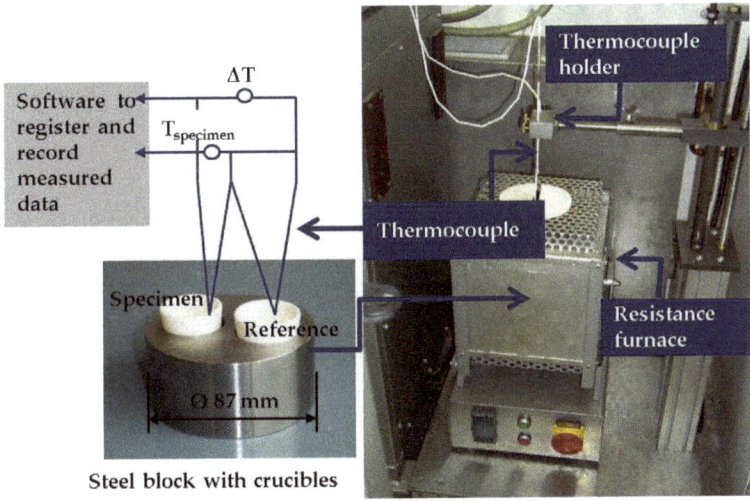

Figure 4. Equipment for large scale Differential Thermal Analysis (DTA) at IME.

3. Results and Discussions

3.1. DTA Experimental Results

Figure 5 illustrates a cooling curve example for the alloy AlSi8Fe2.0Mn1.0 from isopleth AlSi8Fe2.0–Mn. Two curves are indicated: one for the sample alloy and one for the reference (Al_2O_3). The curve of the sample demonstrates two significant effects, whereas the reference curve shows four. This is because of the special bonding of the thermocouples (Figure 4), whereby the reference material becomes very sensitive and can detect changes with lower evolutions of heat, e.g., at the liquidus temperature. Therefore, it was presumed that four phase changes occurred in this alloy. Exemplary DTA results are shown in Table 4 for the isopleths AlSi8Fe0.5-Mn, AlSi8Fe1-Mn, AlSi8Fe1.5-Mn, and AlSi8Fe2.0-Mn; all data are published in reference [17]. After recording and evaluating all cooling curves, the temperature–composition diagrams were created for these isopleths.

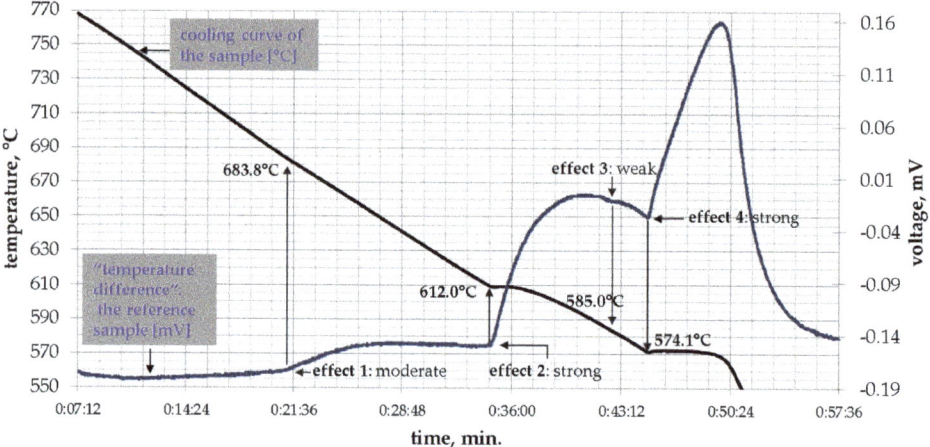

Figure 5. Cooling curve of the alloy AlSi8Fe2.0Mn1.0 from isopleth AlSi8Fe2–Mn.

Table 4. Results of the evaluation of the cooling curve effects of the alloys from isopleths AlSi8Fe0.5–Mn, AlSi8Fe1–Mn, AlSi8Fe1.5–Mn, and AlSi8Fe2.0–Mn.

Alloy (Target)	Mn, wt. %	Effect 1 T, °C	Effect 2 T, °C	Effect 3 T, °C	Effect 4 T, °C
AlSi8Fe0.5	0.00	602.3	586.4	-	574.4
AlSi8Fe0.5Mn0.5	0.54	612.0	599.2	585.0	573.7
AlSi8Fe0.5Mn1.0	1.22	649.9	601.2	597.0	574.7
AlSi8Fe0.5Mn2.0	1.94	675.0	633.0	602.1	574.9
AlSi8Fe1.0	0.00	613.0	598.1	-	574.9
AlSi8Fe1.0Mn0.5	0.47	620.7	606.0	600.0	573.5
AlSi8Fe1.0Mn1.0	0.92	634.2	605.7	580.0	573.0
Al Si8 Fe1.0Mn1.5	1.50	680.5	611.0	607.0	574.3
Al Si8 Fe1.0Mn2.0	1.98	691.4	640.0	612.9	574.8
AlSi8Fe1.5	0.00	614.3	606.4	-	574.4
AlSi8Fe1.5Mn0.5	0.56	645.0	638.0	609.9	572.8
AlSi8 Fe1.5Mn1.0	1.12	672.4	614.0	611.7	573.2
AlSi8Fe1.5Mn1.5	1.56	681.3	612.6	576.0	574.0
AlSi8Fe1.5Mn2.0	2.04	693.1	630.0	613.6	573.8
AlSi8Fe2.0	0.00	616.3	608.4	-	574.4
AlSi8Fe2.0Mn0.5	0.54	657.1	609.2	589.1	573.0
AlSi8Fe2.0Mn1.0	1.09	683.8	612.0	585.0	574.1
AlSi8Fe2.0Mn1.5	1.36	703.6	638.0	612.5	573.8
AlSi8Fe2.0Mn2.0	1.99	710.5	613.2	575.0	573.6

3.2. Precipitated Phases

Figure 6 shows exemplary SEM examination patterns of the alloys AlSi8Fe2Mn0.5(a) and AlSi8Fe2Mn1.0(b) performed by GfE (Gemeinschaftslabor für Electronenmikroskopie) RWTH

(Rheinisch-Westfälische Technische Hochschule) Aachen University. The dark grey crystals are eutectic silicon precipitations. White needle-like precipitations indicate the ternary phase Al$_5$FeSi. The groups of white net-forming precipitations (also known as Chinese script) are clusters of the quaternary phase Al(FeMn)Si. These descriptions of phase shapes were previously accepted, as in references [18,19]. The composition of the precipitations was determined by EDX analysis.

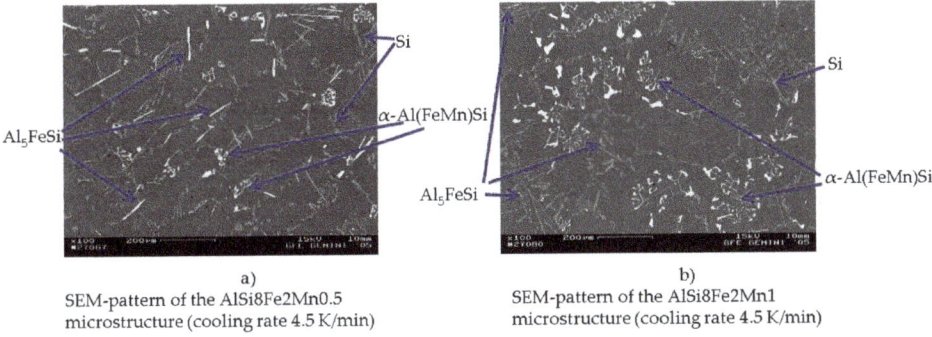

a) SEM-pattern of the AlSi8Fe2Mn0.5 microstructure (cooling rate 4.5 K/min)

b) SEM-pattern of the AlSi8Fe2Mn1 microstructure (cooling rate 4.5 K/min)

Figure 6. SEM pattern of the microstructure.

According to the EDX microanalysis of the investigated alloys, the compositions of the phases precipitated were determined and are shown in Table 5. The appearance of the above-mentioned phases depended on their composition, and the extent varied with the Mn content of the alloy, especially for the precipitation of the Al(FeMn)Si phase. Mn content in the quaternary phase increased from 8.42 to 15.68 wt. %, and Fe content decreased from 18.64 to 12.57 wt. %, correspondingly (Figure 7).

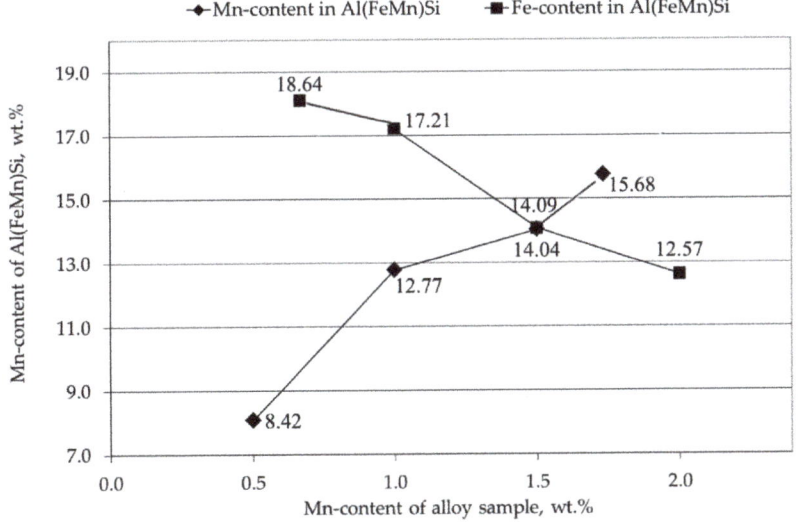

Figure 7. Composition change in the Al(FeMn)Si phase with increasing Mn content in the alloy group AlSi8Fe1.5–Mn.

The ternary Al$_5$FeSi disappeared after a specific Mn content was reached in the alloy, and the formation of Al$_{16}$(FeMn)$_4$Si$_3$ was not as clearly determined as reported by A. Zakharov [12]. This was caused by the fact that the Mn content of our Al(FeMn)Si phases changed with the Mn content of the alloys. On the other hand, the diagram version proposed by L. Mondolfo [17] cannot be accepted as fundamental information for drawing the isopleths because of Al_5FeSi disappearance (see above). Since phase compositions are significantly influenced by the crystallization conditions, the deviations in the Mn content of the Al(FeMn)Si phases, in comparison to A. Zakharov's study, must be considered in consequence of different crystallization conditions.

Table 5. Composition of detected precipitated phases in all investigated alloys.

Phase	Components, wt. %			
	Al	Mn	Fe	Si
Al matrix	98.37–99.66	0.0–0.45	0.0–0.50	0.73–2.55
Al_5FeSi	55.02–56.03	1.92–2.59	23.73–26.21	16.86–17.65
α-Al(FeMn)Si	57.77–61.46	8.07–17.39	12.62–19.85	10.06–13.70
β-Al(FeMn)Si	56.47–62.58	12.63–17.93	11.25–13.44	10.84–11.34
Si	0.30–3.50	-	-	96.50–99.77

3.3. Developing Isopleths from DTA and Phase Analysis Results

Based on the DTA and SEM results, 12 isopleths were drawn (according to Table 4). All isopleths are published in reference [17]. As examples, four isopleths AlSi8Fe–Mn are shown in Figure 8a–d.

The construction of isopleths was based on the following theory as well as on rules of phase boundary drawing:

(1) The quaternary Al(FeMn)Si are differentiated by the Mn/Fe ratio into α-Al(FeMn)Si if Mn/Fe ≤ 1.1 and β-Al(FeMn)Si if Mn/Fe > 1.1. These three systems are formed depending on the Mn/Fe ratio of the alloy: if Mn/Fe < 1.1, after crystallization, the alloys consist of Al–α-Al(FeMn)Si–Si–Al_5FeSi; if Mn/Fe > 1.1, the alloys consist of Al–α-Al(FeMn)Si–Si–β-Al(FeMn)Si; if Mn/Fe = 1.1, only Al–α-Al(FeMn)Si–Si coexist [12].

(2) Crossing the tilted phase boundary line leads to exhaust or precipitation of one phase, whereas passing through the horizontal phase boundary line, where eutectic or peritetic reactions occur, causes exhaust of one phase and precipitation of one phase, respectively. Crossing a point-phase boundary results in either exhaust (precipitation) of two phases or exhaust of one phase and precipitation of the other [20].

In the case of AlSi8Fe1–Mn, α-Al or α-Al(FeMn)Si precipitated primarily, and the liquidus line (marked by ① in Figure 8b) was drawn by fitting the data of the primary precipitation temperature. At the AlSi8Fe1 side, ②, ③, ④ phase boundaries were extended from corresponding points, which indicates, respectively, precipitation of Al5FeSi, Si, and exhaust of the melt. For Mn content from 0.5 to 2 wt. %, the exhaust of melts were caused by two four-phase eutectic reactions:

(1) L + α-Al + α-Al(FeMn)Si + Al_5FeSi = α-Al + α-Al(FeMn)Si + Si + Al_5FeSi and
(2) L + α-Al + α-Al(FeMn)Si + β-Al(FeMn)Si = α-Al + α-Al(FeMn)Si + Si + β-Al(FeMn)Si

Depending on these reactions, ⑤, ⑥ phase boundaries were drawn. At nearly 610 °C, ⑦, ⑧ phase boundaries were drawn because of not only the DTA results, but also of the fact that a three-phase area should occur between of a two-phase area and a four-phase area. For Mn content of 1.0, it was assumed that the precipitation of α-Al(FeMn)Si would lead to a decrease of Mn concentration in the melt, and therefore α-Al was assumed to precipitate prior to that of Al_5FeSi, according to which the phase composition area ⑨ was determined. Lastly, according to rules of phase boundary drawing, ⑩–⑮ phase boundaries were added in the diagram for a complete isopleth.

It is worth noting that in the case of Al Si8Fe1.5–Mn isopleth, ternary phase Al_5FeSi or quaternary α-Al(FeMn)Si precipitated primarily, whereas in the case of AlSi8Fe2-Mn isopleth, ternary phase Al_8Fe_2Si or quaternary α-Al(FeMn)Si precipitated primarily.

In AlSi8Fe2-Mn isopleth, one more phase change occurred before the eutectic equilibrium: L + α-Al + Al_8Fe_2Si + α-Al(FeMn)Si = L + α-Al + Al_5FeSi + α-Al(FeMn)Si (shown as a dotted horizontal line at 591 °C in Figure 8d). Therefore, the ternary Al_8Fe_2Si was absent in the microstructure of solid alloys.

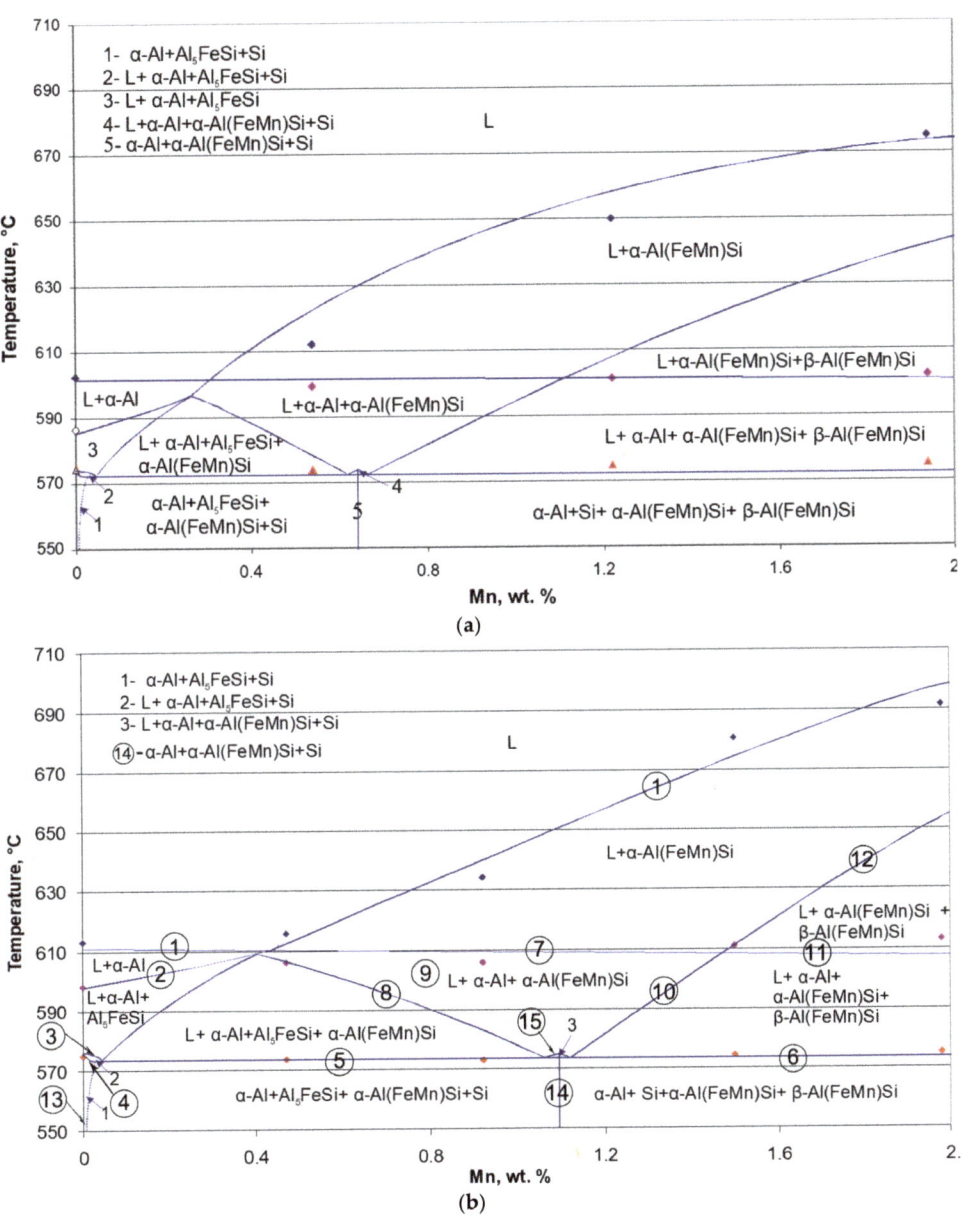

Figure 8. *Cont.*

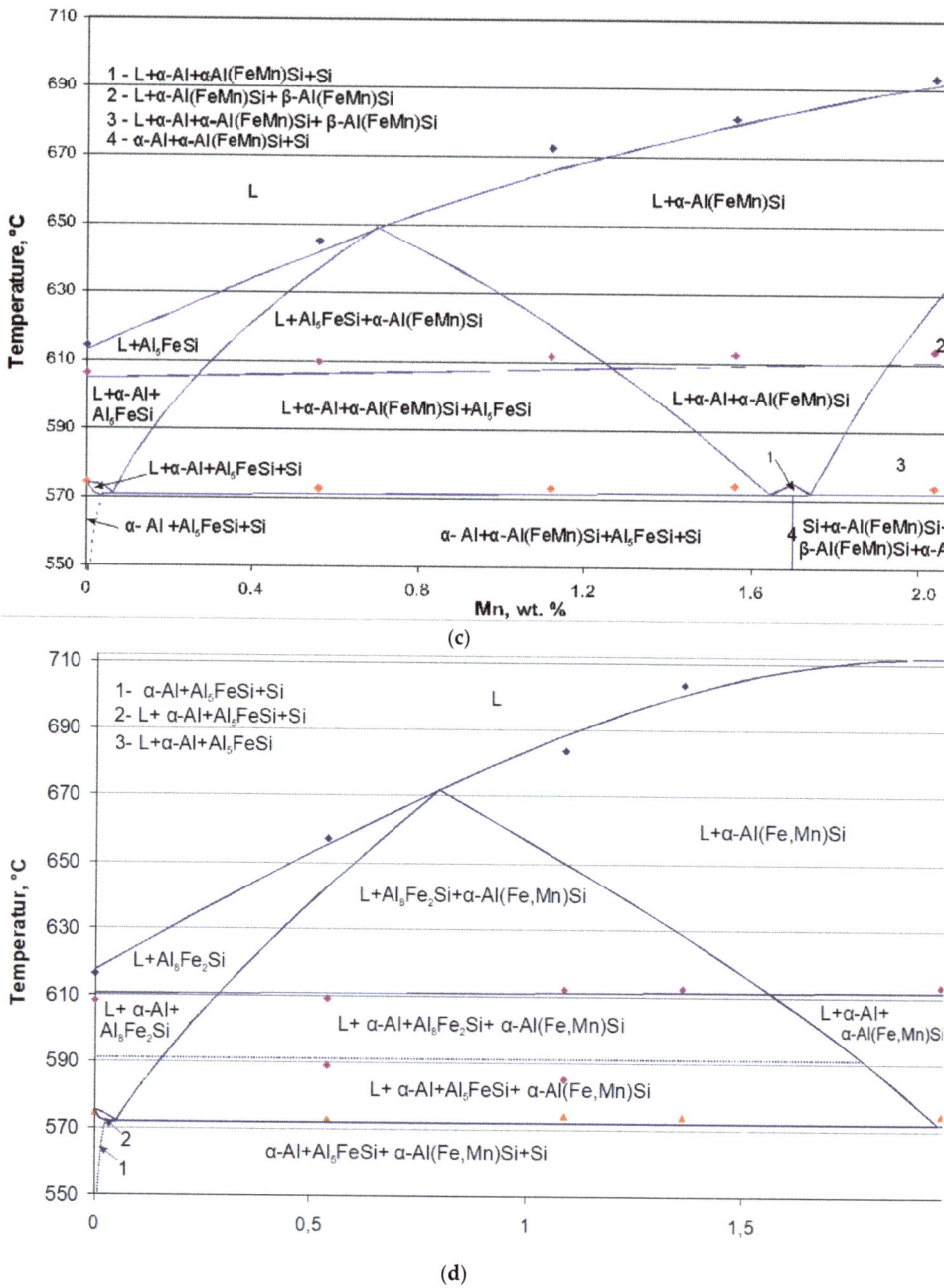

Figure 8. (a) Isopleth AlSi8Fe0.5–Mn. At 0, 0.5, 1, 1.5, and 2.0 wt. % Mn concentrations, three points were set vertically, according to the DTA results shown in Table 4. (b) Isopleth AlSi8Fe1–Mn. At 0, 0.5, 1, 1.5, and 2.0 wt. % Mn concentrations, three points were set vertically according to the DTA results shown in Table 4. (c) Isopleth AlSi8Fe1.5–Mn. At 0, 0.5, 1, 1.5, and 2.0 wt. % Mn concentrations, three points were set vertically according to the DTA results shown in Table 4. (d) Isopleth AlSi8Fe2–Mn. At 0, 0.5, 1, 1.5, and 2.0 wt. % Mn concentrations, four points were set vertically according to the DTA results shown in Table 4.

Serving as a reliable reference for deriving temperature ranges of intermetallic stability in a small continuous range, these isopleths can open part of a process window for the removal of iron from the melt through separation of Fe-enriched intermetallic compounds. For instance, in a melt with 2 wt. % Fe and 1 wt. % Mn (Figure 8d, isopleth AlSi8Fe2.0–Mn), the precipitation of α-Al(FeMn)Si can be controlled by defining the temperature in an interval of 684 °C–643 °C, which could be employed for Fe removal. If the melt is treated in the temperature range of 643 °C–610 °C, the precipitation and segregation of two iron-containing phases, Al_8Fe_2Si and α-Al(FeMn)Si, can be expected. Below 610 °C, α-Al, α-Al(FeMn)Si, Al_8Fe_2Si, and Al_5FeSi will, respectively, crystallize from the melt. However, with a decreasing temperature, the viscosity of the melt increases rapidly because of a more abundant solid/liquid fraction, which makes phases separation difficult.

4. Conclusions

Alloys of the system Al–Si–Fe–Mn were investigated in the concentration range of 6 to 10 Si wt. %, 0.5 to 2.0 Fe wt. %, and 0 to 2.0 Mn wt. % by DTA and SEM analyses. Intermetallics precipitated during solidification in the form of the ternary Al_8Fe_2Si, Al_5FeSi, quaternary Al(FeMn)Si, and Si. With a decreasing temperature, a series of peritectic reactions took place in the melt. Crystallization of the alloys resulted in two four-phase eutectic reactions:

(1) L + α-Al + α-Al(FeMn)Si + Al_5FeSi = α-Al + α-Al(FeMn)Si + Si + Al_5FeSi;
(2) L + α-Al + α-Al(FeMn)Si + β-Al(FeMn)Si = α-Al + α-Al(FeMn)Si + Si + β-Al(FeMn)Si.

In the range of the investigated alloys, solid alloy consisted of α-Al–α-Al(FeMn)Si–Si–Al_5FeSi after crystallization if Mn/Fe < 1.1, of α-Al–α-Al(FeMn)Si–Si–β-Al(FeMn)Si if Mn/Fe > 1.1, and of α-Al–α-Al(FeMn)Si–Si if Mn/Fe = 1.1.

Based on the results and following the rules of phase boundary drawing, isopleths were constructed. It can be inferred from these isopleths that at low Mn content, the melt precipitates primarily the low Fe-containing intermetallics Al_5FeSi or Al_8Fe_2Si. With the rise of Mn content in the melt, quaternary α-Al(FeMn)Si phase becomes the primary phase, thus a better refining effect can be expected.

The isopleths can serve as an informative reference for the purification of secondary recycling aluminium through the precipitation route from an industrial point of view. An initial idea concerning the process design includes: (1) Composition setting by addition of Mn in the melt, (2) Fe-enriched phase precipitation controlling by holding the melt at a specified temperature, and (3) Precipitated phase physical separation by filtration.

The real quantity of precipitated α-Al(FeMn)Si or β-Al(FeMn)Si and Al_5FeSi in the melt at different temperatures is a matter of experimental investigation, which will be presented in future publications.

Author Contributions: A.A. conceptualized the work. B.F. was the principal investigator and supervisor. M.G. performed the experiments. A.A. and M.G. analyzed the data. M.G. wrote and edited the manuscript. C.L. revised the manuscript.

Funding: This research was funded by The DFG (Deutsche Forschungsgemeinschaft)–German Research Society project "Refining of Al–Si melts" FR 1713/5-1.

Acknowledgments: The DFG–German Research Society is greatly appreciated for the support of the project "Refining of Al–Si melts" FR 1713/5-1.

Conflicts of Interest: The authors declare no conflict of interest.

References

1. Ostermann, F. *Anwendungstechnologie Aluminium*, 1st ed.; Springer-Verlag: Berlin/Heidelberg, Germany; New York, NY, USA, 1998; p. 18. ISBN 978-3-662-05788-9.
2. *Liste der Aluminiumgusslegierungen*; Angaben von VAR Verband der Aluminiumrecyclingindustrie e.V.: Düsseldorf, Germany, 2004.

3. Gao, J.; Shu, D.; Wang, J.; Sun, B. Effects of $Na_2B_4O_7$ on the elimination of iron from aluminum melt. *Scr. Mater.* **2007**, *57*, 197–200. [CrossRef]
4. Mohanty, B.P.; Subramnian, S.; Hajra, J.P. Electroslag refining of commercial aluminium. (Retroactive coverage). *Trans. Indian Inst. Met.* **1996**, *39*, 646–651.
5. Zhang, J.; Sun, B.; He, B.; Mao, H.; Chen, G.; Ge, A. Principle and control of new-style purification equipment of 5 N high purity aluminum. *Chin. J. Mech. Eng.* **2006**, *42*, 64–68. [CrossRef]
6. Zhao, H.; Lu, H. The development of 85kA three-layer electrolysis cell for refining of aluminium. *TMS Light Metals.* **2008**, 533–535.
7. Van Der Donk, H.M.; Nijhof, G.H.; Castelijns, C.A.M. The removal of iron from molten aluminum. In Proceedings of the Third International Symposium: Recycling of Metals, and Engineered Materials, Point Clear, AL, USA, 12–15 November 1995; pp. 651–661.
8. De Moraes, H.L.; De Oliveira, J.R.; Espinosa, D.C.R.; Tenorio, J.A.S. Removal of iron from molten recycled aluminum through intermediate phase filtration. *Mater. Trans.* **2006**, *47*, 1731–1736. [CrossRef]
9. Matsubara, H.; Izawa, N.; Nakanishi, M. Macroscopic segregation in Al-11 mass% Si alloy containing 2 mass% Fe solidified under centrifugal force. *J. Jpn. Inst. Light Met.* **1998**, *48*, 93–97. [CrossRef]
10. Mondolfo, L.F. *Aluminium Alloys: Structure and Properties*, 1st ed.; Butterworths: London, UK; Boston, MA, USA, 1976; pp. 282–289, 534–536, 529–530, 661–663. ISBN 0408706805.
11. Petzow, G.; Effenberg, G. *Ternary Alloys*, 1st ed.; VCH Publishers: New York, NY, USA, 1992; Volume 5, pp. 250–264, 394–438. ISBN 0895738953.
12. Zakharov, A.M.; Gul'din, I.T.; Arnol'd, A.A.; Matsenko, Y.A. Phase Equlibria in the Al-Si-Fe-Mn System in the 10–14%Si, 0–3%Fe and 0–4%Mn Concentration Ranges. *Izvest Vyssh Uchebn Zaved Tsvetn Metall.* **1988**, *4*, 89–94.
13. DIN EN 1706:1998-06. Available online: https://www.beuth.de/en/standard/din-en-1706/3359595 (accessed on 19 August 2018).
14. Villars, P.; Prince, A.; Okamoto, H. *Handbook of Ternary Alloy Phase Diagramms*, 1st ed.; ASM International: Geauga Country, OH, USA, 1995; Volume 3, pp. 3501–3514, 3597–3621. ISBN 0871705257.
15. Belov, N.A.; Eskin, D.G.; Aksenov, A.A. *Multicomponent Phase Diagrams: Applications for Commercial Aluminium Alloys*, 1st ed.; ELSEVIER Ltd.: Oxford, UK, 2005; pp. 1–19. ISBN 0080445373.
16. Rhines, F.N. *Phase Diagrams in Metallurgy, Their Development and Application*, 1st ed.; McGraw-Hill Book Company Inc.: New York, NY, USA, 1956; pp. 186, 197, 220–229, 290–302. ISBN 0070520704.
17. Gnatko, M. Untersuchungen zur Entfernung von Eisen aus Verunreinigten Aluminiumgusslegierungen Durch Intermetallische Fällung. Ph.D. Thesis, RWTH Aachen University, Aachen, Germany, 19 May 2008.
18. Hanemann, H.; Schrader, A. *Ternäre Legierungen des Aluminiums, Atlas Metallographicus*, 1st ed.; Band III, Teil 2; Verlag Stahleisen GmbH: Düsseldorf, Germany, 1952; pp. 35–38.
19. Atlas of Microstructures of Industrial Alloys. In *Metals Handbook*, 8th ed.; ASM: Metals park, OH, USA, 1961–1972; Volume 7, pp. 256–263.
20. Zakharov, A.M. *Phase Diagrams of Binary and Ternary Systems*, 3rd ed.; Metallurgia: Moskwa, Russia, 1990; pp. 230–239. ISBN 5229005173.

© 2018 by the authors. Licensee MDPI, Basel, Switzerland. This article is an open access article distributed under the terms and conditions of the Creative Commons Attribution (CC BY) license (http://creativecommons.org/licenses/by/4.0/).

Article

A Holistic and Experimentally-Based View on Recycling of Off-Gas Dust within the Integrated Steel Plant

Anton Andersson [1,*], Amanda Gullberg [2], Adeline Kullerstedt [2], Erik Sandberg [2], Mats Andersson [3], Hesham Ahmed [1,4], Lena Sundqvist-Ökvist [1,2] and Bo Björkman [1]

1. Department of Civil, Environmental and Natural Resources Engineering, Luleå University of Technology, 97187 Luleå, Sweden; hesham.ahmed@ltu.se (H.A.); lena.sundqvist-oqvist@ltu.se (L.S.-Ö.); bo.bjorkman@ltu.se (B.B.)
2. Swerea MEFOS, 97125 Luleå, Sweden; amanda.gullberg@swerea.se (A.G.); adeline.morcel@gmail.com (A.K.); erik.sandberg@swerea.se (E.S.)
3. SSAB Europe, 97437 Luleå, Sweden; mats.andersson@ssab.com
4. Central Metallurgical Research and Development Institute, Cairo 124 22, Egypt
* Correspondence: anton.andersson@ltu.se; Tel.: +46-920-493-409

Received: 4 September 2018; Accepted: 20 September 2018; Published: 25 September 2018

Abstract: Ore-based ironmaking generates a variety of residues, including slags and fines such as dust and sludges. Recycling of these residues within the integrated steel plant or in other applications is essential from a raw-material efficiency perspective. The main recycling route of off-gas dust is to the blast furnace (BF) via sinter, cold-bonded briquettes and tuyere injection. However, solely relying on the BF for recycling implicates that certain residues cannot be recycled in order to avoid build-up of unwanted elements, such as zinc. By introducing a holistic view on recycling where recycling via other process routes, such as the desulfurization (deS) station and the basic oxygen furnace (BOF), landfilling can be avoided. In the present study, process integration analyses were utilized to determine the most efficient recycling routes for off-gas dust that are currently not recycled within the integrated steel plants of Sweden. The feasibility of recycling was studied in experiments conducted in laboratory, pilot, and full-scale trials in the BF, deS station, and BOF. The process integration analyses suggested that recycling to the BF should be maximized before considering the deS station and BOF. The experiments indicated that the amount of residue that are not recycled could be minimized.

Keywords: recycling; cold-bonded briquettes; blast furnace; desulfurization; basic oxygen furnace; dust; sludge; fines

1. Introduction

The production of steel in integrated steel plants generates a considerable amount of solid residues, such as dust, sludges, slags, and scales. Some of these residues have chemical compositions reflecting the raw materials charged to the process, whereas other residues (mainly slags) have properties suitable for external applications. Recycling of the residues within the process or via utilization in other areas is essential for sustainable steel production from the perspective of raw-material efficiency. However, the recycling has to be economically justified and compatible from a process-technical standpoint.

The residues generated within the integrated steel plant differs between sites, depending on things like gas-cleaning equipment, hot metal treatment (e.g., dephosphorization and/or desulfurization), and rolling operation. In crude steel production, the major residues generated in the treatment of off-gases are BF dust, BF sludge, BOF dust, and BOF sludge.

The off-gas dust generated in the production of crude steel contains useful elements, such as iron, carbon, and calcium, as stated in Table 1. Therefore, recycling of these residues within the integrated steel plant has been thoroughly studied, and industrial use has been developed. In pellet-based BF operation, in-plant residues can be included in cold-bonded briquettes that are top-charged into the BF [1]. If the BF operates on sinter, residues can be included in the sintering mix [2]. Furthermore, BF dust injection in the tuyeres has also been reported as an industrial operation practice [1].

Table 1. Typical composition in wt.% of selected off-gas dusts from the blast furnace (BF) and basic oxygen furnace (BOF) [3].

Residue	Fe	C	CaO	SiO$_2$	MgO	Zn	S
BF dust	15–40	25–40	2–8	4–8	0.3–2	0.1–0.5	0.2–1.3
BF sludge	7–35	15–47	3.5–18	3–9	3.5–17	1–10	2.4–2.5
BOF coarse dust	30–85	1.4	8–21	-	-	0.01–0.4	0.02–0.06
BOF fine dust	54–70	0.7	3–11	-	-	3–11	0.07–0.12
BOF sludge	48–70	0.7–4.6	3.0–17	-	-	0.2–4.1	0.03–0.35
BOF primary dedusting	38–85	0.1–6.5	5.7–40	-	-	0.1–1.5	0.02–1.3
BOF secondary dedusting	32–63	1.0–8	3.7–35	-	-	0.5–13	0.1–1.1

Although thoroughly studied, complete recycling of these residues has not been achieved. The challenges of recycling off-gas dusts to the BF arise when levels of tramp elements, mainly zinc, reach undesired levels. Which levels are considered undesirable differs between sites. However, 150–400 g of zinc per ton of hot metal (HM) are typical values reported as acceptable in operations [4]. In the BF, zinc compounds are reduced to metallic zinc vapor by CO-rich gas in the lower regions of the shaft. The zinc vapor follows the ascending gas and is reoxidized to zinc oxide in the colder parts of the furnace. The zinc reoxidizes and condenses on the walls, the burden material, coke, or fines carried by the gas phase. In the latter case, zinc may exit the BF through the off-gas. The zinc deposited on the burden travels down to the lower region where it is reduced and volatilized again, thus forming cyclical behavior. This means that the BF has a circulating load of zinc. The negative effects of high-circulating loads of zinc in the BF includes increased consumption of reducing agents, reduced carbon-brick-lining life, and scaffold formation, which may ultimately lead to disturbances in the burden descent [4].

The main output of zinc from the BF is via the top-gas [5], i.e., the BF dust and sludge. If the dust is recycled internally to the BF, the sludge cannot be recycled, as this would reintroduce the main output of zinc from the BF back to the BF. Furthermore, as there are no external industrial-scale operations utilizing BF sludge, this fraction would be landfilled within the integrated steel plant. This has been recognized and the removal of zinc from BF sludge and recycling of the low-zinc fraction via the sinter [6] or cold-bonded pellets [7] to the BF has been implemented in full-scale operations. However, on-site recycling of the high-zinc fraction generated in the dezincing process has not been reported.

Recycling of the off-gas dust from the BOF to the BF has been successfully achieved using both cold-bonded briquettes [1] and sinter [8]. Again, one of the limiting factors in recycling the dust generated in the BOF process is the zinc content. In the case of BOF dust, the main input of zinc is via the cooling scrap charged to the converter. The zinc content in BOF dust has been addressed by hydrometallurgical approaches [9–12] and by employing a coke breeze-less sintering operation [13]. Also, as zinc evaporation mainly occurs early in the converting process, the possibility of in-process separation of zinc has been suggested [14]. In-process separation of zinc has also been addressed by optimizing the design of the gas-cleaning equipment [15,16]. Another way to enable recycling of a major portion of the BOF dust back to the BF is by avoiding the use of scrap qualities containing zinc by minimizing the zinc input to the BOF [1].

The challenge of zinc mainly applies when considering recycling of off-gas dust to the BF. Thus, if other recycling routes are considered, the raw-material efficiency within the integrated steel plant can be improved. The BOF has been acknowledged as an alternative route for recycling off-gas dust [8,17–20]. In one publication, recycling to the BOF by replacing the sinter coolant was recognized to be limited in tonnage [17]. However, full-scale trials have shown that off-gas dust can successfully be recycled via cold-bonded agglomerates to the BOF in amounts of 23 [18] and 40 kg/tHM [19]. Furthermore, cold-bonded briquettes were shown to be suitable for the recycling of all BOF sludge back to the BOF [20]. In addition, hot briquetting has been employed to recycle the BOF dust back to the BOF in industrial practice [8]. Nonetheless, adopting the BOF recycling route still requires considerations of zinc, especially when BOF dust is recycled to the BF. If the BOF dust is recycled in a closed-loop system to the BOF, zinc can be concentrated in the BOF dust [14]. When the zinc content reaches a certain level, zinc producers can utilize the dust [14].

Based on the above, means for on-site recycling of off-gas dust from the integrated steel plants have already been far-developed. However, there are still residues difficult to recycle, and a holistic view of recycling within the process chain is required in order to find solutions to this issue. The present paper sets out to develop such a holistic approach. Utilizing process-integration analyses, considering the effects on raw materials and energy consumption for steel production, and different recycling scenarios were studied. Based on the results of these analyses, experiments were conducted to analyze recycling approaches that maximize the raw material and energy efficiency while addressing the challenges of zinc. The approach included recycling of cold-bonded agglomerates to the BF, deS station, and BOF.

2. Materials and Methods

2.1. Process Integration Analyses

The reference case used in the process integration analyses was the present scenario of in-plant recycling at the two integrated steel plants in Sweden. The change in energy consumption for a fixed crude steel production was considered in different recycling scenarios. These scenarios are presented as the five cases shown in Figure 1. In the figure, the leftmost column presents the annual generation of non-recycled off-gas dust generated at the BF, deS station, and BOF. The process integration analyses were performed using the Excel spreadsheet-based model, TOTMOD. This model is based on the spreadsheet model Masmod, presented by Hooey et al. [21]. The BF, deS station, BOF, and upgraded method of BF sludge were included in the calculations. In addition, the calorific value of the BF gas and the consumption of gas in the hot stoves were considered. Furthermore, an estimation of the change in energy consumption corresponding to the reduced or increased charging rate of coke and pellets were included.

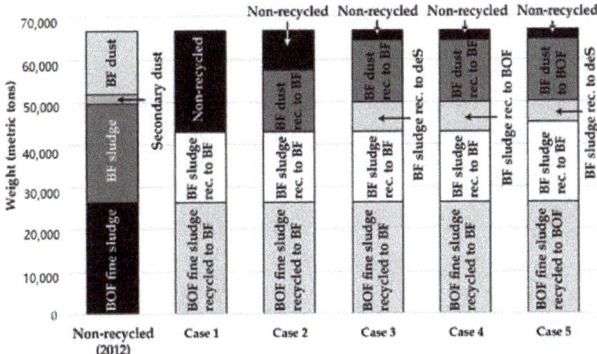

Figure 1. Annual generation of residues not recycled as of 2012, as well as recycling considered in cases 1 through 5. All weights are given in dry weights.

Case 1 through 5 in Figure 1 includes recycling of the BF sludge to the BF. Therefore, the BF sludge has to be dezinced prior to recycling to avoid the accumulation of zinc in the BF. The BFs in Sweden operate on pellets, and use cyclones recovering approximately 80% of the off-gas dust as dry BF dust in the primary gas-cleaning equipment. Research on dezincing of sludges collected under those conditions when only the finest particles are collected in wet gas cleaning were missing. Therefore, experiments aiming at upgrading the BF sludge by generating a low-zinc and high-zinc fraction were performed.

2.2. Upgrading of BF Sludge

In order to upgrade the BF sludge, physical separation methods and a hydrometallurgical approach were tested on a sludge sample, with a d_{90} of 25.0 µm, provided by SSAB Merox. Hydrocycloning and tornado-processing were employed as the physical separation methods. The hydrocycloning of BF sludge has been presented in a previous publication [22]. The tornado process is a high-velocity dry cyclone utilizing pre-heated air, as described by Tikka et al. [23,24]. The hydrometallurgical process employed was leaching in sulfuric acid at different pH levels at 80 °C, as described previously [22]. After generating two fractions of the BF sludge, recycling of the low-zinc fraction to the BF was studied.

2.3. Recycling to the BF

2.3.1. Experiments in Laboratory Scale and Pilot-Plant Scale

Recycling of the low-zinc fraction of upgraded BF sludge to the BF via cold-bonded briquettes was studied in laboratory-scale and pilot-plant scale experiments. The iron, carbon, and zinc content of the low-zinc fraction was 38.1%, 27.1%, and 0.24%, respectively. The different recipes for the briquettes are presented in Table 2. The reference recipe represented a briquette composition used in industrial practice at SSAB in Luleå. Upgraded BF sludge from the tornado process was used in the B1 and B2 recipes.

Table 2. Recipes of the briquettes used in the laboratory scale and pilot-plant scale BF experiments.

Recipe	Upgraded BF Sludge	deS Scrap	BOF Coarse Sludge	BOF Fine Sludge	Briquette Fines	BF Dust	Cement
Ref.	0.0	36.0	18.0	12.0	12.0	10.0	12.0
B1	10.0	31.4	15.7	10.5	10.5	10.0	12.0
B2	20.0	26.8	13.4	8.9	8.9	10.0	12.0

After briquetting, the tumbling index (TI) was determined after 24 h and 28 days of curing in ambient room conditions. The measurements were made in accordance with a modified version of ISO 3271, the modification being the final sieving performed using a 6.0 mm instead of a 6.3 mm sieve.

The reduction of the different briquettes in BF shaft conditions were studied using a laboratory-scale BF shaft simulation experiment. The equipment has been described previously by Robinson [25]. Two programs were run, one representing the descent of the briquette along the wall of the BF and one of the descent in the center. The two programs are depicted in Figure 2. A total gas flow of 50 Nl/min was used in the experiments. The mechanical pressure applied on the sample did not affect the total pressure of the gas phase, as shown in Figure 2.

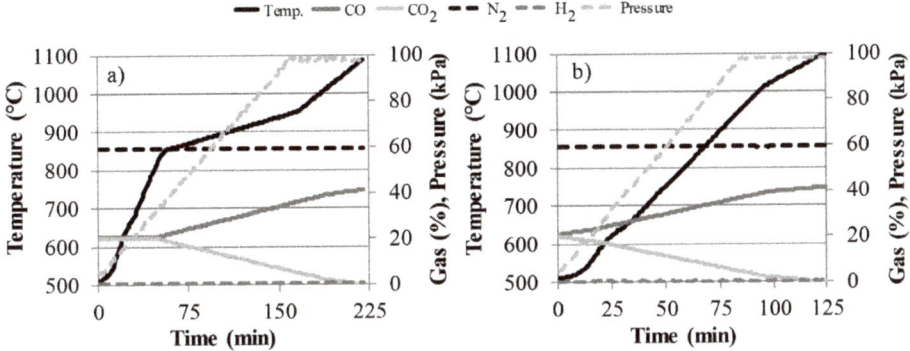

Figure 2. Heating profile, gas profile, and mechanical pressure during the (**a**) wall program and (**b**) center program of the laboratory-scale BF experiment.

Briquettes produced using the recipes of Table 2 were also charged as basket samples to the LKAB (a mining company in Luleå, Sweden) Experimental Blast Furnace (EBF). A thorough description of the LKAB EBF has been provided previously [26]. The weight of the briquettes was recorded when preparing the cylinder-shaped steel-wire baskets. At the end of the campaign, the baskets were charged in eight different coke layers. All baskets contained one briquette of each type. The baskets descended together with the burden material until the EBF was quenched with nitrogen gas. Subsequently, the excavation was carried out by carefully measuring, examining, and photographing the different layers. During the excavation, baskets in six out of the eight layers were retrieved and analyzed.

The laboratory and pilot-plant scale experiments were evaluated by crushing, grinding, and splitting the briquettes and analyzing the sub-samples for the chemical composition and mineralogy. The chemical composition was determined using X-ray fluorescence (XRF) analysis (Malvern Panalytical, Almelo, The Netherlands). Furthermore, the oxidation degree of iron was determined using ISO 2597 and the carbon and sulfur content was determined with a LECO combustion CS444 analyzer (LECO, St. Joseph, MI, USA) with an infrared detector. X-ray diffraction (XRD) (Malvern Panalytical, Almelo, The Netherlands) was used to study the mineralogy.

2.3.2. Full-Scale Trials in the BF

The results of the laboratory-scale and pilot-plant scale experiments were verified in full-scale trials in BF No. 3 at SSAB Luleå. The upgrading of BF sludge, presented in Section 2.2, was done in laboratory-scale experiments. Therefore, non-upgraded BF sludge was utilized in the full-scale trials. Two briquette recipes were studied: one reference briquette (RB), and one briquette containing BF sludge (BSB). In the latter recipe, part of the deS scrap was substituted with BF sludge, Table 3. The briquettes were produced in industrial scale according to the standard method employed at SSAB Luleå. The strength of the briquettes was evaluated after one day and after three weeks using the same TI method as previously described.

After approximately three weeks of curing, the BSBs were charged over three days to BF No. 3 in SSAB Luleå, using a charging rate that averaged at 97.3 kg/tHM. A reference period with three days of stable operation was selected, during which the RBs were charged at an average rate of 99.6 kg/tHM. The evaluation of the full-scale trials was conducted by studying changes in generated sludge and dust amounts and their compositions. Also, the effect on the BF process was analyzed using operational data, direct reduction rate, and mass and energy-balance calculations, deducing things such as the carbon consumed by the process.

Table 3. Recipes of the BF sludge briquette (BSB) and reference briquette (RB) used in the full-scale BF trials (wt.%).

Recipe	Steel Scrap	deS Scrap	BOF Coarse Sludge	BOF Fine Sludge	Briquette Fines	Mill Scale
BSB	10.0	22.5	6.6	6.6	20.6	2.0
RB	10.0	26.3	6.6	6.6	20.6	2.0
Recipe Cont.	BF Dust (Stored)	BF Dust (Fresh)	BF Sludge	Filter Dust	Cement	Water
BSB	4.0	7.6	3.8	1.9	11.6	2.9
RB	4.0	7.6	0.0	1.9	11.6	2.9

2.4. Recycling of Off-Gas Dust to the Steel Shop

2.4.1. Experiments in Laboratory Scale

As only the low-zinc fraction of upgraded BF sludge can be recycled to the BF, the high-zinc fraction has to be recycled in the steel shop. The iron, carbon, and zinc content of this fraction was 29.6%, 19.5%, and 2.18%, respectively. The high-zinc fraction of the tornado-treated BF sludge was incorporated in cold-bonded briquettes and pellets using the recipe presented in Table 4. The mixture was designed to form a self-reducing agglomerate. Screening of the pellets was performed to achieve a narrow fraction between 9.5 and 10 mm.

Table 4. Recipe of the briquettes and pellets used in the laboratory-scale smelting reduction experiments (wt.%).

High-Zinc Fraction of BF Sludge	deS Scrap	Secondary Dust	Cement
25	50	15	10

The briquettes and pellets were subjected to lab-scale smelting reduction experiments to study the melt-in behavior in conditions similar to charging the agglomerates in a ladle with hot metal. The experiments using the briquettes were performed in an induction furnace with 80 kg of hot metal. A smaller induction furnace with 10 kg of hot metal was used in the experiments for testing the pellets. In both cases, the hot metal was taken from SSAB Luleå and the temperature of the melt during the experiments was 1350 °C. The principle of the tests was the same in both setups: an agglomerate was added to the surface of the melt and removed after predetermined times and quenched in nitrogen gas. XRF analysis, titration, and LECO analysis were employed to analyze the chemical composition of the agglomerates. Furthermore, XRD was used to determine the mineralogical composition. The mass loss of the agglomerates was also recorded.

2.4.2. Full-Scale Trials in the deS Station and BOF

After the laboratory-scale experiments were performed, full-scale trials in the deS station and BOF were executed. Again, the upgrading of the BF sludge was made in laboratory scale, meaning that the high-zinc fraction of BF sludge could not be included in the cold-bonded briquettes used in the full-scale experiments. Instead, fine and coarse BOF sludge were used, shown in Table 5. The upgraded BF sludge was assumed to have sufficiently similar characteristics to the BOF sludges to partly replace these residues in future recipes. The steel scrap fines shown in Table 5 comes from the BOF process; it consists of material from the treatment of skulls and material from slopping during the blowing. In order to balance the water content of the mixture prior to briquetting, dry-cast house dust from the BF and water were added.

Table 5. Recipe used for the briquettes used in the full-scale trials in the steel shop (wt.%).

Steel Scrap Fines	BOF Coarse Sludge	BOF Fine Sludge	Mill Scale from Cont. Casting	Cast House Dust	Water	Cement
44	22	18	4	1	1	10

Prior to charging the briquettes to the deS station and BOF, the briquettes were dried to 1.2 wt.% moisture to avoid incidents of smaller explosions. In the deS station, the briquettes were added in ten different trials in amounts ranging from 100 to 300 kg per heat. The additions were made to a ladle holding small amounts of hot metal in the bottom. After adding the briquettes, hot metal from the torpedo car was tapped into the ladle. The melt-in was studied visually and the effect of the addition on the final steel quality was evaluated. The charging of the dried briquettes to the BOF was made together with the steel scrap. Nine trials with an amount of 600 to 1250 kg of briquettes per heat were performed.

The results of the process integration analyses and feasibility for recycling methods based on experimental procedures were used in developing a holistic approach towards the recycling of the off-gas dust.

3. Results and Discussion

3.1. Process Integration Analyses

The category labeled non-recycled in Figure 1 illustrates the annual generation of the residues considered in the analyses. The remaining categories in the figure illustrates the increased raw-material efficiency corresponding to each calculation case. The first two cases consider recycling of BOF fine sludge and upgraded BF sludge to the BF. In addition to these residues, the second case considers increased recycling of BF dust back to the BF as well. Based on the second case, the third and fourth case considers the additional recycling of the high-zinc fraction of upgraded BF sludge to the deS and BOF, respectively. The fifth case considers a different scenario where the majority of the residues are recycled to the BOF. In this case, the BF sludge is not upgraded—instead, part of the sludge is recycled to the deS station while the rest is recycled to the BF.

Case three through five have the highest recycling rates. The fine-grained residue not recycled in these cases is the secondary dust. All materials cannot be recycled from a technical point of view due to the accumulation of tramp elements in the process. The secondary dust is a feasible stream to recycle outside the integrated steel plant, as the tonnage of this residue is, by far, the lowest of these fine-grained residues.

Figure 3a illustrates the change in energy consumption in the process system corresponding to the different calculation cases. By summarizing the effect of each individual process, the net change in energy consumption was calculated for each case, shown in Figure 3b. The net change consistently decreases from case one to case four. The energy savings stem from the decreased specific consumption of coke and iron ore pellets connected to the recycling of the iron and carbon in the residues. The most efficient decrease in the net energy consumption was calculated for case four, where a total decrease of 126 GWh/year was estimated.

Unlike cases one to four, case five mainly considers the recycling of the residues to the steel shop. The calculations suggest that recycling the residues in this manner would generate an increased net energy consumption of 26 GWh/year. This can be explained by the fact that the addition of agglomerates to the BOF will decrease the scrap capacity due to the excess heat required for melting and reduction. To maintain the fixed crude steel production used in the calculations, the hot metal production needs to be increased. This results in a higher energy consumption at the BF, as compared to the reference case.

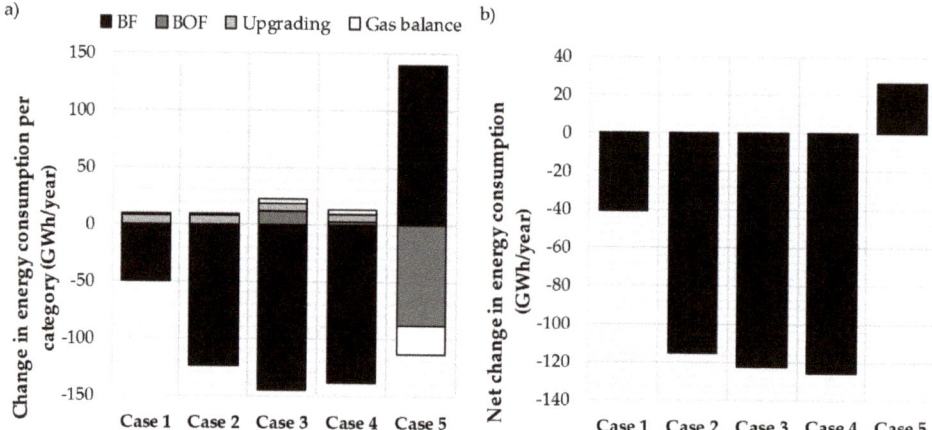

Figure 3. Results of the process integration analyses: (**a**) Change in energy used in each process; (**b**) net change in energy consumption.

The calculations suggest that the residues not recycled as of 2012 should primarily be recycled to the BF prior to considering the steel shop. Case four considers recycling of the remaining BF dust to the BF. Furthermore, BOF fine sludge and the low-zinc fraction of upgraded BF sludge are also considered for the BF. The BOF is considered for the recycling of the high-zinc fraction of the BF sludge.

3.2. Upgrading of Blast Furnace Sludge

In order to achieve recycling for the most promising cases shown in Figures 1 and 3, the BF sludge has to be upgraded, creating a low-zinc and high-zinc fraction. Table 6 presents the results from the upgrading methods applied to the BF sludge of the present study. Considering the performance of the different methods, the leaching in sulfuric acid at pH 1 and 80 °C was most promising in terms of removing zinc. However, leaching in sulfuric acid at pH 3 and 80 °C resulted in a higher recovery of iron and solids in the low-zinc fraction. The leaching time at pH 1 and 3 was 30 min and 6 h, respectively. Nonetheless, the sampling during the leaching process indicated that the zinc was successfully leached within 15 min at pH 1 and 1 h at pH 3. After the leaching process, the zinc in the solution can be precipitated by adding alkali carbonates forming zinc carbonate. Thermal decomposition of the zinc carbonate can be applied to form zinc oxide, which may be used by zinc producers to produce metallic zinc [22].

Table 6. Results of the upgrading of BF sludge.

Method	% of Total Zinc in High-Zinc Fraction	% of Total Iron in Low-Zinc Fraction	% of Total Carbon in Low-Zinc Fraction	% of Total Solids in Low-Zinc Fraction
Leaching, pH 1	95	91	100	86
Leaching, pH 3	80	96	100	93
Hydrocyclone	74	66	37	59
Tornado	81	37	39	31

The results of the leaching experiments suggest that 80% of the zinc in the BF sludge was distributed in weak-acid soluble phases, such as zincite (ZnO) and smithsonite ($ZnCO_3$). The remaining 20% of the zinc was distributed as franklinite ($ZnFe_2O_4$).

Using physical separation methods, the results were promising with regard to the removal of zinc. Both the hydrocycloning and the tornado treatment of the sludge proved to be less efficient as compared to the leaching, with regard to recovering the iron, carbon, and solids in the low-zinc fraction. Using ultrasonic sieving, the sludge was separated into narrow size-fractions and analyzed for zinc, iron, and carbon. The finest fraction, less than 5 microns in size, carried the majority of the zinc; namely, 73.6% of the total zinc content in the sludge. In addition, this size fraction carried 49.2%, 47.2% and 43.6% of the solids, iron and carbon, respectively. Therefore, the efficiency of these methods were limited by the distribution of zinc, iron, and carbon in the different size fractions.

The results of the upgrading experiments presented in Table 6 illustrates that BF sludge generated by a BF operating on 100% pellets as ferrous burden and utilizing a cyclone as the primary gas cleaning equipment can be upgraded, creating a fraction containing the majority of the zinc.

Although superior in performance, leaching has not been reported in full-scale operation. However, a mobile pilot plant utilizing hydrochloric acid as a leaching agent has been developed and tested [27]. Nonetheless, continuation of the present study was made based on the tornado-treated sludge. The choice was made based on three principal reasons: (i) the zinc removal was satisfactory; (ii) during the upgrading, the material was simultaneously dried to below 1 wt.% moisture; and mainly, (iii) this process was the only one handling enough BF sludge required for the subsequent experiments.

3.3. Recycling to the Blast Furnace

3.3.1. Experiments in Laboratory Scale and Pilot-Plant Scale

The cold strength of the cold-bonded agglomerates is essential when determining whether the agglomerates can sustain the conditions inside the BF. Inadequate cold strength leads to breakage during material handling and charging, which results in increased dust formation from the BF. Furthermore, low cold-strength may cause the agglomerates to disintegrate and impair the gas permeability in the furnace. The tumbling indices of the reference B1 and B2 briquettes were determined as 78%, 84%, and 81%, respectively. Thus, including up to 20 wt.% of the low-zinc fraction of the tornado-treated BF sludge to the cold-bonded briquette recipe, shown in Table 2, resulted in briquettes with sufficient tumbling strength to be top-charged into the BF.

The laboratory-scale BF shaft simulation experiments were used to study the reducibility of the three different briquette types. Figure 4a,b shows the diffractograms of each briquette after the wall and center program, respectively. In all cases, metallic iron was the only iron phase, as hematite, magnetite and wüstite were reduced to a level below the detection limit of the XRD. Furthermore, the briquettes with added upgraded BF sludge were consistently less disintegrated and harder to break. These visual observations are in agreement with the tumble indices. Furthermore, the observations are in line with the results presented by Singh and Björkman [28,29], who reported that cold-bonded briquettes with coarser particles had a greater tendency to disintegrate in the LKAB EBF. The d_{50} of the reference, and B1 and B2 recipes in the present study were determined to be 255, 185, and 145 µm, respectively.

The results of the laboratory-scale BF shaft simulation experiments were considered promising and the briquettes were charged as basket samples to the LKAB EBF.

The mass loss of the three different briquettes with respect to the descent in EBF is presented in Figure 5. The reactions presented in the figure were based on the diffractograms of the briquettes at each location. Briquettes descending from the stockline to 1715 mm below the stockline was associated with the reduction of hematite (Fe_2O_3) to magnetite (Fe_3O_4) and magnetite to wüstite (FeO), and it the calcination of calcite ($CaCO_3$) had also started. Reaching 2422 mm below the stockline, the calcination was completed and the reduction of wüstite to metallic iron had finished in all briquettes. All briquettes retrieved at and below 4246 mm below the stockline were partly broken. Also, cementite (Fe_3C) formation was observed in all briquettes at these levels.

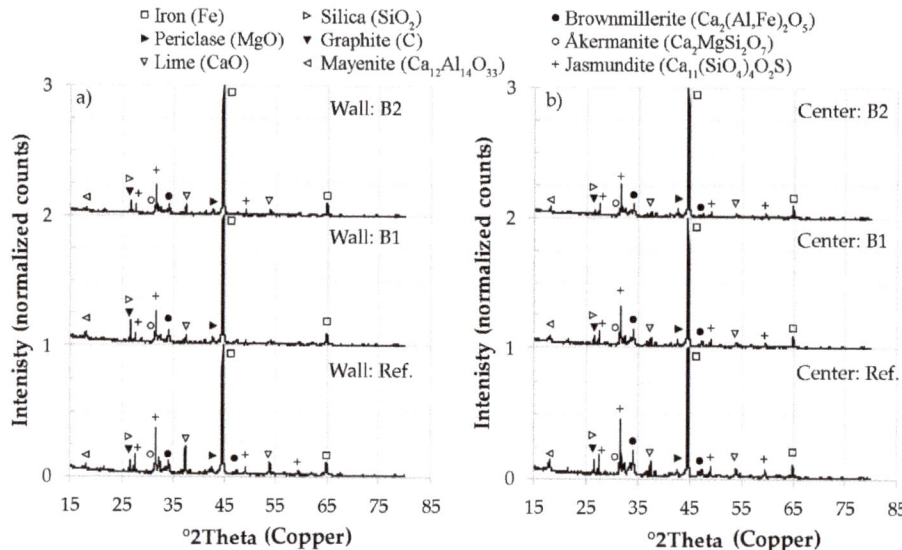

Figure 4. Diffractograms of the briquettes that have gone through the (**a**) wall program and (**b**) center program of the laboratory-scale BF experiments.

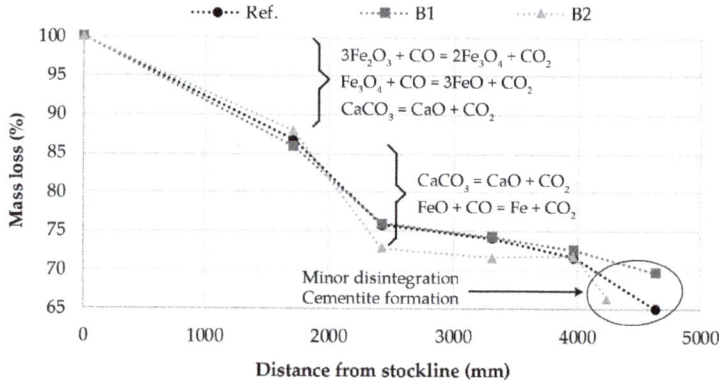

Figure 5. Mass loss and reactions during the descent of the briquettes in the LKAB EBF.

Both the laboratory and pilot-plant scale experiments suggested that adding up to 20 wt.% of upgraded BF sludge to a top-charged cold-bonded briquette is feasible in terms of strength and reduction. Based on these results, a decision to charge briquettes containing 3.8 wt.% of non-upgraded BF sludge, shown in Table 4, to BF No. 3 at SSAB Luleå was made. The non-upgraded BF sludge was used as an approximation of the low-zinc fraction of upgraded BF sludge, as no full-scale upgrading method was available. The more conservative addition of 3.8, as compared to 20 wt.%, was based on the required addition to completely recycle the annual generation of BF sludge.

3.3.2. Full-Scale Trials

The effect of the addition of BF sludge on the cold strength of the briquettes used in the full-scale BF was assessed with regard to the feasibility of top-charging the agglomerates. Table 7 presents the results of the tumbler test experiments. Replacing deS scrap by BF sludge decreased the cold strength both after one day and after three weeks of curing. The general trend during the curing process is an increase in the TI value after the prolonged curing as observed for the RB and the first batch of the BSB. Although the BSB briquettes had lower TI values, the required cold strength for top-charging into BF No. 3 at SSAB Luleå was met. Thus, the briquettes with BF sludge were considered suitable to charge in the full-scale BF.

Table 7. Tumbling strength (TI values in %) of the reference briquettes (RB) and BF sludge briquettes (BSB).

Recipe	1 Day Curing	21 Days Curing
RB	74	81
BSB (first batch)	61	74
BSB (second batch)	69	68

No disturbances could be attributed to the BSB during the full-scale trials, suggesting that the lower cold strength, shown in Table 7, did not affect the operation. Considering these results, the recycling of BF sludge to the commercial-scale BF via the cold-bonded briquettes was achieved without any negative effect on the operation linked to the briquettes. Thus, the trials in laboratory, pilot plant, and full scale showed that cold-bonded briquettes can be used to recycle upgraded BF sludge to the BF.

The zinc contents of the RB and BSB were determined to be 0.076 and 0.081%, respectively. Therefore, when adding 100 kg of briquettes per ton of hot metal, the increased zinc load was 5 g/tHM. The zinc load from the primary raw materials charged to BF No.3 varied between 30 and 41 g/tHM. Thus, as the zinc input to the furnace from the primary raw materials was reasonably low, the increased zinc load of 81 g/tHM for the BSB instead of the reference scenario of 76 g/tHM for the RB was considered acceptable.

Based on the rate of addition of the cold-bonded briquettes and the annual production of hot metal from BF No. 3, 11.4 tons of upgraded BF sludge can be recycled via cold-bonded briquettes each year. This covers the annual on-site generation of BF sludge.

3.4. Recycling to the Steel Shop

3.4.1. Laboratory Scale Experiments

In order to completely recycle the BF sludge, the high-zinc fraction has to be recycled to the steel shop. From an energy-efficiency standpoint, recycling to the BOF is preferred over the deS station, shown in Figures 1 and 3. However, recycling to the BOF is accompanied by sulfur pick-up in the crude steel [19]. This sulfur comes from the cement and residues in the briquettes. Therefore, adding the briquettes to the deS station, prior to the deS of the hot metal, is of interest during the production of steel grades with low sulfur content.

The melt-in behavior of the cold-bonded briquettes of Table 4 was studied in laboratory scale. In the full-scale process, the briquettes would be charged to a ladle with small amounts of hot metal. After the charging of the briquettes, the hot metal from the torpedo car would be poured into the ladle. Thereafter, the ladle would be transported to the deS station. The time required for pouring hot metal from the torpedo and transporting the ladle to the deS station was approximately ten minutes. Therefore, ten minutes was chosen as the longest time the briquettes were in contact with the melt in the laboratory-scale experiments. The propagation of the melt-in of the briquettes during these experiments is presented in Figure 6. A majority of the briquette was still to be melted after ten minutes, suggesting that melt-in problems can be expected in the full-scale process.

Figure 6. Photographs of the briquettes of Table 4 after contact with hot metal at 1350 °C for (**a**) 1 min, and (**b**) 10 min [30].

In order to study the propagation of the reduction in detail, XRD was run on samples within the briquette that were in contact with the melt for 6 min. Five samples from the briquette, distributed perpendicular to the surface of the melt, was analyzed using XRD. The results, provided in-depth in a previous publication [30], showed that as the reduction progressed, the reduced part melted and entered the hot metal. Also, the heat surrounding the rim of the briquette allowed self-reduction of the higher iron oxides.

Four stages need to occur in order for the iron in the cold-bonded briquettes to enter the hot metal: (i) heating, (ii) reduction, (iii) carburization of the iron, and (iv) melting of the carburized iron and slag separation [31]. Based on Figure 6, part of the briquette had gone through all stages. However, the results of the XRD suggested that the middle of the briquette was still undergoing the first stage after 6 min of being in contact with the hot metal.

Considering the indicated slow heat transfer, reduction, and melt-in of the briquettes, the idea of using pellets of the same recipe was to allow these smaller agglomerates to fully reduce and enter the melt. The mineralogy of the pellets being in contact with the melt suggested that the iron oxides were reduced to amounts below the detection limit of the XRD after a time of contact between 4 and 8 min, shown in Figure 7a. Thus, the reduction in the briquettes were limited by a combination of the poor melt-in behavior and limited heat transfer. However, although the pellets were completely reduced and smaller in size as compared to the briquettes, they still had melt-in problems, shown in Figure 7b. These results are in line with the conclusions made by Ding and Warner, who found that the reduction of carbon-chromite composite pellets could be considerably faster than the dissolution when subjected to smelting reduction in high-carbon ferrochromium melts [32]. As the pellets were completely reduced, the poor melt-in suggests that either the carburization of iron or the melting and separation of the slag and carburized iron was the limiting step.

Although the results of the laboratory scale experiments suggested melt-in difficulties, the full-scale trials were considered to be of interest due to the mixing effect during the pouring of hot metal from the torpedo. Also, higher hot metal temperatures than tested in the laboratory-scale experiments are possible, which may facilitate the melting and separation of the slag and carburized iron in the agglomerates. In addition, the internal slag composition of the cold-bonded briquettes used in the full-scale trials was designed to have a lower melting point than that of the laboratory-scale trials.

The upgrading of BF sludge was not made in full-scale. Therefore, the recipe of the briquettes tested in the full-scale trials of the present study did not include any BF sludge. Instead, the briquette recipe included BOF fine and coarse sludge. These two residues can partially be replaced by the high-zinc fraction of the upgraded BF sludge. In such a scenario, the difference in carbon content and oxidation degree of iron between the BF and BOF sludges has to be considered.

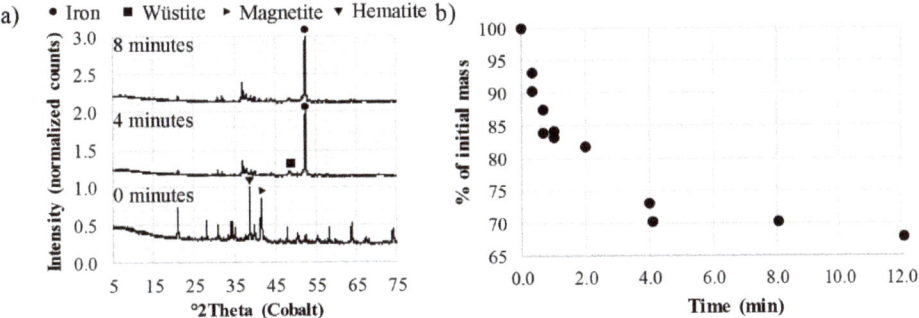

Figure 7. (a) Diffractograms of pellets with 100% peaks of the iron phases denoted; (b) mass loss of the pellets with respect to the time of contact with hot metal at 1350 °C [30].

3.4.2. Full-Scale Trials

Charging the briquettes to the ladle caused minor dusting. However, the moisture content and strength of the briquettes allowed for safe operation without any incidents. The melt-in of the briquettes prior to the deS started was evaluated visually. Charging up to 150 kg of briquettes enabled melting of all added briquettes. In contrast, only partial melt-in was noticed when charging 300 kg per heat. Nonetheless, after the deS process, no briquettes were observed, indicating a successful melt-in. The final steel quality was not compromised in any of the trials, suggesting that up to 300 kg of briquettes was possible to add into the process. This amounts to about 5400 metric tons of briquettes per year. Therefore, the desulfurization station was shown to be a viable recycling route within the steel shop.

Briquettes of the same recipe were charged in amounts of up to 1250 kg per heat in the BOF. The briquettes were charged with the steel scrap, which had several positive outcomes, such as improved slag formation and improved dephosphorization. However, the addition of the briquettes also resulted in increased sulfur content of 6–17 ppm in the crude steel. Therefore, the recycling of these agglomerates to the BOF is restricted to steel qualities that allow slightly higher sulfur content. At the specific plant, 8700 metric tons of briquettes could be added each year.

In total, 14,100 metric tons of briquettes could be recycled annually. The percentage of the high-zinc fraction of BF sludge required to be included in these briquettes to completely recycle the BF sludge depends on the upgrading method employed, shown in Table 6.

3.5. Holistic View on Recycling of Off-Gas Dust

In order to develop a holistic view regarding on-site recycling of off-gas dust, four key aspects should be considered: (i) maximizing the raw-material efficiency, (ii) maximizing the energy-efficiency, (iii) managing tramp elements in the process, and (iv) maintaining the high steel quality and production. The holistic view on recycling which was developed based on the results of the calculations and experimental work of the present study is illustrated in Figure 8. The flowsheet is an extended version of the on-site recycling within the pellet-based integrated steel plant presented by Wedholm [1].

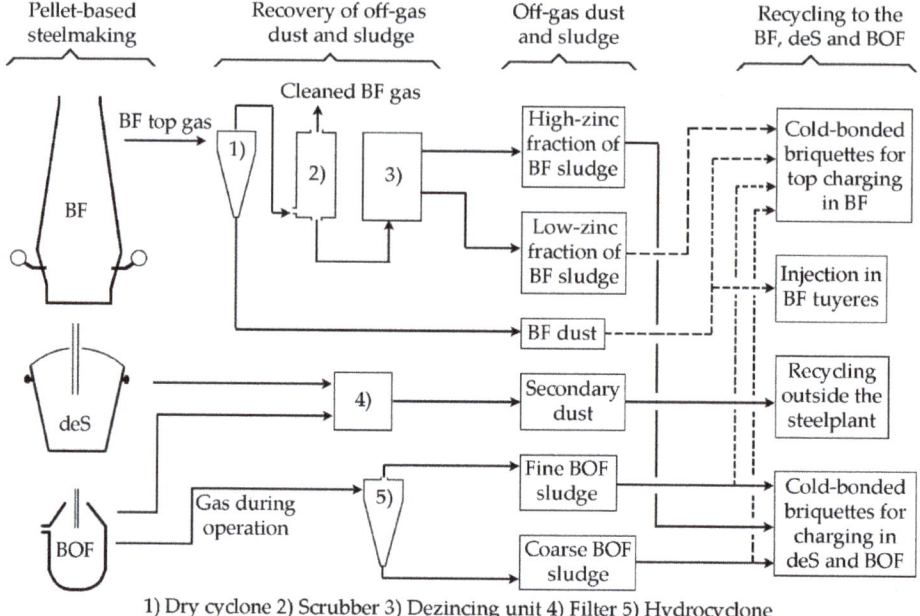

Figure 8. Illustration of the holistic view on recycling of off-gas dust within the integrated steel plant.

In the present study, the process integration analyses suggested that the BF should be utilized as the main recycling route in order to maximize energy efficiency. Therefore, the existing recycling route where the BF dust was generated on-site was completely recycled via the top-charging of cold-bonded briquettes, and injection of BF dust in the tuyeres [1] was maintained in the flowsheet. In order to address the third aspect in the list, the tramp elements, a dezincing step of the BF sludge was introduced. The present study showed that the BF sludge could be upgraded and the low-zinc fraction could be recycled to the BF via the cold-bonded briquettes. This layout allows an outlet of zinc to be introduced from the recycling system of the BF, which mitigates the accumulation and excessive circulating loads of zinc in the BF.

The BOF coarse sludge has previously been successfully recycled via the cold-bonded briquettes to the BF [1]. Also, the BOF fine sludge has recently been included in these briquettes [1,33]. The zinc content of the sludges from the BOF is managed by managing the quality of the cooling scrap [1]. Both of these residues were included in the briquettes used in the experimental work of the present study, Tables 2 and 3, further establishing the possibility of this recycling route.

In order to maximize the raw-material efficiency and manage the tramp elements in the process, part of the off-gas dust has to be recycled to the steel shop. In the present study, residues that are recycled to the BF has been included in the recipes of briquettes that were recycled to the deS station and BOF, shown in Table 5. This contradicts the energy-efficiency maximization, as shown in the results of the process integration analyses in Figure 3. However, including these residues in the briquettes is fundamental to achieve a recipe with a particle-size distribution that is suitable for producing cold-bonded briquettes with adequate properties for handling during recycling.

Recycling via the steel shop using cold-bonded briquettes was shown to be feasible in the present study. By avoiding recycling to steel grades of low sulfur content, the recycling route did not affect the final steel quality. The incorporation of the high-zinc fraction of BF sludge in the briquettes, replacing the BOF sludges, would enable the complete recycling of this residue. The laboratory experiments with the briquettes and pellets containing the high-zinc fraction of the upgraded BF sludge showed melt-in problems in hot metal at 1350 °C. If the addition of this fraction of the BF sludge would facilitate melt-in problems in the deS station, cold-bonded pellets could be used instead of briquettes in order to improve the melt-in. The complete reduction of the pellets, without dissolution in the steel, would still allow them to be recycled as the deS slag is crushed and the magnetic fraction is recycled to the BF via the briquettes.

In the present recycling scenario, the main output of zinc would be the secondary dust. This residue is by far the lowest in tonnage, shown in Figure 1, which poses two benefits: (i) the raw-material efficiency of the on-site recycling is maximized, and (ii) the ability to concentrate tramp elements is easiest. The secondary dust is residue generated in a filter treating the off-gas from the deS station and the off-gas from the BOF prior to the start of blowing. Thus, the zinc being reduced and evaporated from the cold-bonded briquettes charged to the deS station would enter this residue. Furthermore, the zinc in the residues charged together with the cooling scrap to the BOF would at least partly be reduced and evaporated during the charging of desulfurized hot metal to the converter. Thus, an outlet of zinc from the process is created. The efficiency of this outlet depends, to some extent, on the amount of zinc evaporating from the agglomerates prior to the start of blowing. Zinc evaporated after the start of blowing in the BOF would enter the BOF coarse and fine sludge. As these are partly recycled to the BF, the zinc load in the BF would increase. In conclusion, a system analysis is required in order to analyze the effect on the overall zinc load in the integrated steel plant when operating on the proposed recycling scheme presented in Figure 8.

Finally, in order to maximize the recycling, the secondary dust can be recycled outside the process. In that case, the zinc content in the secondary dust has to be concentrated by closed-loop recycling. When the zinc content is sufficiently high, external zinc producers can utilize the residue.

Based on the above, the present paper illustrated the possibility of utilizing process integration analyses to decide the most efficient recycling routes of off-gas dust not being recycled today. Furthermore, the experiments ranging from laboratory scale to pilot-plant scale and full scale showed the feasibility of realizing these efficient recycling routes.

4. Conclusions

In the present paper, process integration analyses and laboratory, pilot plant, and full scale experiments were utilized to develop a holistic view for the recycling of off-gas dust generated in the BF, deS station, and BOF. The holistic approach considered a compromise between energy efficiency and raw-material efficiency for the process system including the BF, BOF, and deS station. Furthermore, the approach accounted for tramp elements, mainly zinc, while maintaining the production of high-quality steel. The study suggested that the off-gas dust could be recycled, minimizing the amount of non-recycled residues. The following findings improved knowledge considering recycling within the integrated steel plant:

- Physical separation or hydrometallurgical approaches were shown to be feasible in upgrading fine-grained BF sludge, although the sludge was low in zinc from the start.
- The low-zinc fraction of BF sludge can be completely recycled to the BF using cold-bonded briquettes.
- Recycling of cold-bonded briquettes to the deS station is feasible but restricted due to melt-in capacity.
- The possible recycling rate to the steel shop is sufficient to completely recycle the high-zinc fraction of upgraded BF sludge, depending on the chosen upgrading method.

A system analysis is required to estimate the increased zinc load in the integrated steel plant when operating the recycling scenario presented in the holistic view.

Author Contributions: Conceptualization, B.B. and L.S.-Ö.; Methodology, A.A., A.G., A.K. and M.A.; Formal Analysis, A.A., A.G. and M.A.; Investigation, A.A., A.G., A.K. and M.A.; Data Curation, E.S.; Writing—Original Draft Preparation, A.A.; Writing—Review & Editing, A.A., H.A., L.S.-Ö. and B.B.; Supervision, H.A., L.S.-Ö. and B.B.; Project Administration, H.A.; Funding Acquisition, L.S.-Ö. and B.B.

Funding: This research was funded by the Swedish Energy Agency and the research program Iron and Steel Industry Energy Use (JoSEn). The work was carried out within CAMM—Centre of Advanced Mining and Metallurgy at Luleå University of Technology.

Conflicts of Interest: The authors declare no conflict of interest.

References

1. Wedholm, A. Briquettes—Taking Advantage of Fine-Grained Residues in a Sustainable Manner. In Proceedings of the SCANMET V, Luleå, Sweden, 12–15 June 2016; Swerea MEFOS: Luleå, Sweden, 2016.
2. Das, B.; Prakash, S.; Reddy, P.S.R.; Misra, V.N. An overview of utilization of slag and sludge from steel industries. *Resour. Conserv. Recycl.* **2007**, *50*, 40–57. [CrossRef]
3. Rasmus, R.; Aguado Monsonet, M.A.; Roudier, S.; Delgado Sancho, L. *Best Available Techniques (BAT) Reference Document for Iron and Steel Production*; European IPPC Bureau: Seville, Spain, 2013; ISBN 978-92-79-26475-7.
4. Geerdes, M.; Chaigneau, R.; Kurunov, I.; Lingiardi, O.; Ricketts, J. *Modern Blast Furnace Ironmaking: An Introduction*, 3rd ed.; IOS Press BV: Amsterdam, The Netherlands, 2015; p. 172, ISBN 9781614994985.
5. Esezobor, D.E.; Balogun, S.A. Zinc accumulation during recycling of iron oxide wastes in the blast furnace. *Ironmak. Steelmak.* **2006**, *33*, 419–425. [CrossRef]
6. Uno, S.; Umetso, Y.; Ohmizu, M.; Munakata, S. Dezincing equipment and operation based on wet classification of wet-cleaned BF dust. *Nippon Steel Tech. Rep.* **1979**, *13*, 80–85.
7. Jeon, J.G.; Jin, S.J. POSCO's Achievement for the Recycling of Sludge. *SEAISI Q.* **2002**, *31*, 53–59.
8. Singh, A.K.P.; Raju, M.T.; Jha, U. Recycling of Basic Oxygen Furnace (BOF) sludge in iron and steel works. *Int. J. Environ. Technol. Manag.* **2011**, *14*, 19–32. [CrossRef]
9. Gargul, K.; Boryczko, B. Removal of zinc from dusts and sludges from basic oxygen furnaces in the process of ammoniacal leaching. *Arch. Civ. Mech. Eng.* **2015**, *15*, 179–187. [CrossRef]
10. Kelebek, S.; Yörük, S.; Davis, B. Characterization of basic oxygen furnace dust and zinc removal by acid leaching. *Miner. Eng.* **2004**, *17*, 285–291. [CrossRef]
11. Cantarino, M.V.; de Carvalho Filho, C.; Mansur, M.B. Selective removal of zinc from basic oxygen furnace sludges. *Hydrometallurgy* **2012**, *111–112*, 124–128. [CrossRef]
12. Trung, Z.H.; Kukurugya, F.; Takacova, Z.; Orac, D.; Laubertova, A.M.; Havlik, T. Acid leaching both of zinc and iron from basic oxygen furnace sludge. *J. Hazard. Mater.* **2011**, *192*, 1100–1107. [CrossRef] [PubMed]
13. Nakano, M.; Okada, T.; Hasegawa, H.; Sakakibara, M. Coke Breeze-less Sintering of BOF Dust and Its Capability of Dezincing. *ISIJ Int.* **2000**, *40*, 238–243. [CrossRef]
14. Fleischanderl, A.; Pesl, J.; Gebert, W. Aspect of recycling of steelworks by-products through the BOF. *SEAISI Q.* **1999**, *28*, 51–60.
15. Ma, N.Y.; Atkinson, M.; Neale, K. In-process separation of zinc from BOF off-gas cleaning system solid wastes. *Iron Steel Technol.* **2012**, *9*, 77–86.
16. Ma, N. Recycling of basic oxygen furnace steelmaking dust by in-process separation of zinc from the dust. *J. Clean. Prod.* **2016**, *112*, 4497–4504. [CrossRef]
17. Makkonen, H.T.; Heino, J.; Laitila, L.; Hiltunen, A.; Pöyliö, E.; Härkki, J. Optimisation of steel plant recycling in Finland: Dusts, scales and sludge. *Resour. Conserv. Recycl.* **2002**, *35*, 77–84. [CrossRef]
18. Su, F.; Lampinen, H.-O.; Robinson, R. Recycling of Sludge and Dust to the BOF Converter by Cold Bonded Pelletizing. *ISIJ Int.* **2004**, *44*, 770–776. [CrossRef]
19. Tang, F.; Yu, S.; Peng, F.; Hou, H.; Qian, F.; Wang, X. Novel concept of recycling sludge and dust to BOF converter through dispersed in-situ phase induced by composite ball explosive reaction. *Int. J. Miner. Met. Mater.* **2017**, *24*, 863–868. [CrossRef]
20. Agrawal, R.K.; Pandey, P.K. Productive recycling of basic oxygen furnace sludge in integrated steel plant. *J. Sci. Ind. Res.* **2005**, *64*, 702–706.
21. Hooey, P.L.; Bodén, A.; Wang, C.; Grip, C.-E.; Jansson, B. Design and Application of a Spreadsheet-based Model of the Blast Furnace Factory. *ISIJ Int.* **2010**, *50*, 924–993. [CrossRef]
22. Andersson, A.; Ahmed, H.; Rosenkranz, J.; Samuelsson, C.; Björkman, B. Characterization and Upgrading of a Low Zinc-Containing and Fine Blast Furnace Sludge—A Multi-Objective Analysis. *ISIJ Int.* **2017**, *57*, 262–271. [CrossRef]

23. Tikka, J.; Lindfors, N.; Bäcklund, E. Learning from nature: The tornado process. In Proceedings of the 6th International Heavy Minerals Conference, Hluhluwe, South Africa, 10–14 September 2007; SAIMM: Johannesburg, South Africa, 2007.
24. Tikka, J.; Hensmann, N.; Lindfors, N.; Bäcklund, E. Utilization of Tornado processed blast furnace gas cleaning sludge in blast furnace injection. In Proceedings of the 6th International Council for Scientific and Technical Information (ICSTI), Rio de Janeiro, Brazil, 14–17 October 2012; ABM: São Paulo, Brazil, 2012.
25. Robinson, R. Studies in Low-Temperature Self-Reduction of By-Products from Integrated Iron and Steelmaking. Ph.D. Thesis, Luleå University of Technology, Luleå, Sweden, 2008.
26. Hallin, M.; Hooey, L.; Sterneland, J.; Thulin, D. LKAB's experimental blast furnace and pellet development. *Revue Métall. Int. J. Met.* **2002**, *99*, 311–316. [CrossRef]
27. Piezanowski, L.; Raynal, S.; Hugentobler, J.; Houbart, M. Selective Hydrometallurgical Extraction of Zn/Pb From Blast Furnace Sludge. In Proceedings of the SCANMET V, Luleå, Sweden, 12–15 June 2016; Swerea MEFOS: Luleå, Sweden, 2016.
28. Singh, M.; Björkman, B. Testing of cement bonded briquettes under laboratory and blast furnace conditions Part 1—Effect of processing parameters. *Ironmak. Steelmak.* **2007**, *34*, 30–40. [CrossRef]
29. Singh, M.; Björkman, B. Effect of reduction conditions on the swelling behaviour of cement-bonded briquettes. *ISIJ Int.* **2004**, *44*, 294–303. [CrossRef]
30. Andersson, A.; Andersson, M.; Kullerstedt, A.; Ahmed, H.; Sundqvist-Ökvist, L. Recycling of the high-zinc fraction of upgraded bf sludge within the integrated steel plant. In Proceedings of the 8th International Council for Scientific and Technical Information (ICSTI), Vienna, Austria, 25–28 September 2018; ASMET: Leoben, Austria, 2018.
31. Wang, G.; Xue, Q.G.; She, X.F.; Wang, J.S. Reduction-melting behaviors of boron-bearing iron concentrate/carbon composite pellets with addition of CaO. *Int. J. Miner. Met. Mater.* **2015**, *22*, 926–932. [CrossRef]
32. Ding, Y.L.; Warner, N.A. Smelting reduction of carbon-chromite composite pellets—Part 2: Dissolution kinetics and mechanism. *Trans. Inst. Min. Metall. Sect. C* **1997**, *106*, C64–C68.
33. Riesbeck, J.; Lundkvist, K.; Brämming, M.; Wedholm, A. Applied Investigation on Waste Minimization in an Integrated Steel Site. In Proceedings of the World Congress on Sustainable Technologies 2015, London, UK, 14–16 December 2015; Infonomics Society: Basildon, UK, 2015.

© 2018 by the authors. Licensee MDPI, Basel, Switzerland. This article is an open access article distributed under the terms and conditions of the Creative Commons Attribution (CC BY) license (http://creativecommons.org/licenses/by/4.0/).

Article

Removal of Tramp Elements within 7075 Alloy by Super-Gravity Aided Rheorefining Method

Lei Guo, Xiaochun Wen, Qipeng Bao and Zhancheng Guo *

State Key Laboratory of Advanced Metallurgy, University of Science and Technology Beijing, Xueyuan Road No. 30, Haidian District, Beijing 100083, China; leiguo@ustb.edu.cn (L.G.); james365183262@163.com (X.W.); baoqipeng123@163.com (Q.B.)
* Correspondence: zcguo@ustb.edu.cn; Tel.: +86-135-2244-2020

Received: 15 July 2018; Accepted: 3 September 2018; Published: 6 September 2018

Abstract: An investigation was made on the super-gravity aided rheorefining process of recycled 7075 aluminum alloy in order to remove tramp elements. The separation temperatures in this study were selected as 609 °C, 617 °C and 625 °C. And the gravity coefficients were set as 400 G, 700 G, 1000 G. The finely distributed impurity inclusions will aggregate to the grain boundaries of Al-enriched phase during heat treatment. In the field of super-gravity, the liquid phase composed of tramp elements Zn, Cu, Mg et al. will flow through the gaps between solid Al-enriched grains and form into filtrate. Both the weight of filtrate and removal ratio of tramp element improved with the increase of gravity coefficient. The total removal ratio of tramp element decreased with the fall of temperature due to the flowability deterioration of liquid phase. The time for effective separation of liquid/solid phases with super-gravity can be restricted within 1 min.

Keywords: super-gravity; rheorefining; aluminum alloy; tramp element; separation

---

1. Introduction

Aluminum and its alloys are important metallic materials in modern industry due to their high specific strength, corrosion resistance and good formability. And they can be considered as sustainable materials due to the little loss of quality when being recycled. The energy consumption of recycling aluminum scrap is only about 5% of that producing primary aluminum from bauxite using molten salt electrolysis method [1]. The production of primary aluminum is energy intensive and causing heavy emission of CO_2. Thus, the production of secondary aluminum from recycled aluminum scrap has both economic and environmental benefits. However, those tramp elements (Fe, Si, Mg, Zn, Cu, Mn, Cr, etc.) exist in the recycled aluminum alloy scraps have to be eliminated or reduced before producing secondary aluminum alloys. The development of sorting technology in solid state can help to adjust the content of aluminum alloy scraps nowadays. Still, the accumulation of trace elements especially as Fe, Mn and Cr after repetitive reuse is the main problem. The removing effect of Fe, Mn, Cr elements was qualitatively proved in this study, and the detailed migration behavior of those three kinds of element will be further investigated in our later works.

The molten metal refining processes can be mainly classified into four types: (1) electrochemical refining (electrolytic refining), (2) physical refining (vacuum refining), (3) chemical refining (refining with fluxes [2,3]), (4) metallurgical refining (refining based on phase diagrams). However, only the metallurgical refining process features high efficient and large quantity. Based on the classification of the metallurgical refining technologies by Ichikawa and Cho [4,5], we think they can be simply divided into two categories depending on the form of tramp element containing phase removed from the raw material: (1) in the form of solid intermetallic compounds [6–8] and (2) in the form of liquid phase [9]. The method used in this study belongs to the later one. Those tramp elements will melt and aggregate to the grain boundaries of solid primary aluminum at high temperatures under the melting point of aluminum. Thus a semi-solid system can be obtained and with certain solid/liquid separation treatment the tramp elements can be removed from the primary aluminum phase. Then the aluminum alloy scraps can be refined and recycled.

Flemings has first introduced the semi-solid processing (also called the partial/fractional solidification or the rheorefining process) into the purification of metal alloy scrap [10]. In the rheorefining process, alloy scraps are heated to the solid/liquid coexisting temperature range. When the volume fraction of solid is small and the solid phases are oxides, nonmetallic inclusions or other particulates with high melting point, the separation can be accomplished effectively by filtration treatment [11,12], electromagnetic force [13–17] or gravitational sedimentation [18]. For example, to separate the Fe-Al-Si or Si solid phases from aluminum melt. However, when the volume fraction of solid is large, then the liquid phase is interspersed in the grain boundaries or interdendritic space, where impurities of low melting point are invaded and accumulated. The liquid phase is so finely dispersed that it is difficult to be removed from the dendritic solid. Ichikawa has investigated the rheorefining process of Al-Sn and Al-Ni alloys to obtain high-purity aluminum assisted by the mechanical squeeze [4]. However, the plunger speed was just about 2.8×10^{-4} mm/s and the total squeezing time had reached to 72 ks. And it has high requirement on the components like plungers. Cho etc. have performed comprehensive investigation on purification of aluminum alloys by backward extrusion process, but we think the refining effect may deteriorate quickly when applied to large scale experiments and the loss of aluminum is high for this method [5,19,20]. In the present work, for the aim of further enhancing the segregation tendency of liquid phase from grain boundaries and improving the separation efficiency, the super-gravity field generated by centrifugation was introduced in the rheorefining process of 7075 aluminum alloys.

Song has investigated the removal of nonmetallic inclusions from liquid aluminum by super-gravity [21,22]. Those nonmetallic inclusions have high melting point and the separation temperature can be easily controlled between the melting points of inclusions and aluminum. While in this study, in order to improve the flowability of liquid intermetallic compounds and keep the aluminum matrix in solid state, the separation temperature should be kept a little lower than the melting point of aluminum. Li has used the super-gravity field to separate valuable components from metallurgical slags [23–25]. The separation temperature in his study was usually above 1300 °C, which called high requirement on apparatus and experiment operation. Moreover, oxide melts usually have high viscosity and low flowability compared to liquid metals. In view of this point, it is more suitable to use the super-gravity technology in the separation treatment of molten metal system. It has been known that in super-gravity field, the interfacial and surface tensions of liquid phase are negligible. Then the flowability of the viscous fluid phase can be improved enormously and the separation efficiency can be improved simultaneously. With the help of super-gravity field, the liquid phase of low melting point will be drained from the Al-enriched solid phase in the form of small liquid drops [26]. Thus, the liquid contaminated with impurities could be separated effectively from the semi-solid alloy through a filter. However, few literatures can be found investigating the super-gravity aided rheorefining of wrought aluminum alloys. The influence of separation temperature/time, gravity coefficient and the detailed removal mechanism of tramp elements in this system are still unknown. Thus, we carried out this investigation and try to clarify above issues.

2. Materials and Methods

In this study, the wrought 7075 aluminum alloy was selected as the experimental material, which has high strength and is usually used for aerospace and structural engineering. The composition of the raw material was determined by ICP-OES analysis and the result is listed in Table 1. The 7075 raw materials used in this study were rods of 17 mm in diameter and 30 mm in height divided from lump material using electric-arc cutting. The experimental apparatus and the rheorefining process are illustrated in Figure 1. The experimental apparatus are mainly composed of an electric furnace and a centrifugal system. The two-stage graphite crucible with a porous support plate was used to complete the filtration process. The graphite felt of 5 mm in thickness 20 mm in diameter and 0.12–0.14 in volume density was used as the filter (the same as reference 25).

Table 1. Chemical composition of 7075 used in this study.

Element	Si	Fe	Cu	Mg	Mn	Cr	Zn	Al
Content/wt %	<0.1	0.15	1.7	2.9	0.9	0.18	6.2	Bal.

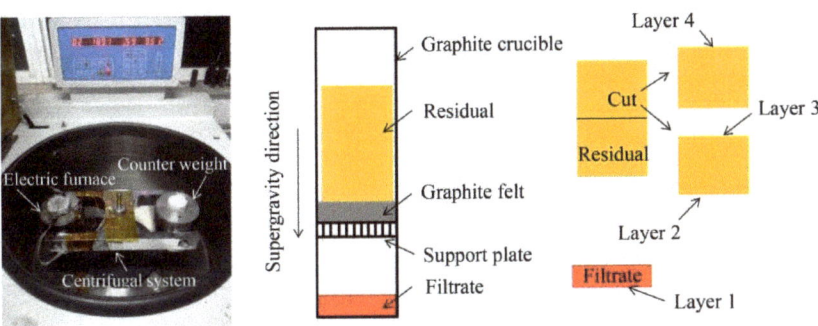

Figure 1. Sketch map of the super-gravity rheorefining apparatus.

The experimental procedure is as follows. First of all, the electric furnace was heated to the preset temperature (the temperatures used in this study are shown in Table 2). A 7075 billet was put into the upper part of the two-stage graphite crucible and kept in the electric furnace for 20 min to achieve a solid/liquid coexisting state. Then the centrifugal system was turned on with a certain gravity coefficient. After the centrifugal treatment with a fixed time, the samples were taken out and cooled in air. The upper sample is the residual and the lower sample is the filtrate. The residual part was equally cut into two parts transversely. Finally, all the samples were grinded and polished for following Scanning Electron Microscope (SEM) and Energy Dispersive Spectrum (EDS) analysis. Four layers were chosen to be observed as illustrate in Figure 1. It was proved that the element content detected by EDS was very close to that determined by ICP-OES analysis after separation processing. Thus, every value of element was derived from the average values of three EDS results in this experiment. Totally 11 trials were carried out in this experiment. The experimental conditions are shown in Table 2. The range of the semi-solid temperature of 7075 was determined to be 475–640 °C [19,27], the separation temperatures in this study were selected as 609 °C, 617 °C and 625 °C. And the gravity coefficients were selected as 400 G, 700 G, 1000 G. The gravity coefficient was calculated as the ratio of super-gravitational acceleration to gravitational acceleration via Equation (1) [28]. Where G is the gravity coefficient, N is the rotating speed of the centrifugal (r/min), R the distance from the centrifugal axis to the center of sample, R = 0.25 m, g = 9.8 m/s². The holding time is the time that the sample was kept in the electric furnace for heating. The separation time is the duration that the sample was subjected in the super-gravity filed.

$$G = \frac{\sqrt{g^2 + (\omega^2 R^2)}}{g} = \frac{\sqrt{g^2 + \left(\frac{N^2 \pi^2 R}{900}\right)}}{g} \tag{1}$$

Table 2. Experimental conditions used in this study.

Trials	Temperature (°C)	Gravity Coefficient	Holding Time (min)	Separation Time (min)
1	609	400	20	5
2	609	700	20	5
3	609	1000	20	5
4	617	400	20	5
5	617	700	20	5
6	617	1000	20	5
7	625	400	20	5
8	625	700	20	5
9	625	1000	20	5
10	625	700	20	1
11	617	1	25	0

3. Results

3.1. Separation Effect of Solid/Liquid Phases

The separation process was carried out according to the introduction in Section 2, after which the filtrate and residual parts were obtained. They were weighed and the result is shown in Figure 2. The total weight was basically constant before and after separation which was about 18.3 g. The weight of filtrates increased both with the enhancement of gravity coefficient and temperature. The minimum weight of filtrate was obtained in trial 1 (1.8 g, 609 °C, 400 G) and the maximum was obtained in trial 9 (6.3 g, 625 °C, 1000 G). In trial 10 the weight of filtrate was 5.2 g and the weight of residual was 12.8 g, which is close to that in trial 8 (filtrate 5.0 g, residual 13.3 g). In trial 11 there was no filtrate obtained and the weight of residual was 18.3 g.

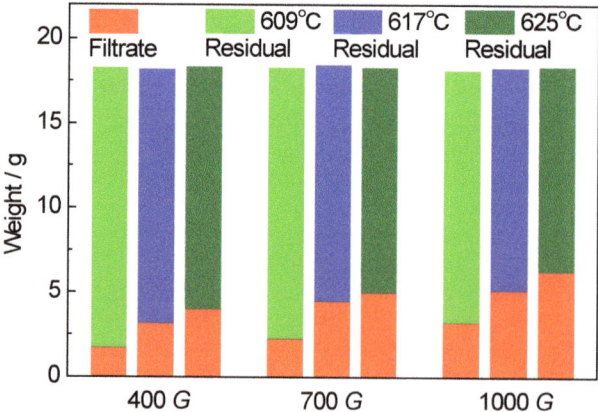

Figure 2. The weights of filtrate and residual in the trials 1–9.

3.2. Morphology in Different Layers

The distribution of alloying elements in the 7075 raw material is shown in Figure 3. It can be seen that those intermetallic compounds are randomly interspersed as small particles in the matrix. The tramp elements in the raw 7075 aluminum alloy are in the form of fine inclusions evenly distributed in the matrix as shown in Figure 4a. Take trail 4 for an example, it can be found from the cross section of filtrate that it is composed of light-colored impurity phase and dark Al-enriched matrix phase. The ellipsoidal columnar crystals of the Al-enriched matrix phase distribute in the dense net shape impurity phase. It can be found that some liquid phase containing tramp elements still remained on the grain boundaries of the residual part after separation treatment. The impurity phase in grain boundaries of the Al-enriched matrix phase decreases from layer 2 to layer 4 gradually as shown in Figure 4d–f. The SEM images in Figure 4b–e showed that the thickness and the number of grain boundaries decreased, and some grain boundaries disappeared due to the compression effect caused by super-gravity. The content of tramp elements from the top of the residual downward decreases slightly, which proves that the separation effect declines from top to bottom of the residual. This phenomenon may result from the increase of inhibition effect on the flow of liquid phase by those Al-enriched matrix grains. However, the removal effects of tramp elements in different positions are all in high level and differ just a little. Thus, the super-gravity field can help to overcome the flowing inhibition effect of Al-enriched matrix grains on the liquid phase.

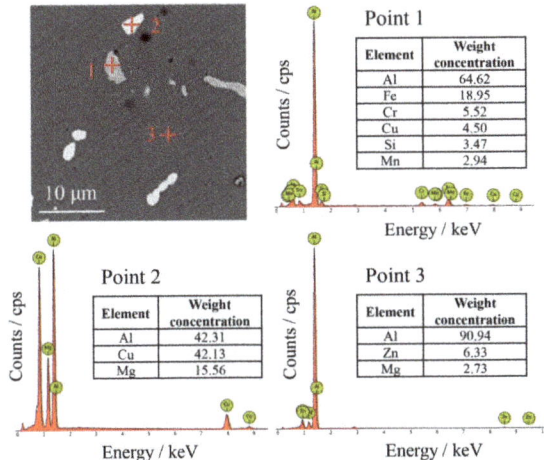

Figure 3. SEM and EDS analysis of the 7075 raw material.

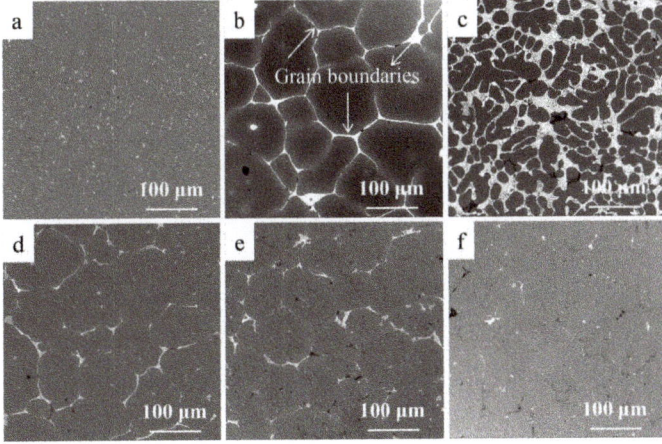

Figure 4. Morphology of raw material (**a**), after heat treatment in trail 11 (**b**) and different layers ((**c**): layer 1, (**d**): layer 2, (**e**): layer 3, (**f**): layer 4) in trail 4 (BSE mode).

3.3. The Composition and Distribution of Residual and Filtrate

The compositions of four kinds of layers as indicated in Figure 1 were analyzed with EDS. The minimum magnification was selected when proceeding the EDS analysis to reduce the error caused by the difference of phase distribution in different areas. 3 different areas in each layer were analyzed and the average values of different elements were chosen as the final composition data. The composition changes of four kinds of elements Al, Zn, Cu and Mg were mainly tracked in this study. The composition data in trial 1–9 is displayed in Figure 5. It can be seen that the contents of Al in all filtrates (layer 1) are lower than the raw material, and the contents of tramp elements Zn, Cu, Mg are all increased. Regardless the gravity coefficient the content of Al reached the lowest value (about 75 wt %) at 609 °C. With the increase of temperature the content of Al in the filtrate increased and the contents of Zn, Cu, Mg decreased. The content of Al in each layer of residual scarcely varied with temperature.

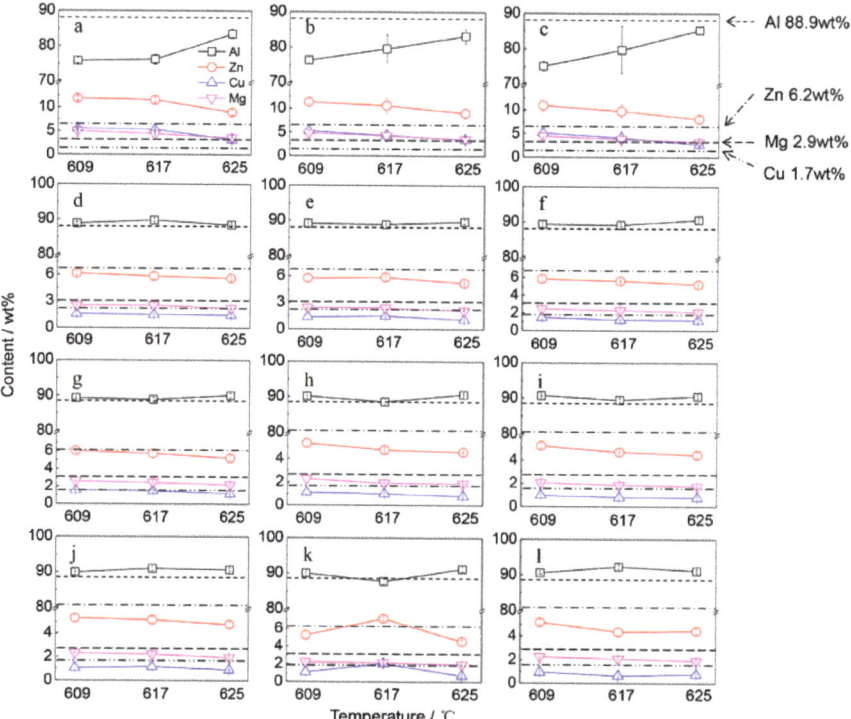

Figure 5. The compositions of different layers in trials 1–9: (**a**–**c**) layer 1, (**d**–**f**) layer 2, (**g**–**i**) layer 3, (**j**–**l**) layer 4, (**a**,**d**,**g**,**j**) 400 G, (**b**,**e**,**h**,**k**) 700 G, (**c**,**f**,**i**,**l**) 1000 G.

To take trial 4 as an example, the EDS mapping results of layer 1 are shown in Figure 6. In the grain boundaries of Al matrix is the net-shaped intermetallic compound zone containing Al-Zn-Cu-Mg (those four elements were mainly investigated in this study. Fe, Mn, Si, Cr may also exist in the intermetallic compound as illustrated in Figure 3). Its composition was determined with EDS spot scan as: Al 17–18 wt %, Zn 35–39 wt %, Cu 24–26 wt %, Mg 15–17 wt %. Iron is the most pervasive impurity element in aluminum alloys. The needlelike primary β-Fe [29] phase was found in layer 1 which composition is Al 70 wt %, Fe 20 wt %, Mn 3.2 wt %, Cr 1.0 wt %. According to the EDS mapping results, the tramp elements Fe, Mn, Cr enriched in the filtrate and it is proved that those elements can also be removed to some extend with this method. Due to the low content of Fe, Mn, Cr and the uneven distribution of needlelike Al-Fe-Mn-Cr phases (Figure 6), which will cause large error by EDS analysis. Thus, we will track their content changes with ICP-OES method in our later study.

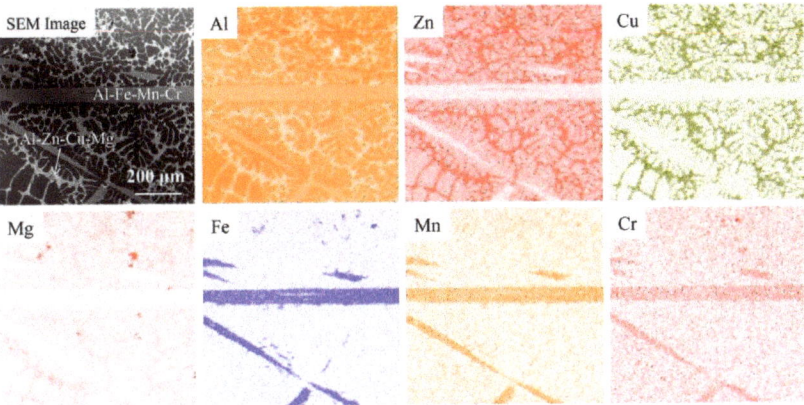

Figure 6. EDS mapping results of layer 1 in trial 4.

It was found that the Mg-Si-Al intermetallic phase exists in some conjunction sites of the net. As shown in Figure 7a, the dark black area as indicated by an arrow is the Mg-Si-Al phase which main composition is Mg 56 wt %, Si 36 wt %, Al 7 wt %. The fine texture of the net shape Al-Zn-Cu-Al intermetallic phase is shown in Figure 7b, which is a typical eutectic structure.

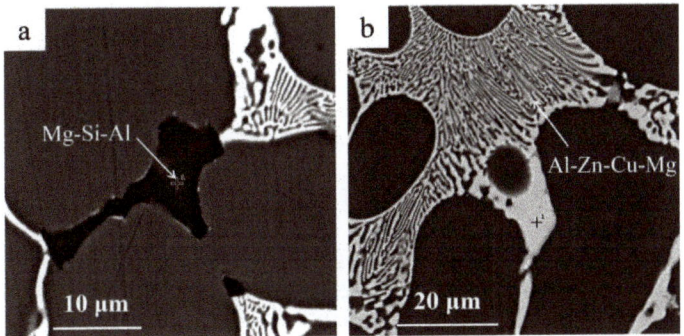

Figure 7. Partial enlarged views of layer 1 in trial 4 under back scattering mode. (**a**) Mg-Si-Al phase; (**b**) Al-Zn-Cu-Mg phase.

3.4. The Removal Ratio of Tramp Elements

The weight of Al, Zn, Cu, Mg in the filtrates can be calculated with the weight of filtrates and the contents of Al, Zn, Cu, Mg. Then the loss ratio of Al and the removal ratio of other tramp elements can be derived as:

$$R_x = \frac{m_f \cdot C_{x_1}}{m_0 \cdot C_x} \quad (2)$$

where R_x: loss ratio or removal ratio; m_f: the weight of filtrate; C_{x1}: the content of Al, Zn, Cu or Mg in filtrate; m_0: the weight of original billet; C_x: the content of Al, Zn, Cu or Mg in the original billet. R_{Al} is the loss ratio of Al, and R_{Zn}, R_{Cu}, R_{Mg} are the removal ratios of Zn, Cu, Mg respectively. The loss ratio of Al and the removal ratio of tramp elements all increase with the enhancement of gravity coefficient.

The driving force for the liquid phase to separate from the grain boundaries increases with the increase of gravity coefficient, then more filtrate can be obtained. Though the content of tramp elements in the filtrate decreases a little with the increase of gravity coefficient as shown in Figure 5a–c, the removal ratio of tramp elements still increases as illustrated in Figure 8b–d. The rise of separation temperature will increase the amount of liquid phase and subsequently more filtrate will be obtained. As the main component in filtrate is aluminum the rise of separation temperature will cause the increase of aluminum loss ratio.

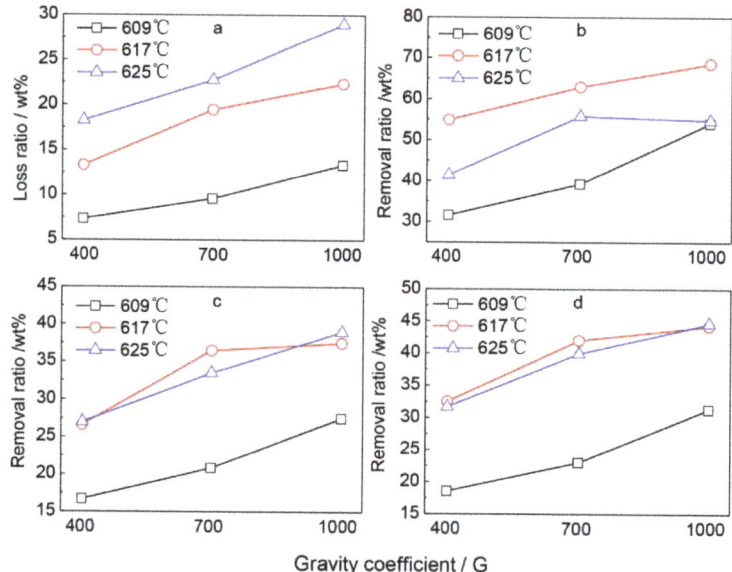

Figure 8. Loss ratio of Al phase (**a**) and removal ratio of tramp elements ((**b**): Cu, (**c**): Mg, (**d**): Zn).

4. Discussion

4.1. Effect of Separation Temperature

The increase of separation temperature will increase the amount of liquid phase and improve its flowability, which will lead to the increase of the filtrate amount. According to the study of ahmad and Cho, the solid fractions corresponding to each temperature in this work are 0.8 (609 °C), 0.72 (617 °C), 0.6 (625 °C) [19,27]. The weight ratio of the residual accounting for the total weight is close to the theoretical solid fraction at the gravity coefficient of 1000 G. Thus, it is proved that the solid/liquid separation ratio with 1000 G is close to the theoretical value of solid/liquid fraction at different experimental temperatures. As shown in Figure 5, the content of aluminum decreases and the content of tramp element increases in filtrate with lower separation temperature. But the amount of liquid phase will decrease and the diffusion speed of tramp elements will drop with lowered separation temperature. Meanwhile, the flowability of liquid phase will decrease. All those above reasons will cause the less amount of filtrate obtained. Thus, the total removal ratio of tramp element decreases with lower separation temperature as shown in Figure 8. With the rise of temperature the solid grains of Al-enriched phase in the residual tended to fuse and left less voids, which can be reflected by comparison of Figures 4 and 9.

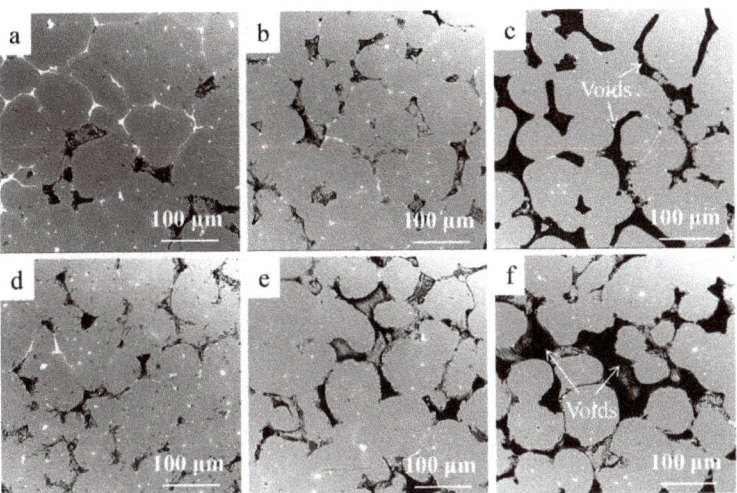

Figure 9. Morphology of layer 3 ((**a**): 400 G; (**b**): 700 G; (**c**): 1000 G) and layer 4 ((**d**): 400 G; (**e**): 700 G; (**f**): 1000 G) in the residual part (609 °C).

4.2. Effect of Gravity Coefficient

The separation rate of solid/liquid phases grows as the increase of gravity coefficient [22], resulting in higher weight of filtrate and dilution of tramp elements in filtrate. Thus, the contents of tramp elements Zn, Cu and Mg decrease slightly with the increase of gravity coefficient as shown in Figure 10. The net shape intermetallic phase in the filtrate becomes sparser with the increase of gravity coefficient as shown in Figure 11, which indicates that the tramp elements are diluted in the filtrate with heightened gravity coefficient. As shown in Figure 9, there were some voids observed in the intercrystalline space on layer 3 and 4 when the separation temperature was 609 °C. Those voids were left after the liquid phase flowed away. The softening trend of the Al-enriched phase was so weak that it kept its original shape and did not fill the voids. The size of the voids increases significantly with the growing of gravity coefficient. The larger size of voids corresponds to larger weight of filtrate.

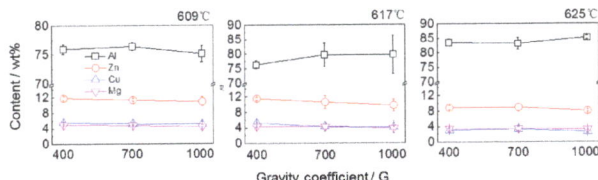

Figure 10. The compositions of filtrates obtained under different gravity coefficients.

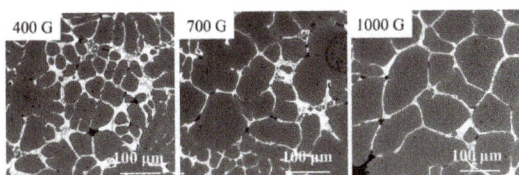

Figure 11. Morphology of layer 1 in the filtrate part (625 °C, backscattered electron mode (BSE)).

4.3. Effect of Separation Time

The experiment with the separation time of 1 min was carried out at 625 °C with the gravity coefficient of 700 G (trail 10) to investigate the influence of separation time on the separation effect. The relevant data is very close to that of trail 8, such as the weight of filtrate, the content of tramp element on layer 1–4, the removal ratio of tramp element. It is indicated that the separation rate of solid/liquid phases has reached the maximum level and the prolonging of separation time dose no good to the separation effect. The gravity coefficient is the crucial factor for separation effect compared to separation time [23,25]. For the liquid phase, its flow trend under super-gravity field equals to the viscous resistance when the separation limitation state is reached. A specific gravity coefficient corresponds to a particular separation limitation state. The separation process under super-gravity field is very fast, which 1 min is proved to be enough in this experiment.

4.4. Removal Mechanism of Tramp Element

7075 aluminum alloy is a kind of cold forged alloy, which features with fine grains and evenly distributed tiny intermetallic inclusions. The eutectic melts with low meting point composed of tramp elements like Zn, Cu, Mg, Fe et al. will migrate to the grain boundaries of Al-enriched phase forming net shape after heat treatment, as shown in Figure 9b. The liquid of impurity phase will flow out of the alloy matrix under super-gravity treatment, as illustrated in Figure 12, then the tramp elements will be concentrated in the bottom filtrate.

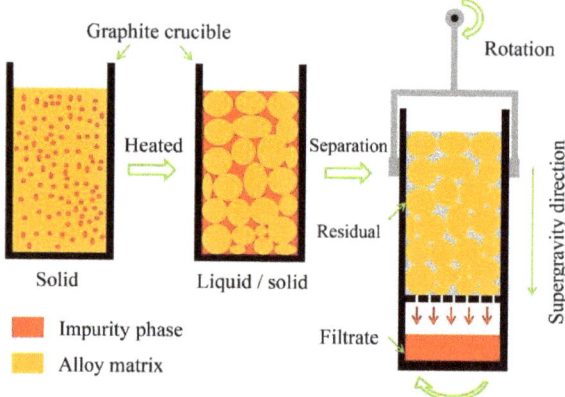

Figure 12. Schematic diagram of the super-gravity aided rheorefining process in this work.

5. Conclusions

11 trails were carried out in this study, influence of gravity coefficient and temperature on the separation effect and removal ratio of tramp elements are all investigated. The experimental results showed that large amount of impurities were enriched in the filtrate. Effective separation of liquid/solid phases with the aid of supergravity can be obtained within 1 min.

(1) It is proved that the tramp elements like Zn, Mg, Cu, Fe, Mn, Cr can be separated partly from the Al-enriched matrix phase using this super-gravity aided rheorefining method. The increase of separation temperature will lead to the increase of filtrate amount and total removal ratio of tramp elements.

(2) The separation process can be extremely accelerated in the super-gravity field, which makes great influence on the flowability of the eutectic melt containing impurities. Larger gravity coefficient will lead to higher weight of filtrate and dilution of tramp elements in filtrate.

(3) The sufficient aggregation of tramp elements in the eutectic phase of low melting point is crucial to improve the removal ratio of impurities. The separation process in the super-gravity field is so short and only 1 min of separation time is enough in this experiment.

Author Contributions: L.G. designed the experiments, analyzed the results and wrote this manuscript; X.W. and Q.B. helped performing most experiments; Z.G. gave some constructive suggestions on this work and helped desigining the experiments.

Funding: This research was funded by the Fundamental Research Funds for the Central Universities FRF-TP-16-037A1 and the Key Projects of the State Key Research and Development Plan of China 2016YFB0601304.

Conflicts of Interest: The authors declare no conflict of interest.

References

1. Moore, J.J. Recycling of non-ferrous metals. *Int. Met. Rev.* **1978**, *23*, 241–264. [CrossRef]
2. Gao, J.W.; Shu, D.; Wang, J.; Sun, B.D. Study on iron purification from aluminium melt by $Na_2B_4O_7$ flux. *Mater. Sci. Technol.* **2009**, *25*, 619–624. [CrossRef]
3. Chen, C.; Wang, J.; Shu, D.; Zhang, S.; Sun, B.D. Removal of non-metallic inclusions from aluminum by electroslag refining. *Mater. Trans.* **2011**, *52*, 2266–2269. [CrossRef]
4. Ichikawa, K.; Katoh, M.; Asuke, F.; Nakazawa, Y. High efficient recovery of pure aluminum from Al-Sn and Al-Ni alloys by rheorefining process. *Mater. Trans.* **1997**, *38*, 622–629. [CrossRef]
5. Cho, T.T.; Sugiyama, S.; Yanagimoto, J. Effect of process parameters on purification of aluminium alloys by backward extrusion process under a semisolid condition. *Mater. Trans.* **2016**, *57*, 404–409. [CrossRef]
6. Kim, S.W.; Im, U.H.; Cha, H.C.; Kim, S.H.; Jang, J.E.; Kim, K.Y. Removal of primary iron rich phase from aluminum-silicon melt by centrifugal separation. *China Foundry* **2013**, *10*, 112–117. [CrossRef]
7. Shu, D.; Li, T.X.; Sun, B.D.; Zhou, Y.H.; Wang, J.; Xu, Z.M. Numerical calculation of the electromagnetic expulsive force upon nonmetallic inclusions in an aluminum melt: Part II. Cylindrical particles. *Metall. Mater. Trans. B* **2000**, *31*, 1535–1540. [CrossRef]
8. Shu, D.; Li, T.X.; Sun, B.D.; Zhou, Y.H.; Wang, J.; Xu, Z.M. Numerical calculation of the electromagnetic expulsive force upon nonmetallic inclusions in an aluminum melt: Part I. Spherical particles. *Mater. Trans. B* **2000**, *31*, 1527–1533. [CrossRef]
9. Hashimoto, E.; Ueda, Y. Zone refining of high-purity aluminum. *Mater. Tran.* **1994**, *35*, 262–265. [CrossRef]
10. Mehrabian, R.; Geiger, D.R.; Flemings, G.A. Refining by partial solidification. *Mater. Trans.* **1974**, *5*, 785–787. [CrossRef]
11. de Moraes, H.L.; de Oliveira, J.R.; Espinosa, D.C.R.; Tenório, J.A.S. Removal of iron from molten recycled aluminum through intermediate phase flitration. *Mater. Trans.* **2006**, *47*, 1731–1736. [CrossRef]
12. Kennedy, M.W.; Akhtar, S.; Bakken, J.A.; Aune, R.E. Electromagnetically modified filtration of aluminum melts-part I: Electromagnetic theory and 30 PPI ceramic foam filter experimental results. *Metall. Mater. Trans. B* **2013**, *44*, 691–705. [CrossRef]
13. Kim, J.H.; Yoon, E.P. Elimination of Fe element in A380 aluminum alloy scrap by electromagnetic force. *J. Mater. Sci. Lett.* **2000**, *19*, 253–255. [CrossRef]
14. Xu, Z.M.; Li, T.X.; Zhou, Y.H. Elimination of Fe in Al-Si cast alloy scrap by electromagnetic filtration. *J. Mater. Sci.* **2003**, *38*, 4557–4565. [CrossRef]
15. Lee, G.C.; Kim, M.G.; Park, J.P.; Lim, J.H.; Jung, J.H.; Baek, E.R. Iron Removal in aluminum melts containing scrap by electromagnetic stirring. *Mater. Sci. Forum* **2010**, *638–642*, 267–272. [CrossRef]
16. He, Y.J.; Li, Q.L.; Liu, W. Separating effect of a novel combined magnetic field on inclusions in molten aluminum alloy. *Metall. Mater. Trans. B* **2012**, *43*, 1149–1155. [CrossRef]
17. He, Y.J.; Li, Q.L.; Liu, W. Effect of combined magnetic field on the eliminating inclusions from liquid aluminum alloy. *Mater. Lett.* **2011**, *65*, 1226–1228. [CrossRef]
18. Zhao, L.X.; Guo, Z.C.; Wang, Z.; Wang, M.Y. Removal of low-content impurities from Al by super-gravity. *Metall. Mater. Trans. B* **2010**, *41*, 505–508. [CrossRef]
19. Cho, T.T.; Sugiyama, S.; Yanagimoto, J. Effect of process parameters of backward extrusion by servo press on purification of A7075 alloy under the semisolid condition. *Mater. Trans.* **2016**, *57*, 1351–1356. [CrossRef]

20. Sugiyama, S.; Meng, Y.; Yanagimoto, J. Refining and recycling of metal scraps by semisolid processing. *Solid State Phenom.* **2012**, *192–193*, 494–499. [CrossRef]
21. Song, G.Y.; Song, B.; Yang, Z.B.; Yang, Y.H.; Zhang, J. Removal of inclusions from molten aluminum by supergravity filtration. *Metall. Mater. Trans. B* **2016**, *47*, 3435–3445. [CrossRef]
22. Song, G.Y.; Song, B.; Yang, Z.B.; Yang, Y.H.; Xin, W.B. Separating behavior of nonmetallic inclusions in molten aluminum under super-gravity field. *Metall. Mater. Trans. B* **2015**, *46*, 2190–2197. [CrossRef]
23. Li, C.; Gao, J.T.; Guo, Z.C. Isothermal enrichment of P-concentrating phase from CaO–SiO$_2$–FeO–MgO–P$_2$O$_5$ melt with super gravity. *ISIJ Int.* **2016**, *56*, 759–764. [CrossRef]
24. Li, C.; Gao, J.T.; Wang, F.Q.; Guo, Z.C. Enriching Fe-bearing and P-bearing phases from steelmaking slag melt by super gravity. *Ironmak. Steelmak.* **2016**, *45*, 1–6. [CrossRef]
25. Li, J.C.; Guo, Z.C.; Gao, J.T. Laboratory assessment of isothermal separation of V containing spinel phase from vanadium slag by centrifugal casting. *Ironmak. Steelmak.* **2014**, *41*, 710–714. [CrossRef]
26. Yang, H.J.; Chu, G.W.; Zhang, J.W.; Shen, Z.G.; Chen, J.F. Micromixing efficiency in a rotating packed bed: experiments and simulation. *Ind. Eng. Chem. Res.* **2005**, *44*, 7730–7737. [CrossRef]
27. Ahmad, A.H.; Naher, S.; Brabazon, D. Thermal profiles and fraction solid of aluminium 7075 at different cooling rate conditions. *Key Eng. Mater.* **2013**, *554–557*, 582–595. [CrossRef]
28. Li, J.C.; Guo, Z.C.; Gao, J.T. Isothermal enriching perovskite phase from CaO-TiO$_2$-SiO$_2$-Al$_2$O$_3$-MgO melt by super gravity. *ISIJ Int.* **2014**, *54*, 743–749. [CrossRef]
29. Cao, X.; Campbell, J. The solidification characteristics of Fe-rich intermetallics in Al-11.5Si-0.4Mg cast alloys. *Metall. Mater. Trans. A* **2004**, *35*, 1425–1435. [CrossRef]

© 2018 by the authors. Licensee MDPI, Basel, Switzerland. This article is an open access article distributed under the terms and conditions of the Creative Commons Attribution (CC BY) license (http://creativecommons.org/licenses/by/4.0/).

Article

Degradation Mechanism of Nickel-Cobalt-Aluminum (NCA) Cathode Material from Spent Lithium-Ion Batteries in Microwave-Assisted Pyrolysis

Fabian Diaz *, Yufengnan Wang *, Tamilselvan Moorthy and Bernd Friedrich

Institute of Process Metallurgy and Metal Recycling IME, RWTH Aachen University, Intzestraße 9, 52056 Aachen, Germany; tamil.moorthy@rwth-aachen.de (T.M.); bfriedrich@ime-aachen.de (B.F.)
* Correspondence: fdiaz@ime-aachen.de (F.D.); yufengnan.wang@rwth-aachen.de (Y.W.)

Received: 19 June 2018; Accepted: 19 July 2018; Published: 24 July 2018

Abstract: Recycling of Li-Ion Batteries (LIBs) is still a topic of scientific interest. Commonly, spent LIBs are pretreated by mechanical and/or thermal processing. Valuable elements are then recycled via pyrometallurgy and/or hydrometallurgy. Among the thermal treatments, pyrolysis is the most commonly used pre-treatment process. This work compares the treatment of typical cathode nickel-cobalt-aluminum (NCA) material by conventional pyrolysis, and by a microwave assisted pyrolysis. In the conventional route, the heating is provided indirectly, while via microwave the heating is absorbed by the microwaves, according to the materials properties. The comparison is done with help of a detailed characterization of solid as well as the gaseous products during and after the thermal treatment. The results indicated at least three common stages in the degradation: Dehydration and evaporation of electrolyte solvents (EC) and two degradation periods of EC driven by combustion and reforming reactions. In addition, microwave assisted pyrolysis promotes catalytic steam and dry reforming reactions, leading to the strong formation of H_2 and CO.

Keywords: Li-ion battery; recycling; pyrolysis; microwave assisted pyrolysis; battery pre-treatment

1. Introduction

The depletion of non-renewable energy source with the growing environmental concern leads to a necessity of using clean energy resources. In addition, the increasing demand of mobile energy resource paved the way for the large requirement of batteries. Li-ion batteries (LIBs) play the most important role in a broad applications, such as portable devices, power tools, hybrid electric vehicles (HEV), plug-in (PHEV), and electric vehicle (EV) market [1], due to their high energy density, better cycle life, lower rate of self-discharge, high average output voltage, reliability, and wide operating temperature range [2]. With the increasing demands of LIBs, not only there will be a shortage of raw material [3–5], but also, growing numbers of LIBs are approaching their end-of-life (EOL). The spent LIBs are harmful to the environment and their disposal or incineration is not allowed by the legislation [4,6] (Directive 2012/19/EU and Restriction of Hazardous Substances (RoHs)). Therefore, recycling is of paramount importance.

LIB consists of cathode (e.g., nickel cobalt aluminum (NCA), lithium iron phosphate (LFP), nickel manganese cobalt (NMC), lithium manganese oxide (LMO), and lithium cobalt oxide (LCO)) and anode (e.g., graphite and carbons) separated by an isolator. Cathode and anode are connected by an organic electrolyte made out of Li-containing salts. Among them, the valuable cathode material account for 42% the cost of Li-ion battery and are the main purpose for recycling [4].

In the recycling process, various methods have been developed for recycling lithium-ion batteries [7,8]. Commonly, spent LIBs are pretreated by mechanical or thermal processing, or are processed with a combined configuration. Valuable elements are then recycled via pyrometallurgical or/and hydrometallurgical treatment. Among all of the pretreatment methods, pyrolysis is the most commonly studied process for LIBs [8,9].

As a new developed pyrolysis method, microwave assisted pyrolysis offer several advantages that are based on its heating phenomena as compared to conventional pyrolysis, such as non-contact heating, energy transfer over heat transfer, high heating rates, easy power control, high selectivity of materials, uniform heating effect, and increased kinetics for the degradations process [10–13]. However, few of them have been applied for recycling of cathode material.

In this work, the degradation mechanism of a selected cathode material (NCA) from commercially used lithium-ion batteries via pyrolysis process is studied. The comparison is drawn between conventional pyrolysis and microwave (MW) assisted pyrolysis. In the microwave pyrolysis, the sample is subjected to different temperatures to understand the characteristics of the material at those temperatures and heating rates. The comparison is done with help of a detailed characterization of solid as well as gaseous products during and after the thermal treatment.

1.1. Theory of Pyrolysis

Pyrolysis is defined as the chemical decomposition of organic materials through the application of heat in the absence of oxygen [14]. Pyrolysis could be represented by the general equation (Equation (1)). In general, it transforms organic materials into three main products: a solid carbonaceous residue, non-condensable gas, and condensable gas called pyrolytic oil [15,16]. There are different parameters that influence the pyrolysis process: temperature of the reaction, residence time, presence of catalysts, gas velocity, particle size, reactor geometry, heating rate, and atmosphere [17].

$$C_nH_mO_p + \text{Heat} \rightarrow \sum\nolimits_{\text{solid}} C + \sum\nolimits_{\text{gas}} C_xH_yO_z + \sum\nolimits_{\text{liquid}} C_aH_bO_c . \quad (1)$$

Generally, pyrolysis undergoes several stages as temperature increases: dehydration (100–200 °C), deoxidation and depolymerisation (~250 °C), cracking of aliphatic boundings (~340 °C), carburation (380 °C), cracking of C-O and C-N boundings (400 °C), generation of bitumen and heavy fuel oils (400–600 °C), cracking of bitumen (600 °C), and generation of aromatic compounds (>600 °C) [18]. Multiple reactions take place once the system enters into a semi-gasification stage during the cracking periods, where carbon dioxide, oxygen, or steam react with each other and produce a combustible gas mixture called syngas. The main reactions involved in this process are indicated in Table 1 [16].

Table 1. Typical reactions during pyrolysis [16].

Type	Reactions		
Combustion of carbon	$C + O_2 \leftrightarrow CO_2$	$\Delta H = -393.5$ kJ/mol	(2)
	$C + 1/2O_2 \leftrightarrow CO$	$\Delta H = -111.4$ kJ/mol	(3)
Combustion of hydrocarbons	$C_xH_y + (x + y/4)O_2 \rightarrow xCO_2 + (y/2)H_2O$		(4)
	$C_xH_y + (x/2)O_2 \rightarrow xCO + (y/2)H_2$		(5)
Gasification reactions	$C + H_2O \leftrightarrow CO + H_2$	$\Delta H = 131.3$ kJ/mol	(6)
	$CO + H_2O \leftrightarrow CO_2 + H_2$	$\Delta H = -41$ kJ/mol	(7)
	$C + 2H_2 \leftrightarrow CH_4$	$\Delta H = -74$ kJ/mol	(8)

The basic heating principle of the traditional pyrolysis is based upon conduction, convection, and radiation, where material is heated completely. The principal features of the conventional pyrolysis are moderate heating rate (~0.17 °C/s), reaction temperature below 600 °C, and residence time between 10 s and 10 min [19,20]. In the traditional pyrolysis, the material is heated from outside to inside and the produced syngas has a relatively high retention time, which permits secondary reactions in the gas phase, which ultimately produces increased amount of condensates [10].

1.2. Thermal Degradation of Organic Materials in Cathode of LIBs

Investigation on thermal degradation of organics via conventional pyrolysis in materials used in LIBs has been studied in the past (see [21–23]). Based on the material composition of a conventional LIB, some of the investigated materials are: Solid-Electrolyte Interphase (SEI) layer, which consists of LiF, Li_2CO_3, $ROCO_2Li$, $(CH_2OCO_2Li)_2$, and/or ROLi; Electrolyte; graphite anode; cathode. The main thermal degradation reactions that are happening in these materials are listed, as indicated in Table 2. The electrolyte degrades with the increasing of temperature, while the SEI starts it degradation at temperatures from 120 °C to 250 °C (see Equations (9) and (10)). After that, graphite anode decomposes with electrolyte following the reactions that are shown in Equations (11)–(18) [24]. In addition, the cathode decomposes when the temperature reaches the onset temperature of their decomposition points (See Equations (19)–(22)). The polyethylene or polypropylene (PE/PP) based separator also undergoes a shrinking and melting step at temperatures above 135–165 °C. In addition, the PVDF binder also decompose itself by lithium reaction with fluorinated binder, as indicated in Equation (23). Apart from these mechanisms, limited information can be found regarding degradation of organics presented in LIBs via microwave pyrolysis [24–29].

Table 2. Main thermal degradation reactions of Li-ion batteries (LIBs) in cathode (based on [24–29]).

Type	Reactions	
SEI	$(CH_2OCO_2Li)_2 \rightarrow Li_2CO_3 + C_2H_4 + CO_2 + 1/2O_2$	(9)
	$2Li + (CH_2OCO_2Li)_2 \rightarrow 2Li_2CO_3 + C_2H_4$	(10)
Electrolyte	$LiPF_6 \rightleftharpoons LiF + PF_5$	(11)
	$LiPF_6 + H_2O \rightleftharpoons LiF + HF + POF_3$	(12)
	$C_2H_5OCOOC_2H_5 + PF_5 \rightarrow C_2H_5OCOOPF_4HF + C_2H_4$	(13)
	$C_2H_5OCOOC_2H_5 + PF_5 \rightarrow C_2H_5OCOOPF_4 + C_2H_5F$	(14)
	$C_2H_5OCOOPF_4 \rightarrow HF + C_2H_4 + CO_2 + POF_3$	(15)
	$C_2H_5OCOOPF_4 \rightarrow C_2H_5F + CO_2 + POF_3$	(16)
	$C_2H_5OCOOPF_4 + HF \rightarrow PF_4OH + CO_2 + C_2H_5F$	(17)
	$C_2H_5OH + C_2H_4 \rightarrow C_2H_5OC_2H_5$	(18)
Decomposition at the cathode	$Li_{0.5}CoO_2 \rightarrow 1/2LiCoO_2 + 1/6Co_3O_4 + 1/6O_2$	(19)
	$Li_{(1-x)}NiO_2 \rightarrow (1-2x)LiNiO_2 + xLiNi_2O_4\ (x \leq 0.5)$	(20)
	$Li_{(1-x)}NiO_2 \rightarrow [Li_{(1-x)}Ni_{(2x-1)/3}][Ni_{(4-2x)/3}]O_{(8-4x)/3} + (2x-1)/3O_2\ (x > 0.5)$	(21)
	$Li_{(1-x)}NiO_2 \rightarrow (2-x)Li_{(1-x)/(2-x)}Ni_{1/(2-x)}O + x/2O_2$	(22)
PVDF binder	$-CH_2-CF_2- + Li \rightarrow LiF + -CH=CF- + 0.5H_2$	(23)

1.3. Theory of Microwave Assisted-Pyrolysis

In comparison to conventional pyrolysis, microwave pyrolysis involves the transfer of energy to the material through the interaction of the molecules inside the material [10]. Microwave behaves in three different levels to materials, according to their dielectric properties, which can be determined by the ratio between the dielectric loss and the dielectric constant of the material. This ratio is also called loss tangent, being high (>0.5), medium (0.1–0.5), and low (<0.1). As an instance, the loss tangent of carbonaceous materials oscillates between 0.1 and 0.8, so carbonaceous materials can be considered as good microwave absorbers. Accordingly, materials could be classified into three categories according to their interactions with microwaves: conductors, insulators, and absorbers [10,30]. Metals are normally considered conductors as they reflect microwaves. Glass/ceramics are classified as insulators since they behave transparent to microwaves [10]. It is understood that metals, being conductors, cannot be heated significantly by microwaves that have penetration depths of few microns, thus they reflect most of the microwaves. However, it is worth noting that there is strong implicit temperature dependency on relative permeability and electrical conductivity. This means that materials can change their dielectric loss factor while heating and become microwaves absorber, which might occur at some specific temperatures. This effect has been seen more in metallic powders and thin metal foils. When metal foils with sharp edges or irregularities are exposed to microwaves, they can also cause electric arcs. These phenomena can be explained due to the fact that electric fields at the sharp edges will become large and the charges on the metal conductor move entirely towards the edges, behaving like an antenna with very high voltages. These electric arcs are called electric discharges [13,31].

Microwave pyrolysis has been used for several organic containing materials like engine oil and various biomasses, printed circuit boards, etc. [10,13,32]. However, the use of microwave pyrolysis in recycling of spent LIBs has rarely been reported. In this work, the degradation of the NCA cathode material through conventional as well as microwave assisted pyrolysis has been investigated. The influence of heating rate, microwave exposure time, and temperature are considered in this study. For both heating methods a detailed characterization of the solid and gas products is performed to understand the fundamental reactions happening during the process.

2. Materials and Methods

Batteries with a cathode material of NCA show a high power reputation and improved performance in terms of safety, having a relevant low market share today, but potential perspective in the future [33,34]. The battery that was used in the experiments is shown in Figure 1a. The material for the experiments is prepared manually by carefully separating the cathode from the anode. The cathode which is comprised of NCA active material bound with binder onto both sides of an aluminum current collector. After dismantle, there is still some portion of the electrolyte solvents (Detected: Propylene carbonate (PC), Diethylcarbonate (DEC)) being distributed in the cathode foil. The cathode foils are softly separated and cut into several pieces, as shown in Figure 1b.

Figure 1. (a) The battery used for the experiment and (b) starting material for the experiment.

After both the conventional and microwave pyrolysis tests, the output material is sieved (250 µm) to identify the separation's degree of aluminum and active mass. Most of the aluminum is present in the unit with the particle size $x > 250$ µm and the active masses are present in the unit with the particle size $x < 250$ µm after sieving. This gives the idea of the feasibility of active mass (also called black mass after pyrolysis) separation after thermal treatment.

The chemical composition of active mass was also detected by X-ray fluorescence and a further surface observation of Al foil was conducted by an electron probe micro-analyzer (EPMA) integrated in a scanning electron microscope (SEM) (JEOL JXA-8530F, Peabody, MA, USA). The gaseous product is analyzed by an integrated Fourier-transform infrared spectroscopy (FTIR) (Gasmet DX4000, Gasmet Technologies Oy, Helsinki, Finland) during the tests.

2.1. Experimental Setup for Conventional Pyrolysis

For conventional pyrolysis, the test was conducted in a programmable resistance furnace (Thermo-Star, Aachen, Germany). It can reach up to 1600 °C with a maximum heating rate of

600 °C/h. For this particular work, a set temperature of 600 °C is selected with a heating ramp of 600 °C/h. The reactor has a volume of 1 L and it is sealed with a water cooling lid to create a fully closed environment. It contained a pressure gauge, a gas sampling vent, a thermocouple, a carrier gas inlet (Ar), and an exhaust. The reactor was hold at this temperature during two hours and then the furnace was stopped and cooled down naturally. The experimental setup for conventional pyrolysis is shown in Figure 2.

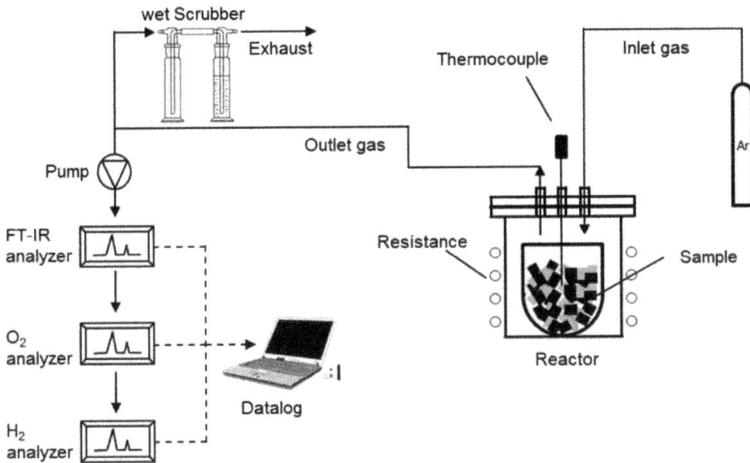

Figure 2. Experimental setup for the conventional pyrolysis test.

Before starting the experiment, a crucible was charged with 30 g of starting material and placed inside the reactor. The thermocouple (type K NiCr-Ni) was adjusted to the center of the reactor to detect the furnace temperature above the crucible during the process. The thermocouple was connected to a thermo-logger to record the temperature every 20 s. The reactor was sealed and placed inside the furnace and the gaps between the reactor and furnace was covered with glass wool. The exhaust was connected to two sequentially connected bottles (As scrubbers), the first bottle was empty and it acted as a safety so that to prevent the liquid in the second bottle entering into the reactor in case of pressure loss in the FTIR pump. The second bottle contained distilled water to clean the produced gas and prevent any oxygen going inside the reactor. The gas sampling vent was connected to the sampling probe to collect samples of gas every 20 s during the process. FTIR analyzer and O_2 analyzer were turned on and left until their measuring cells reached 180 °C and 5 °C, respectively. FTIR analyzer was calibrated manually using Calcmet software (Version 12.16, Gasmet Technologies Oy, Helsinki, Finland). The main pump was adjusted to 3 L/min, whereas, the O_2 analyzer pump was adjusted to 0.5 L/min. The furnace was programmed according to the experiment and turned on while all of the analyzers and thermos-logger were recording.

2.2. Experimental Setup for Microwave Pyrolysis

The microwave reactor that was utilized for this work corresponds to a highly controlled atmosphere microwave with eight microwave (MW) generators of 6 KW each, continuous controlled power, and maximal capacity of 0.033 m³. For this experimental work only six out of eight generators were used with half of its power (3 kW). Each generator was sequentially activated for 30 s to have a homogeneous radiation in the sample. The power supply was stopped after reaching the target temperature of 180, 350, 400 and 450 °C. The higher target temperatures were achieved by prolonging the exposure time of microwaves to the material.

A crucible filled with 30 g of material was placed inside the reactor. The reactor was closed and sealed with resistant screws. In general, the analyzing system that was used in conventional pyrolysis had also been used in the microwave assisted pyrolysis. The true temperature inside the crucible was monitored and recorded with a thermocouple. The setup of the scrubber and the off gas analysis method remained the same. The microwave system includes an infrared camera, which monitors and records the temperature in the surface of the crucible. Before starting the microwave irradiation, argon (6 L/min) was supplied to create the inert atmosphere reducing the amount of oxygen inside the reactor. The microwave assisted pyrolysis plant is shown in Figure 3.

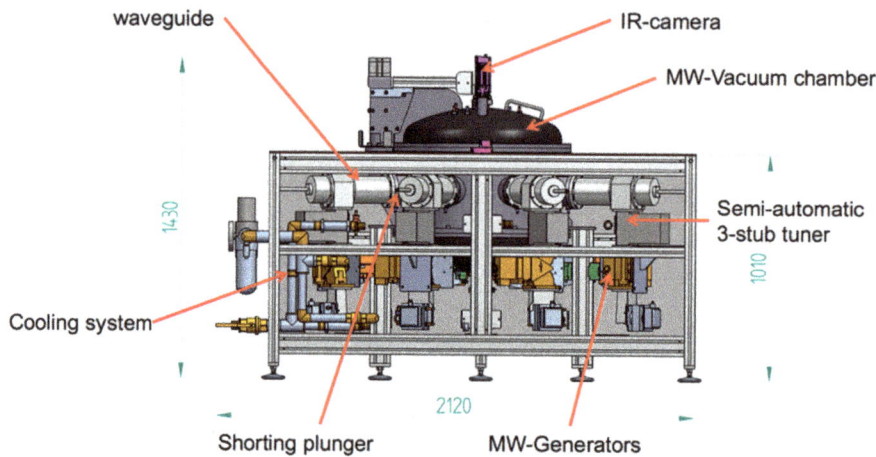

Figure 3. Microwave pyrolysis plant (unit: mm).

3. Results and Discussion

Because of the electric arcs that are formed in microwave, the true temperature of microwaves-assisted pyrolysis is higher than the target temperature. The true temperature of microwaves-assisted pyrolysis were 220, 360, 418, and 501 °C, respectively. The conventional pyrolysis is taken in this discussion as a reference experiment. As it is known, a conventional heating assures the complete removal of the electrolyte and induces the less impact in the metal foils in the absence of oxygen.

3.1. Solid Products Characterizations

The amount of mass loss after conventional as well as microwave pyrolysis, which indicates the content of volatile in starting material, is shown as Figure 4. The amount of mass loss after the pyrolysis in each experiment slightly increases from conventional to the microwave pyrolysis at temperatures lower than 360 °C. For microwave pyrolysis the amount of mass loss increases with the temperature increasing. As well, in Figure 4 it is seen that the amount of aluminum share (particle size $x > 250$ μm) in the output material drastically decreases in the microwave treatment at temperatures higher than 360 °C, whereas the amount of active mass (particle size $x < 250$ μm) increases. These observations can be explained to the following driven phenomena: the microwaves are normally reflected in thick metals. For thin pieces of metal, like aluminum foils, prolonged microwave exposure would lead to rapid heating and generation of electric discharges with very high voltages at the edges, which can easily cause strong metal fragmentation of metal foils. Therefore, after 360 °C by microwave treatment the active mass increases its share in the solid product.

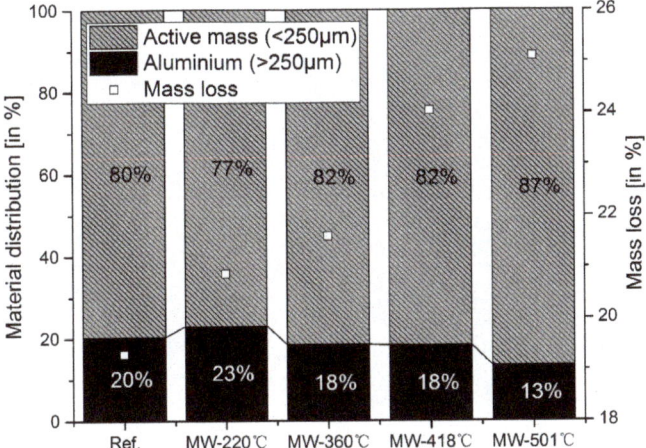

Figure 4. Mass loss and material distribution (active mass vs. aluminum) after pyrolysis.

In addition to the active mass yield ratio, the chemical composition of sieved active mass is also detected, as shown in Figure 5. It could be seen that except for Ni and Al, there is no different for the distribution of elements in the active mass after both conventional and microwave pyrolysis. The distribution of Al and Ni show an exactly opposite relation. The purity of active mass decreases after microwave pyrolysis at temperatures above the 360 °C with increasing content of Al_2O_3 and deceasing content of NiO. To explain nickel mobilization, an electron probe micro-analyzer (EPMA) is conducted for aluminum foil after microwave pyrolysis at 501 °C. It can be seen from Figure 6 that most of elements like P, O, C, and Ti have similar distribution in the sample in the form of highly concentrated spots following the remaining traces of the black mass. On the contrary, Ni element has quite even and fine distribution in the sample in the whole surface, except for the typical concentrated spots, as indicated for other elements. This can suggest some interaction between nickel and the aluminum foils during the microwave treatment.

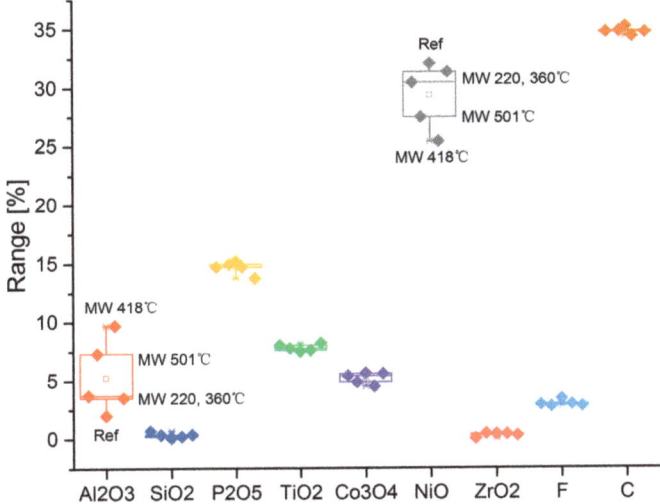

Figure 5. Variation of elements in the active mass during microwave pyrolysis.

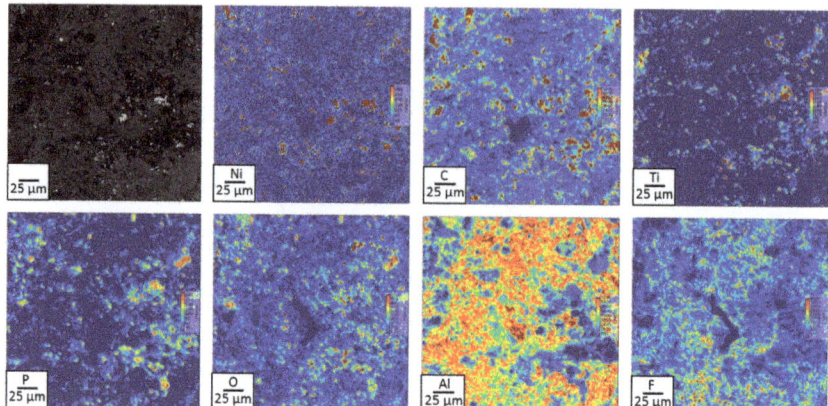

Figure 6. Aluminum foil after microwave (MW) treatment (MW 501 °C): presence of main elements in the surface (electron probe micro-analyzer, EPMA).

In addition to elements analysis, surface observation is also carried out, as demonstrated in Figure 7. When compared with the reference, there is still a lot of active mass attached on the Al foil after microwave-assisted pyrolysis at 220 °C, which accounts for the low active mass yield ratio. When the temperature is increased to 360 °C, the surface characteristic is similar with the reference. Nevertheless, if the material is exposed to longer time to microwaves, some sparks with increased temperature are expected. Therefore, Al foil melts at the edges at measured temperature of 418 °C. This might lead to the increased of alumina content in active mass. When the temperature is increased to 501 °C, it can be observed strong metal fragmentations of the Al foils, which can explain the high mass loss for this trial (see Figure 4).

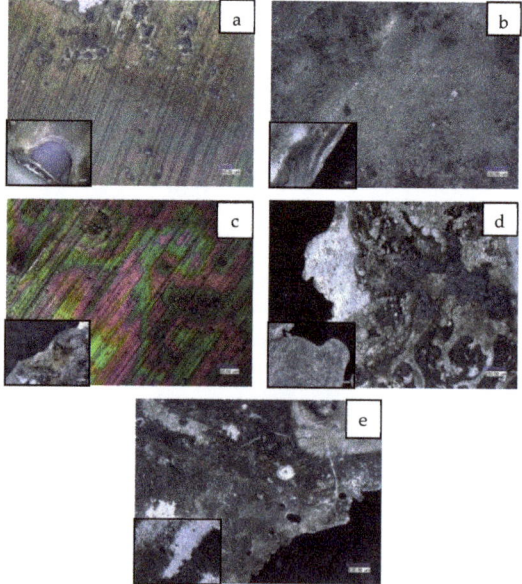

Figure 7. Aluminum foil after treatment: (**a**) Reference; (**b**) MW 220 °C; (**c**) MW 360 °C; (**d**) MW 418 °C; and, (**e**) MW 501 °C.

3.2. Gas products Characterization

3.2.1. Degradation during Conventional Pyrolysis of a NCA Cathode Material

Cracking of organics during pyrolysis can be described as a very complex process and exact formation of component are difficult to define due to the multiple reactions that happen simultaneously. However, the determination of the gas formation gives some hints about the reactions that occurred during the process. The cracking periods that happened during the conventional pyrolysis is identified in Figure 8. During the whole process, at least three periods can be identified, where the organic material degrades into volatile components and char by, dehydration, cracking, partial oxidation, or reforming reactions.

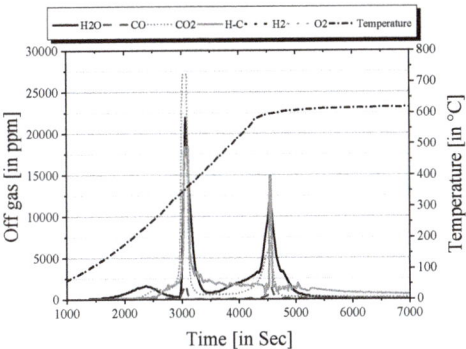

Figure 8. Produced gas analysis in conventional pyrolysis (Reference test).

The first period correspond to a dehydration process, which can be evidenced with the presence of H_2O in the produced gas at temperatures between 100–250 °C. After this, the second period begins with formation of CO_2, hydrocarbons, H_2O, and CO. They are most probably being produced due to combustion of hydrocarbons as indicated in reaction Equation (4). At 320 °C a white smoke appeared in the scrubber, which continued until the temperature reached 360 °C. During this cracking period, there are breakdowns of long chain into short chain molecules, which produces various gaseous compounds, namely CO_2, CO, H_2O, CH_4, phenols, and free carbon. The third cracking period is registered between 400 °C and 600 °C, with the appearance of smoke again and generation of more volatile gases with similar chemical constituents as the second cracking period but in less quantity. In general, it can be observed that combustion of long chain hydrocarbons are probably the main reactions taking place during the second and third cracking period, as is evidenced with the correlation between H_2O, CO_2, and CO formations. However, free hydrogen was not detected during the whole test. This could be explained due to the presence of free hydroxyl radicals that oxidized the hydrogen atoms as concentration of them is not sufficient to form molecules. In addition, secondary reactions in the gas phase can also explain the lack of hydrogen molecules in the pyrolysis gases. For this particular case, free hydrogen could react in the gas phase with other components, like CO, C, and free oxygen radicals before being measured, as indicated in Equation (8).

3.2.2. Degradation during Microwave Pyrolysis of a NCA Cathode Material

The microwave pyrolysis offers a heating rate of around 3.5 °C/s depending on the material and quality of heat transfer to the material, which in contrast to conventional pyrolysis would be about 20 times faster. Due to the high speed pyrolysis, some strong variations are noticed in the generated gas, as indicated in Figure 9. The main gases registered correspond to H_2O, CO, CO_2, C_xH_y, and H_2. Individually, for each test, it can be observed that as long as the material prolong the time that is exposed to microwaves to reach higher temperatures, the concentration of the already defined gases are also increased. In general, the concentration of these gases reaches its maximum and then decreases drastically. It can be observed in Figure 9 that after certain point, as long as the target temperature is increased, some gases, like CO_2 and H–C, decreases. This might be due to the formation of higher concentration of other gases, like CO and H_2. The microwave pyrolysis produced 49–94% of the H–C produced in the conventional pyrolysis (5.71 mL/g of material). The main difference of microwave when compared with conventional pyrolysis is the strong presence of free hydrogen in the produced gas as well as increased presence of short chain molecules.

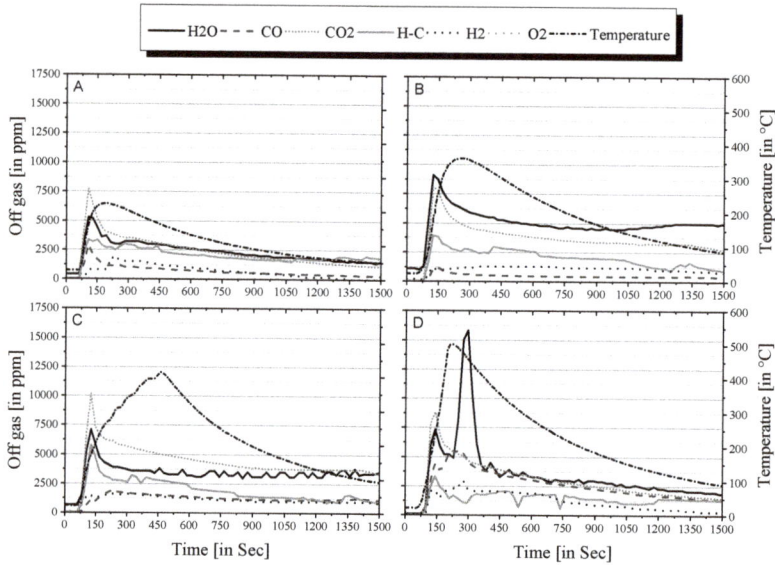

Figure 9. Produced gas analysis in microwave pyrolysis at max. temp.: (**A**) 220 °C; (**B**) 360 °C; (**C**) 418 °C; and, (**D**) 501 °C.

Formation of hydrogen during microwave pyrolysis can be explained due to two mechanisms that might occur separately or together: (I) Equations (5) and (7) take place due to the high energy coupled with the material that can reach higher temperatures than the temperature that is registered by the thermocouple; (II) Hydrogen is formed due to the strong degradation of long chain hydrocarbons, which due to very high cooling rate of the produced gas when leaving the sample hinders major secondary reactions involving hydrogen in the gas phase. Similar to conventional pyrolysis, the presence of H_2O, CO_2, and CO strongly indicated that the combustion/partial oxidation of long hydrocarbons are also playing an important role in the degradation of LIB cathode material by microwave heating.

From Figure 9, it is also possible to notice that all experiments under 360 °C showed similar degradation mechanisms. However, when prolonged exposure to microwave to reach higher temperatures than that, aluminum experience a self-heating effect with the microwaves, reaching very high temperatures in very short period of time, thus leading to partial melting and eventually the formation of sparks (see Figure 7). This can be evidenced by the increased amount of hydrogen at higher temperatures in the form of peaks (Figure 9D). However, it is also noticed that, at the same time, formation of water in the gas phase is taking place while the presence of hydrocarbons is decreased. This can be explained by strong formation of hydroxyl radicals at a short period of time, which represent an important source of oxygen [35], leading to strong oxidation reactions with hydrocarbons.

3.2.3. Comparative Study between Conventional and Microwave Pyrolysis on the Formation of Heavy and Light Molecules, and Toxic Compounds

Formation of aromatic-rich mixtures of heavy hydrocarbons with molecular weight greater than benzene might lead to the formation of tars [36]. These tars condense at relative low temperatures leading to possible clogging in the off-gas system, which bring challenges to the commercialization of pyrolysis. Therefore, the formation of lighter hydrocarbons is considered to be beneficial for the robustness of the process.

In order to perform a fair comparison, the total amount of relevant species is calculated based on the measured results. The amount of off-gas is calculated by integrating the time-concentration records from the FTIR, obtaining total volumes in NL, as shown in Equation (24).

$$V = \frac{1}{d} \sum_{i=0}^{n} \dot{V}(t_{i+1} - t_i) (c_{i+1} - c_i)/2, \qquad (24)$$

where V is the volume of compound (NL), d is the diluted ratio, $\dot{V}$ is flow rate of carried gas (NL/min), c_i is concentration of compound at time t_i (ppm), and c_{i+1} is concentration of compound at time t_{i+1} (ppm).

The cracking of long chain polymers leads to production of linear hydrocarbons. These large hydrocarbons will break when they are exposed for a longer time and at a higher temperature. This cracking is a function of microwave power and working temperature. It is understood that too high temperature, long residence time, and secondary cracking mechanisms will lead to the formation of aromatic compounds [37], which account for the generation of aromatic hydrocarbons, like toluene, ethylbenzene, and styrene. The aromatic hydrocarbons contain sigma bonds, which are difficult to break at lower temperatures. The amount of generated heavy hydrocarbons with molecular weight greater than benzene is shown in Figure 10. It can be seen that in conventional pyrolysis (as referenced in this article) where the rate of breaking bonds is slow when compared with the microwave heating, large amounts of toluene and phenol are expected. In addition, a large amount of electrolyte (EC) is detected in the conventional pyrolysis. This concentration can be explained as a simple volatilization of the electrolyte at lower temperatures that limits the bond breaking in the process. In addition, phenol formation can be explained by the oxidation reactions of hydrocarbons with some free hydroxyl radicals. These secondary reactions can take place due to a prolonged retention time in the reactor when compared to the microwave test.

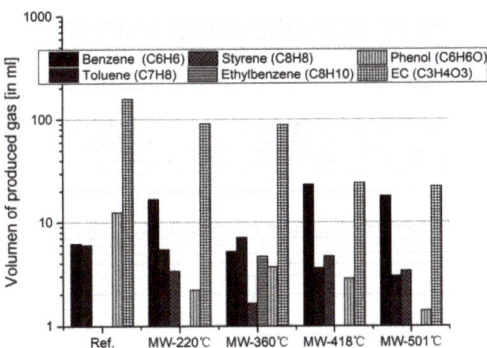

Figure 10. Volume of heavy molecules generated at different target temperatures with microwave heating.

As indicated in Figure 10, the total amount of heavy molecules is calculated by summering the volume of benzene, toluene, styrene, ethylbenzene, phenol, and EC. The results for conventional and microwave pyrolysis are 183 mL, 120 mL, 110 mL, 59 mL, and 49 mL, respectively. In addition to heavy molecules, light molecules are also investigated to study the influence of heating rate. As shown in Figure 11, the volume of light molecules gases, such as CO_2, CO, H_2, H_2O, CH_4, and C_2H_4 produced in conventional method are about 2.14 L in total. It is also observed in Figure 11 that methane and ethene formation remains relatively constant in all of the trials. Generally, in microwave experiment, the volumes of light molecules gases range from 1.11 to 7.11 L. The minimum registered volume is achieved at 221 °C and the maximum at 360 °C. With this evidence, it can be concluded that, when compared with conventional pyrolysis, microwave-assisted pyrolysis decreases the amount of heavy molecules produced. In addition, prolonged exposure to microwave leads to higher temperatures and heating rates, which promotes the strong formation of shorter molecules by degradation of the electrolyte and consequently, leading to lower production of heavy molecules.

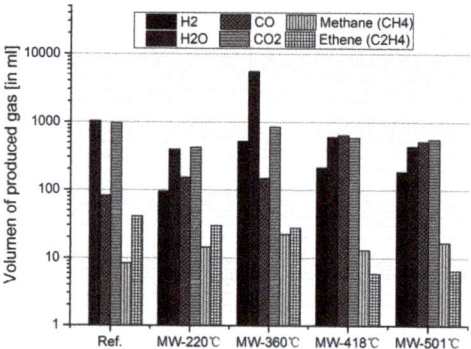

Figure 11. Volume of light molecules generated at different target temperatures with microwave heating.

In general, lots of hydrocarbons are produced, which bring some risk of explosion and promotes the formation of highly toxic fluorides, such as HF and POF_3. Therefore, additional attention should be paid to toxic compounds. The amount of some relevant toxic species (e.g., fluorides and dangerous electrolyte solvents) are indicted in Figure 12. From this figure, it can be concluded that the toxic compounds achieved their minimum volume at 360 °C. Moreover, COF_2 and POF_3 are only detected in the conventional pyrolysis. For this particular group of gases, they are supposed to be produced following the degradation reactions of the electrolyte (Equations (11)–(18)).

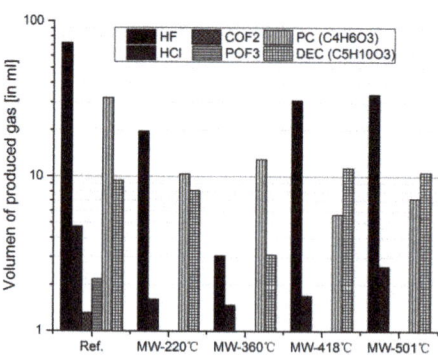

Figure 12. Volume of toxic compounds generated at different target temperatures with microwave heating.

Microwave assisted pyrolysis might follow a different degradation mechanism as formation of H_2 and CO was clearly registered in Figure 12. From this, some observations might bring some hints about some possible mechanism, like the water-aluminum reaction (see Equations (25)–(27)) [38].

This based on the smaller amount of H$_2$O and strong aluminum corrosion in the surface compared with conventional pyrolysis (See Figures 5 and 7). In addition, high content of Nickel (~20%) in the active mass favors the catalytic reactions, which results in increased formation of H$_2$ and CO$_2$ as well. It is worth saying that nickel or nickel-aluminum is a common catalyst used to increase formation of H$_2$ in the pyrolysis process of other organic containing materials and organic gases, like methane [16,39–41]. Some of the most typical catalytic reactions are the so-called steam reforming reaction (Equation (28)) and the "dry" reforming reaction (Equation (29)). These reactions can take place at low temperature only under catalytic condition, which otherwise would require temperatures above 1100 °C [16]. In particular, methane can be reformed into CO and H$_2$ by means of a catalytic steam reforming, as well as dry reforming reactions, as it is indicated in Equations (30) and (31), respectively. These reactions can take place at temperatures above 700 °C [16]. This catalytic reaction is also plausible as temperature in the material is expected to be much higher than that of the registered temperature with the thermocouple, which could be proved by the melted Al-foil shown in Figure 7. In contrast, the conventional pyrolysis—because of the low heating rate volatilization of organics and water—occurs before any catalytic reaction takes place, which explains the neglectable formation of hydrogen in the produced gas.

$$2Al + 6H_2O \rightarrow 2Al(OH)_3 + 3H_2, \tag{25}$$

$$2Al + 4H_2O \rightarrow 2Al(OH) + 3H_2, \tag{26}$$

$$2Al + 3H_2O \rightarrow Al_2O_3 + 3H_2. \tag{27}$$

Catalytic steam reforming reaction [16]:

$$C_nH_m + nH_2O \rightarrow (n + m/2)H_2 + nCO. \tag{28}$$

Catalytic carbon dioxide (or dry) reforming reaction [16]:

$$C_nH_m + nCO_2 \rightarrow (m/2)H_2 + 2nCO. \tag{29}$$

Catalytic steam reforming of methane [16]:

$$CH_4 + H_2O \rightarrow 3H_2 + CO + 206 \text{ kJ/mol}. \tag{30}$$

Catalytic dry reforming of methane [16]:

$$CH_4 + CO_2 \rightarrow 2H_2 + 2CO + 247 \text{ kJ/mol}. \tag{31}$$

4. Conclusions

The fact that microwave assisted pyrolysis involves the transfer of energy to the material through the interaction of the molecules inside the material offers several advantages when compared to conventional pyrolysis, such as non-contact heating, high heating rates, easy power control, high selectivity of materials, uniform heating effect, and increased kinetics for the degradations process [10–13]. The microwave pyrolysis offers a heating rate of around 3.5 °C/s depending on the material and quality of heat transfer to the material, which in contrast to conventional pyrolysis, would be about 20 times faster.

In this work, the degradation of organics from a selected cathode material (NCA) by conventional (600 °C) as well as microwave assisted pyrolysis was studied. For the microwave tests, the sample was subjected to different exposure times resulting in different temperatures and heating rates. The comparison was performed with a detailed solid and gas characterization of products.

As indicated in Figure 13, conventional pyrolysis of cathodes from LIBs undergoes three different steps of volatilization: a dehydration & EC boiling step, which is followed by two cracking and reforming periods.

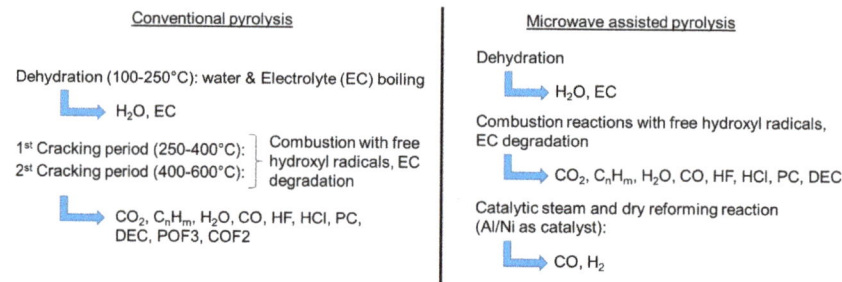

Figure 13. Basic degradation of organics present in commercial cathodes from LIBs during conventional and microwave assisted pyrolysis.

The produced compounds in the gas phase might undergo secondary reactions e.g., combustion of hydrocarbons with free hydroxyl radicals and compounds from the electrolyte solvent degradation.

Microwave assisted pyrolysis introduces an extra degradation mechanism to the conventional pyrolysis. This mechanism is driven by steam and dry reforming reactions, which can take place as a result of catalytic reactions due to the presence of aluminum and nickel in the cathode material. Other factor that benefits this reaction is the high heating rate, which permits reaching high temperatures in the presence of electrolyte and organic gases that can be reformed into CO and H_2.

When prolonged exposure to microwave, aluminum experience a self-heating effect with the microwaves, reaching very high temperatures in very short period of time, as evidenced by partial melting of aluminum foils. Nevertheless, over exposure of the material to microwaves (temperature > 360 °C) leads to metal fragmentations thus contamination of the black mass with aluminum.

From the main findings, the following can be highlighted:

- For microwave assisted pyrolysis, the amount of mass loss increases with the temperature. However, excessive exposure of microwave to the cathode material leads to uncontrollable rapid heating of aluminum, which leads to partial melting and eventually the formation of sparks. Like this, the fragmentation of the metal foil takes place and consequently contamination of the black mass cannot be avoided.
- When compared with conventional pyrolysis, microwave-assisted decrease the amount of heavy molecules gases produced. In addition, prolonged exposure to microwave leads to higher temperatures thus even lower amount of heavy molecules are registered.
- Microwave assisted pyrolysis applied to cathodes from LIBs permits catalytic steam and dry reforming reactions, which is evidenced by strong formation of H_2 and CO.
- Short chain molecules are more likely to be formed in microwave assisted pyrolysis when compared to conventional pyrolysis. This, due to the rapid heating and breaking of long chain molecules into short molecules. In addition, the process experienced limited secondary reactions in the gas phase due to fast cooling of the produced gas after leaving the sample.
- Microwave pyrolysis at 360 °C is suggested in this study after taking mass loss, active mass yield ratio, heavy and light molecules, and toxic compounds into consideration.

Author Contributions: F.D., Y.W., and T.M., Writing-Original Draft Preparation; F.D., Y.W., Formal Analysis, Writing-Review & Editing; F.D., Conceptualization & Methodology; B.F., supervision.

Funding: This research received no external funding.

Conflicts of Interest: The authors declare no conflict of interest.

References

1. McDowall, J. Understanding lithium-ion technology. In Proceedings of the Battcon, Marco Island, FL, USA, 5–7 May 2008.
2. Wu, Y. *Lithium-Ion Batteries: Fundamentals and Applications*; CRC Press: Vancouver, BC, Canada, 2015.
3. Pillot, C. Lithium Ion Battery Raw Material Supply & Demand 2016–2025. In Proceedings of the Advanced Automotive Battery Conference, Mainz, Germany, 30 January–2 February 2017.
4. O'Driscoll, M. Industrial mineral recycling in Li-ion batteries: Impact on raw material supply chain? In Proceedings of the Advanced Automotive Battery Conference, Mainz, Germany, 30 January–2 February 2017.
5. Baylis, R. LIB raw material supply chain bottlenecks: Looking beyond supply/demand/price. In Proceedings of the Advanced Automotive Battery Conference, Mainz, Germany, 30 January–2 February 2017.
6. Friedrich, B.; Träger, T.; Peters, L. Lithium Ion Battery Recycling and Recent IME Activities. In Proceedings of the Advanced Automotive Battery Conference, Mainz, Germany, 30 January–2 February 2017.
7. Ordoñez, J.; Gago, E.J.; Girard, A. Processes and technologies for the recycling and recovery of spent lithium-ion batteries. *Renew. Sustain. Energy Rev.* **2016**, *60*, 195–205. [CrossRef]
8. European Li-Ion Battery Advanced Manufacturing for Electric Vehicles (ELIBAMA). Li-ion BATTERIES RECYCLING. The batteries end of life..., 2014. Available online: http://www.webcitation.org/717f6wRdJ (accessed on 23 July 2018).
9. Zeng, X.; Li, J.; Singh, N. Recycling of Spent Lithium-Ion Battery: A Critical Review. *Crit. Rev. Environ. Sci. Technol.* **2014**, *44*, 1129–1165. [CrossRef]
10. Huang, Y.-F.; Chiueh, P.-T.; Lo, S.-L. A review on microwave pyrolysis of lignocellulosic biomass. *Sustain. Environ. Res.* **2016**, *26*, 103–109. [CrossRef]
11. Sun, J.; Wang, W.; Liu, Z.; Ma, Q.; Zhao, C.; Ma, C. Kinetic Study of the Pyrolysis of Waste Printed Circuit Boards Subject to Conventional and Microwave Heating. *Energies* **2012**, *5*, 3295–3306. [CrossRef]
12. Walkiewicz, J.W.; Kazonich, G.; McGill, S.L. Microwave heating characteristics of selected minerals and compounds. *Miner. Metall. Process.* **1988**, *5*, 39–42.
13. Gupta, M.; Eugene, W.W.L. *Microwaves and Metals*; John Wiley & Sons (Asia) Pte Ltd.: Singapore, 2007.
14. Preto, F. Pyrolysis, Char and Energy. Presentation at the Canadian Biochar Initiative, inaugural Meeting, CAnmetEnergy: December 12, 2008, Ste Anne de Bellevue. Available online: http://www.webcitation.org/717hJZ4pw (accessed on 23 July 2018).
15. Luyima, A. Recycling of Electronic Waste: Pinted Wiring Board. Ph.D. Thesis, Missouri University of Science and Technology, Rolla, MO, USA, 2013.
16. Basu, P. *Biomass Gasification and Pyrolysis. Practical Design and Theory*; Elsevier/AP: Amsterdam, The Netherlands, 2010.
17. Marto, C. Pyrolysis of Peat: An Experimental Investigation. Master's Thesis, University of Padua, Padua, Italy, 2014.
18. Epple, B.; Leithner, R.; Linzer, W.; Walter, H. *Simulation von Kraftwerken und Wärmetechnischen Anlagen*; Springer: Vienna, Austria, 2009.
19. Luda, M.P. *Waste Electrical and Electronic Equipment (WEEE) Handbook, Handbook: Pyrolysis of WEEE Plastics*; Woodhead Publishing: Cambridge, UK, 2012.
20. Diaz, F.; Flerus, B.; Nagraj, S.; Bokelmann, K.; Stauber, R.; Friedrich, B. Comparative Analysis about Degradation Mechanisms of Printed Circuit Boards (PCBs) in Slow and Fast Pyrolysis: The Influence of Heating Speed. *J. Sustain. Metall.* **2018**, *408*, 183. [CrossRef]
21. Sun, L.; Qiu, K. Vacuum pyrolysis and hydrometallurgical process for the recovery of valuable metals from spent lithium-ion batteries. *J. Hazard. Mater.* **2011**, *194*, 378–384. [CrossRef] [PubMed]
22. Träger, T.; Friedrich, B.; Weyhe, R. Recovery Concept of Value Metals from Automotive Lithium-Ion Batteries. *Chem. Ing. Tech.* **2015**, *87*, 1550–1557. [CrossRef]
23. Guo, H.-J.; Li, X.-H.; Zhang, X.-M.; Wang, Z.-X.; Peng, W.-J.; Zhang, B. Optimizing pyrolysis of resin carbon for anode of lithium ion batteries. *J. Cent. South Univ. Technol.* **2006**, *13*, 58–62. [CrossRef]
24. Feng, X.; Ouyang, M.; Liu, X.; Lu, L.; Xia, Y.; He, X. Thermal runaway mechanism of lithium ion battery for electric vehicles: A review. *Energy Storage Mater.* **2018**, *10*, 246–267. [CrossRef]

25. Kawamura, T.; Kimura, A.; Egashira, M.; Okada, S.; Yamaki, J.-I. Thermal stability of alkyl carbonate mixed-solvent electrolytes for lithium ion cells. *J. Power Sources* **2002**, *104*, 260–264. [CrossRef]
26. Spotnitz, R.; Franklin, J. Abuse behavior of high-power, lithium-ion cells. *J. Power Sources* **2003**, *113*, 81–100. [CrossRef]
27. Tobishima, S.-I.; Yamaki, J.-I. A consideration of lithium cell safety. *J. Power Sources* **1999**, *81–82*, 882–886. [CrossRef]
28. Wang, Q.; Ping, P.; Zhao, X.; Chu, G.; Sun, J.; Chen, C. Thermal runaway caused fire and explosion of lithium ion battery. *J. Power Sources* **2012**, *208*, 210–224. [CrossRef]
29. Chen, Y.; Tang, Z.; Lu, X.; Tan, C. Research of explosion mechanism of lithium-ion battery. *Prog. Chem.* **2006**, *18*, 823–831.
30. Sutton, W. Microwave Processing of Ceramic Materials. *Am. Ceram. Soc. Bull.* **1989**, *68*, 376–386.
31. Sun, J.; Wang, W.; Yue, Q. Review on Microwave-Matter Interaction Fundamentals and Efficient Microwave-Associated Heating Strategies. *Materials* **2016**, *9*, 231. [CrossRef] [PubMed]
32. Sun, J.; Wang, W.; Liu, Z.; Ma, C. Recycling of Waste Printed Circuit Boards by Microwave-Induced Pyrolysis and Featured Mechanical Processing. *Ind. Eng. Chem. Res.* **2011**, *50*, 11763–11769. [CrossRef]
33. Blomgren, G.E. The Development and Future of Lithium Ion Batteries. *J. Electrochem. Soc.* **2016**, *164*, A5019–A5025. [CrossRef]
34. Ehsan, R.; Kerstin, S.; Moritz, V. Kompendium: Li-Ionen-Batterien. Available online: http://www.webcitation.org/717pBmpxj (accessed on 23 July 2018).
35. Jackson, W.M.; Conley, R.T. High temperature oxidative degradation of phenol–formaldehyde polycondensates. *J. Appl. Polym. Sci.* **1964**, *8*, 2163–2193. [CrossRef]
36. Gasification. *Chemistry, Processes, and Applications*; Baker, M.D., Ed.; Nova Science Publishers: New York, NY, USA, 2012.
37. Gracida-Alvarez, U.R.; Mitchell, M.K.; Sacramento-Rivero, J.C.; Shonnard, D.R. Effect of Temperature and Vapor Residence Time on the Micropyrolysis Products of Waste High Density Polyethylene. *Ind. Eng. Chem. Res.* **2018**, *57*, 1912–1923. [CrossRef]
38. Setiani, P.; Watanabe, N.; Sondari, R.R.; Tsuchiya, N. Mechanisms and kinetic model of hydrogen production in the hydrothermal treatment of waste aluminum. *Mater. Renew. Sustain. Energy* **2018**, *7*, 4013. [CrossRef]
39. Galdámez, J.R.; García, L.; Bilbao, R. Hydrogen Production by Steam Reforming of Bio-Oil Using Coprecipitated Ni−Al Catalysts. Acetic Acid as a Model Compound. *Energy Fuels* **2005**, *19*, 1133–1142. [CrossRef]
40. Wang, D.; Czernik, S.; Chornet, E. Production of Hydrogen from Biomass by Catalytic Steam Reforming of Fast Pyrolysis Oils. *Energy Fuels* **1998**, *12*, 19–24. [CrossRef]
41. Shah, N.; Panjala, D.; Huffman, G.P. Hydrogen Production by Catalytic Decomposition of Methane. *Energy Fuels* **2001**, *15*, 1528–1534. [CrossRef]

© 2018 by the authors. Licensee MDPI, Basel, Switzerland. This article is an open access article distributed under the terms and conditions of the Creative Commons Attribution (CC BY) license (http://creativecommons.org/licenses/by/4.0/).

Article

Zinc Recovery from Steelmaking Dust by Hydrometallurgical Methods

Piotr Palimąka *, Stanisław Pietrzyk, Michał Stępień, Katarzyna Ciećko and Ilona Nejman

Faculty of Non-Ferrous Metals, AGH University of Science and Technology, 30-059 Krakow, Poland; pietstan@agh.edu.pl (S.P.); mstepien@agh.edu.pl (M.S.); kat.ciecko@gmail.com (K.C.); inejman@agh.edu.pl (I.N.)
* Correspondence: palimaka@agh.edu.pl; Tel.: +48-126-172-671

Received: 17 June 2018; Accepted: 16 July 2018; Published: 18 July 2018

Abstract: Hydrometallurgical recovery of zinc from electric arc furnace dust was investigated on a laboratory scale, using aqueous sodium hydroxide solution as a leaching agent. Special attention was paid to the effect of NaOH concentration, temperature and liquid/solid phase ratio on the zinc leachability. It was found that all tested factors increased the leachability, with the maximum efficiency of 88% obtained in a 6 M NaOH solution at a temperature of 80 °C and the liquid/solid phase ratio of 40. The test results confirmed the high selectivity of the zinc leaching agent. In spite of this, complete recovery of zinc from steelmaking dust has proved to be very difficult due to the occurrence of this element in the form of stable and sparingly soluble $ZnFe_2O_4$ ferrite. Purification of the solution by cementation and electrolysis gave zinc of purity 99.88% in powder form.

Keywords: steelmaking dust; zinc recycling; alkaline leaching; electric arc furnace

1. Introduction

One of the commonly used techniques to protect steel surfaces from corrosion is galvanizing. Almost 50% of the world's zinc production is consumed for this purpose. Worn steel elements are usually sent to metallurgical plants and are subjected to a remelting process in electric arc furnaces (EAF). During this process, zinc is evaporated, oxidized and then, as solid ZnO, transferred to dedusting devices. One ton of smelted steel scrap produces about 15–25 kg of dust, in which the content of zinc is 15–40% [1]. The dusts additionally contain a significant amount of iron (about 20%) and other elements such as cadmium, nickel, chromium, manganese, carbon, tin, antimony and copper. The storage of this type of waste is associated with serious threats to the natural environment, as metals can leach out to surface and groundwater, contaminating the environment [2]. Moreover, following accumulation in organisms such as plants and animals, they reach humans. Since the metals do not decompose, they can bioaccumulate once they are absorbed [3]. Therefore, the waste should be recycled to the steelmaking process, which is not currently possible due to the high content of zinc. On the other hand, a significant iron content eliminates the possibility of treating the steel dust as a raw material for the traditional hydrometallurgical zinc manufacturing process.

Current technologies for steel dust processing are focused on pyrometallurgical methods, and consist mainly of high-temperature reduction of ZnO contained in dust and secondary oxidation of Zn in the gas phase. Figure 1 shows the pyro-and hydrometallurgical methods for steelmaking dust processing currently used on an industrial and pilot scale. The Waelz kiln process remains the predominant method for processing dust, with over 85% of the market [4]. Any other pyrometallurgical process, indicated in Figure 1, results in zinc in the form of ZnO, which is fed to the traditional process of sulfate electrolysis or Imperial Smelting Process (ISP).

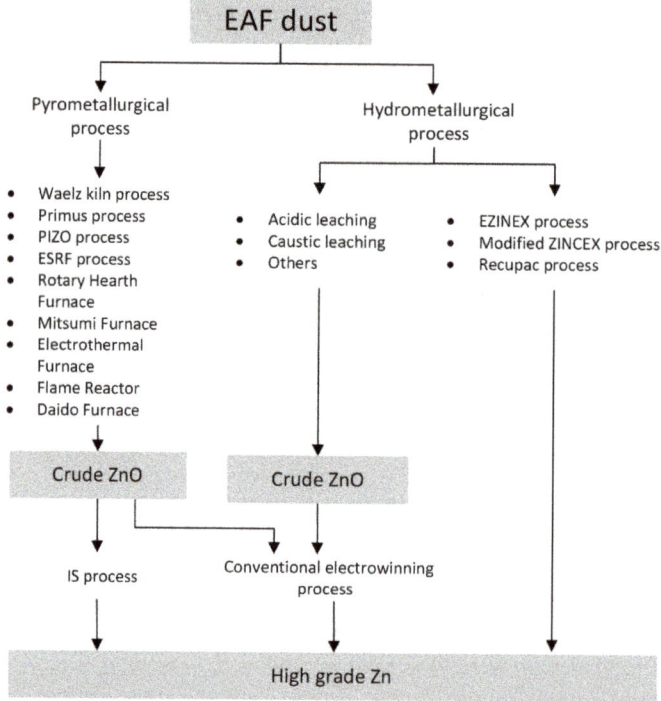

Figure 1. Schematic presentation of electric arc furnace dust treatment [3–7].

The attempt to implement a comprehensive and waste-free process described in References [3,5] was undertaken by a plant in Poland. The pilot installation that has been launched allows two fully usable products to be made in a shaft furnace. The first product is Zn, in the form of ZnO, removed from properly formed briquettes and captured in dedusting devices. The second product includes zinc-free sintered briquettes, which can be used in steelmaking. In this way, no waste is generated in the process.

Hydrometallurgical processes are not as popular as pyrometallurgical methods, but there are several installations converting EAF dust on an industrial scale. One of the currently used technologies is the EZINEX process, which involves ammonia leaching of dust, cementation-based purification of the solution and zinc separation through electrolysis. In recent years, the process has been modified with high-temperature pre-treatment of zinc-bearing material (fuming in an induction furnace). As a result of this process, raw zinc oxide is subjected to a hydrometallurgical treatment, and the integrated technology is called ENDUTEC/EZINEX [6]. Another process (Modified Zincex Process) is a modification and simplification of the former process (Zincex) using atmospheric leaching (H_2SO_4 solution), solvent extraction and conventional electrowinning, in order to resolve impurity difficulties and recover Special High Grade (SHG) zinc ingots [8]. In turn, Recupac has developed a patented recycling process to extract marketable iron and zinc compounds from EAF dust using proven hydrometallurgical technology. The zinc and iron are recovered, and the iron oxide is used to make industrial pigments. Many processes developed on a laboratory, pilot or major scale have not been implemented in industry, or, for indefinite reasons, have been delayed after several years (Cebedeau, Cardif, AMAX) [9]. The biggest problem in hydrometallurgical processes is the maximization of zinc recovery with simultaneous inhibition of iron extraction into the solution [10]. Since zinc occurs in the dust as ZnO and $ZnFe_2O_4$, the key problem is how to extract it from the zinc ferrite and transfer it to the solution. If zinc in the dust is mainly present in the form of ZnO, high leaching efficiency is to be expected. If, in turn, zinc is in large part bound in $ZnFe_2O_4$-a phase of very high stability-then zinc recovery may be low [10]. Generally, hydrometallurgical processes involve leaching of steelmaking dust and extraction of zinc from the solution in a metallic form or as zinc compounds. The main goals are to obtain the end product of the highest purity, develop a method for the management of intermediate products formed during hydrometallurgical processing, and reduce zinc content in the processed EAF dust to a level that would allow its recycling for use in the steelmaking industry.

Until now, most of the attention in studies of the leaching process has been paid to the use of sulfuric acid and caustic soda as leaching agents. The advantages of using sulfuric acid include high dissolution kinetics, availability and a modest price, as well as the possibility of using low concentrations and a well-known process of electrolysis to obtain metallic zinc, with regeneration of the acid. Conversely, the biggest disadvantage of using sulfuric acid as a leaching agent is the lack of selectivity towards other metals found in steelmaking dust. This applies mainly to iron, which can sometimes have a content in steelmaking dust much higher than that of zinc. This results in high consumption of the leaching agent combined with the need to purify the leach solution from significant amounts of Fe, which is both complex and difficult. Iron hinders the cementation of copper, cadmium and cobalt [9]. It is also necessary to control pH to avoid hydrolysis and $Zn(OH)_2$ precipitation.

The main advantage of using an alkaline leaching agent is its selectivity towards zinc over iron, which practically eliminates the difficult and complicated process of iron removal from the electrolyte. It is also possible to separate lead from zinc, which may be used as an additional intermediate product and raw material for the recovery of this metal. Therefore, the use of NaOH as a leaching agent seems to be a promising alternative for the processing of steelmaking dust, although alkaline leaching agents also have some disadvantages. The most important include the high concentration necessary to obtain significant zinc leaching efficiency, a higher price compared to sulfuric acid, and the difficult recovery of the solution.

Despite many problems, the hydrometallurgical methods of steelmaking dust processing are much more attractive due to lower operating costs and energy expenditure. Problems related to the high zinc leaching rate and effective extraction from the solution are still a challenge.

Results of research on EAF dust leaching in NaOH solutions that have been published in recent years indicate a strong dependence of some parameters on leaching efficiency. Mordoğan et al. [10] investigated leaching in solutions with a low concentration of NaOH (3.75 M) at ambient temperatures. The effects were poor; only 21.45% of zinc was extracted to the solution. Xia and Pickles [11] and Dutra et al. [12] obtained leaching efficiency of about 74% at higher concentrations of NaOH (6 M) and temperature (90 °C). Orhan [13] attempted to use even more concentrated solution and a higher temperature (10 M NaOH, 95 °C), and achieved up to 85% Zn recovery. Previous results show that the sodium hydroxide concentration and leaching temperature have significant influence on the dissolution of zinc from the EAF dust. Other parameters such as liquid to solid ratio, particle size and agitation speed were also investigated in the above research, but it was found that these parameters had relatively insignificant influence on the extraction process.

Therefore, this study was undertaken to determine the optimal conditions for zinc extraction from EAF dust using caustic soda solutions. The results of the cementation process and electrolysis of the chosen solution, after leaching of dust, were also obtained.

2. Methods

2.1. Leaching

The leaching process was carried out in two stages. The aim of the first stage, involving leaching in water, was to remove chlorine from the material. This is an indispensable step because, during the main leaching process, the presence of chlorine (and also of Na, K, Ca) in the dust causes high consumption of the leaching agent and risk of harmful chlorine being released during electrolysis. The liquid/solid ratio during leaching was 20 and the temperature was kept at 60 °C. The process lasted 120 min using mechanical stirring at 400 rpm. After water leaching, the material was vacuum filtered. Then, after washing and drying, the residue was subjected to alkaline leaching. The aim of this operation was to transfer as much zinc as possible from the EAF dust to the solution. Sodium hydroxide was used as the leaching agent. The variable parameters investigated for the leaching process included NaOH concentration, temperature and the liquid/solid ratio (L/S). These parameters are summarized in Table 1.

Table 1. The parameter values for alkaline leaching of Electric Arc Furnace (EAF) dust.

Parameter	Value
NaOH Concentration, M	2, 4, 6
Temperature, °C	20, 50, 80
Liquid–solid ratio, cm^3/g	10, 20, 30, 40
Stirring speed, rpm	400
Time, min	120

The experiments were carried out in an 800 mL glass beaker placed on a heating plate. The temperature was controlled to the nearest 1 °C and kept constant using a sensor, protected by a quartz shield, placed in the solution. The duration of each leaching experiment was 120 min. During this process, 4 mL samples of the solution were taken at 5, 15, 30, 60 and 120 min. Each sample taken was filtered and then subjected to chemical analysis using X-ray fluorescence (XRF) (MiniPal 4, PANalytical B.V., Espoo, Finland). Analysis of the concentration of metals in the solution was based on a multi-point calibration curve. The efficiency of zinc extraction into the solution was calculated from Equation (1):

$$L_e = (m_{Zn,t}/m_{Zn,0}) \times 100\% \qquad (1)$$

where: L_e (%) is the zinc extraction efficiency, $m_{Zn,t}$ (g) is the zinc content in the solution after leaching time t, and $m_{Zn,0}$ (g) is the zinc content in the sample before leaching.

2.2. Cementation

The cementation-based purification of the solution was carried out with a 99.95% pure zinc metal plate, keeping the process temperature at a level of 60 °C and stirring the solution with a magnetic stirrer. The cementation time was 70 min. During the experiment, samples for analysis were taken every 3 min (during the first 15 min of the experiment), then every 5 min (up to 30 min of the experiment) and at 50 and 70 min.

2.3. Electrolysis

The electrolysis was carried out for 90 min at a current density of 250 A/m^2 and a temperature of 25 °C. Graphite was used as the material of the anode, while the cathode was made of stainless steel. The current efficiency was calculated from Equation (2):

$$\eta = (m_{Zn}/(k \times I \times t)) \times 100\% \qquad (2)$$

where: η (%) is the current efficiency, m_{Zn} (g) is the mass of zinc after electrolysis, k is the electrochemical equivalent, I (A) is the current and t (sec) is the electrowinning time.

3. Materials

The steelmaking dust used in the study was obtained from a Polish steel plant. All the experiments were carried out on dust in the form of finely divided fractions (less than 100 µm) with a moisture content of 5% (estimated from a 24 h drying test at 105 °C). The chemical and mineralogical compositions of the dust were determined by Microwave Plasma Atomic Emission Spectroscopy

(MP-AES) (MP-AES 4200, Agilent, Penang, Malaysia), and Scanning Electron Microscopy (SEM) (Hitachi, Tokyo, Japan) and X-ray diffraction (XRD) (MiniFlex II, Rigaku, Tokyo, Japan), respectively. The composition and morphology of the dust strongly depends on the quality of the scrap and on the mode of operation of the electric furnace [3]. Typical EAF dust particles obtained by scanning electron microscopy with analysis of a selected particle, are presented in Figure 2.

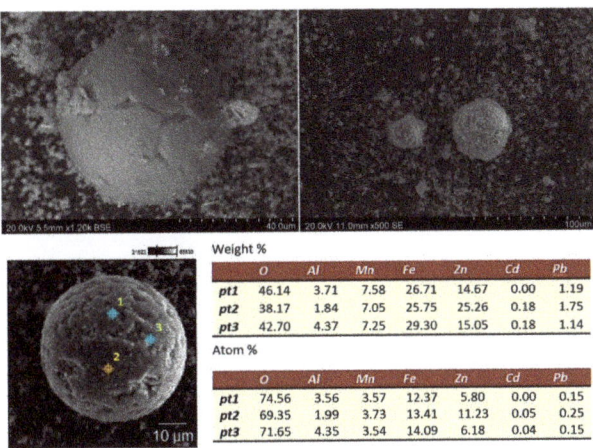

Figure 2. Spherical Electric Arc Furnace (EAF) dust particles and chemical analysis using Scanning Electron Microscope (SEM).

The presence of spherical particles rich in iron, oxygen and zinc was observed. The results of the chemical composition of EAF dust are given in Table 2. The X-ray diffraction pattern of the EAF dust is shown in Figure 3.

Table 2. Chemical composition of the main elements present in the dust.

Element	Zn	Cr	Mn	Fe	Cd	Pb	Al
wt %	33.00	0.30	2.58	21.85	0.05	1.63	0.33

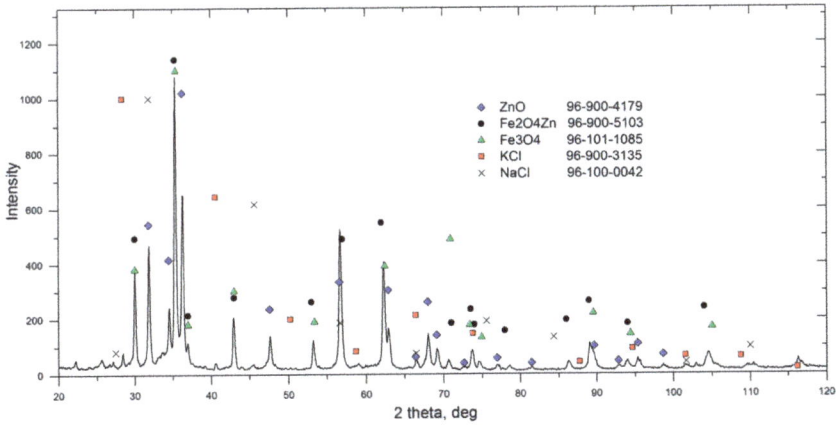

Figure 3. X-ray diffraction analysis of EAF dusts.

Five major phases were identified: franklinite, magnetite, zincite, and sodium and potassium chlorides. The identified chlorides were a residue generated from chlorinated rubber coatings present in plastic admixtures (including polyvinyl chloride and synthetic rubber) in the charge melted to generate the EAF dust. These chlorides are volatile under the process conditions used and evaporate easily, which explains their presence in the dust. Many other compounds have also been identified in the dust [11,12,14,15], but being present in insufficient quantities, these were difficult to detect.

4. Results

4.1. Leaching

In this study, the effect of 2, 4 and 6 M NaOH on the zinc extraction efficiency was investigated. The tests were limited to a concentration of 6 M, because the viscosity of solutions with higher NaOH content increased too much, hindering further processing operations.

The main reactions occurring during leaching of dust can be written in the form of reactions (3) and (4):

$$ZnO_{(s)} + 2NaOH_{(aq)} \rightarrow Na_2ZnO_{2(aq)} + H_2O_{(l)} \tag{3}$$

$$PbO_{(s)} + 2NaOH_{(aq)} \rightarrow Na_2PbO_{2(aq)} + H_2O_{(l)} \tag{4}$$

The high stability of ferrites means that only some of them can dissolve in NaOH solutions, according to reaction (5) [10]:

$$ZnFe_2O_4 + 2NaOH_{(aq)} \rightarrow Na_2ZnO_{2(aq)} + H_2O_{(l)} + Fe_2O_{3(s)} \tag{5}$$

Sample leaching curves for all NaOH concentrations, limit values for temperature and L/S conditions (20 and 80 °C and L/S of 10 and 40) are shown in Figure 4.

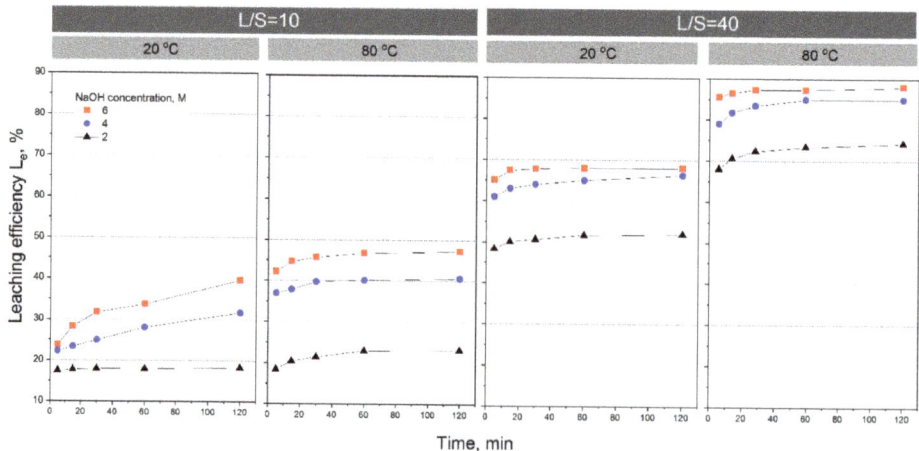

Figure 4. Leaching efficiency for zinc at different leaching conditions of NaOH concentration (2, 4, 6 M), temperature (20 and 80 °C) and liquid/solid (L/S) ratio (10 and 40).

The leaching results obtained for the limit parameter values lead to the conclusion that leaching in 2 M NaOH and L/S = 10 offers a very modest process efficiency, giving a maximum of 24%. During leaching at 20 °C in 4 and 6 M solutions (L/S = 10), a continuous increase in the zinc extraction was observed, which means that these solutions have the potential to dissolve more Zn, but in a time much longer than that used in the experiments. Increasing the temperature to 80 °C increases the zinc extractions to 40 and 45% for 4 and 6 M NaOH, respectively, after only 30 min. Further increases in the leaching time do not result in higher zinc extractions. For L/S = 40, the extractions are much higher, giving 68 and 88%, after 120 min in 6 M NaOH, at 20 °C and 80 °C, respectively. Table A1 in Appendix A summarizes all the results of zinc extraction and its concentration obtained after 120 min for each leaching test. Figures 5 and 6 show these results in graphical form.

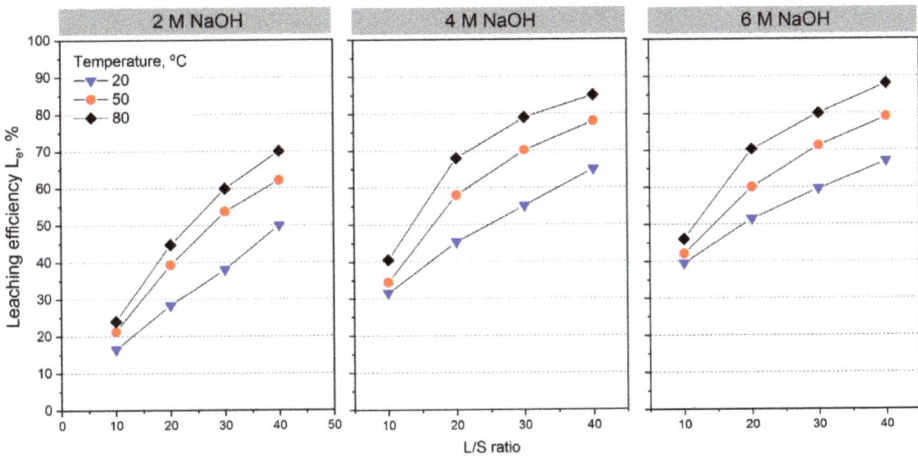

Figure 5. Leaching efficiency for zinc as a function of L/S ratio for different NaOH concentrations and temperature.

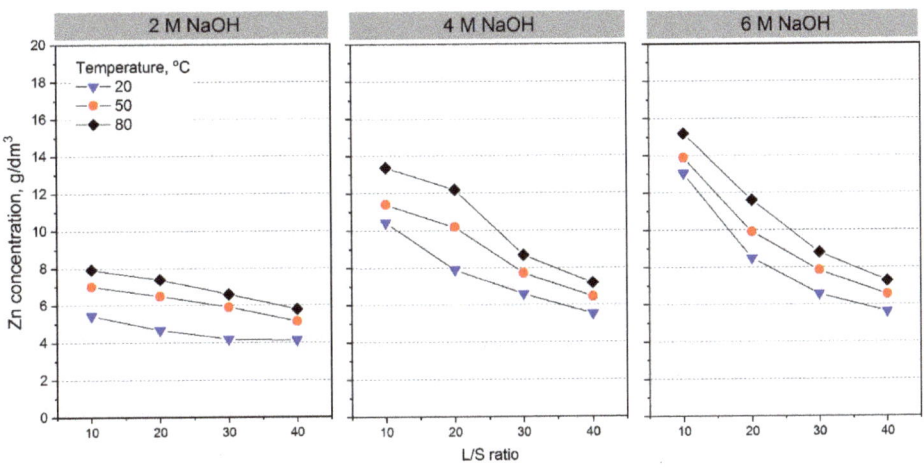

Figure 6. Zinc concentration as a function of L/S ratio for different NaOH concentrations and temperature.

The results obtained show that zinc leaching efficiency depends on all the examined factors, i.e., time, temperature, NaOH concentration and the solid phase content in the leaching solution. At each temperature, the increase in concentration from 2 M to 4 M NaOH increased zinc extraction by a dozen percentage points (13.3–23.4). For further increase in NaOH concentration, the efficiency gains were much lower (0.9–7.9 pp). This may be due to either higher viscosity of the solution and the resultant difficulty in transporting the reactants to and from the reaction surface, or from achieving the maximum extraction under the conditions employed. Additionally, the smallest effect of temperature was observed in solutions with a high solid phase content (L/S = 10). In this case, raising the temperature by 30 °C increased the efficiency of the zinc leaching process by a maximum of 6 pp. In each subsequent leaching test, in which the liquid phase content was increased with respect to the solid phase, the effect of temperature was stronger, and the maximum increase in leaching efficiency was about 15.6 pp for L/S = 30 and 2 M NaOH. Although increasing the liquid phase content in the leached slurry had a beneficial effect on the dust leaching efficiency in terms of metal extraction into the solution in general, the concentration of zinc in the solution actually decreased with increased L/S. The increase of L/S from 10 to 40 in the case of leaching in 2 M NaOH at 80 °C resulted in the drop of zinc concentration from 7.93 g/dm^3 to 5.78 g/dm^3, i.e., about 27%. At the highest NaOH concentration of 6 M, the decrease in zinc concentration associated with dilution of the solution was 52% (from 15.19 to 7.26 g/dm^3). Erdem et al. [16] observed that waste materials containing zinc in the form of $ZnFe_2O_4$ were difficult to leach because franklinite is very stable and poorly soluble in most types of alkaline solutions. They achieved a maximum leachability of 21.47% during leaching at 25 °C for 120 min, using a stirring speed of 300 rpm, L/S = 5 and NaOH concentration of 15%. In the present study, it was found that high solid phase content results in low leaching efficiency of Zn, therefore the results obtained by Erdem et al. [16] appear to be lower than those which can be achieved. The maximum zinc extraction (88.0%) in this study was obtained for 6 M NaOH at 80 °C and L/S = 40. Figure 7 presents a diffractogram of EAF dust samples before and after leaching under selected process conditions.

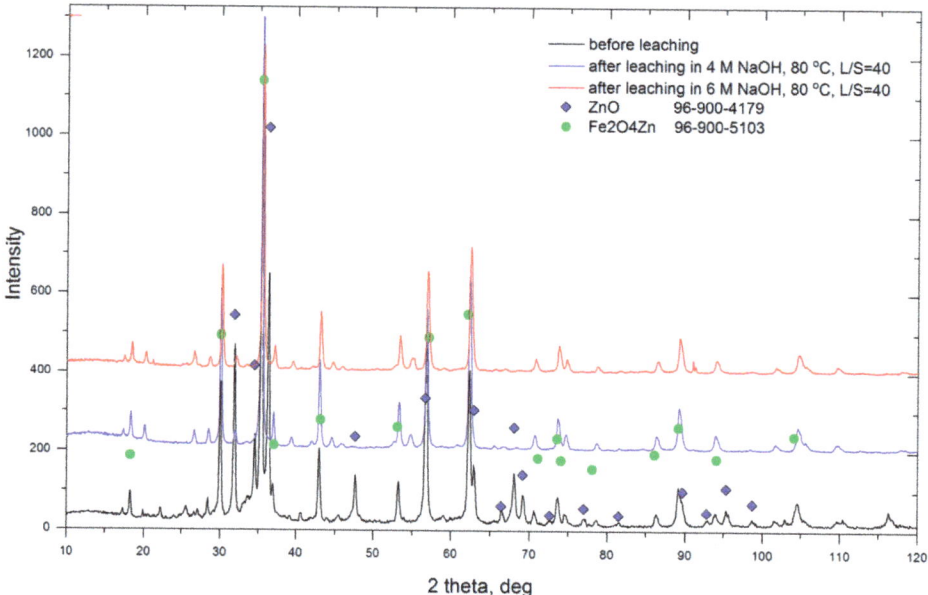

Figure 7. X-ray diffraction analysis of EAF dusts before and after leaching (in 4 and 6 M NaOH, L/S = 40, temp. = 80 °C, time = 120 min).

XRD analysis of the residue obtained after leaching in 4 M NaOH at 80 °C with L/S = 40 showed that virtually all zinc, present in the form of ZnO, was removed from the dust. An increase in NaOH concentration from 4 to 6 M increased the leaching efficiency by a very small extent only, i.e., by 3 pp. It was, however, an indication that zinc in the form of $ZnFe_2O_4$ was being dissolved according to reaction (5), although this process was slow. The dissolution of franklinite was suggested by the lower intensity of franklinite peaks in the diffractogram for the sample leached in 6 M NaOH at 80 °C with L/S = 40 (Figure 7). Analysis of the concentrations of other elements in the leach solution confirmed good selectivity of NaOH towards zinc. In none of the conducted leaching tests did the concentration of iron in the solution exceed 20 mg/dm^3. The only dust component that could dissolve in larger amounts according to reaction (4) is lead. Its concentration under the optimal for zinc leaching conditions (6 M NaOH at 80 °C with L/S = 40) did not exceed 550 mg/dm^3, though even this concentration is quite significant and might affect the process of electrolysis (i.e., contaminate the zinc cathode). The concentration of other metals (Cr, Mn, Cd) in all leaching tests was low and generally below 40 mg/dm^3, or much lower. The tested leaching parameters had no specific effect on the concentrations of these metals. Table 3 outlines the results of the solution analysis after leaching at 80 °C (6 M NaOH; L/S = 10). This particular solution was selected for further treatment, as it had the highest concentration of zinc.

Table 3. Leach solution composition after 120 min (6 M NaOH; 80 °C; L/S = 10).

Element	Zn	Cr	Mn	Fe	Cd	Pb
mg/dm^3	15190	40	12	20	12	420

4.2. Cementation

The removal of undesirable elements from the electrolyte is necessary for the final stage of processing: Electrolysis. In the electrolysis process, any metal more noble than zinc will be deposited on the cathode, contaminating the zinc cathode and reducing the current efficiency. Therefore, it is important that the solution used for electrolysis contains the least amount of undesirable metals. The cementation process can be written in the general form of Equation (6):

$$n_2 \text{Me}^{n_1+} + n_1 \text{M}^0 = n_2 \text{Me}^0 + n_1 \text{M}^{n_2+} \tag{6}$$

where: n_1, n_2 is the ion charge; Me is the cementated metal; and M is the cementing metal.

According to the above reaction it is possible to remove more noble metals with zinc, and in this particular case, mainly a significant amount of lead. Figure 8 shows the changes in the concentration of zinc and lead throughout the 70-min cementation process.

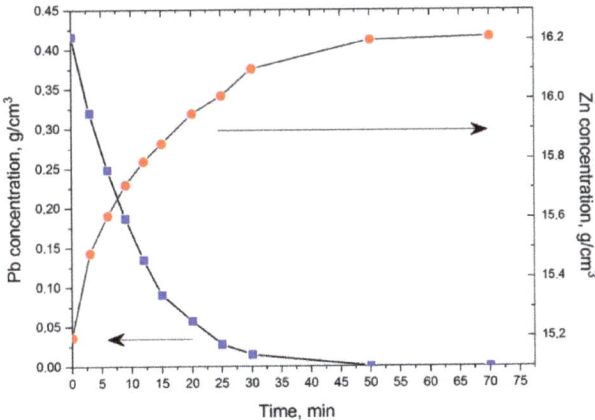

Figure 8. Concentration of Pb (blue) and Zn (red) during the cementation process.

As can be seen from Figure 8, practically all lead was removed in just 50 min. The concentration of other metals remained unchanged during the cementation process. The lead concentration in the solution after cementation (<10 mg/dm^3) was found to be low enough to not affect zinc electrolysis. It should be noted here for comparison that Orhan [13] studied the effectiveness of the cementation process using zinc powder. Despite the selection of optimal process parameters, even after 3 h of cementation, lead was removed from the solution to a level of 100–120 mg/dm^3. Furthermore, Mordogan et al. in Reference [10], who also used zinc powder, removed lead from the solution to the level of about 210 mg/dm^3 using zinc powder.

The composition of the post-cementation solution which was used for electrolysis is presented in Table 4.

Table 4. The composition of the solution after the cementation process.

Element	Zn	Cr	Mn	Fe	Cd	Pb
mg/dm^3	16200	40	12	20	12	<10

4.3. Electrolysis

The reactions which occur during the process of electrolysis can be written in the following form:

$$A(+): 2OH^- \rightarrow 1/2 O_2 + H_2O + 2e \tag{7}$$

$$K(-): Zn(OH)_4^{2-} + 2e \rightarrow Zn + 2OH^- + H_2O \tag{8}$$

In total, the process of metallic zinc deposition from the solution can be written as:

$$Na_2ZnO_2 + H_2O \rightarrow Zn + 2NaOH + 1/2 O_2 \tag{9}$$

Since the concentration of elements in the electrolyte was low (Table 4), the process of electrolysis was straightforward. The result was a spongy powder, having a morphology as shown in Figure 9.

Figure 9. Images of zinc powder deposited on cathode; SEM.

The process efficiency with the applied current density of 250 A/m^2 was 86%. A similar value under similar experimental conditions was obtained by Zhang et al. [17], but there was no mention of the structure of the cathode deposit and purity of the obtained zinc. The chemical composition of the zinc deposited on the cathode from the present study is given in Table 5.

Table 5. Chemical composition of the cathodic zinc.

Compound	Concentration, %
Fe	0.064
Cd	0.051
Zn	99.88

The deposit obtained in this study was easy to remove from the cathode surface and, as shown by the analysis in Table 5, had a high degree of purity. The deposit was repeatedly rinsed with distilled water before the analysis was free of sodium.

5. Discussion

Currently, in terms of development, the hydrometallurgical technologies are lagging behind the pyrometallurgical processes used in most industrial installations for the processing of steelmaking dust. Nevertheless, studies of hydrometallurgical techniques for zinc recycling from this type of waste are still ongoing and have become important for several reasons:

- Hydrometallurgical processes are environmentally friendly, generating practically no air pollution.
- Installations are much smaller in size compared to pyrometallurgical plants, and as such require lower investment and operating costs.
- Hydrometallurgical processes are applicable in the recovery of other valuable metals.
- The product obtained from leaching is an iron-bearing material that can be recycled and reused in the steelmaking process (if the zinc content will be at the minimum level).

As shown by this study, leaching with NaOH solution is selective towards zinc, and the process proceeds with a high yield of up to 88.0%. However, the presence of zinc in the form of $ZnFe_2O_4$ ferrite poses some problems in regards to its complete recovery from steelmaking dust. Studies have shown that even high concentrations of NaOH (6 M), high temperature (80 °C), and adequate excess of leaching solution in relation to the solid phase (L/S = 40) does not remove all of the zinc present. Although the results of the XRD analysis of the leach residue, generated in 4 and 6 M NaOH at a temperature of 80 °C, indicated that zinc ferrite dissolution does occur, the rate of this reaction is low. Xia and Pickles [11] found that even in a highly aggressive leaching environment (10 M NaOH, temperature of 93 °C, leaching time of 3 h), the extraction yield of zinc from the zinc ferrite is only 9%. A solution to this problem might be sintering of dusts with Na_2CO_3 before the leaching process, in order to decompose the zinc ferrite according to reaction (10), as suggested by Holloway et al. [18].

$$ZnFe_2O_4 + Na_2CO_3 \rightarrow ZnO + 2NaFeO_2 + CO_2 \qquad (10)$$

Another solution may be to grind the EAF dust with iron powder (mechanochemical reduction). This also enables the breakdown of $ZnFe_2O_4$, as demonstrated by Zhang et al. [19]. Such approaches generate additional costs due to the expansion of the installations to include aggregates for material processing, such as mills, furnaces and dryers, but the end product—i.e., an iron-bearing material—is free from zinc (or has only a very low content of this metal), and thus can be returned to the steelmaking process. The results of cementation and electrolysis presented in this study showed that both processes run smoothly to give a high purity final product.

6. Conclusions

1. Leaching of EAF dust using NaOH solutions confirmed high zinc selectivity.
2. All parameters tested (time, temperature, L/S ratio and NaOH concentration) had a positive effect on the efficiency of zinc leaching.
3. Increasing the leachate content in relation to the solid phase content increases the leaching efficiency, but also decreases the zinc concentration in the solution. This means that it is necessary to handle larger volumes of solution to obtain the required EAF dust throughput.
4. The maximum zinc leaching efficiency of 88.0% was achieved in a 6 M NaOH solution at a leaching temperature of 80 °C and L/S = 40. Complete leaching of zinc from steelmaking dust under the tested conditions was not possible due to the presence of this element in the form of $ZnFe_2O_4$.

5. The cementation process using metallic zinc enables the removal of lead from the solution to a very low level of concentration (<10 mg/dm^3).
6. The cathode zinc was obtained as spongy-powder and is characterized by high purity (99.88%).

Author Contributions: Conceptualization, writing the manuscript, investigation—Piotr Palimąka; Data analysis—S.P.; Investigation, MP-AES analysis—M.S.; Leaching test—K.C.; SEM analysis—I.N. All authors have discussed the results, read and approved the final manuscript.

Acknowledgments: This paper is supported by the Ministry of Science and Higher Education (Grant No.11.11.180.959).

Conflicts of Interest: The authors declare no conflict of interest.

Appendix A

Table A1. Results of leaching experiments after 120 min.

L/S, cm^3/g	Temp., °C	C$_{NaOH}$, M	L$_e$, %	C$_{Zn}$, g/cm^3	C$_{NaOH}$, M	L$_e$, %	C$_{Zn}$, g/cm^3	C$_{NaOH}$, M	L$_e$, %	C$_{Zn}$, g/cm^3
10	20		16.5	5.4		31.6	10.4		39.5	13.0
	50		21.2	7.0		34.5	11.4		42.0	13.9
	80		24.0	7.9		40.6	13.4		46.0	15.2
20	20		28.3	4.7		45.4	7.9		51.4	8.5
	50		39.3	6.4		58.1	10.2		56.0	9.9
	80	2	44.8	7.4	4	68.2	12.2	6	70.3	11.6
30	20		38.1	4.2		55.2	6.6		59.6	6.6
	50		53.6	5.9		70.1	7.7		71.2	7.8
	80		59.8	6.7		79.1	8.7		80.0	8.8
40	20		50.0	4.1		65.1	5.5		68.1	5.6
	50		62.2	5.1		78.1	6.4		79.1	6.5
	80		70.1	5.8		85.0	7.2		88.0	7.2

References

1. Guézennec, A.G.; Huber, J.C.; Patisson, F.; Sessiecq, P.; Birat, J.P.; Ablitzer, D. Dust formation in Electric Arc Furnace: Birth of the particles. *Powder Technol.* **2005**, *157*, 2–11. [CrossRef]
2. Ostrowska, P.; Mierzwa, K. Recovery of zinc from selected metallurgical waste. *Hutnik-WH* **2007**, *64*, 369–373.
3. Woźniacki, Z.; Telejko, T.K. Sintering as the method of utilization of steelmaking dusts with a high content of zinc oxides. *Hutnik-WH* **2014**, *81*, 166–171.
4. Stevart, C. Sustainability in Action: Recovery of Zinc from EAF Dust in the Steel Industry, 2015 Intergalva Conference, Liverpool, England, June 2015. Available online: http://www.icz.org.br/upfiles/arquivos/apresentacoes/intergalva-2015/5-2-Stewart.pdf (accessed on 24 April 2018).
5. Palimaka, P.; Pietrzyk, S.; Stepien, M. Recycling of Zinc from the Steelmaking Dust in the Sintering Process. In *Energy Technology 2017, The Minerals, Metals & Materials Series*; Zhang, L., Drelich, J.W., Neelameggham, N.R., Eds.; Springer International Publishing AG: Cham, Switzerland, 2017; pp. 181–189, ISBN 978-3-319-52191-6.
6. Maccagni, M.G. INDUTEC/EZINEX Integrate Process on Secondary Zinc-Bearing Materials. *J. Sustain. Metall.* **2016**, *2*, 133–140. [CrossRef]
7. Nakamura, T.; Shibata, E.; Takasu, U.T.; Itou, H. Basic consideration on EAF dust treatment using hydrometallurgical processes. *Resour. Process.* **2008**, *55*, 144–148. [CrossRef]
8. Díaz, G.; Martín, D. Modified Zincex process: The clean, safe and profitable solution to the zinc secondaries treatment. *Resour. Conserv. Recycl.* **1994**, *10*, 43–57. [CrossRef]
9. Xia, K. Recovery of Zinc From Zinc Ferrite and Electric Arc Furnace Dust. Ph.D. Thesis, Department of Materials and Metallurgical Engineering, Queen's University, Kingston, ON, Canada, Septemper 1997.
10. Mordoğan, H.; Çiçek, T.; Işik, A. Caustic soda leach of electric arc furnace dust. *Turkish J. Eng. Environ. Sci.* **1999**, *23*, 199–207.
11. Xia, D.K.; Pickles, C.A. Kinetics of zinc ferrite leaching in caustic media in the deceleratory period. *Miner. Eng.* **1999**, *12*, 693–700. [CrossRef]
12. Dutra, A.J.B.; Paiva, P.R.P.; Tavares, L.M. Alkaline leaching of zinc from electric arc furnace steel dust. *Miner. Eng.* **2006**, *19*, 478–485. [CrossRef]

13. Orhan, G. Leaching and cementation of heavy metals from electric arc furnace dust in alkaline medium. *Hydrometallurgy* **2005**, *78*, 236–245. [CrossRef]
14. Stefanova, A.; Aromaa, J. Alkaline Leaching of Iron and Steelmaking Dust. Research Raport, Aalto University publication series SCIENCE + TECHNOLOGY 1/2012. Available online: https://aaltodoc.aalto.fi/handle/123456789/3570 (accessed on 24 April 2018).
15. Havlik, T.; Kukurugya, F.; Miskufova, A.; Parilak, L.; Jascisak, J. Leaching of EAF Steelmaking Dust in Sulfuric Acid Dilute Solution: Case of Calcium. In Proceedings of the 2015 Sustainable Industrial Processing, Antalya, Turkey, 4–9 October 2015; pp. 81–88.
16. Erdem, M.; Yurten, M. Kinetics of Pb and Zn leaching from zinc plant residue by sodium hydroxide. *J. Min. Metall. Sect. B Metall.* **2015**, *51*, 89–95. [CrossRef]
17. Zhang, Y.; Deng, J.; Chen, J.; Yu, R.; Xing, X. The electrowinning of zinc from sodium hydroxide solutions. *Hydrometallurgy* **2014**, *146*, 59–63. [CrossRef]
18. Holloway, P.C.; Etsell, T.H.; Murland, A.L. Roasting of La Oroya zinc ferrite with Na_2CO_3. *Metall. Mater. Trans. B Process. Metall. Mater. Process Sci.* **2007**, *38*, 781–791. [CrossRef]
19. Zhang, C.; Zhuang, L.; Wang, J.; Bai, J.; Yuan, W. Extraction of zinc from zinc ferrites by alkaline leaching: Enhancing recovery by mechanochemical reduction with metallic iron. *J. South. African Inst. Min. Metall.* **2016**, *116*, 1111–1114. [CrossRef]

© 2018 by the authors. Licensee MDPI, Basel, Switzerland. This article is an open access article distributed under the terms and conditions of the Creative Commons Attribution (CC BY) license (http://creativecommons.org/licenses/by/4.0/).

Article

The Scrap Collection per Industry Sector and the Circulation Times of Steel in the U.S. between 1900 and 2016, Calculated Based on the Volume Correlation Model

Alicia Gauffin * and Petrus Christiaan Pistorius

Department of Materials Science and Engineering, Carnegie Mellon University, 5000 Forbes Avenue Wean Hall 3325, Pittsburgh, PA 15213, USA; pistorius@cmu.edu
* Correspondence: agauffin@andrew.cmu.edu or alicag@kth.se; Tel.: +1-412-268-2700

Received: 12 March 2018; Accepted: 25 April 2018; Published: 10 May 2018

Abstract: On the basis of the Volume Correlation Model (VCM) as well as data on steel consumption and scrap collection per industry sector (construction, automotive, industrial goods, and consumer goods), it was possible to estimate service lifetimes of steel in the United States between 1900 and 2016. Input data on scrap collection per industry sector was based on a scrap survey conducted by the World Steel Association for a static year in 2014 in the United States. The lifetimes of steel calculated with the VCM method were within the range of previously reported measured lifetimes of products and applications for all industry sectors. Scrapped (and apparent) lifetimes of steel compared with measured lifetimes were calculated to be as follows: a scrapped lifetime of 29 years for the construction sector (apparent lifetime: 52 years) compared with 44 years measured in 2014. Industrial goods: 16 (27) years compared with 19 years measured in 2010. Consumer goods: 12 (14) years compared with 13 years measured in 2014. Automotive sector: 14 (19) years compared with 17 years measured in 2011. Results show that the VCM can estimate reasonable values of scrap collection and availability per industry sector over time.

Keywords: lifetime of steel; steel scrap; circulation; industry sector; dynamic material flow model; recycling rate; material flow analysis

1. Introduction

Information on the lifetime of products and applications is important to evaluate the availability of recyclable metals. The metal contained in products and applications becomes available as scrap at the end of product life. Metal scrap reserves are a substitute for natural resources (iron ore) that are used for ironmaking and steelmaking. By optimizing the collection of recyclable metals, it is possible to preserve natural resources, save energy, and lower the CO_2 emissions associated with steelmaking [1].

The availability of recyclable metals can be estimated with empirical survey-based methods or with model-based methods. The model-based methods are referred to as dynamic material flow models (DMFMs) and are based on both the inflow and outflow of metals as well as the lifetime of products and applications in society [2–8]. Product lifetimes can be obtained from surveys of products and applications; lifetimes can be estimated with the Volume Correlation Model (VCM), a model-based method.

Surveys of measured lifetimes have been published for product groups such as household appliances, automotive and transportation applications, buildings, and bridges [9–18]. These studies have been published by bodies such as NACE International, U.S. Department of Transport (National Bridge Inventory), Nevada Department of Taxation, U.S. Department of Commerce, National

Association of Home Builders, U.S. Army Corps of Engineers, U.S. Department of Transportation, and Amtrak. An overview of the reported lifetimes of products in the U.S. is given in Appendix A [9–18], which is summarized in Table 1. These analyses are based on the average age of products in-use and depreciation periods and do not state the steel tonnages in the products.

Table 1. Previously reported average lifetimes of steel by industry sector and product groups in the U.S. between 1925 and 2014.

Industry Sector	Methods	Typical Product Groups	Average Lifetime of Steel Products in Years [1]
Construction	Average age of buildings and bridges in-use in US	Housing, industrial buildings, mobile offices, office buildings, warehouses, bridges	39 (2015) [9] 44 (2013 & 2014) [9] 42 (2011) [9] 50 (2010) [10,11] 17–48 (1925–1997) [9] 28 (1980) [9] 24 (1973) [9]
Industry goods	Service lives of machinery and depreciation estimates	Machinery for metalworking, construction, agriculture, and special industry	7–30 (2010) [12] 10–16 (1925–1999) [13] 16–25 (1925–1997) [13]
Consumer goods	Average age of appliances in-use in US	Household appliances, video and audio products, computers	5–20 (2011) [14]
Automotive	Average age of automobiles and trucks in operation in the United States	Transportation by air, automobiles, locomotives, ferry boats	16–18 (1990–2014) [15] 11–16 (1985–2014) [16] 11–26 (1972–2015) [17] 15–20 (1960–1997) [18]

[1] with date range shown in parentheses.

Another model-based method to estimate the lifetime of products and applications is the Volume Correlation Model (VCM) [19–23]. The VCM estimates the in-use lifetime of steel. Two different lifetimes are calculated: the scrapped and apparent lifetimes (previously termed "true" and "full" lifetimes) [19–23]. The scrapped lifetime corresponds to the service lifetime, similar to the average in-use lifetime of steel estimated with empirical survey-based studies.

The VCM relies on annual data on steel consumption and scrap collection. Because end-of-life (EOL) products are normally not identified by origin or application, information on scrap collection per industry sector is not generally available [24]. In this study, scrap collection per industry sector in the United States was estimated from a scrap survey conducted by the World Steel Association for the year 2014 (see Appendix B) [25]. The sectors considered were construction, appliances, industrial goods, and automotive products. The survey accounted for approximately 29% of all scrap collected in the U.S. in 2014 and was assumed to be representative. In this work, the results of the scrap survey were extrapolated from zero recycling to the 2014 values for the period 1900 to 2016. Steel consumption per industry sector for this period (1900 to 2016) was based on an analysis conducted by the World Steel Association [26].

In this study, it was investigated whether it was possible to verify the input data on scrap collection per industry sector between 1900 and 2016 based on the VCM method. More specifically, this study (1) investigated the use of the VCM method to estimate the service lifetimes of steel in the US by industry sector on an annual basis and (2) compared model results with lifetime measurements of products and applications in the US. The VCM results (calculated steel lifetimes) for the US were also compared with calculated lifetimes for Sweden and the world between 1900 and 2010.

2. Input Data

2.1. Scrap Collection

The input data of the VCM are the tonnages of steel consumption and scrap collection for the construction, industrial goods, automotive, and consumer goods (appliances) sectors. A simple material flow chart of steel is shown in Figure A1 (see Appendix C). Scrap collection in the United States was estimated according to U.S. Geological Survey (USGS) data [27]. Total collected steel scrap in the US was calculated based on reported scrap consumption in ironmaking and steelmaking, scrap exports and imports, and subtracting internal scrap (years 1938 to 2016). For the years 1900 to 1934, there were no data available for the reported scrap consumption and an approximate mass balance was used instead, as follows: Total raw material usage per ton of crude steel production between 1934 and 1954 was assumed to have been the same from 1900 to 1933. Raw material usage includes scrap consumption, the iron content in the apparent consumption of iron ore (including iron ore used to produce direct reduced iron), and the net trade of pig iron. The average mass ratio of total metal usage to crude steel production was calculated to have been 1.28 between 1934 and 1954. This time base was used because the ratio appeared to be constant during this period. The mass ratio of total scrap usage to crude steel production was estimated by subtracting the iron content in the reported consumed iron ore and the net use of pig iron in crude steel production between 1900 and 1933 (yielding a ratio of 0.74). The resulting estimated mass ratio of scrap consumption (home and purchased scrap) to crude steel production was 0.54. Variations in the accuracy of this necessary approximation would not significantly affect the overall calculation because the tonnage of steel consumed in the period 1900–1933 was small compared with the overall period (see Figure 2).

The purchased steel scrap tonnages for the years 1905 to 1933 were taken from USGS data and the values were extrapolated to 4 M ton in 1900 [28]. The internal scrap was then estimated by subtracting the purchased steel scrap from total scrap usage. Half of the purchased steel scrap was assumed to be obsolete scrap for the years 1900 to 1955. Prompt steel scrap values were taken from a Nathan Associates report covering the years 1956 to 2009 [29]. According to the USGS report, internal scrap also included obsolete scrap from old steel plant equipment and buildings, which amounted to approximately 2% of scrap in 1980 and in 2014 [27]. The obsolete scrap tonnage was subtracted from the internal scrap tonnage by using the same ratio of 0.02 (fraction of internal scrap that is obsolete scrap) for the period 1900 to 2014. Consumption of old scrap at foundries was estimated from a 36% ratio of scrap usage to total casting production between 1972 and 2013 [27]. The total domestic steel scrap collection was the sum of purchased steel scrap at steel mills, scrap consumption at foundries, and exports, from which imported steel scrap was then subtracted [27].

Prompt steel scrap is generated when the finished steel from steel mills and foundries is used to fabricate products at other facilities. Prompt steel scrap was included in the total calculation for steel products and for steel scrap collection. Theoretically, adding this mass to steel consumption and to scrap collection should not significantly affect the model results. Prompt steel scrap was calculated to be 27% of the total scrap collection in 2014 in the U.S., based on the World Steel Association scrap survey [25]. This ratio was similarly calculated to be 26% in 2009 in the US according to a separate study performed by Nathan Associates [29].

Scrap collection per industrial sector was estimated according to the 2014 scrap survey conducted by the World Steel Association [25]; see Table 2 and Appendix B. The proportion of scrap originating from the four sectors was assumed to be equal to the 2014 values for the years 2011 to 2016. In the absence of other information, the ratios were extrapolated to earlier years (see Figure 1a). The proportion of scrap from consumer goods was assumed to increase from 0.05% in 1931 to 0.6% in 1944, subsequently increasing to 11.5% in 2011. The relative contribution from the automotive sector was assumed to increase from 0.001% in 1911 to 1.7% in 1944 and then to 31.7% in 2011. The assumed increase in automotive scrap collection after 1944 appears reasonable on the basis of the following: (1) the relatively low consumption of steel by these sectors in earlier years (see Figure 2), and (2) the reported history of automotive recycling, with shredders having only been widely available from the 1950s as well as "graveyards" of automobile hulks persisting until the 1970s [30]. The small "tails" before 1945 for automotive and consumer goods were added to reduce edge effects (discussed below).

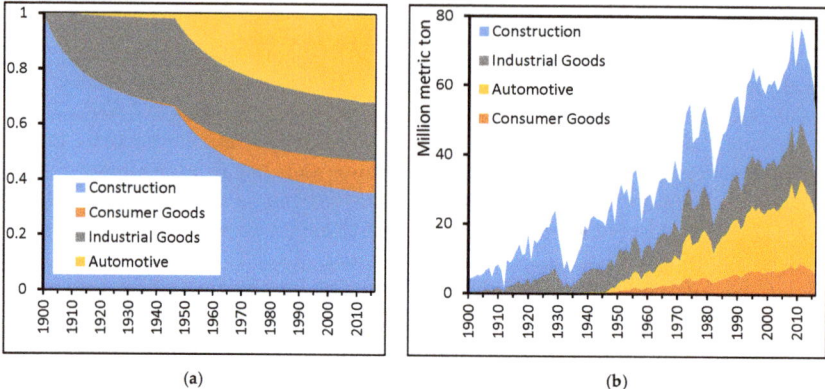

Figure 1. Input data on U.S. scrap collection by industry sector between 1900 and 2016: (**a**) ratios of scrap collected and (**b**) tonnages of scrap collection.

The proportion of scrap from industry goods was assumed to increase from zero before 1900 to 31.4% in 1945 and then decrease to 21.1% in 2011. The assumed zero ratio before 1900 for industry goods appears reasonable based on the rapid industrial development during the late 1800s and early 1900s, when new technologies and specialized machines replaced manual work. Examples include (1) the implementation of large scale factories with new standardized production methods and (2) the upscaling in farming due to new agricultural technology such as the mechanical reaper [31].

The relative scrap collection per industry sector (from the World Steel Association scrap survey; see Table 2) was adjusted slightly to achieve better agreement with the lifetime values in Table 1. The largest change was for consumer goods. It appears that the scrap survey overestimated the contribution of this sector, or that the World Steel Association analysis underestimated the consumption by this sector. Using these values as-is would have resulted in a recycling rate greater than 100% (the calculations for which are presented below; see Figure 5). Figure 1b shows the estimated tonnage of steel scrap from the four sectors based on the estimates of obsolete scrap collected and of the relative contributions of the different sectors. Figure 1a shows the ratios of scrap collection per industry sector between 1900 and 2016. These ratios were assumed to be representative for the total steel scrap collection in the U.S. including prompt steel scrap.

Table 2. Relative contribution of scrap from different industry sectors to obsolete steel scrap collection for the year 2014. Results of the scrap survey [25] and adjusted values used as input data in the calculations.

Industry Sector	Reported	Input Data
Construction	32.4%	35.7%
Consumer Goods	15.6%	11.5%
Industrial Goods	20.4%	21.1%
Automotive	31.6%	31.7%
Prompt scrap	27.3%	-

Figures 1 and 2 show large variations in steel consumption and scrap collection, with the business cycles lasting several years. The scrap collection rate follows the steel consumption rate, likely because of the sensitivity of scrap collection to price. Increased steel demand increases the scrap price, which increase the scrap collection rate [32]. Concurrently, steel consumption is highly sensitive to economic cycles [33]. Major economic events can be recognized in Figure 2, including the downturns of the 1930s and 2008–2009 as well as the rapid growth associated with the period 1940–1950. Over the century of data considered in this work, the share of consumption in the automotive and consumer-goods sectors has increased, while that of industrial goods has remained approximately constant.

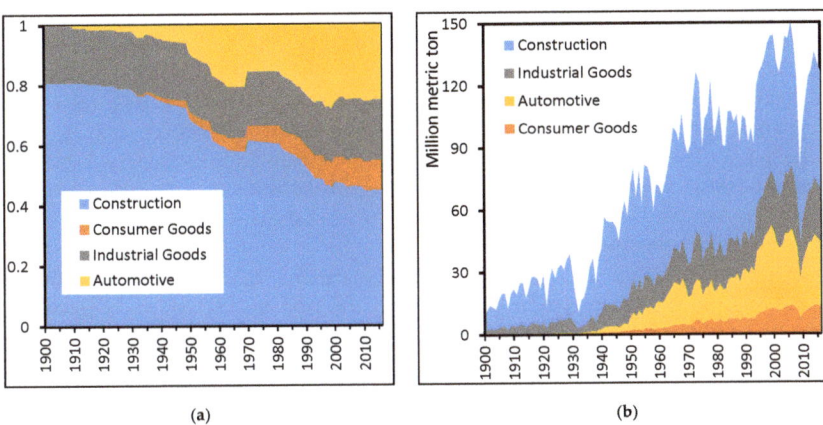

Figure 2. Input data on U.S. steel consumption by industry sector between 1900 and 2016: (**a**) ratios of relative steel consumption per industry sector and (**b**) tonnages of steel consumed.

2.2. Steel Consumption

The total steel consumption represents the net utilization of steel as finished products, including steel that was imported (as semi-finished or finished products as well as manufactured goods). The lifetime of steel was assumed to last from the start of utilization as a finished product until collection and processing into recyclable scrap (ready for use in steel mills and foundries).

Steel consumption in the United States was calculated from USGS data [27] as the sum of apparent finished steel consumption between 1900 and 2016 as well as foundry shipments between 1972 and 2016 (adjusted for exports and imports). No earlier data on foundry shipments were available. The average foundry shipments were, on average, 15% of the total shipments (from steel mills and foundries) for the years 1971 to 1981. The net indirect trade of steel in manufactured goods was taken from World Steel Association data for the years 1970 to 2016 [34,35]. The average net indirect trade of steel was calculated to be nine million metric tons per annum during the period, which made the U.S. a net importing country of steel in manufactured goods. Prompt steel scrap was included in the steel consumption because prompt scrap is generated after finished steel products have been shipped from steel mills and foundries.

Steel consumption ratios in the four industrial sectors in the U.S. were taken from World Steel Association data for the years 1900 to 2013 [9] and assumed to be the same for 2014–2016 as for 2013; see Figure 2a. The ratios were multiplied by the calculated total steel consumption to yield the estimated consumption by sector, shown in Figure 2b.

Figure 2b shows that, by far, the largest steel consumer in the U.S. was the construction sector, accounting for approximately 45–80% of the total steel consumption in US between 1900 and 2016. The remaining steel consumption in US in 2016 was in the automotive sector (25%), industry goods (20%), and consumer goods (10%).

3. Calculation Procedure

The model used in this study, the Volume Correlation Model (VCM), has been described in detail in other studies [19–23]. The model was implemented in Excel®spreadsheets in combination with MATLAB® [36,37]. The VCM can be used to evaluate the time difference between reported data on steel consumption and scrap collection. The model calculates the scrapped and apparent lifetimes of steel. The scrapped lifetime of steel is the (estimated) actual service lifetime of steel; the apparent lifetime is defined below.

In a previous study, the scrapped and apparent lifetimes of steel were calculated for Sweden and the world [19,20]. In this study, the lifetimes of steel were recalculated with Swedish and global steel data for 1900 to 2010, for direct comparison with the calculations for the U.S. as presented here.

The apparent lifetime is predominantly used to calculate moving averages of scrap recovery rates, as described below. The apparent lifetime is calculated by assuming full recovery (as scrap) of all steel in final products, i.e., every ton of steel that entered service as a finished product is assumed to be eventually recoverable as scrap. Clearly, this is not always the case, and incomplete recovery is considered when calculating the scrapped lifetime.

The apparent lifetime of steel (λ_{app}) has been found as the time difference between a given year t_x (with a known cumulative tonnage of scrap collected up to this time: the left term in the equation below) and the number of years (in the past) required to consume the same tonnage of steel:

$$\sum_{0}^{t_x} \Delta m_{scrap}(t) = \sum_{0}^{t_x - \lambda_{app}} \Delta m_{consumed}(t) \qquad (1)$$

In this expression, $\Delta m_{scrap}(t)$ is the mass of scrap collected during year t (years since 1900), $\Delta m_{consumed}(t)$ is the total steel consumption in year t, and λ_{app} is the apparent lifetime of steel. This expression is used to find the apparent lifetime (λ_{app}).

Not all consumed steel is recycled. The apparent recovery rate $\eta(t)$ is found as the ratio of steel recycled in a given year to the moving average steel consumption over a longer period, taken to be $2\lambda_{app}$.

$$\eta(t_x) = \frac{\Delta m_{scrap}(t_x)}{\left[\sum_{t_x - 2\lambda_{app}}^{t_x} \Delta m_{consumed}(t)\right] / 2\lambda_{app}} \qquad (2)$$

At the beginning of the time period considered in the calculations, the time period (for averaging the recovery rate) could be longer than the input data available. For calculating the recovery rate, the moving average was instead calculated over the time period zero to t, in cases where $2\lambda_{app} > t$. For the input data used here, the shorter integration period was used for the first seven years of data for the automotive sector, four years for consumer goods, one year for the construction sector, and the first six years of industrial goods.

Given the long lifetime of steel, the non-recirculated amount of steel is a reserve that is potentially available for future collection. This reserve may become a loss if the steel is not collected after a significant time period. Since the annual apparent recovery rate is based on the moving average of steel consumption, it would not necessarily be equivalent to the recycling rate in any given year; however, the longer-term averages (weighted by tonnage of steel) of the apparent recovery rate and recycling rate would be equal. Using the (time-varying) recovery rate, the amount of potentially recyclable steel added from each year's consumption is estimated as follows:

$$\Delta m_{recyclable}(t) = \eta(t) \times \Delta m_{consumed}(t) \qquad (3)$$

Here, potentially recyclable steel refers to previously consumed steel that has been recycled within an apparent lifetime of steel. The tonnage of recyclable steel depends not solely on the availability of a reserve of steel but also strongly on scrap price; this formulation helps to highlight this effect.

The scrapped lifetime of steel is calculated as the time difference ($\lambda_{\text{scrapped}}$) between a given year t_x (with a known cumulative tonnage of scrap collected) and time required to have accumulated the same tonnage of potentially recyclable steel in products:

$$\sum_{0}^{t_x} \Delta m_{\text{scrap}}(t) = \sum_{0}^{t_x - \lambda_{\text{scrapped}}} \Delta m_{\text{recyclable}}(t) = \sum_{0}^{t_x - \lambda_{\text{scrapped}}} \eta(t) \times \Delta m_{\text{consumed}}(t) \qquad (4)$$

In this expression, $\lambda_{\text{scrapped}}$ is the scrapped lifetime of steel. As Equation (4) shows, the principle suggests that the cumulative mass of scrap collected up to a given year equals a fraction (the recovery rate) of the steel used in products up to a specific date $\lambda_{\text{scrapped}}$ years earlier. Use of an annual recovery rate (rather than a constant ratio) helps to account for some of the variations in scrap collection.

The integration period used to calculate the moving average steel consumption rate affects the results. The effect of the integration period arises, in part, from fluctuations in steel consumption and recycling (in response to economic cycles) as well as the increase in steel consumption over the period considered. To test the effect of the integration period, the scrapped lifetime for total steel in the U.S. for the years 1900 to 2016 was calculated based on three different integration periods: λ_{app}, $2\lambda_{\text{app}}$, and $3\lambda_{\text{app}}$ (as used in Equation (2)). The resulting calculated scrapped lifetimes of steel for the different integration periods are shown in Figure 3. The difference between the apparent lifetime and scrapped lifetime was larger with a shorter integration period. If this large difference were real, it would imply that a large proportion of the steel would be lost and never recycled, which does not appear to be realistic, given reported recycling rates. The results were also compared with weighted averages of the reported lifetime of steel (the box in Figure 3) for 2010–2015. The reported sector lifetimes (listed in Table 1) were weighted using either the relative amount of steel consumed by each sector (yielding an average lifetime of 29 years) or the relative amount of scrap recovered from each sector (yielding an average lifetime of 25 years). The proportions of steel consumed were as given in Figure 2 and scrap ratios as in Table 2. The calculated average lifetime was close to the scrapped lifetime of steel, calculated with $2\lambda_{\text{app}}$ as the integration period. On the basis of these results, the integration period was chosen to be $2\lambda_{\text{app}}$.

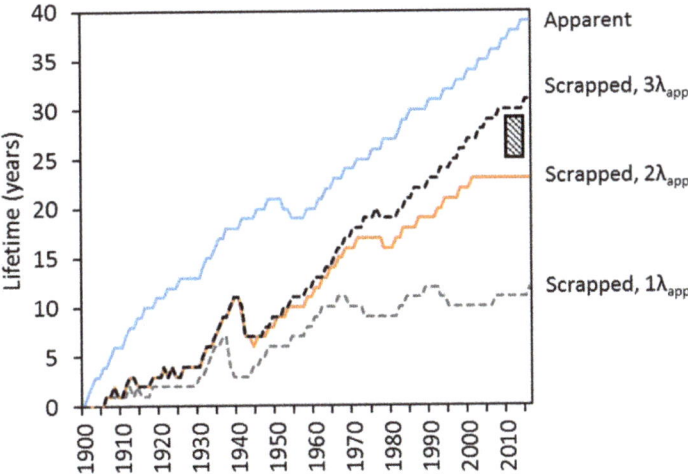

Figure 3. The apparent lifetime of all steel consumed in the U.S., with the scrapped lifetime calculated with integration periods of λ_{app}, $2\lambda_{\text{app}}$, and $3\lambda_{\text{app}}$. The shaded box shows the weighted average of the reported lifetime of steel for 2010–2015.

4. Results and Discussion

4.1. Calculated Lifetimes per Industry Sector in the U.S.

Calculated annual steel consumption and scrap collection tonnages per industry sector in the U.S. between 1900 and 2016 are shown in Figure 4, with apparent recovery rates in Figure 5. On the basis of the low recovery rate, the construction sector had the largest tonnage of non-recirculated steel, with approximately 55% of the total steel consumption collected within the apparent lifetime of steel. The results also show that the apparent recovery rate could temporarily be above 100%, as seen in Figure 4b–d, in years when the tonnage of steel scrap collected was larger than the moving average tonnage of steel produced. In those years, scrap was collected from accumulated scrap reserves. From simple mass balance considerations, the weighted average apparent recovery rate cannot be greater than 1 (100% recovery) over time. However, for all sectors apart from construction, the apparent recovery rate was close to 1, resulting in similar scrapped and apparent lifetimes and indicating low losses of unrecycled steel from the system (see Figure 5b–d).

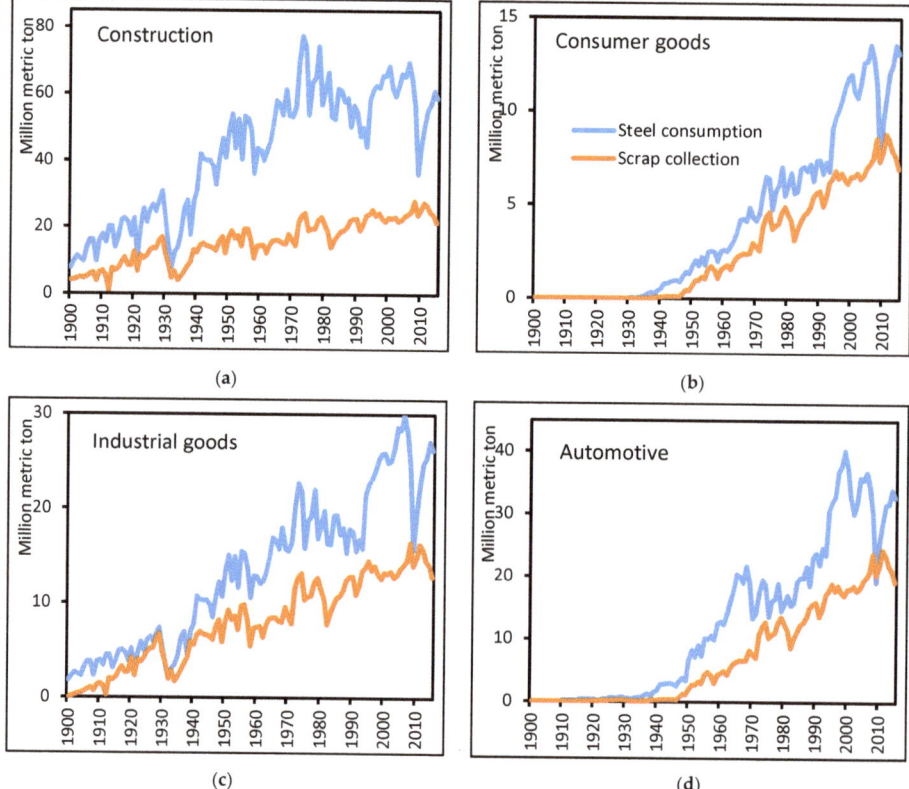

Figure 4. Input data on steel consumption and scrap collection for the following industry sectors: (**a**) construction, (**b**) consumer goods, (**c**) industry goods, (**d**) automotive sector between 1900 and 2016 in the U.S.

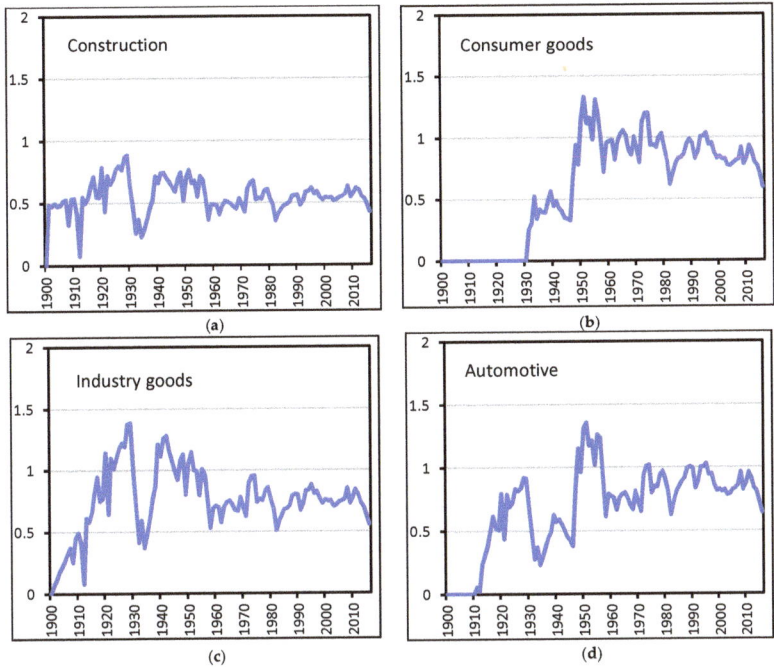

Figure 5. Apparent recovery rates of steel $\eta(t)$ per industry sector in the U.S. between 1900 and 2016: (**a**) construction, (**b**) consumer goods, (**c**) industry goods, and (**d**) automotive sector.

The end-of-life recycling rate (EOL-RR) of metals is defined as the ratio of the tonnage of old steel scrap collected to the steel tonnage in EOL products [38]. A 2011 United Nations Environment Programme report on the EOL-RR of steel in different countries showed the value to be between 70% and 90% [38]. In a USGS study, the EOL-RR of steel in the U.S. in 1998 was calculated to be 47%, taking the lifetime of steel as 19 years [39]. In the present study, the collected steel scrap included both old scrap and prompt scrap. Although the calculated apparent recovery rates fluctuated due to the input data, the longer-term average appeared to be constant and similar to the previously reported EOL-RR. In this work, the average apparent recovery rate for the years 1998 to 2010 for total steel in the U.S. was calculated to be 73%, with a lowest value of 69% and a highest value of 83%, similar to the values of the end-of-life recycling rate quoted above.

The calculated scrapped and apparent lifetimes of steel per industry sector in the U.S. between 1900 and 2016 are shown in Figure 6a–d. For comparison, the figure includes data on measured in-use lifetimes of products and applications in the U.S. over certain periods; these are shown as individual data points or ranges of measure lifetime and period of measurement (lines and boxes). These measured lifetimes are also summarized in Table 1.

The scrapped (and apparent) lifetimes of end-of-life steel products in 2016 in the U.S. were calculated to be as follows: a scrapped lifetime of 30 years (apparent lifetime, 53 years) for the construction sector; 17 (28) years for industrial goods; 17 (21) years for the automotive sector; and 13 (15) years for consumer goods.

The results show edge effects at the beginning of the calculation period for the automotive and consumer goods sectors, resulting in a peak in the calculated scrapped and apparent lifetimes (marked in Figure 6b,d). These edge effects stemmed from an inconsistency between the input data on steel consumption and scrap collection and resulted in part from the assumption that scrap was increasingly collected from these sectors from 1945. This appeared to cause the calculated lifetime of steel at the start of the period of substantial scrap collection to be excessively large, resulting in a peak in the calculated lifetime of steel. However, the peak might be real. In the absence of large-scale recycling, older scrap would have accumulated before 1945. It does not appear to be possible to test this calculation with the available data. Although the in-use lifetime of automobiles did not change for automobiles that were deregistered over this period [40], the time lapse between deregistration and recycling (for this period) is not known.

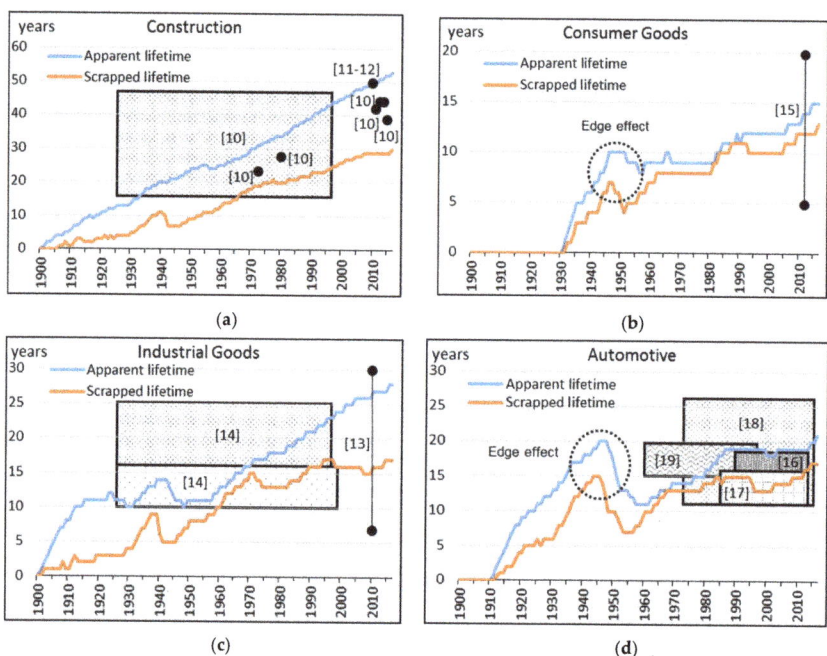

Figure 6. The scrapped and apparent lifetimes of steel in the U.S. between 1900 and 2016 calculated with the Volume Correlation Model (VCM), compared with previously reported lifetime measurements of steel products in-use in the U.S. (literature values are shown as filled circles for single years and boxes for studies referring to multiple years, with reference numbers indicated): (**a**) construction, (**b**) consumer goods, (**c**) industrial goods, and (**d**) automotive sector.

A similar but less obvious effect applies to the calculated lifetimes of steel construction and industrial goods, at the start of the period considered (from 1900 onwards). The scrapped lifetime is incorrectly shown as zero. This inevitably arose from the arbitrary choice of a starting date, subject to the limitations of the available data. However, for all sectors, the calculated scrapped lifetimes did correspond to the previously reported lifetimes for later years (Figure 6). Additionally, the calculated values at least fell within the rather wide range of reported lifetimes. On the basis of that agreement, the estimated lifetimes appeared to be reliable for 1950 onwards.

The average lifetime of steel generally showed an increasing trend for all sectors over the century of data considered, likely indicating that the quality of steel products has improved over time. The comparison between the model results and the lifetime measurements of steel showed that the calculated values based on the VCM were within the range of lifetime measurements of steel in the U.S. (Figure 6). Exceptions included two estimates of the lifetimes of steel in the industrial and consumer appliance sectors: the calculated lifetime (from the VCM in this work) was significantly shorter than the reported (measured) lifetime. However, these sectors cover a wide range of products (with a wide range of lifetimes, as indicated by Figure 6) and the disagreement may have simply reflected the difficulty of obtaining a representative sample when measuring in-use lifetimes.

4.2. The Lifetimes of Steel in the U.S., Sweden, and the World

The lifetime of steel (total for all sectors) was estimated with the VCM based on steel consumption and recycling in the U.S., Sweden, and the world between 1900 and 2010. The scrapped and apparent lifetimes of steel in the different regions are compared in Figure 7a,b. The estimated scrapped lifetime of steel was similar in all the regions, reflecting the global nature of steel markets. There were larger differences between the apparent lifetimes, especially around the middle of the 20th century. Since the difference between the apparent and scrapped lifetime was related to the recovery rate, these results indicate the relatively large proportions of non-recycled steel in the U.S. and the world. The estimated scrapped (and apparent) lifetimes of steel in 2010 were as follows: a scrapped lifetime of 32 years (apparent lifetime: 35 years) in Sweden, 29 (39) years on a global scale, and 23 (37) years in the U.S.

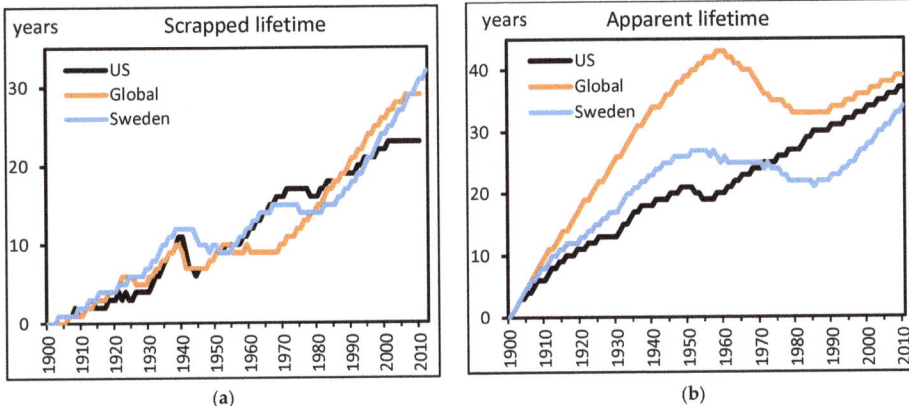

Figure 7. Steel lifetimes (total for all sectors) in the U.S., Sweden and the world, calculated with the VCM for 1900 to 2010: (**a**) scrapped lifetime; (**b**) apparent lifetime.

5. Conclusions

The scrapped and apparent lifetimes of steel per industry sector in the U.S. were calculated on an annual basis for the years between 1900 and 2016, using the Volume Correlation Model (VCM) method. The calculated steel lifetimes were within the range of previously reported lifetime measurements for products and applications in all industry sectors. The apparent lifetime of steel varied between countries and between different industry sectors in the U.S. The difference between the scrapped and apparent lifetimes of steel was significant and should be considered when forecasting the availability of steel scrap based on Dynamic Material Flow Models (DMFMs). In previous DMFMs the lifetime of steel was based on lifetime measurements on products and applications [2–8]. On the basis of DMFMs, it is possible to estimate steel scrap generation at national and industry sector levels. The information on the availability of steel scrap per industry sector can assist in the planning of waste management and steelmaking facilities.

Supplementary Materials: The following data are available online at http://www.mdpi.com/2075-4701/8/5/338/s1, Excel document with input data and results.

Author Contributions: A.G. did the literature survey, data gathering and calculations, analyzed the data, wrote the paper; P.C.P. did literature survey, data gathering, analyzed the data, and wrote the paper.

Acknowledgments: The authors would like to thank the Swedish Steel Producers Association, the Hugo Carlssons Foundation, and the Axel Hultgrens Foundation for financial support of AG. The authors would also like to thank Carnegie Mellon University Libraries for its financial support for the publication fee.

Conflicts of Interest: The authors declare no conflict of interest.

Appendix A

Survey-based estimates of the in-use lifetime of products and applications in the United States.

Table A1. Average lifetime per unit in the construction sector in the United States.

Product groups	References	Method	Timeline	Average Lifetime
American Housing Survey, General Housing Data, all Housing Units.	U.S. Census Bureau, Current Housing Reports, Series H150/11, American Housing Survey for the United States: 2011, Table C-01-AH. [9]	Lifetime of buildings used in Unites States	2015 2013–2014 2011	39 years 44 years 42 years
American Housing Survey, General Housing Data, all Housing Units.	U.S. Department of Commerce. Bureau of Economic Analysis. Table A-1. Characteristics of the housing inventory 1973, 1980 and 1970 page 6 or page 1. [9]	Lifetime of buildings used in Unites States	1973 1980	24 years 28 years
Industrial buildings, mobile offices, office buildings, commercial warehouses, other commercial buildings, religious buildings, educational buildings, hospital and institutional buildings, hotels and motels, amusement and recreational buildings, all other nonfarm buildings.	U.S. Department of Commerce. Bureau of Economic Analysis. Fixed Assets and Consumer Durable Goods in the United States, 1925–1999. [9]	Lifetime estimates	1925–1997	17–48 years
Bridges	National Bridge Inventory [10,11]	Lifetime of bridges in-use, maximum age distribution. Most bridges were built for a 50-year design life.	2010	50 years

Table A2. Average lifetime of industrial-goods products and applications in the United States.

Product Groups	Reference	Method	Timeline	Average Lifetime
Construction machinery and equipment, metalworking machinery and equipment, general purpose machinery and equipment.	U.S. Department of Commerce. Bureau of Economic Analysis. Fixed Assets and Consumer Durable Goods in the United States, 1925–1999. [13]	Service lives and depreciation estimates.	1925–1999	10–16 years
Agricultural and different machinery.	Division of Assessment Standards, Nevada Department of Taxation. [12]	Service lives.	2010	7–30 years
Metalworking machinery, durable machinery, special industry machinery.	U.S. Department of Commerce. Bureau of Economic Analysis. Fixed Assets and Consumer Durable Goods in the United States, 1925–1997. [13]	Service lives.	1925–1997	16–25 years

Table A3. Average in-use lifetime per unit in the automotive sector in the United States.

Product Groups	Reference	Method	Timeline	Average Lifetime
Boats and vessels—dry cargo, tanker, towboat, passenger, offshore support/crew-boats, dry barge, tank/liquid barge, (figures include vessels available for operation)	U.S. Army Corps of Engineers, Waterborne Transportation Lines of the United States, Volume 1, National Summaries, Table 4, available at http://www.navigationdatacenter.us/veslchar/pdf/ as of 21 June 2016. [15]	Age is based on the year the vessel was built or rebuilt.	1990–2014	18–16 years
Locomotives, passenger and other train cars.	Amtrak Annual Report, Statistical Appendix. [17]	Fiscal year-end average (30 September of stated year). Active units less backshop units undergoing heavy maintenance, less back-ordered units undergoing progressive maintenance and running repairs.	1972–2015	11–26 years
Commuter rail locomotives, commuter rail passenger coaches, commuter rail self-propelled passenger cars, heavy-rail passenger cars, light rail vehicles (streetcars), articulated full-small size trolley vans, ferry boats.	U.S. Department of Transportation, Federal Transit Administration, National Transit Database. National Transit Summaries and Trends, Table 25. [16]	Average Age of Urban Transit Vehicles. Locomotives used in Amtrak intercity passenger services are not included.	1985–2014	11–16 years
Aircraft: Transportation by air, depository institutions and business services.	U.S. Department of Commerce. Bureau of Economic Analysis. Fixed Assets and Consumer Durable Goods in the United States, 1925–1999. [18]	Average age of aircraft.	1960–1997	15–20 years

Table A4. Average in-use lifetime of appliances in the United States.

Product Groups	Reference	Method	Timeline	Average Lifetime
Mobile phones, cordless telephones, answering machines, fax machines, personal computers, laptops, printers, computer monitors, computer mice, keyboards (Metal content 8–69%).	Study of Life Expectancy of Home Components. Prepared by the Economics Group of NAHB. [14]	Current lifetime	2011	5–11 years
Household appliances—air conditioners, dishwashers, dryers, freezers, microwave ovens, ranges, refrigerators, clothes washers, water heaters, trash compactors (metal content in all units between 46–96%).	National Association of Home Builders/Bank of America Home Equity. [14]	Life expectancy is based on first-owner use.	2011	5–20 years
Video and audio products—projection TVs, plasma, LCD, and color TVs, TV/VCR combinations, videocassette players, VCR decks, DVD players, camcorders, home and portable audio products (Metal content 21–30%).	National Association of Home Builders/Bank of America Home Equity. [14]	Life expectancy is based on first-owner use.	2011	9–15 years

Appendix B

Table A5 shows the summary of the World Steel Association U.S. scrap survey [25] based on the answers from the scrap-producing companies in the United States in 2014 (excluding phone interviews). The category of "Other" scrap was estimated to be predominantly construction material (92.83%),

0.86% packaging, 4.17% mechanical machinery, 2.14% prompt scrap. The "total production of scrap" represents the coverage of scrap the survey.

Table A5. Results from 2014 World Steel Association scrap survey [25].

Scrap Type	Percentage of Yearly Scrap (%)
Appliances	11.37
Vehicles	19.78
Tires	0.02
Packaging	4.2
Construction	19.16
Mechanical machinery	4.57
Electrical and Electronic products	5.8
Transport	3.16
Prompt scrap	27.23
Other	4.71
Total production of scrap	19,095,547 ton

Appendix C

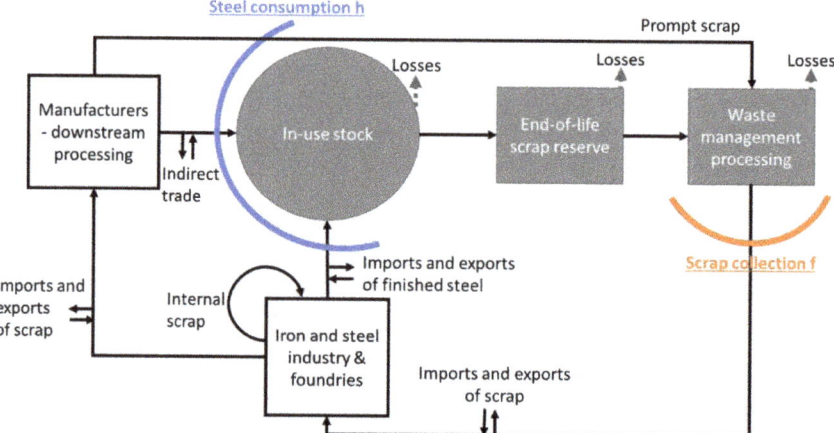

Figure A1. Material flow of steel in society, showing the extent of the system considered in this study and the model input data.

Table A6. Nomenclature on stocks and flows, with definitions.

Term	Definition
Steel consumption (h)	Net consumption of steel used for its application purpose plus prompt steel scrap. Marked with a thick blue curved line in Figure A1.
Scrap collection (f)	Net collection of obsolete and prompt steel scrap in the US. Domestic collected steel scrap which is commercially available. Marked with a thick orange curved line in Figure A1.
Purchased steel scrap	Net receipt of scrap in US iron and steel mills and foundries, including imports and excluding exports of scrap.
Internal scrap	Processing scrap generated at iron and steel mills and foundries.
Prompt scrap	Processing scrap generated at external manufacturers (during downstream processing); also termed "new scrap".
Obsolete scrap	Old scrap which has been collected and processed from end-of-life products and applications.
Indirect trade	Imports and exports of steel in further manufactured goods; steel contained in products.

References

1. Grimes, S.; Donaldson, J.; Gomez, G.C. *Report on the Environmental Benefits of Recycling*; Bureau of International Recycling (BIR), Centre for Sustainable production & Resource Efficiency (CSPRE), Imperial College London: London, UK, 2008.
2. Müller, E.; Hilty, L.M.; Widmer, R.; Schluep, M.; Faulstich, M. Modeling metal stocks and flows: A review of dynamic material flow analysis methods. *Environ. Sci. Technol.* **2014**, *48*, 2102–2113. [CrossRef] [PubMed]
3. Müller, D.B.; Cao, J.; Kongar, E.M.A.; Weiner, P.-H.; Graedel, T.E. Service Lifetimes of Mineral End Uses. Minerals Resources External Research Program. Available online: https://minerals.usgs.gov/mrerp/reports/Mueller-06HQGR0174.pdf (accessed on 7 May 2018).
4. Müller, D.B.; Wang, T.; Duval, B. Patterns of iron use in societal evolution. *Environ. Sci. Technol.* **2001**, *45*, 182–188. [CrossRef] [PubMed]
5. Pauliuk, S.; Milford, R.L.; Müller, D.B.; Allwood, J.M. The steel scrap age. *Environ. Sci. Technol.* **2013**, *47*, 3448–3454. [CrossRef] [PubMed]
6. Cooper, D.R.; Skelton, A.C.H.; Moynihan, M.C.; Allwood, J.M. Component level strategies for exploiting the lifespan of steel in products. *Resour. Conserv. Recycl.* **2014**, *84*, 24–34. [CrossRef]
7. Reck, B.K.; Chambon, M.; Hashimoto, S.; Graedel, T.E. Global stainless steel cycle exemplifies China's rise to metal dominance. *Environ. Sci. Technol.* **2010**, *44*, 3940–3946. [CrossRef] [PubMed]
8. Hatayama, H.; Daigo, I.; Matsuno, Y.; Adachi, Y. Outlook of the world steel cycle based on the stock and flow dynamics. *Environ. Sci. Technol.* **2010**, *44*, 6457–6463. [CrossRef] [PubMed]
9. U.S. Census Bureau, Current Housing Reports. *American Housing Survey for the United States: 2015–2013, 2011, 1997, 1980, 1973*; U.S. Government Printing Office: Washington, DC, USA, 2016–2014, 2012, 1998, 1981, 1974.
10. Emily, Yu. Analysis of National Bridge Inventory (NBI) Data for California Bridges. Master's Thesis, California Polytechnic State University, Pomona, CA, USA, April 2015.
11. NACE. Corrosion Control Plan for Bridges. A NACE International White Paper. Available online: https://www.nace.org/Newsroom/Press-Releases/NACE-International-White-Paper-Corrosion-Control-Plan-for-Bridges-Now-Available-Online/ (accessed on 3 May 2018).
12. *Personal Property Manual 2011–2012*; Division of Assessment Standards, Department of Taxation: Carson City, NV, USA, 2010.
13. U.S. Department of Commerce. *Fixed Assets and Consumer Durable Goods in the United States, 1925–1997 and 1925–1999*; U.S. Government Printing Office: Washington, DC, USA, 2003; pp. M-29–M-33.
14. *National Association of Home Builders/Bank of America Home Equity, Study of Life Expectancy of Home Components*; NAHB: Washington, DC, USA, 2007.
15. U.S. Army Corps of Engineers. Waterborne Transportation Lines of the United States, Volume 1, National Summaries; Table 4. Available online: http://www.navigationdatacenter.us/veslchar/pdf/wtlusvl1_04.pdf (accessed on 21 June 2016).
16. U.S. Department of Transportation, Federal Transit Administration, National Transit Database (Washington, DC: Annual Reports). National Transit Summaries and Trends, Table 25. Available online: https://www.transit.dot.gov/ntd/annual-national-transit-summaries-and-trends (accessed on 3 May 2018).
17. *Amtrak Annual Report*; Statistical Appendix and Personal Communications, Tables 1–33: Age and Availability of Amtrak Locomotive and Car Fleets; Amtrak: Washington, DC, USA, 1972–2015.
18. Survey of current business, U.S. Department of Commerce. *Fixed Assets and Consumer Durable Goods in the United States, 1925–99*; U.S. Government Printing Office: Washington, DC, USA, 2000.
19. Gauffin, A.; Andersson, N.Å.I.; Storm, P.; Tilliander, A.; Jönsson, P.G. Use of volume correlation model to calculate lifetime of end-of-life steel. *Ironmak. Steelmak.* **2015**, *42*, 88–96. [CrossRef]
20. Gauffin, A.; Andersson, N.Å.I.; Storm, P.; Tilliander, A.; Jönsson, P.G. Time-varying losses in material flows of steel using dynamic material flow models. *Resour. Conserv. Recycl.* **2017**, *116*, 70–83. [CrossRef]
21. Gauffin, A.; Andersson, N.; Storm, P.; Tilliander, A.; Jönsson, P. The Global Societal Steel Scrap Reserves and Amounts of Losses. *Resources* **2016**, *5*, 27. [CrossRef]
22. Gauffin, A. Improved Mapping of Steel Recycling from an Industrial Perspective. Ph.D. Thesis, Royal Institute of Technology, Stockholm, Sweden, November 2015.
23. Gauffin, A.; Ekerot, S.; Tilliander, A.; Jönsson, P. KTH steel scrap model—Iron and Steel Flow in the Swedish Society 1889–2010. *J. Manuf. Sci. Prod.* **2013**, *13*, 47–54. [CrossRef]

24. Fenton, M.D. *2015 Minerals Yearbook—Iron and Steel Scrap (Advanced Release)*; U.S. Geological Survey (USGS): Reston, VA, USA, 2014.
25. *Scrap Survey (Answers from Scrap Dealers Excluding Phone Interviews)*; World Steel Association: Brussels, Belgium, 2014.
26. Ciftci, B. *Statistical Data and Analysis on the World Steel Flow*; World Steel Association: Brussels, Belgium, 2016.
27. *Minerals Yearbook (1932–2016) Iron and Steel Scrap Statistics*; U.S. Geological Survey: Reston, VA, USA, 1933–2017.
28. Pehrson, E.W. *Minerals Yearbook Review of 1940, Iron and Steel Scrap Statistics*; Figure 1; U.S. Geological Survey: Reston, VA, USA, 1941; p. 502.
29. Damuth, R.J. *Iron and Steel Scrap—Accumulation and Availability as of December 31, 2009*; Institute of Scrap Recycling Industries: Washington, DC, USA, 2010.
30. Zimring, C.A. The complex environmental legacy of the automobile shredder. *Technol. Cult.* **2011**, *52*, 523–547. [CrossRef]
31. Bensel, R.F. *The Political Economy of American Industrialization, 1877–1900*; Cambridge University Press: Cambridge, UK, 2000; ISBN-13: 978-0521776042.
32. Bever, M.B. The recycling of metals—I. Ferrous metals. *Conserv. Recycl.* **1976**, *1*, 55–69. [CrossRef]
33. Zheng, X.; Wang, R.; Wood, R.; Wang, C.; Hertwich, E.G. High sensitivity of metal footprint to national GDP in part explained by capital formation. *Nat. Geosci.* **2018**, *11*, 269–273. [CrossRef]
34. *Report on Indirect Trade in Steel (1970–2013)*; World Steel Association: Brussels, Belgium, 2015; p. 39.
35. *Steel Statistical Yearbook (2014–2017), Indirect Trade of Steel*; Tables 55–57; World Steel Association: Brussels, Belgium, 2014–2017.
36. *Microsoft Office Excel Toolbox Release 2013 (Office15)*; Microsoft Redmond Campus: Redmond, WA, USA, 2013.
37. *MATLAB and Statistics Toolbox Release 2012a*; The Math Works, Inc.: Natick, MA, USA, 2018.
38. Graedel, T.E.; Buchert, M.; Reck, B.K.; Sonnemann, G. *Assessing Mineral Resources in Society: Metal Stocks & Recycling Rates*; United Nations Environment Programme: Nairobi, Kenya, 2011; ISBN 978-92-807-3182-0.
39. Fenton, M.D. *Iron and Steel Recycling in the United States in 1998*; U.S. Geological Survey: Reston, VA, USA, 1998.
40. Sawyer, J.W. *Automotive Scrap Recycling: Processes, Prices and Prospects*; Johns Hopkins University Press: Baltimore, MA, USA, 1974.

© 2018 by the authors. Licensee MDPI, Basel, Switzerland. This article is an open access article distributed under the terms and conditions of the Creative Commons Attribution (CC BY) license (http://creativecommons.org/licenses/by/4.0/).

Article

Low-Waste Recycling of Spent CuO-ZnO-Al$_2$O$_3$ Catalysts

Stanisław Małecki and Krzysztof Gargul *

Faculty of Non-ferrous Metals, AGH University of Science and Technology, 30-059 Krakow, Poland; stanmal@agh.edu.pl
* Correspondence: krzygar@agh.edu.pl; Tel.: +48-12-617-2646

Received: 22 January 2018; Accepted: 7 March 2018; Published: 12 March 2018

Abstract: CuO-ZnO-Al$_2$O$_3$ catalysts are designed for low-temperature conversion in the process of hydrogen and ammonia synthesis gas production. This paper presents the results of research into the recovery of copper and zinc from spent catalysts using pyrometallurgical and hydrometallurgical methods. Under reducing conditions, at high temperature, having appropriately selected the composition of the slag, more than 66% of the copper can be extracted in metallic form, and about 70% of zinc in the form of ZnO from this material. Hydrometallurgical processing of the catalysts was carried out using two leaching solutions: alkaline and acidic. Almost 62% of the zinc contained in the catalysts was leached to the alkaline solution, and about 98% of the copper was leached to the acidic solution. After the hydrometallurgical treatment of the catalysts, an insoluble residue was also obtained in the form of pure ZnAl$_2$O$_4$. This compound can be reused to produce catalysts, or it can be processed under reducing conditions at high temperature to recover zinc. The recovery of zinc and copper from such a material is consistent with the policy of sustainable development, and helps to reduce the environmental load of stored wastes.

Keywords: sustainable development; recycling; spent catalysts; zinc; copper

1. Introduction

Spent CuO-ZnO-Al$_2$O$_3$ catalysts are very important secondary resources for metal recovery and could be highly usable for copper and zinc recycling [1]. This type of catalyst is used in low-temperature processes of carbon monoxide conversion with steam and to obtain hydrogen as well as a synthesis gas to produce ammonia or methanol [2–5]. Manufacturers of catalysts define the content of the basic components in new products as follows: CuO—min 50%, ZnO—min 25%. Multiple variations of these are available on the market. They are different in terms of their Cu, Zn and Al oxide content. The copper to zinc mass ratio in industrial catalysts manufactured using the co-precipitation method is usually 7:3 [6]. The presence of spinel structures is a feature specific to the catalysts that are analyzed in the paper. In the literature, CuAl$_2$O$_4$ spinels on the surface of the catalysts and the ZnAl$_2$O$_4$ stoichiometry spinel structures throughout their volume are the most frequently reported [7,8]. X-ray diffraction studies in the literature [9] indicate that the ZnAl$_2$O$_4$ compound is present at a temperature of 1100 °C. Therefore, processing of the spent catalyst with the infusible spinel structures present in them, and the complete recovery of zinc from this type of material, can be difficult using the pyrometallurgical method. Due to the toxic metal content (Zn, Cu), these materials can be considered to pollute the natural environment. In Poland, the volume of CuO-ZnO-Al$_2$O$_3$ waste catalysts is estimated at around 2000 tons per year. The current method for their recycling consists of pyrometallurgical processing together with other waste containing zinc and/or copper. In the literature, there is only quite scarce information related to the processing of this type of catalysts. This is limited to the patents [10–12] describing the hydrometallurgical processing of spent CuO-ZnO-Al$_2$O$_3$ catalysts, which consists of leaching the spent catalysts in a solution of nitric acid after high-temperature roasting or leaching in ammonia solutions. From the resulting solutions, copper and zinc compounds can be selectively precipitated and then recycled for the production of catalysts. In this work, it was decided to analyze the use of pyro and hydrometallurgical methods for the processing of catalysts in terms of maximum recovery of metals and minimization of waste.

2. Properties of Examined Material

The material tested in this work comprises spent CuO-ZnO-Al_2O_3 catalysts available on the Polish market. These occur in the form of identical rolls with a diameter of 4.5 mm and a height of 3.5 mm. The chemical analysis (AAS) of the examined samples of spent catalysts indicates different content of copper and zinc in the materials analyzed. With the detailed chemical analysis of the catalyst sample selected to be examined, the following results were obtained: 35.1% of Cu, 29.9% of Zn and 9.2% of Al. A sample of the material was also examined using X-ray diffraction (XRD) and scanning electron microscope observations combined with qualitative energy dispersion chemical analysis (EDS). The results of the phase analysis are shown in Figure 1. CuO and ZnO oxides are the main phase components. Additionally, small quantities of Cu_2O and hydroxy-carbonate complexes of zinc and copper are present.

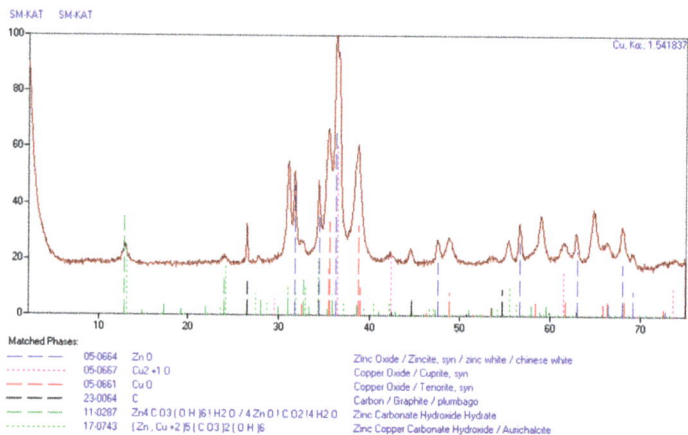

Figure 1. XRD pattern of the grounded catalyst sample.

In Figures 2 and 3, respectively, an image of a sample of fragmented catalyst material and the EDS analysis of the area presented are shown. The elements identified are copper, zinc, aluminum, carbon and oxygen. This is only a confirmation of the phase analysis.

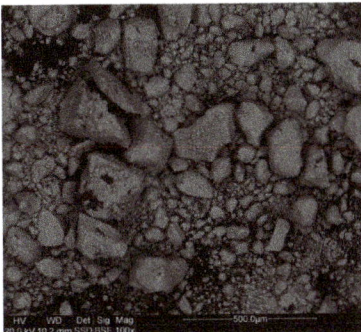

Figure 2. The microscopic image of the sample of the fragmented catalyst material.

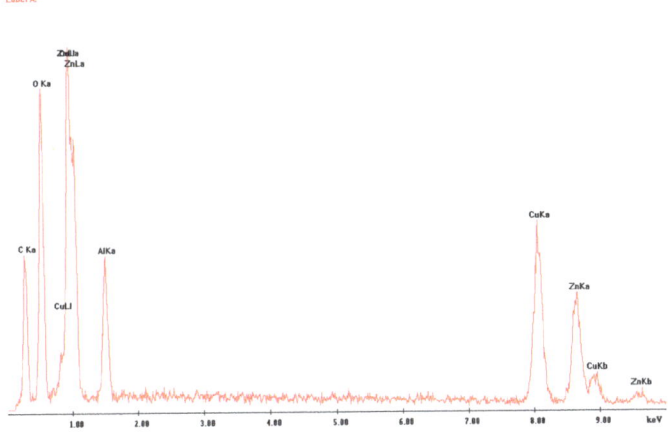

Figure 3. EDS analysis of the area of the sample seen in Figure 2.

3. Pyrometallurgical Processing of Spent Catalysts

With information about the content of the most valuable components in the material having been obtained, it was decided to perform the test melting in order to extract copper and zinc in the form of zinc oxide from spent catalysts and to optimize the process for maximum yields of Cu and ZnO, as mentioned above. Bearing in mind the oxidic nature of catalysts, in order to lower the melting point (softening point) of the slag created and, if possible, to decrease its viscosity, it was decided to carry out the recovery process under reducing conditions and to apply slag-forming additives (e.g., CaO, SiO_2, Na_2O). Therefore, in order to determine the amount of additives that would ensure the lowest melting point of the slag, the relevant binary and ternary systems were analyzed. In the case of a binary ZnO-SiO_2 system, the liquid phase occurs at a temperature of 1432 °C, with a content of about 52% ZnO by weight [13]; while in the Na_2O-SiO_2 system, the liquid phase occurs in a wide range of concentrations and temperatures close to 1000 °C [14]. Therefore, it can be assumed that in the Na_2O-ZnO-SiO_2 ternary system, compositions are likely in which the melting point of the respective ternary phase will be low enough to melt the catalysts processed. The above hypothesis is confirmed by reports in the literature [15], which indicate that once the concentration level of (weight percent) 21.5% Na_2O, 12.0% ZnO, 66.5% SiO_2 in this system is reached, the minimum melting point is 680 °C. On the basis of the available information about the CaO-ZnO-SiO_2 and CaO-ZnO-Al_2O_3-SiO_2 systems [16,17], it was assumed that the lowest melting points could be obtained when the relevant components were at similar concentration levels.

3.1. Experimental Procedure of Pyrometallurgical Processing

The catalysts were processed in an induction furnace, operating at a temperature range between 1100–1300 °C. This temperature was to ensure maximum stripping of zinc and the obtaining of liquid copper by reducing its oxide. The amount of coal for the reduction of CuO and ZnO was determined assuming the formation of CO and CO_2. As a result, the amount of reductant added varies between 6 and 12 wt %. For further studies, the mean value was taken and increased by 10% due to the ash content in coal. Initial testing made it possible to determine that at 10% of added reductant, the degree of zinc stripping was at a level of approx. 65%. The remainder of the zinc goes into the slag and metal phase. It should be noted that tests were carried out in a graphite crucible, which additionally improved the reduction conditions. Based on these results, it was possible to determine the test plan to optimize the amount of slag-forming additives.

Four variants of laboratory tests were performed. The tests were different from each other in terms of quantity of slag-forming components added. The optimal process parameters were sought in order to obtain the lowest melting points of slag, and thereby to minimize the loss of metals extracted from the catalysts to slag. 100 g of uncrushed catalysts were used for each of the tests. The weights of copper and zinc contained in these were 35.1 and 29.9 g, respectively. Reducing conditions of the process were secured by adding 10 g of coal and by the fact that the melting was carried out in a graphite crucible. Once the melting was complete, the liquid products contained in the crucible were cooled, separated, and weighed, and the chemical analyses of the materials obtained were performed.

3.2. Results and Discussion of Pyrometallurgical Tests

3.2.1. Test No. 1

100 g of catalysts were mixed with 10 g of carbon, 50 g of SiO_2 and 15 g of CaO. The whole feed was placed in a graphite crucible and heated until a temperature of 1300 °C was reached. At the beginning, intensive reduction of ZnO to Zn was observed, and the metal was lifted in the gas phase and re-oxidized. After about 45 min, this process had definitely halted (no emission of ZnO white films), and hence the decision was taken to terminate the test at this time. After cooling the crucible, its contents were separated into three fractions, and 26 g of metal, 62 g of glassy slag and about 17.5 g of unmelted fine fraction were yielded. This phase is, most likely, unreacted slag-forming components and unmelted $ZnAl_2O_4$ spinel structures. Additionally, in the experimental system, it was impossible to identify the amount of dust produced. It should be added that in the slag phase, no sedimented tiny metallic copper inclusions were visible.

3.2.2. Test No. 2

100 g of catalysts, 10 g of carbon, 60 g of SiO_2 and 20 g of NaOH were the feed for smelting. The temperature of the process was 1300 °C, and the duration was 50 min. Like in the previous test, for the first 45 min of the process, the stripping of the zinc in the form of ZnO was very intense. Upon its completion, after product cooling and separation, 25 g of metal, 95 g of glassy slag and 7.5 g of free-flowing phase (slag-forming components and spinels) were found. Additionally, in this test, metallic copper inclusions were seen in the slag.

3.2.3. Test No. 3

100 g of catalysts, 10 g of carbon, 31 g of Na_2CO_3 and 55.4 g of SiO_2 were the feed for the process. The feed was melted at a temperature of 1250 °C, and the melting time was 90 min. After this period, stripping of zinc drastically decreased, and hence the decision to terminate the smelting was taken. 25 g of metal and 107 g of glassy slag with minor copper inclusions were yielded.

3.2.4. Test No. 4

100 g of catalysts, 10 g of carbon, 16 g of Na_2CO_3 and 28 g of SiO_2 were melted. The melting time was 90 min and the process temperature was 1250 °C. 20 g of metal and 61 g of glassy slag with a certain amount of copper drops were yielded.

The smelting products, namely slag and metallic alloy, were subjects of chemical analysis for the content of copper and zinc. The summary of the test results in Table 1 takes into account the fact that the unbalanced portion of zinc is transferred to the dust phase.

Table 1. List of parameters for processing the spent catalysts conducted at laboratory scale according to smelting variants.

Process Parameters	Smelting Variant			
	I	II	III	IV
FEED				
Mass of catalysts (g)	100	100	100	100
Mass of silica (g)	50	60	55.4	28
Mass of carbon (g)	10	10	10	10
Mass of CaO (g)	15	-	-	-
Mass of Na_2CO_3 (g)	-	-	31	16
Mass of NaOH (g)	-	20	-	-
Smelting time (min)	45	50	90	90
Process temperature (°C)	1300	1300	1250	1250
PRODUCTS				
The overall mass of the alloy (g)	26	25	25	20
Mass of Cu in the alloy (g)	23.2	22.1	22.0	17.7
Mass of Zn in the alloy (g)	2.5	2.7	2.8	2.3
Mass of slag (g)	62	95	107	61
Mass of copper in the slag (g)	6.4	7.5	7.6	11.9
Mass of zinc in the slag (g)	6.0	6.1	6.7	6.3
The copper yield in the alloy (%)	66.1	63.0	62.7	50.4
The estimated yield of zinc in the dust (%)	71.6	70.6	68.2	71.2

Slag-forming additives have a significant effect on the recovery of Cu to alloy and Zn to dust. The use of CaO (variant I) results in the best recovery of copper and zinc. The use of other additives (variants II and III) results in similar effects with a much larger quantity of waste slag produced. In the last variant of the remelting (IV), a small addition of Na_2CO_2 was used and an unsatisfactory degree of copper extraction to metallic alloy was obtained. However, the results of the pyrometallurgical test of catalyst recycling do not provide grounds for optimism. The too-low yield of copper may cause the processing to be less cost-effective. This is due to the problem of obtaining a low-viscosity slag. It is probable that the presence of zinc aluminate is responsible for the high viscosity of the slags. Correction of the slag composition results in the formation of a large amount of slag, and even at a lower copper content, total losses are significant.

4. Processing of Catalysts Using the Hydrometallurgical Method

4.1. Procedure of Hydrometallurgical Tests

The results of using the described pyrometallurgical method to recycle the spent catalysts are not fully satisfactory. Therefore, the decision was taken to use a hydrometallurgical method for their processing. With knowledge about the structure of catalysts, their chemical composition and their phase composition, as well as being familiar with how they are produced [6], an innovative method for processing them was developed. Zinc aluminate ($ZnAl_2O_4$), which is present in the catalysts, is a compound highly resistant to both acids and alkalines [18]. Therefore, it has been recognized that after leaching, copper oxides and zinc oxides will be left as insoluble residue. First, catalysts in the form of pellets were fragmented to reach sizes of less than 90 µm. In order to separate zinc and copper, zinc oxide and copper oxide leaching processes were selectively carried out, consisting of two stages:

- Leaching in NaOH solution (temperature 75 °C, process duration 120 min, NaOH concentration = 200 g/dm^3, l/s = 10),
- Leaching in H$_2$SO$_4$ solution (temperature 60 °C, process duration 120 min, H$_2$SO$_4$ concentration = 180 g/dm^3, l/s = 10).

The process conditions were adjusted based on previous experience in alkaline leaching [19], and pilot tests. After each leaching process, the slurry was filtered (filter Munktell & Filtrak, Stockholm, Sweden, type 392) to separate the deposit. The filtration process is difficult because the deposit consists of very fine grains. During the leaching using NaOH, only zinc was transferred to the solution. In fact, after the acidic leaching, the solution contained just copper. The amount of zinc in the solution was 100 to 150 times less than the amount of copper. Additionally, in order to facilitate transferring the copper to the solution, as acidic leaching was conducted, small amounts of hydrogen peroxide solution were added. A small quantity of this may in fact be present in metallic form.

4.2. Results and Discussion of Hydrometallurgical Tests

The above-described procedure made it possible to obtain the following products from 100 g of recycled catalysts (information based on AAS analysis):

- Zn solution—0.8 dm^3 (Zn-23 g/dm^3),
- Cu solution—0.8 dm^3 (Cu-43 g/dm^3),
- ZnAl$_2$O$_4$ deposit in the amount of 33 g.

After leaching, the solutions still contain highly concentrated leaching agent, and should be returned to the initial leaching of subsequent batches. In order to recover as much metal as possible, the final leaching must be carried out using highly concentrated leaching agent. The zinc aluminate deposit was of dark gray color, since it contained a certain amount of carbon. An attempt to burn it out (600 °C, air atmosphere) resulted in weight reduction by about 10%, and the color was changed to light gray. The sample obtained in this way was analyzed using X-ray phase analysis. The results are presented in Figure 4.

Results of the phase analysis indicate the presence of only zinc aluminate and carbon. The presence of carbon results from the fact that during leaching, coal does not pass into the solution and accumulates in the residue.

The balance of the processing performed is as follows (100 g of catalysts):

- Amount of zinc in alkaline solution—18.5 g
- Amount of copper in acidic solution—34.5 g
- Zinc aluminate—30 g
- Carbon—3 g

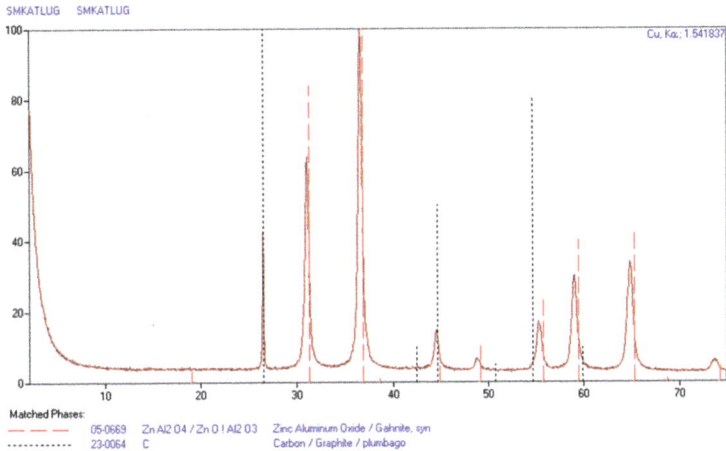

Figure 4. XRD pattern of zinc aluminate obtained from recycling catalysts.

Material balance and analysis of the solutions made it possible to determine the yield of copper in solution at the level of 98%. The yield of zinc to the alkaline solution is 61.9%. Because the residue after leaching is actually pure $ZnAl_2O_4$, it can be treated as a product of the process. In this case, zinc yield increases to 97.5%, and the yield of aluminum is 96%.

5. Conclusions

Laboratory tests carried out indicate that the pyrometallurgical method can be used to recover copper and zinc contained in the spent catalysts processed. However, after conducting the tests, some amounts of the alloy in the slag were noted in the form of small. This may be a result of the high viscosity of the slag and a result of the difficult sedimentation conditions under the test conditions. The process should be carried out at temperatures above 1200 °C, and a reductant should be used in the form of coal in an amount of about 10% by weight of catalysts. Liquid copper and ZnO in the form of dust are products of the process. Silica and CaO or NaOH or Na_2CO_3 are an indispensable technological additive in processing catalysts for slag adjustment. This makes it possible to obtain a glassy slag and to minimize copper losses in the process. Laboratory test conditions made it possible to carry out the process of copper recovery with a yield of up to 66%. The Zn yield for dust was estimated based on the amounts of zinc contained in the alloy obtained and in slag. It can be assumed that under proper process conditions (reducing atmosphere, ~1300 °C, enough time for Zn stripping), transferring to dust more than 70% of zinc contained in the spent catalysts is very possible.

Hydrometallurgical processing based on selective leaching makes it possible to accurately separate the components of spent catalysts. More than 96% degree of metal recovery is reached. In the case of zinc and aluminum, this value is relative to the overall yield. It also includes the content of these metals in the leach residue. This residue may be the product of the process, because it actually contains pure $ZnAl_2O_4$. Solutions obtained as a result of applying acidic and alkaline leaching make it possible to perform the selective extraction of the Cu and Zn they contain. These metals can be recovered from solutions in the form of compounds by precipitation or in a metallic form by electrolysis. The residue (mainly $ZnAl_2O_4$) left after leaching may be returned to produce new catalysts or may be thermally processed to recover the zinc it contains.

Taking into account the ecological side of the proposed processes, it should be noted that during the pyrometallurgical processing, a waste slag is formed with a relatively high content of zinc and copper. For this reason, the waste slag cannot be considered neutral for the natural environment. Hydrometallurgical conversion of catalysts leads to their waste-free management, and thus is completely environmentally friendly. The optimization of the leaching process parameters is the way of developing the proposed method.

Acknowledgments: This paper is supported by the Ministry of Science and Higher Education (Grant No. 11.11.180.726).

Author Contributions: All parts of the work were done jointly by both authors with the exception of laboratory work. pyrometallurgical tests—Krzysztof Gargul; hydrometallurgical tests—Stanisław Małecki.

Conflicts of Interest: The authors declare no conflict of interest.

References

1. Sun, Z.; Xiao, Y.; Agterhuis, H.; Sietsma, J.; Yang, Y. Recycling of metals from urban mines—A strategic evaluation. *J. Clean. Prod.* **2016**, *112*, 2977–2987. [CrossRef]
2. Amphlett, J.; Mann, R.; Weir, R. Hydrogen production by the catalytic steam reforming of methanol: Part 3: Kinetics of methanol decomposition using C18HC catalyst. *Can. J. Chem. Eng.* **1988**, *66*, 950–955. [CrossRef]
3. Agarwal, V.; Patel, S.; Pant, K.K. H_2 production by steam reforming of methanol over $Cu/ZnO/Al_2O_3$ catalysts: Transient deactivation kinetics modelling. *Appl. Catal. A* **2005**, *279*, 155–164. [CrossRef]
4. Waugh, K.C. Methanol Synthesis. *Catal. Lett.* **2012**, *142*, 1153–1166. [CrossRef]
5. Riaz, A.; Zahedi, G.; Klemes, J.J. A review of cleaner production methods for the manufacture of methanol. *J. Clean. Prod.* **2013**, *57*, 19–37. [CrossRef]
6. Behrens, M.; Studt, F.; Kasatkin, I.; Kühl1, S.; Hävecker, M.; Abild-Pedersen, F.; Zander, S.; Girgsdies, F.; Kurr, P.; Kniep, B.L.; et al. The Active Site of Methanol Synthesis over $Cu/ZnO/Al_2O_3$ Industrial Catalysts. *Science* **2012**, *336*, 893–897. [CrossRef] [PubMed]
7. Barroso, M.N.; Gomez, M.F.; Gamboa, J.A.; Arrua, L.A.; Abello, M.C. Preparation and characterization of CuZnAl catalysts by citrate gel process. *J. Phys. Chem. Solids* **2006**, *67*, 1583–1589. [CrossRef]
8. Turco, M.; Bagnasco, G.; Costantino, U.; Marmottini, F.; Montanari, T.; Ramis, G.; Busca, G. Production of hydrogen from oxidative steam reforming of methanol: II. Catalytic activity and reaction mechanism on $Cu/ZnO/Al_2O_3$ hydrotalcite-derived catalysts. *J. Catal.* **2004**, *228*, 56–65. [CrossRef]
9. Walerczyk, W.; Zawadzki, M.; Grabowska, H. Glycothermal synthesis and catalytic properties of nanosized $Zn_{1-x}Co_xAl_2O_4$ ($x = 0, 0.5, 1.0$) spinels in phenol methylation. *Catal. Lett.* **2011**, *141*, 592–601. [CrossRef]
10. Zhou, H.; Huang, Y. Method for Recycling Waste Copper-Based Methanol Waste Catalyst. Patent CN103 495426A, 18 January 2014.
11. Wang, B.; Meng, Y. Recovery Method of Copper and Zinc Oxide from Waste Copper-Zinc Catalyst. Patent CN1258752A, 30 December 1998.
12. Ling, H.; Liu, J.; Zhang, X.; Xu, X.; Li, Q.; Hu, Z.; Hu, G.; Qiu, C.; Huang, H. Application Method of Waste Copper Based Catalyst to Preparing Catalyst for Preparing Hydrogen from Methanol. Patent CN102125851A, 20 April 2011.
13. Bunting, E.N. Phase equilibria in the system SiO_2-ZnO. *J. Am. Ceram. Soc.* **1930**, *13*, 5–10. [CrossRef]
14. Kracek, F.C. The system sodium oxide-silica. *J. Phys. Chem.* **1930**, *34*, 1583–1598. [CrossRef]
15. Holland, A.E.; Segnit, E.R. The ternary system Na_2O-ZnO-SiO_2. *Aust. J. Chem.* **1966**, *19*, 905–913. [CrossRef]
16. Segnit, E.R. The System CaO-ZnO-SiO_2. *J. Am. Ceram. Soc.* **1954**, *37*, 273–277. [CrossRef]
17. Segnit, E.R. Three planes in the quaternary system CaO-ZnO-Al_2O_3-SiO_2. *J. Am. Ceram. Soc.* **1962**, *45*, 600–607. [CrossRef]
18. Tang, Y.; Shih, K.; Wang, Y.; Chong, T. Zinc stabilization efficiency of aluminate spinel structure and its leaching behaviour. *Environ. Sci. Technol.* **2011**, *45*, 10544–10550. [CrossRef] [PubMed]
19. Gargul, K.; Jarosz, P.; Małecki, S. Alkaline leaching of low zinc content iron-bearing sludges. *Arch. Metall. Mater.* **2016**, *61*, 43–50. [CrossRef]

© 2018 by the authors. Licensee MDPI, Basel, Switzerland. This article is an open access article distributed under the terms and conditions of the Creative Commons Attribution (CC BY) license (http://creativecommons.org/licenses/by/4.0/).

Article

Anodic Lodes and Scrapings as a Source of Electrolytic Manganese

Daniel Fernández-González [1,*], José Sancho-Gorostiaga [2], Juan Piñuela-Noval [1] and Luis Felipe Verdeja González [1]

[1] Laboratorio de Metalurgia, Dpto. de Ciencia de los Materiales e Ingeniería Metalúrgica, Escuela de Minas, Energía y Materiales, Universidad de Oviedo, 33004 Oviedo, Asturias, Spain; 120195juan@gmail.com (J.P.-N.); lfv@uniovi.es (L.F.V.G.)
[2] Departamento de Medioambiente, Gobierno del Principado de Asturias, 33005 Oviedo, Asturias, Spain; jsanchog1@yahoo.es
* Correspondence: fernandezgdaniel@uniovi.es; Tel.: +34-985-104-303

Received: 26 December 2017; Accepted: 1 March 2018; Published: 7 March 2018

Abstract: Manganese is an element of interest in metallurgy, especially in ironmaking and steel making, but also in copper and aluminum industries. The depletion of manganese high grade sources and the environmental awareness have led to search for new manganese sources, such as wastes/by-products of other metallurgies. In this way, we propose the recovery of manganese from anodic lodes and scrapings of the zinc electrolysis process because of their high Mn content (>30%). The proposed process is based on a mixed leaching: a lixiviation-neutralization at low temperature (50 °C, reached due to the exothermic reactions involved in the process) and a lixiviation with sulfuric acid at high temperature (150–200 °C, in heated reactor). The obtained solution after the combined process is mainly composed by manganese sulphate. This solution is then neutralized with CaO (or manganese carbonate) as a first purification stage, removing H_2SO_4 and those impurities that are easily removable by controlling pH. Then, the purification of nobler elements than manganese is performed by their precipitation as sulphides. The purified solution is sent to electrolysis where electrolytic manganese is obtained (99.9% Mn). The versatility of the proposed process allows for obtaining electrolytic manganese, oxide of manganese (IV), oxide of manganese (II), or manganese sulphate.

Keywords: manganese; Zinc; electrolytic lodes and scrapings; electrolytic manganese; metallurgy; hydrometallurgy; recycling

1. Introduction

Some secondary products, such as muds, collected powders, and slags, are considered as wastes in different industries, especially in metallurgy. They are sent to controlled disposal as sometimes they contain hazardous substances, being an economical and environmental problem for the factory. In some cases, these wastes are mixed/recycled with the raw materials, as, for instance, in the iron metallurgy (in the iron ore sintering process mill scale, LD (Linz and Donawitz, LD) slag, sludges, and refractory oxides [1], are recycled in the sintering process, others, such as blast furnace slags, are used in the manufacture of cement, and certain gases are burnt in power stations) [2–6] or in the ferroalloys industry (ultrafine oxidized dust from the ferromanganese and silicomanganese production is mixed with cement and recycled in the process) [7]. Mixing is not always an option, even though some of these wastes are produced in small amounts. This is the case of the waste produced in the zinc electrowinning, known as anode manganese-lead waste, which is produced in amounts of 15–35 kg per 10^3 kg of zinc [8]. This waste cannot be mixed with other manganese ores in the ferroalloys industry as a consequence of the presence of lead (the incompatibility of lead in the metallurgy of manganese is related with the easily reducibility of lead compounds and the volatility of this metal at the temperatures that are involved in the carbothermic reduction furnace employed in the ferromanganese industry, see Sancho et al., pages 386–394 [9]). As said, some of these wastes are considered toxic-hazardous and their deposit is regulated and carried out in controlled areas with high costs, especially due to preparation, inertization, and the control of the area. On another note, sustainability is deeply rooted in developed countries policies leading to take advantage of the metallic content of the wastes produced in metallurgical plants, becoming them in by-products of interest for the metallurgical industry, as is the case of anode manganese-lead waste.

The origin of the waste used in this process is found in the zinc metallurgy. Zinc is produced in large plants (0.20–0.55 × 10^9 kg of zinc) through electrolysis (the full description of the process can be read in Sancho et al., pages 354–369 [9]). First, blende is lixiviated with sulfuric acid solution in a return cell. After the purification of the solution from the nobler metals than zinc, the solution is electrolyzed, and high-quality zinc is produced [9]. Jarosite precipitation for iron elimination from the solution requires iron in solution as 3+. This can be guaranteed by using $Mn\,(IV)(s)$ or $Mn\,(VI)(aq)$ in solution, which is achieved by adding the corresponding manganese compounds. The existence of $Mn\,(II)(aq)$ in the rich solution produces, when electrolyzing zinc, some deposition of $MnO_2(s)$ on the anode's surface because of the reaction (1).

$$2H_2O(l) + Mn^{2+}(aq) \rightarrow MnO_2(s) + 4H^+(aq) + 2e^- \tag{1}$$

This layer is beneficial as avoids the contamination of the electrodes with lead. Manganese is also present in zinc ores (depending on the rock containing the zinc deposit, and consequently, manganese can appear as $MnCO_3$ or $MnO \cdot SiO_2$ [10]). Mn is not harmful for the process, as it was previously mentioned, because deposits on the surface of the anodes (Pb-0.5Ag anodes), avoiding its corrosion, and the correspondingly contamination in the cathode. The lode falls and forms the anodic lodes sweeping along some Cu, As, Sb, and Co, and for that reason, facilitates de purification of the electrolyte. Mn content in the electrolyte is around 7 g/L (maximum considered reasonable 25 g/L) [9]. The problem is when the crust of MnO_2 is too deep and this causes the increase of voltage, making the removal of both the lodes and the crust (scrapings) necessary. The granulometry of the lode is mainly lower than 0.1 mm, while the granulometry of the crust (scrapings) depends on the removal system, usually mechanic, forming the scraps with a granulometry higher than 1 mm. Homogenization is, for that reason, required, partially achieved by milling the material (milling also facilitates de leaching [11]) and the screening. For that reason, there are two products susceptible of being treated for manganese recovery: anode scraps and bottom cell lodes. Both the anode scraps and the cell lodes are mainly formed by $Mn\,(IV)$ oxides, and there is an important mechanical contamination of lead sulphate with strontium and little silver [8].

Manganese finds its main applications in steel production since 85–90% of all manganese is consumed in the steel industry [12], mainly as ferromanganese and silicomanganese. These products are typically produced by pyrometallurgical methods and using metallurgical grade manganese ores (>40%) [12,13]. The development of processes to recover low-grade manganese ores and other secondary sources has taken emphasis in the last decades as the manganese demand has grown rapidly [12,13], and the depletion of high manganese ore sources. Several processes have been studied to recover low grade manganese ores (20–30% Mn) by using different methods: leaching of manganese carbonate in ammonium sulfate solution [14]; recovery of manganese from electric arc furnace dust of ferromanganese by using sulfuric acid as leaching agent, and oxalic acid, hydrogen peroxide, and glucose as reducing reagents [15]; reduction-roasting of low-grade manganese dioxide ores by using sulfuric acid as leaching solvent and cornstalk as reducing reagent [16]; sulfuric acid leaching of ocean manganese nodules using phenols as reducing agents [17]; sulfur-based reduction roasting-acid leaching of low-grade manganese oxide ores [18]; reductive leaching of low-grade manganese ores, using cane molasses as reducing reagent and sulfuric acid as solvent [19]; reduction-acid leaching of low grade manganese ores using CaS as reductant [20]; recovery of manganese from spent batteries [21];

reuse of anode slime from the zinc electrolysis [22]; and, recovering manganese from treated sludge of the exhaust gases of ferroalloy production furnaces [23].

As mentioned, manganese finds its main applications in the steel industry; for instance, high strength steels contain more than 1% manganese, representing 3–4% of the tonnage of steel produced worldwide [24,25], but also in the modern TRIP/TWIP (Transformation Induced Plasticity/Twining Induced Plasticity) steels with >20% Mn [26–30]. Other steels, such as stainless steels, also contain important amounts of manganese [24,25]. Electrolytic manganese (99.9% Mn) is used in the production of aluminum (as improves corrosion resistance) and copper (manganese bronzes are strengthened by small additions of manganese) alloys, special grades of stainless steels and other special steels, and for electronic applications [13]. Other non-metallurgical applications include potassium permanganate, which is used in chemistry and medicine as a disinfectant agent [31], and, manganese dioxide in dry cell batteries [32].

The treatment of these two by-products (anode manganese-lead waste and scrapings), but also low-grade manganese ores and other by-products of the manganese industry, could be performed following Jacobs patent [33], the inventor of the electrolytic process, by a combination of a pyrometallurgical treatment, to reduce using carbon or hydrocarbides $Mn\ (IV)$ (*solid solution, s. s.*) to $Mn\ (II)$ (*s. s.*) compound, mainly MnO. Once it has been done, the material can be etched with strong acids, like sulfuric acid water solution, below boiling temperature in a typical hydrometallurgical process. Doing that, $Mn\ (II)\ (aq)$ is dissolved but also some other heavy metals (Zn, Cu, Ni, Cd) that is necessary to eliminate. Finally, the electrolytic process could be carried out to obtain electrolytic manganese.

The objective of this work is the production of electrolytic manganese from anode lodes and scrapings obtained in the zinc electrolysis cells. The obtaining of other manganese products, such as oxide of manganese (IV), oxide of manganese (II), or manganese sulphate would also be possible.

2. Materials, Methods, and Results

2.1. Manganese Residue

Treated residues/by-products were, as previously mentioned, anodic lodes and scrapings from the zinc electrolysis process. First of all, the characterization of the anodic lodes and scrapings, which are recovered from the zinc electrolysis cells as a single product, is carried out.

The chemical composition of the by-product (anodic lode and scrapings) is given in Table 1, from which we can find that both manganese and lead are the most important elements. X-ray fluorescence was used to analyze the by-product. X-ray fluorescence measurements were performed with wavelength dispersive X-ray fluorescence (WDXRF) spectrometer (Axios, PANalytical, Faculty of Materials Science and Ceramics, AGH University, Krakow, Poland) equipped with an Rh-anode X-ray tube with maximum power 4 kW. The samples were measured in vacuum with 15–50 eV energy resolution. For quantitative analysis of the spectra, the PANalytical standardless analysis package Omnian was used. Manganese exists as MnO_2, as shown in Equation (1), because it is the result of an oxidation electrochemical reaction in the anode of the zinc electrolysis cell.

Table 1. Compositions of anodic lodes and scrapings from zinc electrolysis process. Determined by X-ray fluorescence (mass %).

Mn (%)	O (%)	Pb (%)	S (%)	Ca (%)	Si (%)	Sr (%)	Cd (%)
34.45	25.47	24.53	5.992	2.862	1.975	1.120	1.107
Zn (%)	Al (%)	K (%)	Fe (%)	Ag (%)	Cl (%)	Cu (%)	Others (%)
0.9183	0.3871	0.2803	0.2604	0.2552	0.2064	0.1081	0.034

As shown in Table 1, the residue is rich in manganese (>30% Mn). It could be considered as a manganese source if compared with traditional manganese ore deposits (pyrolusite, 63.2% Mn; braunite, 48.9–56.1% Mn; manganite, 62.5%; etc. [13]), near to the contents of manganese carbonates (around 47.6% Mn, [13]). The main difficulty that makes unusable this residue is the high lead content (Table 1). If this residue was briquetted and used in the production of ferroalloys, lead would be volatilized at the temperatures of the furnace [11], and this question would be inadmissible since the environmental point of view. In this way, hydrometallurgical processes are more suitable to treat this residue/by-product, in order to separate lead from manganese, but mainly because the anodic lode and scraps are very reactive facing to acids (such as dissolved H_2SO_4) similar to manganese carbonates (manganese carbonates are susceptible to be used as neutralizing agents instead of the lime in the proposed process). The amount of silver is also important, and for that reason these anodic lodes and scrapings could be considered as a silver source.

2.2. Industrial Process

In our project we propose treating anodic lodes and scrapings from the zinc electrolysis. It would be also possible to recover other raw materials/residues of manganese, classified in: chemically refractories (difficultly lixiviated, such as slags, powders, or other residues coming from the pyrometallurgy of manganese) and reactive (easily lixiviated, such as anodic lodes and scrapings of the zinc electrolysis and manganese carbonates). We designed a process that we have divided into four sub-processes (see Figure 1): acid leaching at low temperature (50 °C), acid leaching at high temperature (150–200 °C), neutralization with CaO (s), and chemical purification and electrolysis process. In this way, a conceptual design of plant was carried out while considering a production of 1000 kg of Mn (metal) every day. Even when the objective is obtaining electrolytic manganese, widely used in the production of aluminum and copper alloys, for special grades of stainless steel and other special steels, and for electronic applications [13], the process described could be applied for the obtaining of manganese sulphate, which could be sold as $MnSO_4$ or used in the production of manganese oxide (IV) and manganese oxide (II).

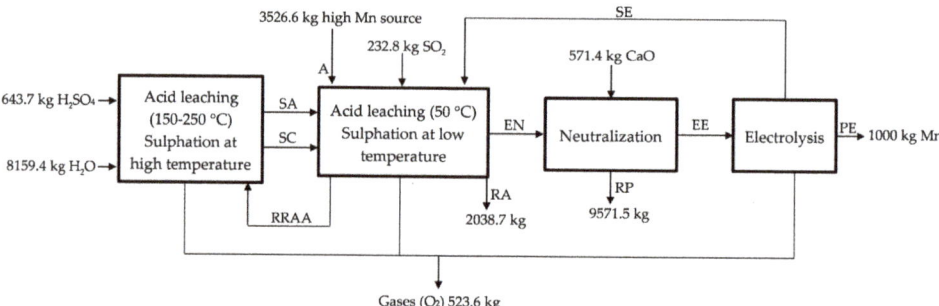

Figure 1. Process used for treating anodic lodes and scrapings from the zinc electrolysis process. SA, exit of acid leaching at high temperature; SC, hot sulfated; RRAA, recycled residue of acid leaching; A, high MnO_2 source; SE, exit of electrolysis; EN, entrance to neutralization; RA, residue low MnO_2; RP, residue of the neutralization process; EE, entrance to electrolysis; and, PE, product of electrolysis.

The basic characteristics of the process for obtaining manganese aqueous solutions used in the manufacture of electrolytic manganese are:

- aqueous medium of sulphate basis (SO_4^{2-} (aq));
- proposal of combined leaching: pyrometallurgical (T < 300 °C) and hydrometallurgical (T = 50 °C); and,

- utilization of non-conventional raw materials-wastes: wastes with high Mn contents as those coming from the zinc hydrometallurgy; powders and slags (of oxidized characteristics) generated in the manganese pyrometallurgy; powders of metallic characteristics, which are generated in the ferromanganese or silicomanganese production; and, other raw materials, which nowadays are not considered by the manganese pyrometallurgic industry, with low grade of Mn and high reactivity facing to the sulfuric acid, such are the manganese carbonates.

As it was previously mentioned, the proposed industrial process was divided into four stages or sub-processes. We will describe each one of the stages separately, being each of them supported by laboratory trials.

2.2.1. Sulphation at Low Temperature (50 °C)

The objective of this stage of the process is reducing MnO_2 in presence of a reductant agent, SO_2 (g) (other reducing reagents have been used in treating manganese ores, such as oxalic acid, hydrogen peroxide and glucose [15], cornstalk [16], phenols [17], cane molasses [19], CaS [20], carbon [23] or waste tea [34]) and obtaining a solution of $MnSO_4$ (aq) to be sent to the neutralization stage, and then to electrolysis. The mixed leaching allows for recovering almost all of the manganese of the anodic lodes and scrapings (95% is supposed in the calculations). The reactivity of the waste is exploited in this stage as a kind of neutralization of the sulfuric acid coming from other stages, supported by the utilization of SO_2 as reductant reagent. The low manganese waste contains valuable elements, such as lead and silver (see Table 2), coming from both the low and high temperature leaching processes.

The laboratory scale process begins with the drying of the lodes and scrapings in stove at 110 °C, and homogenization of the by-product. It was then mixed with Na_2SO_3, H_2SO_4 and water to obtain a solution containing manganese as sulphate and impurities, and leaving a solid product that contained lead sulphate and impurities (Table 2). The presence of Na_2SO_3 (s) allows for the generation of SO_2 (g) that acts as reductant agent of the by-product.

Tests were carried out in a hastelloy reactor (Laboratorio de Metalurgia, Dpto. de Ciencia de los Materiales e Ingeniería Metalúrgica, Universidad de Oviedo, Oviedo (Asturias), Spain) with different entries, allowing the introduction of a thermocouple to control the temperature and the feeding of the reagents. Besides, the reactor has an agitation system to homogenize and mix the different reagents. As the reaction of decomposition of the Na_2SO_3 (s) to generate SO_2 (g) is exothermic, the reactor is insulated with refractory wool with the purpose of minimizing the heat loss (this heat is used to make more favorable the process). In the proposed industrial process, the SO_2 (g) can be supplied directly as gas instead of using Na_2SO_3 (s) in the decomposition of H_2SO_4 (aq) to obtain SO_2 (g) used as reducing agent.

Table 2. Compositions of the filtration by-product (mass %). Determined by X-ray fluorescence.

Pb (%)	O (%)	S (%)	Mn (%)	Sr (%)	Si (%)	Ca (%)	Zn (%)
55.48	23.03	9.743	3.553	3.546	1.129	0.995	0.6289
Ag (%)	Al (%)	Fe (%)	Cl (%)	Hg (%)	K (%)	Ba (%)	Others (%)
0.6289	0.4730	0.2408	0.2164	0.1191	0.09077	0.08992	0.0682

Once the amounts of the different reagents previously mentioned were weighted, they were loaded in the reactor. The feeding process had a sequence that was: with the reactor open the Na_2SO_3 (s) was previously loaded with the manganese by-product and 2/3 of the water; the reactor is then closed and the agitation system is connected (at low agitation speed, around 30 rpm), while the H_2SO_4 (aq) is loaded into the vessel in small amounts; finally, 1/3 of the water is loaded into the reactor, it is completely closed and from this moment and the process has a duration of 30 min. A thermocouple was used to control the temperature, and it is observed that it was kept at 40–60 °C because of the exothermic reaction previously mentioned. Once the process finished, the solution is filtrated and an aqueous solution of $MnSO_4$ and a solid by-product, containing $PbSO_4$ (lead sulphate is poorly soluble in water) and other impurities, are obtained, being the solution analyzed by atomic absorption spectroscopy to evaluate the amount of manganese in the solution. The by-product coming from the filtration process was analyzed by using X-ray fluorescence and the results are shown in Table 2. Lead, as well as manganese and zinc, appear in this by-product as sulphate. This by-product (RA in Figure 1) contains significant amounts of lead and silver that could be treated with the purpose of recovering both of them (see pages 378–404 in [9]), thus making the process much more economically profitable.

The amounts of reagent in this process at low temperature were calculated by means of the following chemical reaction considering 15 g. of anodic lode as base of calculus:

$$Na_2SO_3(s) + MnO_2(s) + H_2SO_4(aq) \rightarrow MnSO_4(aq) + Na_2SO_4(aq) + H_2O(l) \qquad (2)$$

This reaction is thermodynamically favorable ($\Delta G^0 < 0$), according to the software HS5.1 (Outokumpu Research Oy, 5.11, Pori, Finland) even at room temperature. However, the real process includes other reactions of importance like that one:

$$Na_2SO_3(aq) + H_2SO_4(aq) \rightarrow Na_2SO_4(aq) + H_2SO_3(aq) \qquad (3)$$

As H_2SO_3 (aq) is unstable, it decomposes and:

$$H_2SO_3(aq) \rightarrow SO_2(g) + H_2O(l) \qquad (4)$$

And the simplified reaction for the reduction of the MnO_2 in the anodic lode is:

$$MnO_2(s) + SO_2(g) \rightarrow MnSO_4(aq) \qquad (5)$$

Being the SO_2 (g) the reductant agent used in the process (see Equation (5)). For that reason, when the amount of reductant agents was calculated (verifying the last equations) the amount of water and lode were kept constant. In this way, anodic lode and scrapings were a constant value of 15 g. To evaluate the effect of milling two situations were considered after 30 s milling and after 90 s milling (a finer granulometry will increase the leaching, but also a product with a more homogenous granulometry is obtained as lodes and scrapings have different initial granulometries). The results are shown in Tables 3 and 4.

Table 3. Low temperature process after 30 s milling.

Test	Na_2SO_3	H_2SO_4	Mn Extraction (%)
Condition 1	25% defect	11% excess	10.11
Condition 2	15% defect	18.5% excess	29.03
Condition 3	Stoichiometric	30% excess	39.82
Condition 4	15% excess	55% excess	79.92

Table 4. Low temperature process after 90 s milling.

Test	Na_2SO_3	H_2SO_4	Mn Extraction (%)
Condition 1	25% defect	11% excess	55.33
Condition 2	15% defect	18.5% excess	45.77
Condition 3	Stoichiometric	30% excess	71.04

It should be considered that the manual introduction of H_2SO_4 and the nearly simultaneous generation of SO_2 (g) cause the release of this gas, and consequently the loss of reducing gas. If the supply of H_2SO_4 had carried out automatically, then the manganese extraction would be better.

Moreover, the direct supply of SO_2 (g) would also increase the manganese extraction. This is the reason of a lower recovery in Condition 2 (in the industrial scale process the SO_2 supply will be automatic). It is also significant the improvement in the manganese extraction when the higher the milling, as the lower the particle size the easier the chemical lixiviation as the surface is increased and the reactions solid-gas became more favorable [11].

The residue obtained after filtering was analyzed by X-ray diffraction (PANalytical X'Pert Powder, Servicios Científico-Técnicos, Universidad de Oviedo, Oviedo (Asturias), Spain) and the following phases were obtained (see Figure 2, which is consistent with the information provided in Table 2): anglesite ($PbSO_4$) as the main phase, and sulphates of manganese and strontium. Lead is recovered in the solid residue obtained after the filtration as lead sulphate. This lead sulphate could be used in the manufacture of lead [9]. The presence of certain amount of manganese in the solid residue is always unavoidable, as we did not reach a full extraction of manganese. Other impurities also end in the solid residue as calcium, strontium, and potassium, but also silver. The presence of silver and lead in the solid residue of filtration makes it economically interesting. The solution containing most of the manganese (as $MnSO_4$) should be purified before being used to produce electrolytic manganese.

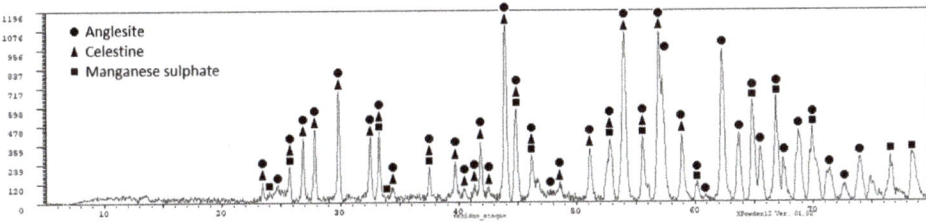

Figure 2. X-ray diffraction pattern for the residue of filtration.

The acid leaching/sulphation at low temperature (50 °C, temperature reached because of the exothermic reactions) is described in Figure 3 for the industrial scale process. The by-product/waste is loaded in a reactor (A, high MnO_2 source), where it is leached with SO_2 (g) (like that proposed in a laboratory scale but replacing Na_2SO_3 (s), used in the generation of SO_2 (g), by using directly SO_2 (g)) (see Equation (5)) and H_2SO_4 (aq) (see Equation (7)). MnO_2 is reduced in presence of both SO_2 (g) (20% of the initial manganese, according to Equation (5)) and H_2SO_4 (aq) (10% of the initial according to Equation (7)) from $Mn\ (IV)$ to $Mn\ (II)$, and a solution of $MnSO_4$ (aq) is obtained (a neutralization also happens in this stage as H_2SO_4 (aq) is used in the lixiviation of the manganese source). The presence of a reducing agent is always necessary for the complete extraction of the manganese, for instance, in the process described by Sanchez-Recio and Sancho [23], they take advantage of the presence of carbon in the powders used as manganese source [23]. The acid solution of $MnSO_4$ (aq) is sent with the $MnSO_4$ (aq) solution obtained in the acid leaching at high temperature (SC, hot sulfated) to neutralization (EN, entrance to neutralization). 70% of the initial manganese is treated in the acid leaching at high temperature (RRAA, recycled residue of acid leaching). The manganese source (anodic lodes and scrapings) contains significant amounts of impurities (lead, strontium, silver, etc.) that leave the process as a product, which we call RA (Residue low MnO_2). Most of the impurities enter and leave the process as sulphate so there are not H_2SO_4 (aq) losses, but ~3.5% Mn is lost in this residue that leaves the process wet. Figure 3 describes the acid leaching at low temperature.

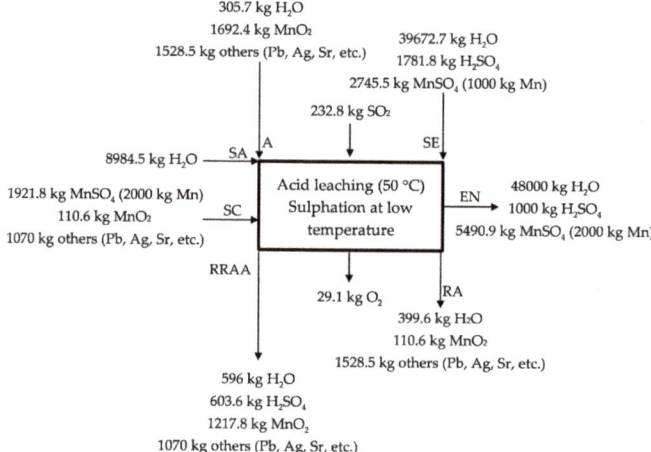

Figure 3. Description of the acid leaching at low temperature (50 °C).

2.2.2. Sulphation at High Temperature (150–200 °C)

The high temperature leaching was also considered in our research project with the idea of recovering all manganese from anodic lodes and scrapings by using a combination of high (70% of the raw material) and low (30% of the raw material) temperature leaching processes. First of all, a study of temperatures, where leaching with carbon and without carbon takes place, was carried out. The reactions that were considered were:

$$MnO_2(s) + C(s) + H_2SO_4(aq) \rightarrow MnSO_4(aq) + H_2O(g) + CO(g) \tag{6}$$

$$MnO_2(s) + H_2SO_4(aq) \rightarrow MnSO_4(aq) + H_2O(g) + 1/2O_2(g) \tag{7}$$

Tests were carried out in thermal balance and it was observed that all of the reactions take place at temperatures below 250 °C, and for that reason experiments were carried out at temperatures of 200 and 225 °C. Other authors worked at higher temperatures [22], but as observed in the thermal analysis, it is not necessary. We have as other variables: milling, the presence of reducing agents and time.

The procedure was: samples were mixed in crucibles (H_2SO_4 was added with 33% excess with the purpose of facilitating the extraction of manganese as sulphate) and then loaded into the furnace. After that, the samples were heated up to the reaction temperature slowly, and then held at that temperature for the considered time (45 or 60 min). With the purpose of facilitating the removal of the sample from the crucible, crucibles were removed at 100 °C and the product was filtered in presence of water, being the solution analyzed by atomic absorption spectroscopy in order to know the extraction of manganese. The extraction of manganese depending on the conditions is shown in Tables 5–7. As in the low temperature leaching, milling increases the extraction of manganese. Extractions would improve with a better control of the furnace temperature avoiding the risks of sulfuric acid evaporation (not pure sulfuric acid was used).

Table 5. Test carried out at 200 °C for 45 min.

Test	Mn Extraction (%)
After milling 90 s with C	85.55
Without milling with C	63.01
After milling 90 s without C	46.11
Without milling without C	33.81

Table 6. Test carried out at 200 °C for 60 min.

Test	Mn Extraction (%)
After milling 90 s with C	55.23
Without milling with C	58.30
After milling 90 s without C	38.90
Without milling without C	33.70

Table 7. Test carried out at 225 °C for 45 min.

Test	Mn Extraction (%)
After milling 90 s with C	68.60
Without milling with C	66.10
After milling 90 s without C	42.80
Without milling without C	33.70

It is observed that increasing the time does not improve the extraction, the same as increasing the temperature. In thermal analysis, it was observed that reactions took place at 200 °C. As happened in low temperature leaching, there were two products after the process: the solid residue containing lead sulphate and other impurities, and the solution containing manganese as $MnSO_4$ to be used in the production of electrolytic manganese (water was added to put $MnSO_4$ in solution, while $PbSO_4$ remained as solid residue because lead sulphate is poorly soluble in water).

The acid leaching at high temperature is described in Figure 4 for the industrial scale proposed process. The process would be like that described in Equation (7), but when considering the industrial scale. It is necessary to supply water and sulfuric acid in this stage to obtain the solution of manganese sulphate. 70% of the initial manganese is lixiviated in this stage by following the reaction presented in the Equation (7). A $MnSO_4$ (aq) solution is sent with the $MnSO_4$ (aq) solution obtained in the low temperature lixiviation to the neutralization process. The residue containing lead, silver, strontium, etc., leaves the process as RA in Figure 1.

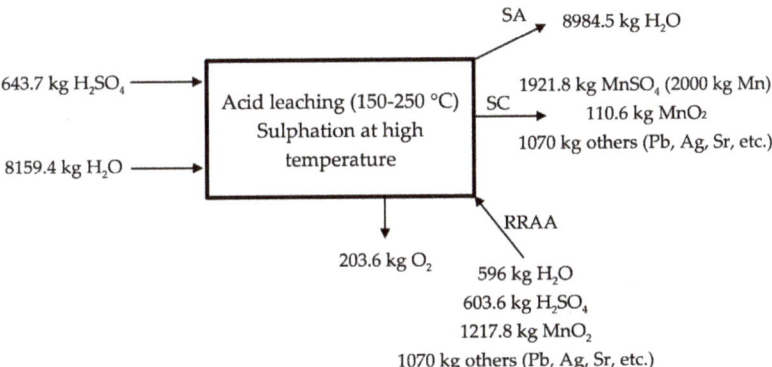

Figure 4. Description of the acid leaching at temperatures <300 °C.

2.2.3. Neutralization with CaO

The objective of the neutralization, apart from the purification of the solution from the elements that could compete with manganese in the electrolytic process, is changing the pH from the acid of the previous stage to the almost neutral that is required in the entrance of the electrolysis process.

Free acidity (pH < 1) is assumed at the entrance of the neutralization process (EN), while at the end of the neutralization pH = 6 is considered (~0 kg free H_2SO_4) (EE). The neutralization process is described in Figure 5. As neutralizing agent, CaO (s) is used, leaving a residue containing $CaSO_4$ (s) hydrated and most of the impurities (Zn, Co, Cu and Ni) (RP) that could make impure the solution used in the obtaining of manganese by electrolysis. Lime (CaO) is proposed in Figure 5 as neutralizing agent, however, manganese carbonates and even lodes could be used as neutralizing agents. The $MnSO_4$ (*aq*) solution that is enriched in Mn (50 g Mn/kg H_2O) is sent to electrolysis. Part of the water would be recirculated in the process.

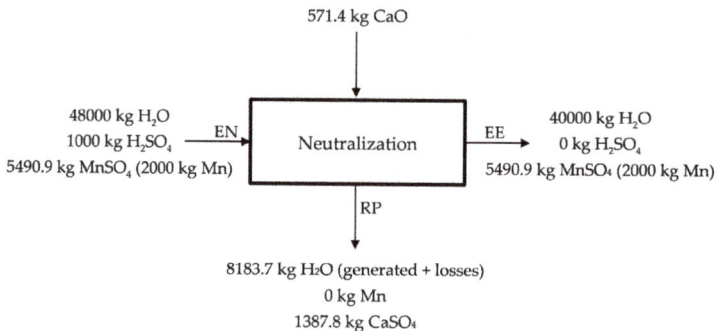

Figure 5. Description of the neutralization stage.

2.2.4. Chemical Purification. Electrolytic Process

The electrolysis process is the final stage for obtaining electrolytic manganese. However, depending on the market conditions, it could be interesting the production of oxide of manganese (IV), oxide of manganese (II), or manganese sulphate, and, for that reason, this stage is optional. Anyway, the process is described in Figure 6.

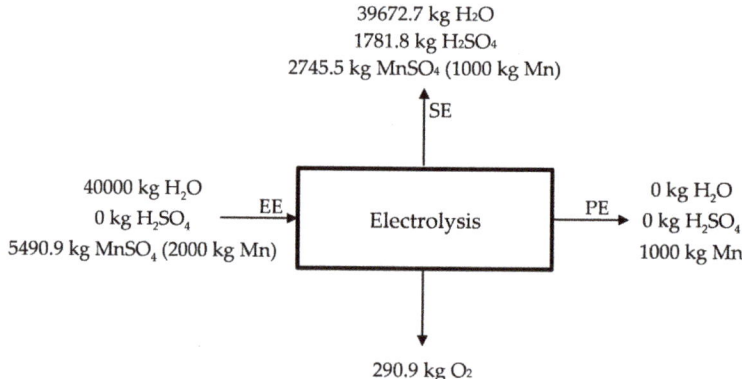

Figure 6. Manganese electrolytic process.

The reactions involved in the electrolysis process are:

$$Catodic\ reaction:\ Mn^{2+}(aq) + 2e^- \rightarrow Mn\ (s)$$

$$Anodic\ reaction:\ H_2O(l) - 2e^- \rightarrow 1/2 O_2(g) + 2H^+(aq)$$

$$Global\ reaction:\ Mn^{2+}(aq) + H_2O(l) \rightarrow 2H^+(aq) + 1/2 O_2(g) + Mn(s)$$

The result of the process is electrolytic manganese (Figure 7), although other products are obtained as it is possible to see in the previous equations (H_2SO_4, H_2O, and O_2). H_2SO_4 and H_2O are sent back to the previous stages of lixiviation (SE), the same as the poor liquor (25 g Mn/kg H_2O) (see Figure 6) (SE). Manganese deposits on the surface of the cathode as flakes with small amounts of manganese dioxide on the anode (see Figure 7) acting as catalyst [35]. The deposited manganese plate is then removed from the cathode surface by mechanical means, and is prepared to be sold, obtaining manganese qualities of 99.9%. Conditions for the electrolysis process are deeply described by Sánchez-Recio and Sancho [23].

Figure 7. Electrolytic manganese flakes.

As previously mentioned, several metals are separated in previous stages, for instance, the waste identified as RA in Figure 1 contains most of the lead and the silver present in the initial waste (this residue could be treated with the purpose of recovering both lead and silver). However, certain undesirable metals remain in the solution, and as a consequence, it should be purified before the electrolysis process. The first stage is based on separating several impurities by pH control following the Pourbaix diagrams (potential-pH) keeping the manganese in solution. Iron and aluminum, as well as other less problematic contaminants (such as cobalt and nickel), are removed during this stage. The pH is increased by adding lime to the pulp while it is stirred. The precipitate containing the impurities is separated by filtration. Finally the solution is passed through an active carbon filter.

The second stage of purification is applied to contaminants that cannot be removed by pH control. In this group of contaminants are included those that are nobler than manganese, being the most important the zinc. The removal is achieved by their precipitation as sulphide at slightly acid pH (see that pH at the end of the neutralization process is slightly acid). The problem is the formation of manganese sulphide and for that reason the precipitation requires a sufficient residence time allowing for the manganese sulphide to redissolve, but avoiding the re-dissolution of the other impurities that we wanted to remove.

The introduction in special electrolysis tanks (diaphragm cells, anolyte and catholyte are separated by a membrane) of the purified liquor requires a pH almost neutral and for that reason is conditioned by addition of a base. Finally, the pulp with a manganese concentration of around 30–50 g Mn/L is passed through a crystallizer, where calcium and magnesium are removed as ammonium salts. Ammonium sulphate and hydroxylamine sulphate are added as manganese stabilizer and buffering agent, and antioxidant, respectively. Electrolysis process is carried out in a diaphragm cells with separated anolyte and catholyte. The consumed electrolyte is then recirculated to the process.

The metallic manganese is deposited on the surface of the cathode in the form of flakes (99.9% Mn; 0.015% C; 0.05% S; 0.002% P; 0.001% Ti; 0.004% Mg; 0.006% Fe; 0.004% K; 0.002% Si; 0.003% Ca; 0.004% Zn; 0.001% Cu; 0.002% Co; 0.004% Ni), which is then removed by mechanical means (manganese oxide deposited in the anode acts a catalyst for the desired anodic reaction [35]).

2.2.5. Mass Balance of the Process

The process that was described in previous pages was summarized in Figure 1. The objective of the process is the obtaining of electrolytic manganese, as shown in Figure 7, from a high manganese source (now considered a waste), such are the anode lodes and scrapings (it could be also applied to other wastes containing high manganese). The amounts of reagents, sub-products, and raw materials are calculated considering 1000 kg of electrolytic manganese at the end of the process produced every day, and the results are supported by the laboratory tests. These amounts are summarized in Table 8, considering an efficiency of the 95% in the recovery of manganese.

Table 8. Summary of the amounts of reagents, raw materials and sub-products involved in the industrial process (1000 kg of electrolytic manganese, daily).

Raw Materials	
High Mn source: MnO_2 waste (~35.5% Mn; ~10% H_2O; others (Pb, Sr, etc.), ~50%) (kg)	3526.6
Reagents	
H_2SO_4 (kg)	643.7
H_2O (kg)	8159.4
SO_2 (kg)	232.8
CaO (kg)	571.4
Sub-Products	
Residue low MnO_2 (~3.5% Mn; ~16.5% H_2O; balance others (Pb, Ag, etc.)) (kg)	2038.7
Residue of neutralization (H_2O + $CaSO_4$ + H_2SO_4) (kg)	9571.5
Gases (O_2 (kg))	523.6
Final Product	
Electrolytic Mn (kg)	1000

3. Discussion

In this paper, we proposed a mixed leaching (low temperature and high temperature) to treat anodic lodes and scrapings from the zinc electrolysis process. The objective of the mixed leaching is taking advantage of the reactivity of the lodes and scrapings facing to sulfuric acid and SO_2 in solid solution (low temperature leaching; partial neutralization of the acidity) and achieve a complete the extraction of the manganese by using H_2SO_4 at high temperature. Even when in the paper, the process could seem divided into four stages independents; in fact, the proposed process would operate in a continuous way. As we see from Section 2.2.2., Sulphation at high temperature, manganese is lixiviated in presence of sulfuric acid without any kind of reducing agent at temperatures of around 200 °C, while from Section 2.2.1., Sulphation at low temperature, we see that manganese extraction is better than in the other case at room temperature (temperature will rise to nearly 50 °C as a consequence of the exothermic behavior of reaction [2]). We also see from the studies of lixiviation at high temperature that the addition of a reducing agent improves the manganese extraction, the same as the milling (the finer the granulometry the higher the manganese extraction). Manganese extractions are more or less two times higher in the presence of carbon or SO_2 than those that wer obtained without any kind of reducing agent. For that reason, a combined leaching allows for reaching efficiencies of around 95% in manganese extraction. Better experimental procedure (control of the temperature in the furnace and supply of SO_2) would have improved the manganese extractions that are shown in the tables.

Under the premise of a double leaching, a proposal for industrial upscaling was presented for the treatment of anodic lodes and scraping from the zinc electrolysis (this process could be applied in the treatment of other wastes/raw materials, as described in the Section 2.2). Even when the calculations of the main raw materials consumed in the process (water, sulfuric acid, residue and lime) were carried out under the objective of obtaining 1000 kg of electrolytic manganese with recirculation of the sulfuric acid that is generated in the electrolysis, it could be applied in the obtaining of manganese oxide (II), manganese oxide (IV), or manganese sulphate, depending on the market and economic conditions (for instance, high price of the electricity or strong demand of the each one of the manganese products). In that case, the process could be stopped after the mixed leaching (sulfuric acid supply would be increased). The obtaining of manganese oxides (II) and (IV) would be carried out through the following chemical reactions. First of all, manganese is precipitated by using ammonia ($NH_3 + H_2O \rightarrow NH_4OH$):

$$MnSO_4\ (aq) + 2NH_4OH(aq) \rightarrow Mn(OH)_2(s) + (NH_4)_2SO_4(aq)$$

Ammonium sulphate could be recycled by adding lime (NH_3), and manganese hydroxide (II) is obtained. Manganese hydroxide (II) could be roasted, and manganese oxide (II) would be obtained:

$$Mn(OH)_2(s) \rightarrow MnO(s) + H_2O \qquad (8)$$

And finally, manganese oxide (II) could be oxidized to obtain manganese oxide (IV):

$$MnO(s) + O_2(g) \rightarrow MnO_2(s) \qquad (9)$$

Aluminothermy of the manganese oxides could be performed with the purpose of obtaining metallic manganese (with lower quality than that obtained through the electrolysis process), but that could be used in the aluminum industry.

The treatment of the anodic lodes and scraping would mean for zinc plants an improvement in the competitiveness as the current practice with this material is disposing it in controlled areas with increasing costs. The treatment of this residue implies finishing with the costs of disposing the waste, but it also implies an economic value for the factories, as they will produce a valuable material that could be sold in a market with a strong demand of manganese coming from the steel and aluminum metallurgies. The versatility of the process would allow adapting to the market requirements.

4. Conclusions

Solid residues coming from metallurgical industries are being considered secondary sources of metals in a context of depletion of rich ores and environmental protection awareness. In this way, zinc production by electrowinning produces a waste/by-product of great interest, especially for the manganese industry, known as anodic lodes and scrapings. The interest of this waste/by-product comes from its high manganese content (>30%), and by means of the process that is described in this paper, electrolytic manganese could be obtained. Electrolytic manganese has interest in the aluminum industry (as it provides resistance and ductility to the alloys), in the steel industry (as desulphurizing and fine alloying agent for high performance stainless steels and HSLA (High Strength Low Alloy) steels), in copper and nickel alloys industry and other applications including the production of manganite or zinc-manganese ferrites.

The process described in this paper includes a mixed leaching at low temperature and high temperature, allowing for an optimization of costs and operating conditions (taking advantage of the lodes and scrapings would: reduce costs of storage for this waste; produce an economic impact derived from the sale of the manganese compound (metal, oxide, or sulphate) and the high lead and silver residue; and, optimize the zinc production process as less invaluable products would be generated) for zinc plants. In this way, low temperature lixiviation is supported by the addition of $SO_2\ (g)$ directly or through $Na_2SO_3\ (s)$ (used in the generation of $SO_2\ (g)$ as described in the paper), while high temperature lixiviation is based on the lixiviation achieved with H_2SO_4 at 150–250 °C, as described in the thermal analysis. The combination of both processes allows reaching the almost full leaching of the waste/by-product proposed.

The described process would have special interest for zinc factories, as they are the generators of anodic lodes and scrapings, but also because they produce H_2SO_4 that is used as a sulphation agent. They would also avoid the cost and dangers of storing this residue in specially prepared places. Economically, the profitability of the process depends on the capacity of the plant, the amortization costs, the electricity price, and electrolytic manganese price.

The described process is considered for anodic lodes and scraping obtained during the zinc electrolysis process, but it could be applied to other wastes/raw materials that we classified in: chemically refractories (low reactivity, slags, powders or other residues coming from the pyrometallurgy of manganese) and reactive (high reactivity, anodic lodes, and scrapings of the zinc electrolysis and manganese carbonates ores). The same as for the raw materials we could talk for the final products, as it would be possible to obtain manganese oxides (MnO and MnO_2) and manganese sulphate ($MnSO_4$).

Acknowledgments: The authors want to thank the advices and help of José Pedro Sancho Martínez during the research. We could not forget the valuable information provided by María Teresa Suárez Rodríguez. We thank Juan Carlos Sánchez Recio and Alberto Fuentes for their cooperation during the research. Finally, we want to thank Janusz Prazuch of the AGH University for his help during the research. The research was also supported by the Spanish Ministry of Education, Culture, and Sports via an FPU (Formación del Profesorado Universitario) grant to Daniel Fernández González (FPU014/02436).

Author Contributions: Luis Felipe Verdeja and José Sancho-Gorostiaga designed the work and supervised the experiments. Daniel Fernández-González performed the tests and wrote the manuscript. Juan Piñuela Noval helped in the revision of the manuscript.

Conflicts of Interest: The authors declare no conflict of interest.

References

1. Verdeja, L.F.; Sancho, J.P.; Ballester, A.; González, R. *Refractory and Ceramic Materials*, 1st ed.; Editorial Síntesis: Madrid, Spain, 2014; ISBN 9788490775837.
2. Fernández-González, D.; Ruiz-Bustinza, I.; Mochón, J.; González-Gasca, C.; Verdeja, L.F. Iron ore sintering: Raw materials and granulation. *Miner. Process. Extr. Metall. Rev.* **2017**, *38*, 36–46. [CrossRef]
3. Fernández-González, D.; Ruiz-Bustinza, I.; Mochón, J.; González-Gasca, C.; Verdeja, L.F. Iron ore sintering: Process. *Miner. Process. Extr. Metall. Rev.* **2017**, *38*, 215–227. [CrossRef]
4. Fernández-González, D.; Ruiz-Bustinza, I.; Mochón, J.; González-Gasca, C.; Verdeja, L.F. Iron ore sintering: Quality indices. *Miner. Process. Extr. Metall. Rev.* **2017**, *38*, 254–264. [CrossRef]
5. Fernández-González, D.; Ruiz-Bustinza, I.; Mochón, J.; González-Gasca, C.; Verdeja, L.F. Iron ore sintering: Environment, automatic and control techniques. *Miner. Process. Extr. Metall. Rev.* **2017**, *38*, 238–249. [CrossRef]
6. Fernández-González, D.; Martín-Duarte, R.; Ruiz-Bustinza, I.; Mochón, J.; González-Gasca, C.; Verdeja, L.F. Optimization of sínter plant operating conditions using advanced multivariate statistics: Intelligent data processing. *JOM-J. Miner. Met. Mater. Soc.* **2016**, *68*, 2089–2095. [CrossRef]
7. Ordiales, M.; Iglesias, J.; Fernández-González, D.; Sancho-Gorostiaga, J.; Fuentes, A.; Verdeja, L.F. Cold agglomeration of Ultrafine Oxidized Dust (UOD) from ferromanganese and silicomanganese industrial process. *Metals* **2016**, *6*, 203. [CrossRef]
8. Chandra, N.; Amritphale, S.S.; Pal, D. Recovery of manganese and lead values from zinc industry anode mud. *J. Solid Waste Technol. Manag.* **2010**, *36*, 116–125.
9. Sancho, J.P.; Verdeja, L.F.; Ballester, A. *Metalurgia Extractiva: Procesos de Obtención*, 1st ed.; Editorial Síntesis: Madrid, Spain, 2008; Volumen II, pp. 319–375; ISBN 84-7738-803-2.

10. Bateman, A.M. Geology of zinc deposits. In *Zinc the Science and Technology of the Metal, Its Alloys and Compounds*, 1st ed.; Mathewson, C.H., Ed.; Reinhold Publishing Corporation: New York, NY, USA, 1959; pp. 36–37.
11. Ballester, A.; Verdeja, L.F.; Sancho, J. *Metalurgia Extractiva. Fundamentos*, 1st ed.; Editorial Síntesis: Madrid, Spain, 2000; ISBN 84-7738-802-4.
12. Zhang, W.; Cheng, Z. Manganese metallurgy review. Part I: Leaching of ores/secondary materials and recovery of electrolytic/chemical manganese dioxide. *Hydrometallurgy* **2007**, *89*, 137–159. [CrossRef]
13. Olsen, S.E.; Tangstad, M.; Lindstad, T. *Production of Manganese Ferroalloys*, 1st ed.; Tapir Academic Press: Trondheim, Norway, 2007; ISBN 978-82-519-2191-6.
14. Lu, J.; Dreisinger, D.; Glück, T. Electrolytic manganese metal production from manganese carbonate precipitate. *Hydrometallurgy* **2016**, *161*, 45–53. [CrossRef]
15. Ghafarizadeh, B.; Raschchi, F.; Vahidi, E. Recovery of manganese from electric arc furnace dust of ferromanganese production units by reductive leaching. *Miner. Eng.* **2011**, *24*, 174–176. [CrossRef]
16. Cheng, Z.; Zhu, G.; Zhao, Y. Study in reduction-roast leaching manganese from low-grade manganese dioxide ores using cornstalk as reductant. *Hydrometallurgy* **2009**, *96*, 176–179. [CrossRef]
17. Zhang, Y.; Liu, Q.; Sun, C. Sulfuric acid leaching of ocean manganese nodules using phenols as reducing agents. *Miner. Eng.* **2001**, *14*, 525–537. [CrossRef]
18. Zhang, Y.; You, Z.; Li, G.; Jiang, T. Manganese extraction by sulfur-based reduction roasting-acid leaching from low-grade manganese oxide ores. *Hydrometallurgy* **2013**, *133*, 126–132. [CrossRef]
19. Su, H.; Wen, Y.; Wang, F.; Sun, Y.; Tong, Z. Reductive leaching of manganese from low-grade manganese ore in H_2SO_4 using cane molasses as reductant. *Hydrometallurgy* **2008**, *93*, 136–139. [CrossRef]
20. Li, C.; Zhong, H.; Wang, S.; Xue, J.; Wu, F.; Zhang, Z. Manganese extraction by reduction-acid leaching from low grade manganese oxide ores using CaS as reductant. *Trans. Nonferrous Met. Soc.* **2015**, *25*, 1677–1684. [CrossRef]
21. Buzatu, M.; Saceanu, S.; Petrescu, M.I.; Ghica, G.V.; Buzatu, T. Recovery of zinc and manganese from spent batteries by reductive leaching in acidic media. *J. Power Sources* **2014**, *247*, 612–617. [CrossRef]
22. Ayala, J.; Fernández, B. Reuse of anode slime generated by the zinc industry to obtain a liquor for manufacturing electrolytic manganese. *JOM-J. Miner. Met. Mater. Soc.* **2013**, *65*, 1007–1014. [CrossRef]
23. Sanchez-Recio, J.C.; Sancho, J. Method of Obtaining Electrolytic Manganese from Ferroalloy Production Waste. U.S. Patent 8,911,611 B2, 4 December 2014.
24. Pero-Sanz, J.A. *Aceros Metalurgia Física, Selección y Diseño*, 1st ed.; Editorial CIE Dossat: Madrid, Spain, 2004; ISBN 84-89656-54-1.
25. Pero-Sanz, J.A.; Quintana, M.J.; Verdeja, L.F. *Solidification and Solid-State Transformations of Metals and Alloys*, 1st ed.; Elsevier: Amsterdam, The Netherlands, 2017; pp. 255–324; ISBN 978-0-12-812607-3.
26. Grässel, O.; Krüger, L.; Frommeyer, G.; Meyer, L.W. High strength Fe-Mn-(Al, Si) TRIP/TWIP steels development—Properties—Application. *Int. J. Plast.* **2000**, *16*, 1391–1409. [CrossRef]
27. Frommeyer, G.; Brüx, U.; Neuman, P. Supra-Ductile and High-Strength Manganese-TRIP/TWIP Steels for High Energy Absorption Purposes. *ISIJ Int.* **2003**, *43*, 436–446. [CrossRef]
28. Yang, P.; Xie, Q.; Meng, L.; Ding, H.; Tang, Z. Dependence of deformation twinning on grain orientation in a high manganese steel. *Scr. Mater.* **2006**, *55*, 629–631. [CrossRef]
29. Sipos, K.; Remy, L.; Pineau, A. Influence of austenite predeformation on mechanical properties and strain-induced martensitic transformations of a high manganese steel. *Metall. Mater. Trans. A* **1976**, *7*, 857–864. [CrossRef]
30. Ueji, R.; Tsuchida, N.; Terada, D.; Tsuji, N.; Tanaka, Y.; Takemura, A.; Kunishige, K. Tensile properties and twinning behavior of high manganese austenitic steel with fine-grained structure. *Scr. Mater.* **2008**, *59*, 963–966. [CrossRef]
31. Ordiales, M.; Fernández, D.; Verdeja, L.F.; Sancho, J. Potassium permanganate as an alternative for gold mining wastewater treatment. *JOM-J. Miner. Met. Mater. Soc.* **2015**, *67*, 1975–1985. [CrossRef]
32. Compton, T.R. *Battery Reference Book*, 3rd ed.; Newnes Reed Educational and Professional Publishing Ltd.: Oxford, UK, 2000; ISBN 978-0-7506-4625-3.

33. Carosella, M.C.; Culbertson, J.B.; Jacobs, J.H. Electrolytic Manganese. U.S. Patent 2,805,195 A, 3 September 1957.
34. Tang, Q.; Zhong, H.; Wang, S.; Li, J.; Liu, G. Reductive leaching of manganese oxide ores using waste tea as reductant in sulfuric acid solution. *Trans. Nonferrous Met. Soc.* **2014**, *24*, 861–867. [CrossRef]
35. Wiechen, M.; Berends, H.-M.; Kurz, P. Water oxidation catalyzed by manganese compounds: From complexes to "biomimetic rocks". *Dalton Trans.* **2012**, *41*, 21–31. [CrossRef] [PubMed]

© 2018 by the authors. Licensee MDPI, Basel, Switzerland. This article is an open access article distributed under the terms and conditions of the Creative Commons Attribution (CC BY) license (http://creativecommons.org/licenses/by/4.0/).

Review

Refining Principles and Technical Methodologies to Produce Ultra-Pure Magnesium for High-Tech Applications

Seifeldin R. Mohamed [†], Semiramis Friedrich [*,†] and Bernd Friedrich

IME Institute of Process Metallurgy and Metal Recycling, RWTH Aachen University, 52056 Aachen, Germany; sraslan@ime-aachen.de (S.R.M.); bfriedrich@ime-aachen.de (B.F.)
* Correspondence: SFriedrich@ime-aachen.de; Tel.: +49-241-80-95977
† These authors contributed equally to this work.

Received: 23 November 2018; Accepted: 10 January 2019; Published: 15 January 2019

Abstract: During the last decade, magnesium-based medical implants have become the focal point of a large number of scientific studies due to their perceived favorable properties. Implants manufactured from magnesium alloys are not only biocompatible and biodegradable, but they are also the answer to problems associated with other materials like stress shielding (Ti alloys) and low mechanical stability (polymers). Magnesium has also been a metal of interest in another field. By offering superior technical and economic features in comparison to lithium, it has received significant attention in recent years as a potential battery anode alternative. Natural abundancy, low cost, environmental friendliness, large volumetric capacity, and enhanced operational safety are among the reasons that magnesium anodes are the next breakthrough in battery development. Unfortunately, commercial production of such implants and primary and secondary cells has been hindered due to magnesium's low corrosion resistance. Corrosion investigations have shown that this inferior quality is a direct result of the presence of certain impurities in metallic magnesium such as iron, copper, cobalt, and nickel, even at the lowest levels of concentration. Magnesium's sensitivity to corrosion is an obstacle for its usage not only in biomedical implants and batteries, but also in the automotive/aerospace industries. Therefore, investigations focusing on magnesium refinement with the goal of producing high and ultra-high purity magnesium suitable for such demanding applications are imperative. In this paper, vacuum distillation fundamentals and techniques are thoroughly reviewed as the main refining principles for magnesium.

Keywords: magnesium; refining; recycling; ultra-high purity; vacuum distillation; condensation

1. Introduction

In today's high-tech, application-oriented world, the demand for materials with superior quality suiting these applications is on the rise. Moreover, such materials should also be resource-saving and environmentally friendly. That is the reason for the recent significantly growing attention to magnesium, the lightest of the structural metals. Magnesium's extremely low density in comparison to iron and even aluminum results in lower energy consumption, which has made it attractive for automotive and aerospace industries in the last few decades. Not only this, but its high thermal and electrical conductivity, high alkaline resistivity, dimensional stability in electronic housing, as well as the machinability and formability of magnesium also make it a favorable metal for various applications [1,2].

Typically, magnesium is found in the Earth's crust in the form of dolomite ($CaCO_3 \cdot MgCO_3$), magnesite ($MgCO_3$), and magnesium oxide (MgO), among others. Regarding liquid resources, magnesium is usually found in form of carnallite ($KCl \cdot MgCl_2 \cdot 6H_2O$) and bischofite ($MgCl_2 \cdot 6H_2O$) [3].

For primary magnesium production there are two routes: electrolytic or thermal processing. USA, Canada, and some EU countries have dominated magnesium production from the 1970s to the 1990s, using electrolytic methods such as the Dow process. However, since the start of the 21st century, the growth in Chinese magnesium production capacity has led to the almost complete decline of western magnesium and the electrolytic route. China's adoption of the Pidgeon process (thermal route) enabled it to be the world's biggest producer, due to the relative easiness, versatility, and reduced capital cost of this process. Despite these great advantages, the Pidgeon process consumes a significant amount of energy, requires high labor costs, and suffers from low productivity (a batch process): aspects that make this process unfeasible to utilize in any other country [4]. The Pidgeon and the Dow processes are not the only examples of their respective processing routes. Processes like Magnetherm and Mintek have also emerged over the years as substitutes for the Pidgeon process.

It should be noted that all of these processes are primary processes to extract magnesium from its various ores and are not appropriate for extreme refining/purification purposes, as they are only capable of producing metallurgical grade pure Mg. Typically, electrolytic processes can produce Mg with 99.8% purity [4]. Even to this day, US Magnesium LLC (one of the few remaining magnesium producers using the electrolytic route) offers its pure magnesium in this grade. Regarding thermal processes, they can also produce comparable Mg purities; 99.97%, 99.93%, and 99.02% by the Pidgeon, Magnetherm, and Mintek processes respectively, with these figures obtained by considering only the main impurities (Al, Si, Ca, and Fe) [4,5]. The magnesium produced by such processes is classified as 'pure' by the ASTM, as they use this term for Mg with purities between 99.8% and 99.98% (3N) [6]. This level of purity is typically enough when talking about most conventional industrial applications of magnesium, as shown with their consumption rates in Figure 1 [7].

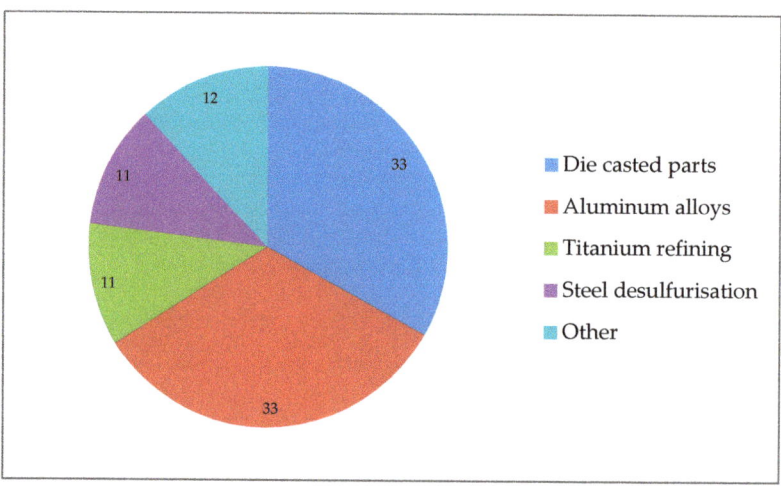

Figure 1. Global consumption of major Mg applications in 2012 (in %).

However, a quite trendy application of magnesium is Mg-based implants, which has recently become one of the most researched areas in the world of magnesium, and for which the commercial purity grade is not sufficient. For almost a decade, magnesium alloys were and still are under extensive investigation to test their feasibility as a superior replacement for titanium alloys and polymer-based implants. Magnesium alloys solve the problems associated with the other materials like stress shielding and mechanical stability for titanium and polymer based implants respectively [8,9]. Moreover, magnesium has antibacterial properties, is biocompatible and biodegradable, and eliminates the need for further surgical intervention to remove the implant after healing [10–12].

Last but not least, another trendy application of magnesium in recent years is magnesium-based anodes in batteries. Since magnesium is an abundant, light, environmentally friendly metal and it is significantly cheaper than lithium, it is a highly feasible alternative for lithium. Additionally, it eliminates the toxic effects related to discarding or recycling of lithium batteries [13]. Moreover, magnesium possesses almost double the volumetric energy density of lithium, and offers improved operational safety due to the lack of non-dendritic deposition and lower air sensitivity [14].

For these sorts of high-tech applications, magnesium purity must be upgraded to high (4N) and ultra-high (5N or more) purity graded Mg [15,16], because their development is hindered by the weak corrosion resistance of magnesium. Extensive research has found that magnesium corrosion is extremely sensitive to impurity elements such as iron, nickel, copper, and cobalt at even ppm levels, as illustrated in Figure 2 [17]. These impurities generally tend to be highly cathodic in a Mg matrix, thus forming micro-galvanic cells that result in severe corrosion [18]. In the case of biomedical applications, this corrosion behavior compromises the mechanical integrity of the implant and shortens its estimated service life [19]. An optimum corrosion rate accompanied by a symmetric laminar degradation can be achieved by using high purity magnesium [20]. Even after achieving tolerable levels of the above mentioned hazardous impurity elements to control corrosion, their toxic effect is another matter that drastically decreases the permissible limit [21]. When it comes to batteries, the corrosion of the anode is undesired as it results in the formation of a passive layer that blocks the electrode, which leads to a decrease in efficiency and unstable anodic dissolution, as well as a decrease in the output voltage [22].

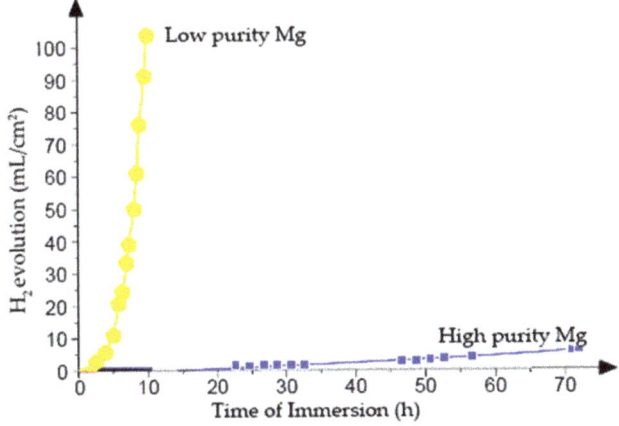

Figure 2. High purity Mg vs. low purity Mg corrosion rates.

The focus of this paper is to introduce the most important magnesium refining processes used to achieve high and ultra-high purity magnesium, with the aim of giving an extensive overview on the principles and the devised techniques and setups. Various methodologies have been investigated over the years to test their feasibility in producing ultra-high purity 'XHP' magnesium, such as electro-refining by using fluorides, chlorides, or oxides as the electrolytes. Impurity precipitation was also investigated as a potential methodology for this task. Electro-refining was distinguished to be a complex trial with high energy consumption and large operation costs, while impurity precipitation has not been successful in effectively and economically producing XHP magnesium. Therefore, vacuum distillation is still the only reliable and cost-effective process for this purpose [15,23]. Before introducing different methodologies based on vacuum distillation, the fundamentals of this principle will be briefly explained in the following section.

2. Fundamentals of Vacuum Distillation

Vacuum distillation offers several advantages when it comes to purification of metals. Besides reliable prevention of any contact to atmosphere, dramatic evaporation and sublimation rates at much lower temperatures can be achieved in a vacuum in comparison to the normal atmospheric or inert conditions. This, of course, decreases the reaction time, energy consumption, and operation cost, hence increasing the overall profitability of the process. Furthermore, vacuum distillation is environmentally friendly, as it hinders the formation of exhaust gases, waste water, or slags [24,25].

Distillation in general is based on the fact that different chemical substances (elements or compounds) have different boiling points, hence different vapor pressures. It means that upon heating any given feed material, the more volatile substances evaporate, leaving the less or non-volatile substances in the initial container. It is a fact that each substance has a certain vapor pressure at any given temperature (the only variable), independent of the mass or the volume of that substance [26,27]. In order to represent the vapor pressures of materials depending on their temperature, the following equation is used [28]:

$$\log p^* = AT^{-1} + B\log T + CT + D \qquad (1)$$

where p^* is the saturated vapor pressure of pure substances (Pa) and T is the absolute temperature (K). Regarding A, B, C, and D, they are evaporation constants for different elements, for which the exemplary values are shown in Table 1.

Table 1. Values of evaporation constants for Mg and its typical impurities [28].

Element	A	B	C	D	Temperature Range (K)
Ca	−9350	−1.39	0	14.94	298–1112
Cu	−17770	−0.86	0	14.42	298–1356
Fe	−21080	−2.14	0	19.02	298–1809
Mg	−7780	−0.855	0	13.54	298–923
Mg	−7550	−1.41	0	14.915	923–1363
Mn	−14920	−1.96	0	18.32	298–1517
Pb	−10130	−0.985	0	13.28	600–2013
Zn	−6620	−1.255	0	14.465	692–1773

By solving Equation (1) with the evaporation constants mentioned in Table 1, the vapor pressures of these impurities (among others) can be compared to that of Mg. Preliminary results indicate the feasibility of the distillation process, as illustrated in Figure 3, where all common magnesium impurities except zinc have a lower vapor pressure than magnesium [28]. Therefore, in a magnesium vacuum distillation process, most of the impurities will be expected to stay in the crucible as residue, while Mg and Zn will be collected on a condensing surface.

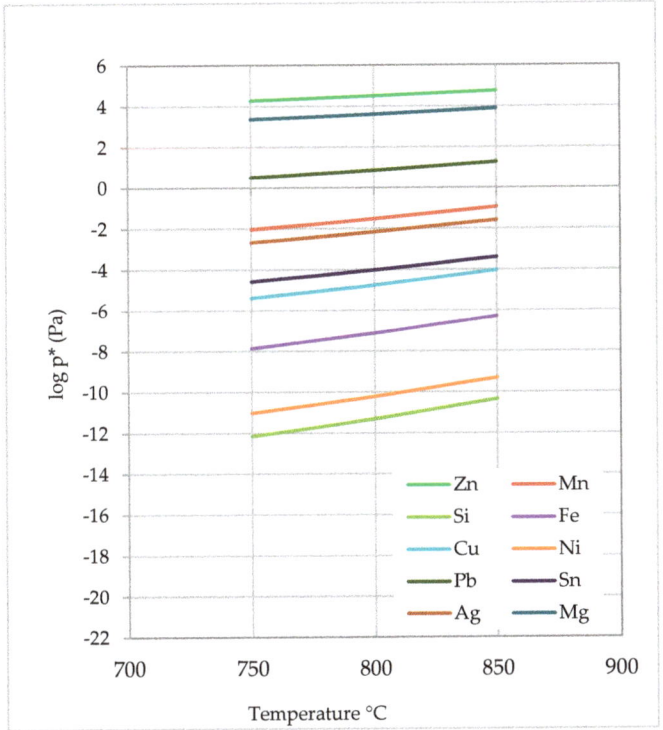

Figure 3. The variation of log p^* versus temperature for Mg and selected impurities.

It should be noted that the vapor pressure of pure substances is only a preliminary indicator as to the feasibility of the separation process; therefore, a more accurate parameter is devised to show the extent of separation between an impurity element and liquid magnesium. That parameter is the so called separation coefficient (β), which depends on the vapor pressures of both investigated elements as well as their activity coefficients (γ). The separation factor of an exemplary impurity element (i) is to be calculated as shown in Equation (2) [23,29,30].

$$\beta_i = \frac{\gamma_i}{\gamma_{Mg}} \cdot \frac{p_i^*}{p_{Mg}^*} \qquad (2)$$

where γ_{Mg} is taken approximately as unity (due to the almost pure raw Mg). Obtaining (γ_i) is the main challenge, because it varies with the temperature of the system and the initial concentration of each impurity. In order to theoretically predict the activity coefficients, the molecular interaction volume model (MIVM) was developed on the basis of statistical thermodynamics [31]. Various studies have used this model to explain the behavior of impurity elements in Pb- and Sn-based alloys during vacuum distillation, reporting comparable results to their experimental trials, hence proving the feasibility of this model for vacuum distillation [32–34].

Another approach to calculate the separation coefficients is the use of an empirical formula Equation (3) to obtain the evaporation coefficient (α_i) [30].

$$y_i = 100 - 100(1 - \frac{x_i}{100})^{\alpha_i} \qquad (3)$$

where x_i and y_i are the evaporated wt. % from the main metal and an impurity respectively, calculated from the results of the process. The obtained evaporation coefficient is then substituted into Equation (4) to calculate the separation factor [30].

$$\beta_i = \alpha_i \sqrt{\frac{M_i}{M_{Mg}}} \qquad (4)$$

where M_i and M_{Mg} are the molar masses of the investigated impurity and magnesium, respectively. The farther the separation coefficient is from unity, the better the separation between the given impurity and magnesium or vice versa. Table 2 shows the calculated coefficients of some impurities at various temperatures and their (with the exception of zinc) easy separation possibility from magnesium. In this trial-dependent model, the calculated separation coefficient is then used to calculate the activity coefficient of the corresponding impurity through Equation (2).

Table 2. Separation coefficients between Mg and impurities at different temperatures [23].

T (K)	βCa	βCu	βPb	βZn
873	1.08×10^{-5}	3.98×10^{-12}	4.74×10^{-7}	1.03
923	1.31×10^{-5}	1.66×10^{-11}	6.58×10^{-7}	0.85
973	1.72×10^{-5}	6.58×10^{-11}	9.74×10^{-7}	0.79
1023	2.12×10^{-5}	2.20×10^{-10}	1.34×10^{-6}	0.72

It is worth noting that calculation of the activity coefficients is also important for the prediction of the impurity mass fraction in the gas phase, with the result also depending on the initial concentration of the impurity in the liquid and the vapor pressures of Mg and the investigated impurity [23].

$$i_g = \left[1 + \left(\frac{Mg_l}{i_l}\right) \cdot \left(\frac{\gamma_{Mg}}{\gamma_i}\right) \cdot \left(\frac{p^*_{Mg}}{p^*_i}\right)\right]^{-1} \qquad (5)$$

where l and g stand for liquid and gas phases, and Mg_l, i_l, and i_g are the mass fractions of Mg and any given impurity in the liquid and gas phase respectively.

It should be noted that for vacuum distillation, the kinetic aspect is as important as the thermodynamic one, but since the kinetics of this process are highly dependent on the geometry of the used reactor, it is very difficult to obtain information applicable to the process in general, and thus the kinetic aspect will not be discussed in this review.

3. Vacuum Distillation Setups for Producing Ultra-High Purity Magnesium

Vacuum distillation has been investigated as a way to refine magnesium to unprecedented purities throughout the years. In 1978, Revel et al. [35] created a vertical retort (see Figure 4) for magnesium refining. The equipment is composed of a graphite crucible and a diaphragm holding a stainless steel condenser that contains baffles. The system is heated via a resistance heater situated between two sintered alumina walls, with the external wall isolating the equipment.

The design of the system allows a homogeneous temperature at the bottom part (crucible to diaphragm) and a decreasing temperature throughout the condenser's compartments. The apparatus is able to distill a 50 g load of magnesium.

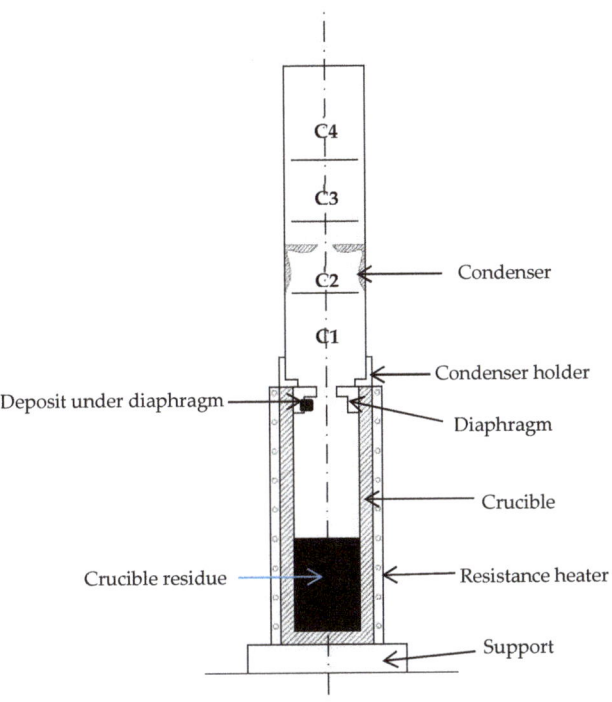

Figure 4. Illustration of Revel's distillation column (C1–C4: compartments of condenser).

The aim of their study was to find the optimum parameters and the effect of various process parameters on the purification degree of impurities with higher volatility than magnesium (e.g. Na, Zn) as well as with low volatility (e.g. Cu, Fe, and Mn). They investigated the effect of the load and condensation temperatures, residual gas pressure, distillation rate, and distilled quantity. The optimum load temperature for this process is 700 °C, and no benefit for increasing the temperature over this value was observed. Regarding the residual gas pressure in the retort, no significant effect on the process was noticed at low pressure levels (0.013 to 0.0013 mbar). The purest deposits of magnesium were found in the lower compartment of the condenser (C1), which had a temperature varying between 400 and 450 °C, having a noticeable difference with deposits condensed below 350 °C. Several distillation rates were investigated (3.5, 24, and 50 g/h) at the optimum temperature by changing the number and size of holes in the diaphragm.

The lowest rate was proven to be the optimum, as it simplifies the control and the reproducibility of the conditions. Table 3 shows a comparison between the impurity concentrations of selected impurities after different stages of distillation. The results shown are of a trial conducted with the mentioned optimum parameters, and the sample is collected from the C1 compartment.

Table 3. Impurity concentration (in ppm) in samples with varying distillation quantity [35].

Element	Before Distillation	After 55% Distillation	After 72% Distillation	After 83% Distillation
Cu	3	0.8	0.6	0.05
Fe	97	<0.07	<0.04	<0.6
Mn	17–19	0.2	0.05	0.01
Na	4.3	0.4	0.3	0.6
Zn	52–58	17	16	17

Despite the extremely high volatility of sodium and zinc, and thus the difficulty to separate them from Mg in a distillation process, the results of Table 3 show a decrease in their concentration. This can be attributed to the multi-chambered condenser, where these elements tend to condense in the upper, low-temperature chambers, while most of the Mg distillate (almost 90%) condenses/solidifies in the lowest chamber [35].

In 1996, Lam and Marx devised a distillation column able to produce Mg with 5N metallic purity [15]. This column (depicted in Figure 5) comprises a crucible and a vertical condenser with multiple baffles, all made out of highly pure graphite. This setup is maintained in a resistance furnace, including three temperature-controlled zones. Through optimization of the baffle positions and temperature control, impurities with higher condensation temperatures than magnesium, such as iron and copper, condense on the lower baffles, while magnesium deposits on the middle ones. Finally, elements such as zinc (as well as sodium and potassium, if present) with lower condensation temperatures condense on the higher positioned baffles. The preferred parameters here are a strong vacuum of 1.3×10^{-6} mbar, a crucible temperature of approximately 700 °C, and condenser temperatures of approximately 600 and 450 °C for the lower and upper parts of the condenser respectively [15].

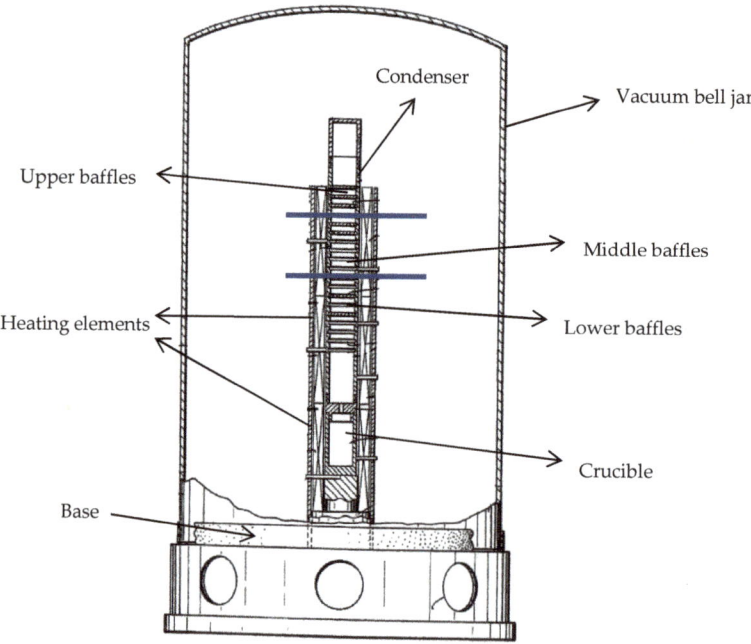

Figure 5. Vacuum distillation column as devised by Lam and Marx.

In 2003, Tayama and Kimura [16] developed a purification method with a newly devised apparatus. Their main concern was to produce XHP magnesium through overcoming some disadvantages they thought present in the Lam and Marx's version of the distillation column. They argued that since the concept of the old apparatus was based on selectively condensing different elements in different temperature zones, it would be difficult to exclude impurities with small differences in solidification temperatures between them and magnesium, especially on an industrial scale. Of course, in order to ensure the exclusion of such elements, the specified recovery zone must have a very small temperature range, which will result in a low metal yield.

This apparatus is also a vertical distillation column, comprising a feed heating zone at the bottom and a condensation/passage way zone thereupon. A cooled chamber is then located above for the condensation and collection of the purified magnesium. At the top, there is an entrapment zone to collect any vapor that is still not solidified. As shown in Figure 6 [16], there are special vapor passage plates, some downwardly convex plates with a centered hole, and some upwardly convex plates with peripheral holes.

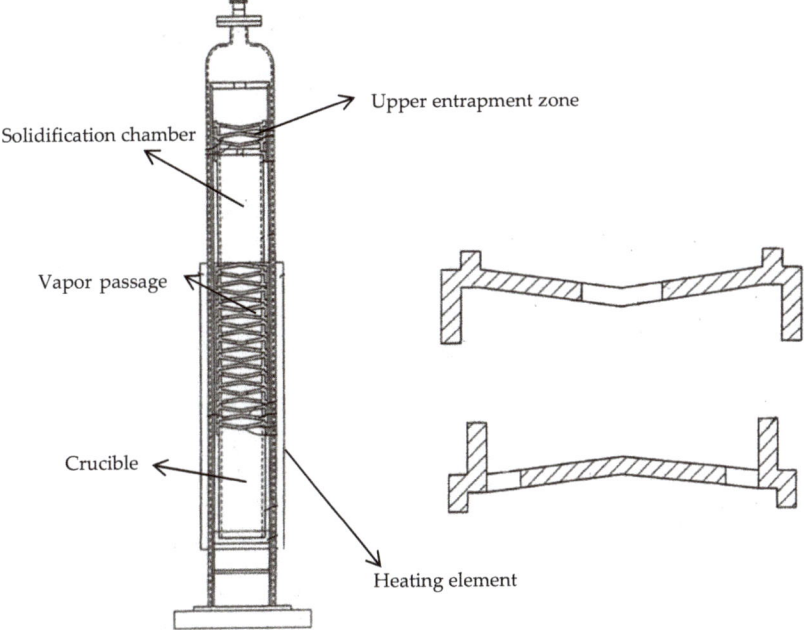

Figure 6. Vacuum distillation column as devised by Tayama and Kimura (**left**) and vapor passage plates (**right**).

The intention of these types of plates is to force the generated magnesium vapor to follow a zigzag path that sufficiently decreases its flow rate. The continuous collisions between the vapor and the plates result in energy losses, thus decreasing the vapor temperature and allowing part of the evaporated Mg to condense on the vapor passage plates. The design and the installation order of these plates allow magnesium vapor to pass upwardly, and at the same time allow the condensed melt (that still contains some impurities) to flow downwards until it eventually reaches the feed crucible. Through this countercurrent mechanism, fractionation of high vapor pressure elements (i.e., magnesium) and low vapor pressure elements (i.e., impurities except zinc) occurs. The Mg vapor escaping the passage then solidifies on the walls of the solidification chamber, while the remaining fumes (Zn–Mg mixture) are collected by the upper entrapment zone of the reactor. The optimum parameters reported were a vacuum atmosphere of 1.3×10^{-3} mbar along with controlled temperatures of 750 °C and 700 °C for the feed crucible and the condensation vapor passage plates respectively. Magnesium with an impurity content of 0.75 ppm (6N purity) can be obtained. If the obtained material from the first run is purified again through the same process, this impurity level is able to be reduced to 0.38 ppm (6N6). It should be noted that these values exclude zinc, as the amount of zinc in the final material after two runs is still 1.7 ppm, which increases the total impurity content to 2.08 ppm, resulting in Mg with 5N8 purity grade being produced [16].

In all the above setups, the condensed magnesium is always collected as fine powdery crystals with high surface/volume ratio. Due to the extremely high reactivity of magnesium, these fine particles can be easily oxidized, leading to problems upon re-melting to produce a semi-finished product, because the formed oxide skins on the fine crystals contaminate the pure melt and remain in the solidified metal in the form of non-metallic inclusions. This, of course, can have severe consequences not only on the mechanical properties of the finished product, but also on its corrosion resistance. In order to overcome this problem, Uggowitzer et al. [36] developed a quite different setup (see Figure 7) that allows the liquid condensation of pure magnesium, so that upon solidification it is collected as a considerable amount of bulk. In this setup, magnesium is heated in a resistance-heated stainless steel retort until evaporation. Since the produced vapor reacts weakly with the stainless steel of the walls, minor contamination can be detected in the condensed molten magnesium droplets. In order to prevent these contaminated droplets from entering the collection crucible, a 'cover' is placed to direct the falling droplets back into the feed. By controlling and keeping the temperature constant via an independent heating element, under the boiling point but over the melting temperature of magnesium, the evaporated magnesium that enters the collection crucible is to be condensed in a liquid state [36].

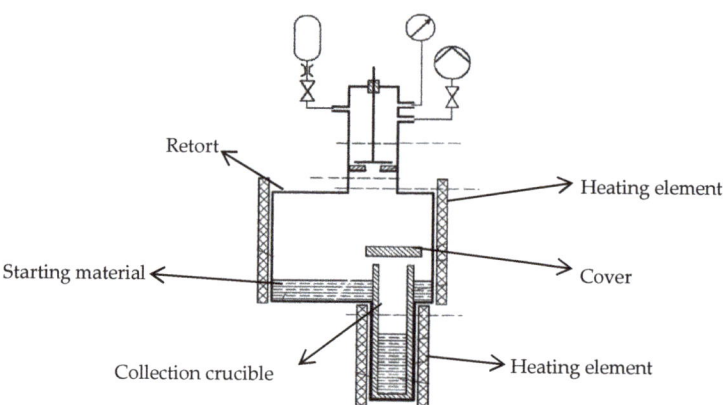

Figure 7. Vacuum distillation retort as devised by Uggowitzer et al.

4. Summary and Outlook

In the never ending search for better materials and technologies to reduce production—as well as environmental—costs, researchers have given special attention to magnesium in previous years for its exceptional qualities. This has led to a significant rise in the production of magnesium, to meet the increasing demand. The major reason hindering the extensive use of magnesium (i.e., weak corrosion resistance) comes from the presence of impurities like iron, cobalt, copper, and nickel, even at minor quantities. To overcome this issue and unlock magnesium's full potential as a biodegradable metal and a battery anode, various methodologies for ultra-purification of magnesium have been investigated over the years.

Throughout this paper, the major refining process for magnesium (i.e., vacuum distillation) was deeply reviewed, discussing the refining principle and the most important setups devised over the years, especially the ones focusing on ultra-purification. The review shows that magnesium with purities up to 5N8 can be achieved by vacuum distillation, with zinc being the main reason for not achieving purity values over 6N.

Despite the efforts of the past years, there is still a big gap in the thermodynamic data. Except for the limited information mentioned in Section 2, no data for the activity coefficients or the separation factors of impurities in magnesium are available. A MVIM study to theoretically predict the behavior of such impurities and confirm the results with experimental trials would drastically enhance our understanding for the process. An extensive kinetic study showing the effect of the evaporation rate on the purification levels of the various impurities is also an interesting topic for further research. For elements difficult to separate from Mg (e.g., zinc), investigating the addition of certain elements to the melt to form a hardly evaporable intermetallic compound could also be a feasible technique. Finally, in an effort to find further successful magnesium refining techniques, the authors are keen on testing the feasibility of fractional crystallization-based methods too.

Author Contributions: B.F. was the principal investigator. S.R.M. and S.F. wrote and edited the manuscript.

Funding: This research received no external funding.

Conflicts of Interest: The authors declare no conflict of interest.

References

1. Avedesian, M.M.; Baker, H. Magnesium and Magnesium Alloys. In *ASM Speciality Handbook*; ASM International: Materials Park, OH, USA, 1999.
2. Engh, T.A. *Principles of Metal Refining*; Oxford University Press: Oxford, UK, 1992; ISBN 9780198563372.
3. Pekguleryuz, M.; Kainer, K.; Kaya, A. (Eds.) *Fundamentals of Magnesium Alloy Metallurgy*; Elsevier: New York, NY, USA, 2013.
4. Wulandari, W.; Brooks, G.; Rhamdhani, M.; Monaghan, B. Magnesium: Current and Alternative Production Routes. In *Australian Conference on Chemical Engineering*; University of Wollongong Australia: Wollongong, Australia, 2010.
5. Pekguleryuz, M.; Mackenzie, L. (Eds.) Pilot Plant Demonstration of the Mintek Thermal Magnesium Process. In Proceedings of the Conference of Metallurgists, Montréal, QC, Canada, 1–4 October 2006.
6. Charles, M. (Ed.) *Engineering Properties of Magnesium Alloys*; ASM International: Novelty, OH, USA, 2017.
7. Magnesium.com. Dynamics of Worldwide Magnesium Metal Consumption, 2008–2013. Available online: http://www.magnesium.com/w3/data-bank/index.php?mgw=241 (accessed on 15 September 2018).
8. Beevers, D.J. Metal vs. bioabsorbable interference screws: Initial fixation. *Proc. Inst. Mech. Eng. Part H J. Eng. Med.* **2003**, *217*, 59–75. [CrossRef]
9. Breen, D.J.; Stoker, D.J. Titanium lines: A manifestation of metallosis and tissue response to titanium alloy megaprostheses at the knee. *Clin. Radiol.* **1992**, *47*, 274–277. [CrossRef]
10. Witte, F.; Hort, N.; Feyerabend, F. Degradable biomaterials based on magnesium corrosion. *Curr. Opin. Solid State Mater. Sci.* **2008**, *12*, 63–72. [CrossRef]
11. Staigera, M.P.; Pietaka, A.M.; Huadmaia, J.; Dias, G. Magnesium and its alloys as orthopedic biomaterials: A review. *Biomaterials* **2006**, *27*, 1728–1734. [CrossRef] [PubMed]
12. Hort, N.; Huangand, Y.; Feyerabend, F. Magnesium alloys as implant materials—Principles of property design for Mg–RE alloys. *Acta Biomater.* **2010**, *6*, 1714–1725. [CrossRef] [PubMed]
13. Richey, F.W.; McCloskey, B.D.; Luntz, A.C. Mg Anode Corrosion in Aqueous Electrolytes and Implications for Mg-Air Batteries. *J. Electrochem. Soc.* **2016**, *163*, A958–A963. [CrossRef]
14. Crowe, A.J.; Bartlett, B.M. Solid state cathode materials for secondary magnesium-ion batteries that are compatible with magnesium metal anodes in water-free electrolyte. *J. Solid State Chem.* **2016**, *242*, 102–106. [CrossRef]
15. Lam, R.; Marx, D.R. Ultra High Purity Magnesium Vacuum Distillation Ultra High Purity Magnesium Vacuum Distillation Purification Method. U.S. Patent 5,582,630, 1996.
16. Tayama, K.; Kimura, S. High Purity Metals, Process and Apparatus for Producing Them by Enhanced Purification. EP 1,335,030 A1, 2003.
17. Liu, M.; Uggowitzer, P.J.; Nagasekhar, A.V.; Schmutz, P.; Easton, M.A.; Song, G.; Atrens, A. Calculated phase diagrams and the corrosion of die-cast Mg–Al alloys. *Corros. Sci.* **2009**, *51*, 602–619. [CrossRef]
18. Narayanan, T.S.N.S.; Park, I.; Lee, M. (Eds.) *Surface Modification of Magnesium and Its Alloys for Biomedical Applications: Mechanical Integrity of Magnesium Alloys for Biomedical Applications*; Elsevier: New York, NY, USA, 2015.
19. Xin, Y.; Huo, K.; Tao, H.; Tang, G.; Chu, P.K. Influence of aggressive ions on the degradation behavior of biomedical magnesium alloy in physiological environment. *Acta Biomater.* **2008**, *4*, 2008–2015. [CrossRef]

20. Hofstetter, J.; Martinelli, E.; Pogatscher, S.; Schmutz, P.; Povoden-Karadeniz, E.; Weinberg, A.; Uggowitzer, P.; Löffler, J. Influence of trace impurities on the in vitro and in vivo degradation of biodegradable Mg–5Zn–0.3 Ca alloys. *Acta Biomater.* **2015**, *23*, 347–353. [CrossRef]
21. Cipriano, A.F.; Sallee, A.; Guan, R.-G.; Zhao, Z.Y.; Tayoba, M.; Sanchez, J.; Liu, H. Investigation of magnesium-zinc-calcium alloys and bone marrow derived mesenchymal stem cell response in direct culture. *Acta Biomater.* **2015**, *12*, 298–321. [CrossRef] [PubMed]
22. Höche, D.; Lamaka, S.V.; Vaghefinazari, B.; Braun, T.; Petrauskas, R.P.; Fichtner, M.; Zheludkevich, M.L. Performance boost for primary magnesium cells using iron complexing agents as electrolyte additives. *Sci. Rep.* **2018**, *8*, 7578. [CrossRef] [PubMed]
23. Wang, Y.C.; Tian, Y.; Qu, T.; Yang, B.; Dai, Y.; Sun, Y. Purification of Magnesium by Vacuum Distillation and its Analysis. *MSF* **2014**, *788*, 52–57. [CrossRef]
24. Bauer, R. Vakuumdestillation von Magnesium aus Aluminium-Magnesium-Legierungen und Aluminium-Magnesium-Silizium-Legierungen. Ph.D. Thesis, RWTH Aachen, Aachen, Germany, 1970.
25. Samanidis, K. Untersuchung der Destillation von Metallen im Vakuum. Ph.D. Thesis, RWTH Aachen, Aachen, Germany, 1989.
26. Zhu, T.; Li, N.; Mei, X.; Yu, A.; Shang, S. Innovative Vacuum Distillation for Magnesium Recycling. *Magnesium Technol.* **2001**, 55–60. [CrossRef]
27. Akbari, S.F. Minimizing Salt and Metal Losses in Mg-Recycling through Salt Optimization and Black Dross Distillation. Ph.D. Thesis, RWTH Aachen, Aachen, Germany, 2011.
28. Yong, D.; Bin, Y. *Vacuum Metallurgy of Nonferrous Metal Materials*; Metallurgical Industry Press: Beijing, China, 2000.
29. Liu, D.C.; Yang, B.; Wang, F.; Yu, Q.C.; Wang, L.; Dai, Y.N. Research on the Removal of Impurities from Crude Nickel by Vacuum Distillation. *Phys. Procedia* **2012**, *32*, 363–371. [CrossRef]
30. Kong, X.; Yang, B.; Xiong, H.; Liu, D.; Xu, B. Removal of impurities from crude lead with high impurities by vacuum distillation and its analysis. *Vacuum* **2014**, *105*, 17–20. [CrossRef]
31. Tao, D.P. A new model of thermodynamics of liquid mixtures and its application to liquid alloys. *Thermochimica Acta* **2000**, *363*, 105–113. [CrossRef]
32. Kong, L.; Yang, B.; Xu, B.; Li, Y.; Liu, D.; Dai, Y. Application of MIVM for Pb–Sn–Sb ternary system in vacuum distillation. *Vacuum* **2014**, *101*, 324–327. [CrossRef]
33. Yang, H.; Yang, B.; Xu, B.; Liu, D.; Tao, D. Application of molecular interaction volume model in vacuum distillation of Pb-based alloys. *Vacuum* **2012**, *86*, 1296–1299. [CrossRef]
34. Wang, A.; Li, Y.; Yang, B.; Kong, L.; Liu, D. Process optimization for vacuum distillation of Sn–Sb alloy by response surface methodology. *Vacuum* **2014**, *109*, 127–134. [CrossRef]
35. Revel, G.; Pastol, J.-L.; Rouchaud, J.-C.; Fromageau, R. Purification of Magnesium by Vacuum Distillation. *MTB* **1978**, *9*, 665–672. [CrossRef]
36. Löffler, J.; Uggowitzer, P.; Wegmann, C.; Becker, M.; Feichtinger, H. Process and Apparatus for Vacuum Distillation of High-Purity Magnesium. WO 2013107644 A1, 2013.

© 2019 by the authors. Licensee MDPI, Basel, Switzerland. This article is an open access article distributed under the terms and conditions of the Creative Commons Attribution (CC BY) license (http://creativecommons.org/licenses/by/4.0/).

Article

Kinetic Investigation of Silver Recycling by Leaching from Mechanical Pre-Treated Oxygen-Depolarized Cathodes Containing PTFE and Nickel

Jil Schosseler [1,*], Anna Trentmann [1], Bernd Friedrich [1], Klaus Hahn [2] and Hermann Wotruba [2]

1. Institute of Process Metallurgy and Metal Recycling IME, RWTH Aachen University, Intzestraße 3, 52056 Aachen, Germany; atrentmann@metallurgie.rwth-aachen.de (A.T.); bfriedrich@metallurgie.rwth-aachen.de (B.F.)
2. Unit of Mineral Processing AMR, RWTH Aachen University, Lochnerstraße 4-20 Haus C, 52064 Aachen, Germany; hahn@amr.rwth-aachen.de (K.H.); wotruba@amr.rwth-aachen.de (H.W.)
* Correspondence: jschosseler@metallurgie.rwth-aachen.de; Tel.: +49-241-80-90855

Received: 26 November 2018; Accepted: 31 January 2019; Published: 5 February 2019

Abstract: This paper focuses on the recycling of silver from spent oxygen-depolarized cathodes through an innovative combination of pre-treatment methods and leaching. A silver- and polytetrafluorethylene (PTFE)-rich fraction was produced by cryogenic milling, screening, and magnetic separation. In order to understand the kinetic leaching mechanism, the silver-rich fraction was leached by different concentrations of nitric acid and hydrogen peroxide. Results showed that nickel influences the silver leaching. This leads to complex reaction systems, which cannot be described by the Arrhenius law.

Keywords: oxygen-depolarized cathodes; silver leaching; cryogenic pre-treatment; negative activation energy

1. Introduction

The utilization of oxygen-depolarized cathodes (ODCs) for chlorine-alkaline electrolysis is becoming interesting as an industrial process due to its lower energy consumption and associated CO_2 emission compared to mercury, diaphragm, or conventional membrane cell electrolysis [1]. The power consumption for chlorine production can be reduced to 30% compared to that of the common membrane process [2].

Oxygen-depolarized cathodes consist of a grid material made of nickel, silver powder as a catalyst, and polytetrafluorethylene (PTFE). As the technology becomes more relevant in the future, the development of a recycling process for used cathodes is an important step towards a sustainable process chain. This publication focuses on the recovery of silver from ODCs with a combination of mechanical pre-treatment and hydrometallurgical methods. The main objective is to extract the silver from the ODC as silver nitrate and produce an optimal silver nitrate solution that is suitable for subsequent silver electrolysis. The main challenge is to leach with a relatively low acid concentration, so that electrolysis can run at pH values of 5–6 without requiring strong neutralization. Furthermore, the nickel concentration in the electrolyte has to be below 1 g/L as nickel influences the silver powder morphology and its quality for reuse in ODCs. The influence of nickel in the silver electrolysis was investigated by a project partner and is not a part of this paper [3].

To develop a zero-waste recycling process for used oxygen-depolarized cathodes, the combination of mechanical pre-treatment and leaching was necessary for the following reasons:

- Pyrometallurgical treatment is not suitable because of the complex material composition and the high amount of PTFE. The combustion of PTFE produces harmful off-gases and requires a complex off-gas cleaning [4]. Furthermore, a pyrometallurgical separation of silver and nickel causes high silver losses, which are investigated in other trials.

- The direct leaching of non-pre-treated oxygen-depolarized cathodes is technically feasible but requires a complex separation of dissolved nickel in order to recover catalytically active silver powder via silver electrolysis. Single pre-trials (5% nitric acid) with non-pre-treated cathodes revealed a higher dissolved nickel content (8 g/L Ni) compared to pre-treated ODCs (0.08 g/L Ni). For this reason, further experiments were carried out with crushed cathodes.
- The mechanical pre-treatment of the ODCs leads to the liberation of the enclosed silver particles, which improves the contact of acid and silver. A visual analysis of the pre-treated material showed a significant exposure of silver particles. In addition, handling crushed material is much easier compared to dealing with complete cathodes.
- Due to the relatively low acid concentration required, the influence of an oxidizing agent (hydrogen peroxide) and its influence on leaching (thermal instability [5] and catalytic decomposition [6]) should be investigated.

A zero-waste process is possible because the pre-processing of ODCs separates magnetic nickel, which can be melted directly after one washing step. Silver will be regained as silver powder for new ODCs from the obtained silver nitrate solution after leaching. Due to its chemically inert behavior, a PTFE fraction is received after leaching and filtration. After an additional leaching with concentrated acid, the PTFE should theoretically be clean enough to be transferred to a new recycling process.

The schematic of the developed recycling procedure, which should be able to deal with the challenges presented, is shown in Figure 1. The mechanical treatment started with cryogenic grinding in a cutting mill under liquid nitrogen. Temperatures below −50 °C increased embrittlement and supported the comminution of the material in the mill, which was necessary for the following processing steps. Separation was achieved by screening and magnetic separation, producing both an enriched Ni and a PTFE–Ag fraction. The enriched PTFE–Ag fraction was the input material for the following leaching step with nitric acid. The silver nitrate solution after leaching was used to produce silver powder by electrolysis, which, however, is not a part of this publication.

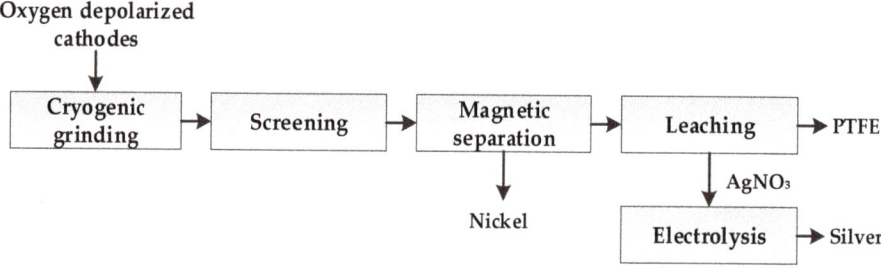

Figure 1. Schematic procedure for recycling silver from oxygen-depolarized cathodes (ODCs).

As shown in the flow sheet for the developed recycling process, the main objective was to devise a pre-treatment for used ODCs and to specify leaching conditions to obtain a suitable $AgNO_3$ solution for subsequent silver electrolysis.

The mechanical processing of the material uses mechanisms that are already well established in the processing of several mineral resources. Whereas screening is normally used to achieve a defined feed grain size for the subsequent separation processes, it can also be used to separate materials with different shapes in combination with selective comminution. There have been various approaches to selective comminution and subsequent screening for materials, such as concrete and mineral construction and demolition wastes in order to recycle their different components [7,8].

Magnetic separation has been increasingly used since the beginning of the 20th century. Its most important field of application is iron ore processing, as strong magnetic magnetite is enriched almost exclusively in this way. Due to its high susceptibility, metallic nickel is characterized as a ferromagnetic material that behaves similarly to magnetite, whereas Ag–PTFE is non-magnetic. This makes the application of a dry low-intensity magnetic separation (LIMS) suitable for the investigated material [9]. Modern dry LIMS aggregates are mainly designed as drum separators with permanent magnets, providing magnetic field strengths below 2000 Gauss (0.2 Tesla) [10]. For small-scale applications, rotary magnetic belt separators with higher magnetic field strengths (up to 5000 Gauss) can be applied.

Silver leaching with nitric acid is a well-known procedure with a complex reaction between metal and acid with a dependency on purity, formed reaction products, and temperature. The chemical and physical state of silver also influences the reaction [11]. Furthermore, the acid concentration can change the reaction order. In a concentration range between 3.0 and 4.9 mol/L, the reaction follows the second order [12]:

$$4HNO_3 + 3Ag \longrightarrow 3AgNO_3 + NO + 2H_2O; \tag{1}$$

at higher acid concentrations (4.9–7.1 mol/L) silver reacts with nitric acid according to the first order [12]:

$$2HNO_3 + Ag \longrightarrow AgNO_3 + NO_2 + H_2O. \tag{2}$$

Also, the reaction system presented in this paper had to be extended because of the partially enclosed silver particles in a PTFE matrix, which certainly influenced diffusion.

A non-catalytic heterogeneous solid–fluid reaction of uniform spherical particles can be expressed with the shrinking core model. Here, the rate-controlling step can be expressed differently under two different scenarios. If the chemical reaction is the dominant step, the expression can be indicated as [13]:

$$1 - (1-\alpha)^{1/3} = k \cdot t, \tag{3}$$

where α is the conversion, k is the reaction rate constant, and t is time.

If the diffusion of the reagent through a solid porous layer dominates the reaction, then the shrinking core model can be expressed, e.g., by an equation proposed by Jander [14]:

$$\left((1+\alpha)^{1/3} - 1\right)^2 = k \cdot t. \tag{4}$$

A plot of $[1 - (1-\alpha)^{1/3}]$ with respect to $[((1+\alpha)^{1/3} - 1)^2]$ versus time should give a straight line with a slope k. Furthermore, the reaction rate constant is a function of temperature and the activation energy of the system can be determined by the Arrhenius equation (Equation (5)) [15]:

$$k = Ae^{-E_a/RT}, \tag{5}$$

where A is the pre-exponential factor, E_a the activation energy, R the gas constant, and T the temperature.

The Arrhenius equation is applicable for elementary reactions between reactants, which are in an endothermic equilibrium. Furthermore, the reaction pathway from the initial to the final state does not influence the activation energy. The Arrhenius law suggests that the activation energy is a constant, which is not always a given, as E_a can show temperature and reaction-phase dependence [16,17].

2. Materials and Methods

To provide a sufficient liberation of the main components, the material was crushed by a cutting mill under cryogenic conditions. The low temperature of liquid nitrogen led to the embrittlement of the PTFE and nickel grid. Thus, a slight comminution of ODCs was possible without major material loss. A rough separation was seen after treatment (Figure 2a). The nickel grid matrix was disassembled into threads and the PTFE with the silver into grains (Figure 2b,c). The grain size was controlled by using different discharge screens (mesh sizes: 1, 2, and 4 mm). Five different batches were generated, whereby the first two batches (fractions 1 and 2) were used to generate enough material for the leaching trials with a discharge screen of 4 mm. In order to investigate the liberation grade, three fractions were generated with different discharge screens. The influence of using different mesh sizes during cryogenic milling on chemical composition, grain size distribution, and leaching behavior was investigated separately. The different fractions obtained underwent the same processing steps afterwards.

Figure 2. Samples after cryogenic treatment. (**a**) Macroscopic image, (**b**) microscopic image of the nickel fraction, and (**c**) microscopic image of the Ag–polytetrafluorethylene (PTFE) fraction.

The main objective of mechanical processing was to generate a high-grade Ag–PTFE concentrate by rejecting the metallic Ni threads. This was achieved by dry magnetic separation. Due to the fact that the Ni threads tended to stick together and agglomerate with Ag–PTFE, the feed had to be deagglomerated by short-time screening. As a positive side effect, the screening process not only ensured deagglomeration but also separated an adequate amount of Ni threads. A simplified flow sheet of the discontinuous process is shown in Figure 3.

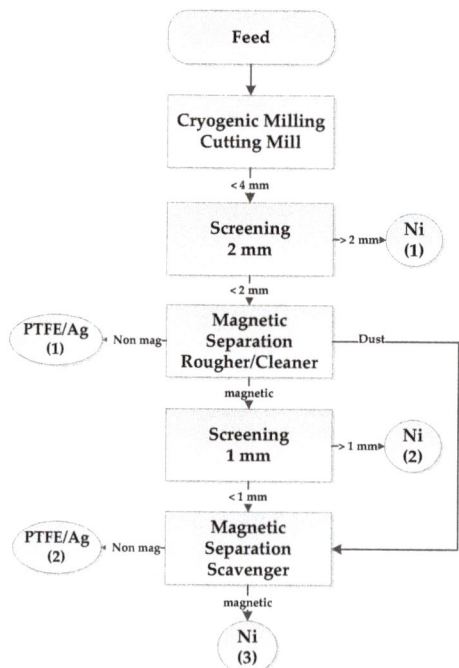

Figure 3. Simplified flow sheet of the discontinuous mechanical treatment process.

Screening was carried out using a laboratory-scale sieve, whereas magnetic separation was carried out with a rotary magnetic belt separator (Steinert, Cologne, Germany) with a permanent magnetic strength of 3500 Gauss (0.35 Tesla). To ensure a clean Ag–PTFE product and to speed up the recovery, a two-staged magnetic separation was performed followed by another short-time screening and a subsequent magnetic separation step.

The analysis of the processed ODCs was conducted with inductively coupled plasma atomic emission spectroscopy (ICP-AES) (Spectro Ciros Vision, Spectro, Kleve, Germany) after a microwave digestion (Multiwave PRO, Anton Paar, Graz, Austria) in concentrated nitric acid. By back weighing, the amount of insoluble PTFE was determined.

Leaching trials of the processed ODCs were carried out in a magnetic stirred beaker with a stirring speed of 300 rpm. To minimize fluid loss during the experiment, a watch glass was placed on the opening of the beaker. The solid to liquid ratio (50 g/L) and the temperature of 50 ± 1 °C was constant for all the leaching experiments. A reaction volume of 0.750 L was chosen and the material was homogenized for every trial to reduce the influence of material inhomogeneity.

In the first step, two different nitric acid concentrations were tested (5% HNO_3 and 10% HNO_3). Further, the influence of the addition of hydrogen peroxide was investigated. For this purpose, the required amount of H_2O_2 was calculated from the silver content and 1:1.5 and 1:3 $Ag:H_2O_2$ stoichiometries were chosen. The overstiochiometrical addition was chosen because a loss of hydrogen peroxide was anticipated due to thermal decomposition and reaction with nickel. At 120 min and at 240 min a liquid sample of 10 mL was taken and filtered with a syringe filter unit (mesh size 0.45 µm). The trials were repeated three times to prove the reproducibility and to reduce the influence of inhomogeneity of the material.

For the kinetic trials a temperature range from room temperature to 85 ± 0.5 °C was chosen. Over a time frame of 240 min, every 30 min a small sample of 10 mL was taken and filtered with syringe filter units (mesh size 0.45 µm). The agitation speed and the solid to liquid ratio were adopted from the leaching tests. The experiments were carried out with 5% nitric acid and a 1:3 stoichiometry was chosen to investigate the influence of hydrogen peroxide. The acid concentration of 5% and 1:3 stoichiometry delivered good results for the required selectivity and silver recovery for a possible recycling process. The results of the kinetic study provided information about the activation energy as well as details about the temperature and diffusion behavior in order to refine leaching parameters. The kinetic trials were repeated two times.

The silver concentration was determined with titration against chloride and nickel using ICP analysis. The liquid loss due to sampling and evaporation was taken into account for the calculation of the mean yields.

3. Results and Discussion

The results of five different test samples are shown in Table 1. Coarse- and medium-grained samples showed similar behavior in the first magnetic separation stage, while the recovery of Ag–PTFE from the fine-grained sample was less effective. The second magnetic separation stage increased the mass recovery of Ag–PTFE in all five samples.

As mentioned above, the process was not designed continuously because of the short-time screening stages between the milling and the magnetic separation stages. To realize an industrial and dry continuous process with a higher throughput, the application of an air jig or air tables could be considered due to the higher density of the Ag-bearing PTFE material compared to the Ni threads [18].

Table 1. Mass recoveries of mechanical processing tests.

Fraction	1 (Coarse)	2 (Coarse)	3 (Coarse)	4 (Middle)	5 (Fine)
Ni > 2 mm (1)	7.8%	8.4%	6.9%	8.2%	5.3%
PTFE–Ag magnetic separation rougher/cleaner (1)	73.2%	70.2%	74.7%	72.5%	61.2%
Ni > 1 mm (2)	8.4%	8.5%	8.5%	6.7%	11.3%
PTFE–Ag magnetic separation scavenger (2)	5.5%	10.6%	7.8%	7.5%	13.2%
Ni magnetic separation scavenger (3)	5.1%	2.3%	2.2%	5.1%	9.0%
Ag–PTFE product	78.8%	80.8%	82.4%	80.0%	74.4%
Ni fraction	21.2%	19.2%	17.6%	20.0%	25.6%

For the following leaching trials the non-magnetic fractions of 1 and 2 were taken and homogenized. Table 2 shows the material composition of the obtained Ag–PTFE fraction.

Table 2. Composition of processed ODCs used for leaching trials.

Ag	Ni	PTFE
91%	0.25%	5.7%

Figure 4 shows the metal extraction for silver and nickel after leaching with 5% and 10% HNO_3 and the addition of different amounts of hydrogen peroxide. By leaching with 5% nitric acid the silver yield reached 63.9% after 120 min; after 240 min the metal extraction increased to 85.2% (Figure 4a). The addition of hydrogen peroxide improved the silver recovery, as the yield reached 83.6% with 1:1.5 or 93.1% with 1:3 stoichiometry. A longer reaction time resulted in a slight increase up to 98.7% (1:3 stoichiometry). The silver dissolution reaction took place at the very beginning and almost reached a steady state after 120 min with the addition of hydrogen peroxide. For the extraction of nickel, a yield of 66.9% was obtained with the lower acid concentration after 120 min. The addition of the oxidizing agent decreased the nickel recovery to 47.1% with 1:3 stoichiometry. Similar to the dissolution of silver, the nickel dissolution also reached a saturation value after 120 min.

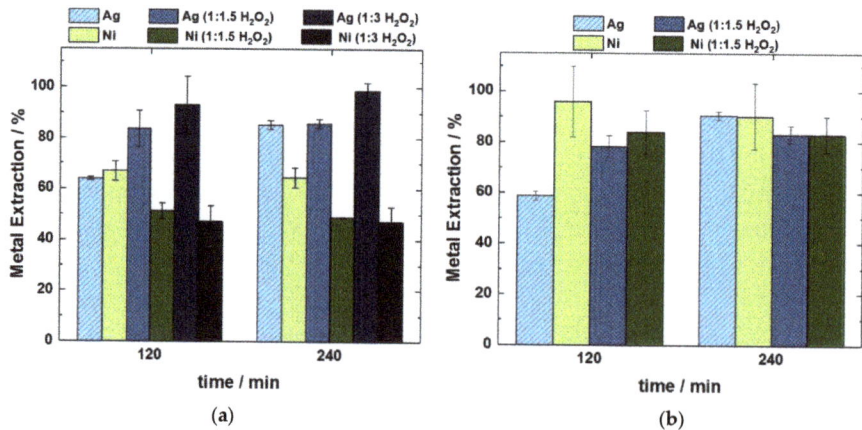

Figure 4. Recovery rate for silver and nickel for leaching trials with (a) 5% HNO_3 and (b) 10% HNO_3.

By leaching with 10% nitric acid, a slightly lower silver yield of 58.7% was obtained compared to the leaching trials with 5% HNO_3 (Figure 4b). The addition of hydrogen peroxide increased the silver recovery to 78.1%. A higher acid concentration led to a higher amount of nickel in the solution compared to leaching trials with 5% HNO_3. Additionally, the influence of the oxidizing reagent was significantly smaller and resulted in a lower silver-to-nickel selectivity. The lower acid concentration was more suitable for producing a silver nitrate solution ideal for electrolysis. Furthermore, the addition of hydrogen peroxide decreased the dissolved nickel content and reduced the reaction time, as shown in Figure 4.

A recovery of 100% was not achieved due to the following observations: Some silver particles were partly or completely surrounded by the PTFE matrix. Consequently, direct contact between the nitric acid and the silver particles was impeded. Furthermore, the presence of metallic nickel threads led to the cementation of metallic silver on the nickel surface. Through this cementation reaction, nickel was passivated as soon as sufficient silver dissolved to cover the nickel surface. This reaction between the dissolved silver and metallic nickel grains results in a useful nickel saturation because of a desired low nickel concentration in the solution on the one hand but consequently leads to silver losses on the other hand.

The influence of different temperatures is shown in Figure 5. The dissolution of silver in 5% HNO_3 at room temperature was slower compared to that at higher temperatures. At room temperature, a delay in the dissolution of silver can be observed for less than 60 min. Due to previous experiences it can be assumed that silver dissolution is dominated by the silver cementation.

At 45 °C, the yield of silver reached a maximum at 80.6%. Higher temperature increases led to reduced silver extraction. The dissolution of silver with the H_2O_2 additive was significantly faster than without it. This can be attributed to the higher standard potential of H_2O_2 compared to that of HNO_3, meaning that silver oxidation occurred more easily. The temperature behavior of the HNO_3–H_2O_2 leaching curves was similar to the HNO_3 curves. The yield increased from 88.0% to 93.7% with a temperature rise of 15 °C after 60 min. However, a further increase in temperature led to lower silver yields.

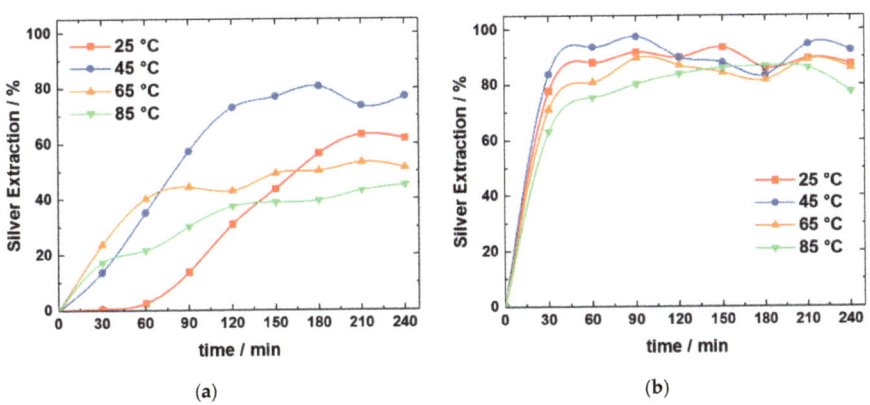

Figure 5. Silver extraction at different temperatures (a) without and (b) with the addition of hydrogen peroxide.

By applying the kinetic models stated above over a time range of 60–180 min, the following diagrams were created. The time range was chosen to exclude the range, where the silver extraction reached a saturation state. The differences between the models for reaction-controlled and diffusion-controlled leaching were only very marginal. Using other forms of the models also made no improvement. The main concern of the results obtained was the unusual behavior as a function of the temperature. The relationship between $1 - (1 - \alpha)^{1/3}$ (reaction-controlled) and $((1 + \alpha)^{1/3} - 1)^2$ (diffusion-controlled) and the time for silver leaching at various temperatures are shown in Figure 6. Comparing the two models, the quality of the linear regressions decreased with increasing temperature. In addition, based on the available data, no decision could be made as to whether the reaction is more diffusion- or reaction-controlled.

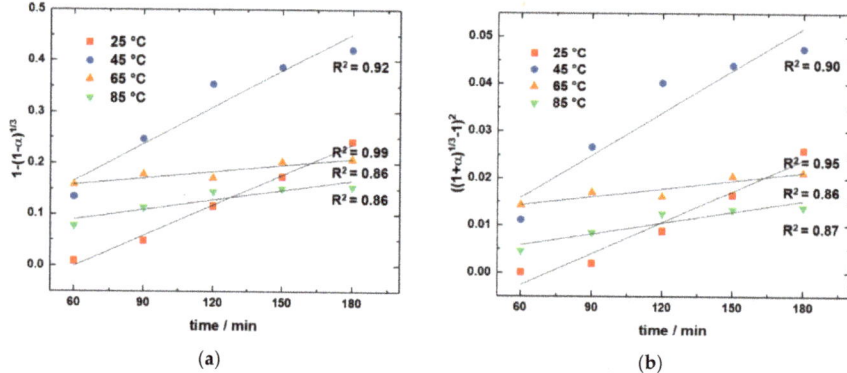

Figure 6. The relationship between (a) $1 - (1 - \alpha)^{1/3}$ (reaction-controlled) and (b) $((1 + \alpha)^{1/3} - 1)^2$ (diffusion-controlled) and the time for silver leaching at various temperatures.

After determining the reaction rate constants from the slopes, a decrease in the k values was observed with temperature increase. By plotting $\ln(k)$ versus $1/T$ in order to determine the activation energy following the Arrhenius equation, a negative activation energy was calculated for the dissolution of silver regardless of the chosen model. This outcome contradicts the literature values for the activation energy of metallic silver in nitric acid [12].

As mentioned in the introduction, the Arrhenius law depends on several factors [16,17]. Considering the observations of the reaction behavior, the non-Arrhenius behavior can be explained by the presence of a complex reaction system [19]. One assumption for the validity is a simple elementary reaction shown in Equation (6).

$$A + B \rightarrow C \tag{6}$$

One indication of the presence of a complex reaction is the leaching behavior of nickel. Leaching trials with 5% nitric acid have shown that it is impossible to achieve a complete dissolution of nickel in spite of its solubility in diluted nitric acid [20]. One possible explanation for this phenomenon is a passivation of the nickel threads after a certain reaction time and a subsequent hindrance for the dissolution of nickel. On examining the nickel threads after leaching by optical enlargement, a silver layer on the nickel surface was obvious from silver cementation (Figure 7).

Figure 7. Silver cementation on nickel threads.

Considering this cementation, it was necessary to extend the model for the dissolution of silver with a consecutive reaction (Equation (7)) which is a complex elementary reaction.

$$[Ag]_{PTFE} \xrightarrow{HNO_3} Ag^+_{(aq)} \underset{HNO_3}{\overset{Ni}{\rightleftarrows}} [Ag]_{Ni}. \tag{7}$$

Furthermore, nickel shows a complex behavior, as it is dissolved in diluted acid and with silver cementation, which results in a parallel reaction.

$$Ni \begin{array}{c} \xrightarrow{HNO_3} Ni^{2+}_{(aq)} \\ \xrightarrow{Ag^+} Ni^{2+}_{(aq)} \end{array} \tag{8}$$

Given the fact that both silver and nickel show complex dissolution reactions in this system, the requirements of the Arrhenius law are not fulfilled and the negative activation energy can be explained because of these phenomena. The three necessary reaction rate constants for the silver reaction include three different temperature dependences. Nickel dissolution can be expressed by two reaction rate constants. In addition, silver and nickel interact with each other, making the leaching model even more complicated. The overall reaction rate constant at one temperature, which is determined by the leaching model, is consequently a combination of several constants and depends on the concentration of the other metal, especially at the beginning of the reaction. By adding hydrogen peroxide, the reaction system will be extended at least for one reaction for each metal and even more reaction rate constants have to be considered. As a result, the kinetics of this system cannot be evaluated with the concepts of a shrinking core model and the Arrhenius law.

4. Conclusions

The kinetic investigation in the range of 25 °C to 85 °C led to a negative activation energy for the dissolution of silver. In the small temperature range from 25 °C to 45 °C, a positive activation energy was possible, but the data was insufficient to prove this. For higher leaching temperatures, a change in the leaching mechanism was possible, which was dominated by complex elementary reactions. In the future, further experiments should be conducted over smaller temperature and time ranges to determine a possible change in the reaction mechanism. The analysis of the data is difficult since little can be found in the literature about negative activation energy [21]. To work in the validity of the Arrhenius law, silver dissolution must be investigated without the influence of nickel. Due to the utilization of a recycled material, a pre-treatment was necessary.

For future kinetic investigations, it will be useful to study all possible chemical reactions. This is shown schematically in Figure 8 for the HNO_3–H_2O_2 system. The silver dissolution can occur on the one hand with nitric acid, where besides Ag^+ and water, nitrogen monoxide will also be formed. On the other hand, hydrogen peroxide can also react with silver to form Ag^+ and water. As the silver particles on the PTFE surface are in the micrometer range, H_2O_2 will decompose catalytically on the metal surface, leading to O_2 and H_2O formation and a slight temperature increase due to the exothermic reaction. Nickel can undergo reactions with nitric acid analogous to silver dissolution and with already dissolved silver to form nickel cations. Additionally, the oxidation of NO, which is formed by the metal–nitric acid reaction, to nitric acid should be mentioned as a side reaction.

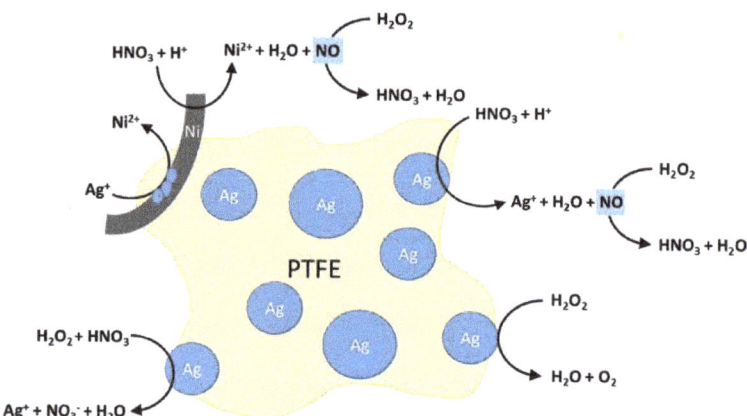

Figure 8. Schematic diagram for the possible reactions between silver, nitric acid, nickel, and hydrogen peroxide.

Author Contributions: Conceptualization and methodology: J.S., A.T., and K.H.; Writing—original draft preparation: J.S., A.T., and K.H.; Writing—review and editing: J.S., A.T., and K.H.; Supervision: B.F. and H.W.

Funding: The project upon which this publication is based was funded by the German Federal Ministry of Education and Research (BMBF) under Project Number 033R144.

Acknowledgments: Special thanks to Prof. Quicker and the Unit of Technologies of Fuels RWTH Aachen University, Germany for cryogenic milling and support. Another thanks goes to all project partners (Covestro Deutschland AG, Siegfried Jacob Metallwerke GmbH & Co KG, Department of Technical Thermodynamics RWTH University, and Institute for Non-Ferrous Metallurgy and Pure Materials TU Bergakademie Freiberg) for their support.

Conflicts of Interest: The authors declare no conflict of interest.

References

1. Sudoh, M.; Arai, K.; Izawa, Y.; Suzuki, T.; Uno, M.; Tanaka, M.; Hirao, K.; Nishiki, Y. Evaluation of Ag-based gas-diffusion electrode for two-compartment cell used in novel chlor-alkali membrane process. *Electrochim. Acta* **2011**, *56*, 10575–10581. [CrossRef]
2. Jörrisen, J.; Turek, T. Chlorherstellung mit Sauerstoffverzehrkathoden. *Chem. Unserer Zeit Bd* **2011**, *45*, 172–182. [CrossRef]
3. Schosseler, J.; Trentmann, A.; Alkan, G.; Friedrich, B.; Dressler, A.; Stelter, M.; Brück, S.; Dörner, H.; Bulan, A.; Douzinas, K.; et al. Silber-Recycling von Gasdiffusionselektroden aus der Chlor-Alkali-Elektrolyse. In *Berliner Recycling-und Rohstoffkonferent*; TK Verlag Karl Thomé-Kozmiensky: Neuruppin, Germany, 2018; pp. 233–251.
4. Arito, H.; Soda, R. Pyrolysis products of polytetrafluoroethylene and polyfluoroethylenepropylene with reference to inhalation toxicity. *Ann. Occup. Hyg.* **1977**, *20*, 247–255. [PubMed]
5. Rice, F.O.; Reiff, O.M. The thermal decomposition of hydrogen peroxide. *J. Phys. Chem.* **1927**, *31*, 1352–1356. [CrossRef]
6. Weiss, J. The catalytic decomposition of hydrogen peroxide on different metals. *Trans. Faraday Soc.* **1935**, *31*, 1547–1557. [CrossRef]
7. Seifert, S.; Thome, V.; Karlstetter, C. Elektrodynamische Fragmentierung—Eine Technologie zur Effektiven Aufbereitung von Abfallströmen, Berliner Recycling-und Rohstoffkonferenz, TK Verlag Karl Thomé-Kozmiensky, Neuruppin, Germany. 2014. Available online: http://www.vivis.de/phocadownload/Download/2014_rur/2014_RuR_431_438_Seifert.pdf (accessed on 20 December 2018).
8. Florea, M.V.A.; Ning, Z.; Brouwers, H.J.J. *Smart Crushing of Concrete and Activation of Liberated Concrete Fines*; University of Eindhoven, Department of the Built Environment, Unit Building Physics and Services: Eindhoven, The Netherlands, 2012.

9. Schubert, H. *Aufbereitung Fester Mineralischer Rohstoffe, Band 2, 3*; Auflage, VEB: Leipzig, Germany, 1996; pp. 125–172.
10. Mular, A.; Halbe, D.; Barratt, D. Mineral Processing Plant. *Des. Pract. Control* **2002**, *1*, 1069–1095.
11. Stansbie, J.H. The reaction of metals and alloys with nitric acid. *J. Soc. Chem. Ind.* **1913**, *XXXII*, 311–319. [CrossRef]
12. Martinez, L.L.; Segarra, M. Kinetecs of the dissolution of pure silver and silver-gold alloys in nitric solution. *Metall. Trans. B* **1993**, *24B*, 827–835. [CrossRef]
13. Levenspiel, O. *Chemical Reaction Engineering*, 2nd ed.; John Wiley & Sons, Int.: New York, NY, USA, 1972; pp. 357–373.
14. Jander, W. Reactions in the Solid State at High Temperatures. *Z. Anorg. Allgem. Chem.* **1927**, *163*, 1–30. [CrossRef]
15. Arrhenius, S. Über die Reaktionsgeschwindigkeit bei der Inversion von Rohrzucker durch Säuren. *Z. Phys. Chem.* **1889**, *4U*, 226–248. [CrossRef]
16. Smith, I.W.M. The Temperature-dependence of elementary reactions rates: Beyond Arrhenius. *Chem. Soc. Rev.* **2008**, *37*, 812–826. [CrossRef] [PubMed]
17. Vyazovkin, S. On the phenomenon of variable activation energy for condensed phase reactions. *New J. Chem.* **2000**, *24*, 913–917. [CrossRef]
18. Wotruba, H.; Weitkämper, L.; Steinberg, M. *Development of a New Dry Density Separator for Fine-Grained Materials*; The Australian Institute of Mining and Metallurgy (XXV IMPC): Carlton, Australia, 2010.
19. Muench, J.L.; Kruuv, J.; Lepock, J.R. A Two-Step Reversible-Irreversible Model Can Account for a Negative Activation Energy in an Arrhenius Plot. *Cyrobiology* **1996**, *33*, 253–259. [CrossRef] [PubMed]
20. Holleman, A.F.; Wilberg, N.; Wilberg, E. *Anorganische Chemie, Band 2, 103*; Auflage, De Gruyter: Berlin, Germany; Boston, MA, USA, 2016; p. 2025.
21. McKibben, M.A.; Tallant, B.A.; del Angel, J.K. Kinetics of inorganic arsenopyrite oxidation in acidic aqueous solutions. *Appl. Geochem.* **2008**, *23*, 121–135. [CrossRef]

© 2019 by the authors. Licensee MDPI, Basel, Switzerland. This article is an open access article distributed under the terms and conditions of the Creative Commons Attribution (CC BY) license (http://creativecommons.org/licenses/by/4.0/).

Article

Behavior of Waste Printed Circuit Board (WPCB) Materials in the Copper Matte Smelting Process

Xingbang Wan [1], Jani Fellman [1], Ari Jokilaakso [1,*], Lassi Klemettinen [1] and Miikka Marjakoski [2]

[1] Department of Chemical and Metallurgical Engineering, School of Chemical Engineering, Aalto University, P.O. Box 16100, 00076 Aalto, Finland; xingbang.wan@aalto.fi (X.W.); jani.fellman@aalto.fi (J.F.); lassi.klemettinen@aalto.fi (L.K.)

[2] Boliden Harjavalta, Teollisuuskatu 1, 29200 Harjavalta, Finland; miikka.marjakoski@boliden.com

* Correspondence: ari.jokilaakso@aalto.fi; Tel.: +358-50-313-8885

Received: 28 September 2018; Accepted: 29 October 2018; Published: 31 October 2018

Abstract: The amount of waste electrical and electronic equipment (WEEE) in the world has grown rapidly during recent decades, and with the depletion of primary ores, there is urgent need for industries to study new sources for metals. Waste printed circuit boards (WPCB) are a part of WEEE, which have a higher concentration of copper and precious metals when compared to primary ore sources. PCB materials can be processed using pyrometallurgical routes, and some industrial processes, such as copper flash smelting, have utilized this type of waste in limited amounts for years. For the purpose of recycling these materials through smelting processes, this work studied the behavior of WPCB scrap when dropped on top of molten slag. A series of experiments was carried out during this research at a temperature of 1350 °C, in an inert atmosphere with different melting times. The time required for complete melting of the PCB pieces was 2–5 min, after which molten alloy droplets containing Cu, Pb, Sn, Ni, Au, and Ag formed and started descending toward the bottom of the crucible. The ceramic fraction of the PCB material mixed with slag and the polymer fraction was pyrolyzed during the high-temperature experiments. The results give an understanding of PCB melting behavior and their use as a part of the smelting furnace feed mixture. However, more research is needed to fully understand how the different elements affect the process as the amount of PCB in the feed increases. The physical behavior and distribution of PCB materials in fayalite slag during the smelting process are outlined, and the results of this work form a basis for future studies about the chemical reaction behavior and kinetics when PCB materials are introduced into the copper smelting process.

Keywords: WPCB; melting behavior; flash smelting

1. Introduction

Recent advancements in electronic and electrical technologies have caused rapid growth in the amount of waste electrical and electronic equipment (WEEE) in the last two decades. This is largely credited to global trends in the number of users, technological advances, efficiency, as well as social and economic development. Additionally, the increase in average disposable income has reduced the cycle times of electronic equipment. Globally generated e-waste amounted to 44.7 million metric tons in 2016 [1]. The amount of collected e-waste is expected to surge rapidly, which would further increase the need for efficient recycling solutions. The plentiful amount of precious metals (PMs) in WEEE are especially becoming of greater importance and focus [2]. In today's circular economy, an essential challenge is how to incorporate the rising amount of WEEE into existing industrial processes as a secondary feed material with additional value [3,4].

With the continuous growth of demand for metals, and parallel degradation and depletion of primary ores, there is an increasing need for secondary raw materials. Some studies have been conducted regarding the behavior and recovery possibilities of different metals in WPCBs with different kinds of metallurgical processes, such as leaching in different media [5,6]. During the last decade, the practice of using WPCB as a feed material in copper smelting has also become a focus of research and development. WPCBs' high PM content has incensed commercial smelter operators toward developing existing processes for accepting WEEE and WPCB. Their use has already been implemented in several integrated processes around the world, such as the Umicore Hoboken plant in Belgium [7] or the Boliden plant in Sweden [8]. In Japan and Korea, the Mitsubishi smelting process has been developed for adapting WEEE as a secondary material source [9]. A few commercial smelting operations have moved towards charging WEEE into a copper flash-smelting furnace (FSF) as an added material; one example is the Horne smelter in Canada [10]. This process differs from direct reduction metal-slag systems, as the flash furnace maintains a matte–slag environment. It also has variances to other matte–slag systems, as WEEE material is charged together with the primary copper concentrate.

Experimental investigations on the thermodynamic properties and distributions of minor (WEEE) elements in different oxygen partial pressures have been carried out recently by several researchers, mostly for metal-slag systems [11–18] but also in matte–slag equilibria [19]. Some model predictions for PCB's minor element distributions, including Pb, Sn, Ni and Zn, have also been proposed [20]. The decomposition and oxidation rates of concentrate material in the FSF reaction shaft have been extensively studied by several researchers [21–27], and some modeling works using fluid dynamic software were finished based on these results.

However, the distribution and kinetic behavior of WEEE/WPCB in the flash-smelting process still lack investigation. More research is needed to evaluate the compatibility of WPCB with traditional FS processes, regarding the capacity of simultaneous smelting of WPCB and copper concentrate.

This research work expands the work done previously [28], which studied the matte–slag reaction system. The aim of this research is to find out the behavior and distribution of WPCB scrap in the slag phase when injected on top of the slag from the roof of a settler where the flight time is short and atmosphere is no longer oxidizing.

2. Materials and Methods

2.1. Materials

The WPCB pieces (shown in Figure 1a) were cut from a disassembled mobile-phone motherboard. The size of the pieces was approximately 5 mm × 5 mm × 1 mm. The comparison samples, referred to as "synthetic PCB" (Figure 1b) were comprised of copper–nickel alloy (Alfa Aesar, Karlsruhe, Germany, 0.51 mm thick, 33% copper and 67% nickel) pieces wrapped in aluminum foil (Fisher Scientific, Pittsburgh, PA, USA, 24 µm thick). The size of these comparison pieces was 5 mm × 5 mm × 1 mm as well.

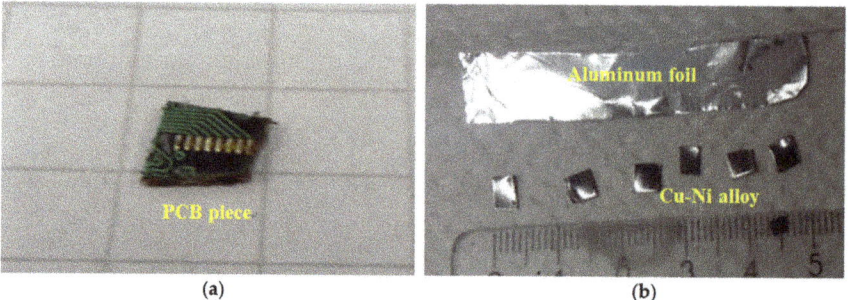

Figure 1. Materials utilized in this study: (**a**) printed circuit-board (PCB) piece; (**b**) synthetic PCB pieces.

The synthetic slag was a mixture of hematite (Alfa Aesar, Kandel, Germany, 99.998% purity), silica (Umicore, Balzers, Liechtenstein, 99.99% purity), and metallic iron (Alfa Aesar, Kandel, Germany, 99.5% purity), with mass ratios of 50% Fe_2O_3, 30% SiO_2, and 20% Fe. The amount of slag used for each experiment was 1.0 g. The slag–powder mixture was placed into a silica crucible and introduced into the furnace.

2.2. Apparatus

The experimental apparatus used in this research is shown in Figure 2. It is comprised of a vertical tube furnace (Lenton LTF 16/-/450, Lenton, Nottingham, UK) which is equipped with 4 silicon carbide (SiC) heating elements and a Eurotherm PID temperature controller (Eurotherm, Ashburn, VA, USA). The furnace working tube is made of impervious pure alumina (Frialit AL 23; Friatec AG, Germany) with 45/38 mm OD/ID. On top of the working tube, a lid equipped with the water-cooling system has three small holes. From the first hole, an S-type Pt/Pt-10Rh thermocouple (Johnson–Matthey Noble Metals, London, UK) inside a protective alumina tube was inserted into the furnace. The output voltage of the thermocouple was measured with a Keithley 2000DMM multimeter (Keithley, Solon, OH, USA). A Pt100 resistant thermometer (SKS-Group, Vantaa, Finland) was connected to a Keithley 2010DMM multimeter for performing cold-junction compensation. Temperature data were logged with LabVIEW software (National Instruments, Austin, TX, USA). From the second hole, a platinum wire (inside a 22 mm alumina guiding tube, see Figure 2) was inserted for raising the sample to the reaction zone of the furnace. The third hole was equipped with a re-sealable rubber cork for dropping the PCB pieces into the alumina guiding tube and eventually into the silica crucible containing the molten slag.

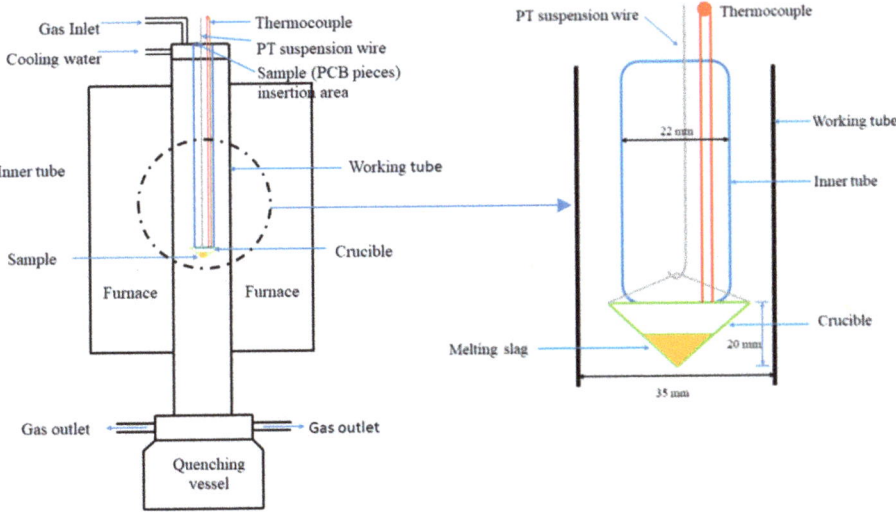

Figure 2. A schematic of the experimental furnace and details of the hot zone. During quenching of the sample, the quenching vessel was lifted to cover the bottom end of the furnace work tube for preserving the inert-gas atmosphere.

2.3. Procedure

Before the experiments, the temperature inside the working tube of the furnace was measured, the area with the highest temperature was defined as the hot zone, and its temperature was set to 1350 °C. The temperature was kept constant throughout the experiments. After introducing the crucible–sample–platinum basket assembly into the cold zone of the furnace and sealing the lower end of the furnace work tube, an argon (Aga Linde, Espoo, Finland, 99.999% purity) gas flow was turned on and given 10 min to fill the tube to exclude oxygen. Then, the crucible with the slag was lifted up to the hot zone with the Pt suspension wire, where it was kept for 5 min for the slag to melt. After this, a piece of WPCB was dropped into the crucible from the top of the furnace through a small resealable hole. The melting time was measured from the moment the piece was dropped through the hole to the moment the sample was quenched in 0 °C ice water in the quenching vessel.

Two different experimental series were conducted. In the first series, actual PCB pieces were used, and the melting times were 25, 60, 120, and 300 s. Each time interval experiment was repeated at least twice to ensure reliable results. In the second series, the behavior of synthetic PCB (copper–nickel alloy wrapped in aluminum foil) pieces was studied with the same procedure.

The quenched samples in their cone-shaped crucibles were mounted in epoxy resin and then cut in half from a location that seemed to contain PCB material in visual inspection. The obtained semicircle-shaped pieces were ground and polished using traditional dry metallographic techniques. Sufficient conductivity on the surface was ensured with a carbon coating (Leica EM SCD050, Leica Microsystems, Wetzlar, Germany). The samples were analyzed with a Tescan MIRA 3 Scanning Electron Microscope (SEM, Tescan, Brno, Czech Republic) equipped with an UltraDry Silicon Drift Energy Dispersive X-Ray Spectrometer and NSS Microanalysis Software (EDS, Thermo Fisher Scientific, Waltham, MA, USA). The acceleration voltage and beam current of the SEM were 15 kV and 10 nA, respectively. The external standard materials used for elemental analysis were olivine (O), hematite (Fe), quartz (Si), and pure metals for Cu, Pb, Sn, Ni, Au, Ag, and PGMs.

3. Results and Discussion

The experiments were conducted in an inert argon atmosphere for verifying the behavior of the unreacted WPCB material in the settler area without free oxygen. During the introduction of the PCB sample pieces into the furnace, tiny amounts of oxygen may have entered the furnace, but the effect of this was considered negligible. The synthetic PCB material was selected for investigating the melting and settling behavior of only the metallic PCB fraction.

3.1. Basic Matte–Slag Reaction during the Flash-Smelting Process

During the flash-smelting process, most of the concentrate particles react with oxygen and become molten due to exothermic reactions. The slag phase forms and separates from the matte phase in the settler area through the following reactions, and, during the processes described in Equations (1)–(4), iron sulfide is continuously removed from the matte phase, the copper content in the matte phase increases, and more slag appears.

$$3\ Fe_3O_4\ (s) + FeS_{(matte)} = 10\ FeO_{(slag)} + SO_2\ (g) \tag{1}$$

$$Cu_2S\ (l) + O_2\ (g) = 2\ Cu_{(metal)} + SO_2\ (g) \tag{2}$$

$$2\ Fe_2O_3 + S_{(matte)} = 4\ FeO_{(slag)} + SO_2\ (g) \tag{3}$$

$$3\ Fe_3O_4\ (s) + FeS_{(matte)} + 5\ SiO_2 = 5\ Fe_2SiO_{4(slag)} + SO_2\ (g) \tag{4}$$

However, during the industrial process, some particles do not react completely and fall to the settler area as partly or fully unreacted. Kim and Themelis [29], Jokilaakso et al. and Ahokainen and Jokilaakso [23,30] found out how larger particles fragment into small ones in the reaction shaft, and the result showed that unreacted particles cannot be ignored. In that case, the decomposition of these particles may start in the settler, where oxygen pressure is low.

The same phenomenon may happen when adding WPCB material into the concentrate during the flash-smelting process, and pieces of WPCB can directly drop into the settler and melt only in the slag phase. In addition, WPCB scrap can be directly added into the settler from the roof of the settler area.

In order to investigate the melting behavior of the WPCB in the slag phase, two series of experiments were done in this study to investigate the physical behavior of the WPCB in the fayalite slag.

3.2. Behavior of WPCB Samples in the Slag Phase

Figure 3a–d shows SEM micrographs of the samples and the structure change of the WPCB pieces as a function of time in argon atmosphere. From the overview of the 25 s WPCB sample, a visible WPCB structure on the top of molten slag can be clearly observed, so it can be concluded that 25 s of contacting time is far too short to melt the whole WPCB sample. With the contacting time of 60 s, the WPCB piece has disintegrated to some extent, and the droplets found in this sample are primarily copper mixed with minor trace metals such as nickel, lead and tin. However, 60 s is not enough to allow the copper alloy droplets to settle through the slag layer towards the bottom.

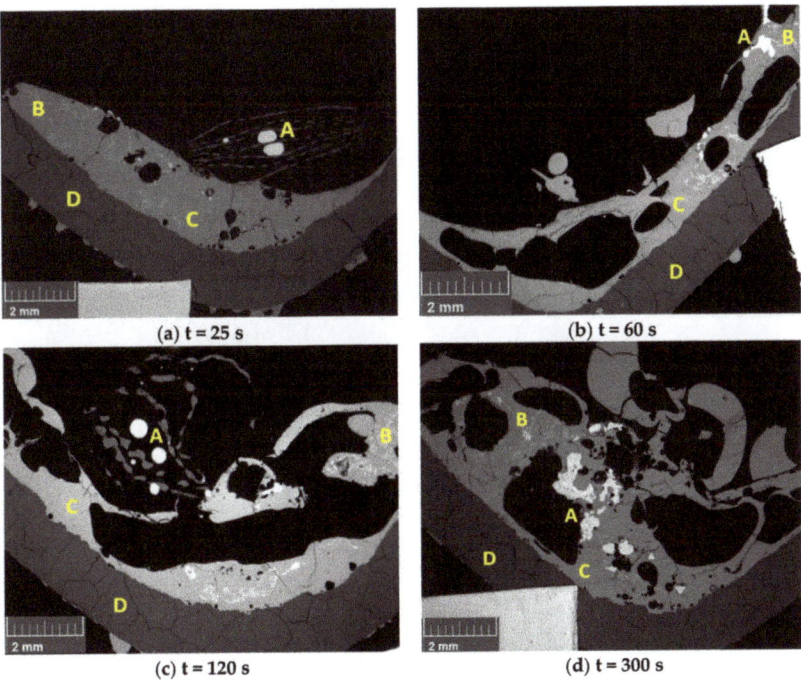

Figure 3. Scanning electron microscope (SEM)—backscattered electron (BSE) micrographs of the waste PCB (WPCB) specimens at different melting times. The letters indicate (**a**) metal droplets from PCB, (**b**) unreacted iron from the slag mixture, (**c**) liquid slag, and (**d**) silica crucible.

After 120 s of melting, shown in Figure 3c, a remainder of the WPCB structure with copper droplets could be found above the slag, and there were also multiple metal droplets inside the slag, which meant the melting and settling process was not finished. Since the plastic and ceramic structure of the WPCB was still visible after 120 s, it was most probably present in the 60 s sample as well; however, this cross-section in Figure 3b does not show it. Figure 3d shows an overview of the 300 s sample, where the WPCB structure has already melted entirely, and the small metal droplets coalesced into bigger-size droplets. At that time, the melting process of the PCB was finished, but the settling process of the metal droplets toward the bottom of the silica crucible was still ongoing.

Due to the short time allowed for the slag mixture to melt before introducing the WPCB samples, some metallic iron is found from the slag (indicated with B in Figures 3a–d and 5a–d). This indicates that the slag was still far from equilibrium. The aim of this study was to investigate the behavior of the WPCB pieces in a molten-slag bath, and therefore the state of the slag was not of great importance, as long as it was molten. When the melting time of the slag–powder mixture was prolonged, a considerable amount of slag dripped through the crucible (visible in Figure 3a), reducing the volume of the slag inside the crucible. If the slag volume were too low, i.e., only a slag "film" remained on the surface of the crucible, observations about the settling behavior of the metal droplets originating from the PCB pieces could not be made.

For some of the metal droplets, SEM-EDS analysis areas are shown in Figure 4a–d. The compositions (in wt%) of analysis areas 1–8 in Figure 4a–d are listed in Table 1.

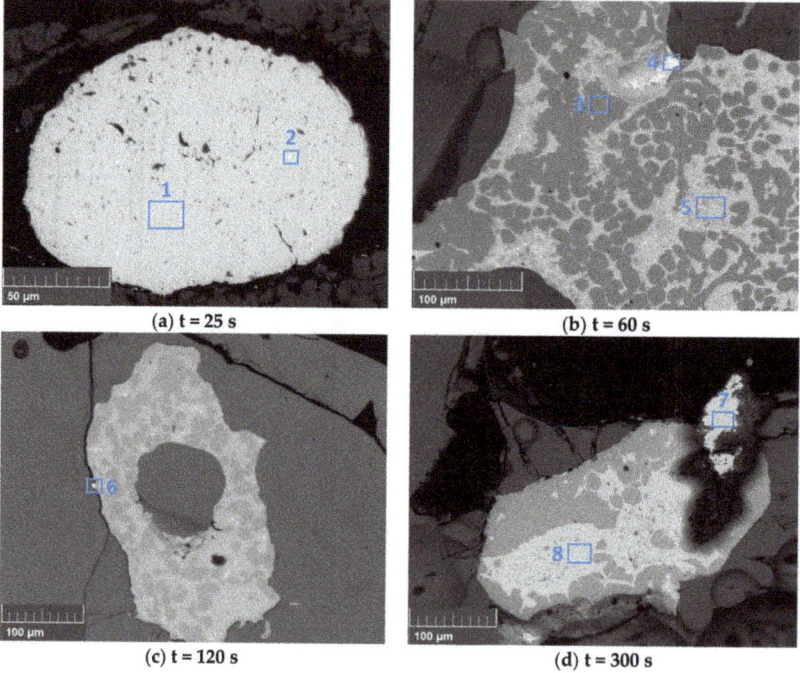

Figure 4. SEM-EDS analysis areas of metal droplets originating from PCB pieces. 1: Cu, 2: Cu–Ni alloy, 3: Fe, 4: Cu–Pb–Sn alloy, 5: Cu–Sn alloy, 6: Cu–Fe–Pb alloy, 7: Pb, 8: Cu–Fe alloy.

The metal-alloy droplets, originating from the WPCB pieces, had different compositions and could be roughly divided into Cu-rich and Fe-rich alloys. According to the SEM-EDS results, these droplets contained copper, nickel, lead, tin, and minor traces of gold, silver, and platinum group metals. Different alloy compositions are commonly found from the same large droplets, as shown in Figure 4b–d. In was observed that Cu-rich areas were more likely to contain valuable and precious metals than Fe-rich areas.

Table 1. EDS analysis results (in wt%) of the metal droplets shown in Figure 4a–d.

Area	Cu	Fe	Ni	Pb	Sn	Ag	Au
1	95.7	<0.1	0.23	0.8	0.1	0.1	0
2	90.9	<0.1	8.1	0.1	0	0	0
3	0	95.7	0	0	0	0	0
4	42.1	0	0	37.3	16.0	0	0
5	80.9	0	0	0	20.5	0.8	0
6	33.2	32.8	0	4.3	0	0	0
7	2.3	0	0	85.4	0	0	<0.1
8	44.4	57.6	0	0	0	0.4	0

When observing the WPCB sample results, the melting of the WPCB structure started at the outer copper layers and moved toward the center of the sample. The WPCB pieces were found to gradually melt in 2–5 min. After 5 min, WPCB structure seemed to be completely molten and broken down. When comparing this time to the theoretical results using a lumped capacitance method [31] for calculating melting time, the experimentally observed melting time of around 5 min seemed appropriate.

3.3. Behavior of Synthetic PCB Samples in the Slag Phase

Figure 5a–d shows micrographs of the synthetic PCB samples. After 25 s, the synthetic sample remained on top of the molten slag and melted more thoroughly than the WPCB sample at the same time. The large metal-alloy droplet already started to mix into the slag phase. After longer contacting times, from 60 to 300 s, basically no further changes took place, but the molten and almost undivided metal droplets settled through the slag without considerable mixing into the slag or disintegration. Without the ceramic and plastic fraction of the WPCB samples, the synthetic PCB melted much faster, and remained as one metal lump settling through the slag. In the WPCB samples, the metallic fractions were scattered throughout the plastic–ceramic support structure of the circuit board, and therefore a larger metal droplet most likely formed only when all the individual droplets settled through the slag layer to the bottom of the crucible.

During the WPCB melting, the complex material layers and their different properties made the melting process very non-uniform. Copper layers and lead–tin solders melt almost instantaneously, while ceramic structures and plastic parts take more time to melt, which increases the total melting time.

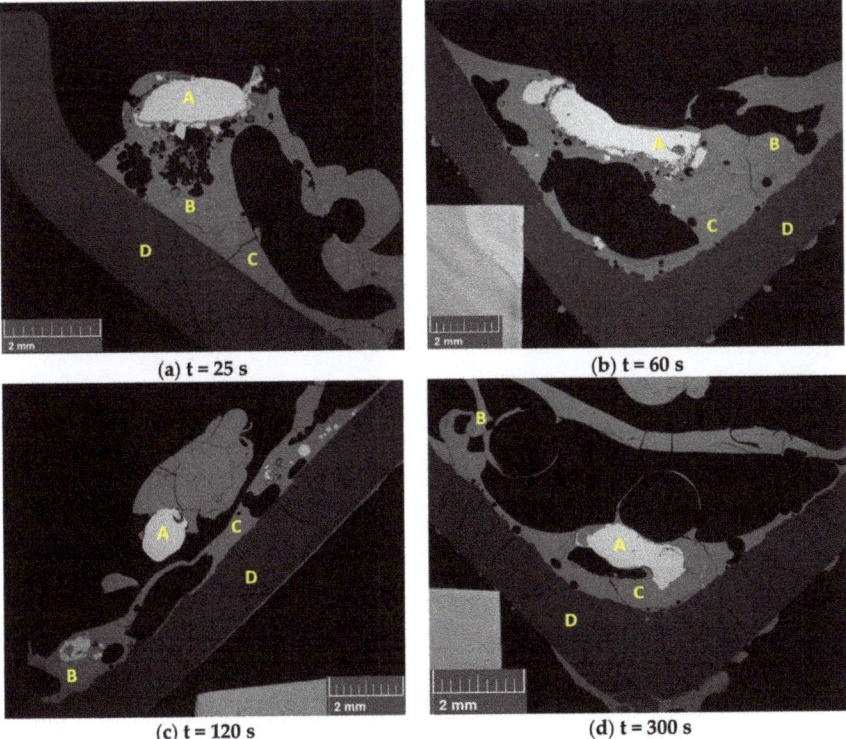

Figure 5. SEM-BSE micrographs of the synthetic PCB specimens after different melting times. The letters indicate (**a**) alloy droplets from PCB, (**b**) unreacted iron from the slag mixture, (**c**) liquid slag, and (**d**) silica crucible.

Furthermore, pyrolysis of plastics is also a factor in melting behavior. The ceramic fraction dissolved into the slag phase, while the plastic fraction was not found in the EDS analyses. Some black, charred residue was observed in the quenching vessel when opening the gas seal for quenching, which suggests that the plastic fraction was pyrolyzed and exited into the gas phase during the process.

4. Conclusions

In this study, the behavior of WPCB and synthetic PCB pieces were examined using silica crucibles that were suspended in a vertical tube furnace. The PCB pieces were dropped on top of a molten slag bath and given time to melt or react for 25, 60, 120, and 300 s. After each time interval, the sample–crucible assembly was quenched in ice water, followed by analysis using SEM and EDS.

The most abundant metal in the WPCB scrap was copper, followed by the base metals of lead, tin, and nickel, plus gold and silver as precious metals; all these metals were found in the metal droplets. The metallic fraction of the WPCB formed droplets that tended to settle toward the bottom of the silica crucible due to the density difference between the metals and slag.

The melting time between the WPCB and synthetic PCB were found to be different, mainly due to the ceramic and plastic parts in the WPCB. The ceramic fraction dissolved into the slag phase, and the plastic fraction experienced pyrolysis and formed gases, solid carbon-based compounds, or liquid oils. The additives in the plastic fraction, mainly halogenic fire retardants and possible carcinogenic compounds, may cause both production and safety problems on an industrial scale and, consequently, have to be treated properly.

Based on the observations from the micrographs and EDS analyses, the following conclusions can be drawn:

(1) In this research, 5 min was enough for the samples to completely melt, but the settling process needs more time. In industrial flash-smelting operations, the amount of WPCB scrap feed is much higher and the required melting time needs further investigation and simulation.
(2) WPCB materials can be added into the FS process, and the metal fraction melts readily and forms droplets that settle through the slag layer and mix into the matte phase. However, the ceramic and plastic fractions of WPCB should be removed before; especially the plastic fraction that may cause both production and safety problems. The possible pyrolysis or combustion behavior in the concentrate suspension and slag reduction in the settler also require thorough investigation before larger quantities of WPCB can be safely used as raw material in an industrial-scale process. If not removed in a pretreatment process, the pyrolysis products of the plastic fraction would change the composition of the flue gas, and the recovery and use of flue gas for sulfuric acid production would be influenced by these organic impurities.
(3) The ceramic fraction of the WPCB dissolves into the slag phase after melting, so the slag volume increases if WPCB materials are added in larger amounts.
(4) The precious metals (for example Au, Ag, and Pt) in the WPCB enter the matte phase in the industrial FS process after settling, and they are finally recovered from the anode slime.

Author Contributions: X.W. wrote most of the manuscript and supervised the experiments with L.K.; J.F. performed the experiments, prepared the samples for analysis, and conducted the SEM-EDS analyses with X.W. and L.K.; L.K. and M.M. revised the manuscript; A.J. conceptualized the research idea, supervised the study, helped write the manuscript, and edited and proofread the draft.

Funding: This paper has been supported by TEKES (Business Finland) (CMEco, grant number 7405/31/2016) and by Boliden Harjavalta Oy, and made use of the Academy of Finland's RawMatTERS Finland Infrastructure (RAMI) based at Aalto University. X.W. also received financial support from the China Scholarship Council.

Conflicts of Interest: The authors declare no conflicts of interest.

References

1. Radulovic, V. Portrayals in Print: Media Depictions of the Informal Sector's Involvement in Managing E-Waste in India. *Sustainability* **2018**, *10*, 966. [CrossRef]
2. Baldé, C.P.; Forti, V.; Gray, V.; Kuehr, R.; Stegmann, P. *The Global E-Waste Monitor 2017: Quantities, Flows and Resources*; United Nations University: Bonn, Germany; International Telecommunication Union: Geneva, Switzerland; International Solid Waste Association: Vienna, Austria, 2017.
3. Ghosh, B.; Ghosh, M.K.; Parhi, P.; Mukherjee, P.S.; Mishra, B.K. Waste Printed Circuit Boards recycling: An extensive assessment of current status. *J. Clean. Prod.* **2015**, *94*, 5–19. [CrossRef]
4. Khaliq, A.; Rhamdhani, M.A.; Brooks, G.; Masood, S. Metal extraction processes for electronic waste and existing industrial routes: A review and Australian perspective. *Resources* **2014**, *3*, 152–179. [CrossRef]
5. Reyes-Valderrama, M.I.; Salinas-Rodriguez, E.; Montiel-Hernandez, J.F.; Rivera-Landero, I.; Cerecedo-Saenz, E.; Hernandez-Avila, J.; Arenas-Flores, A. Urban Mining and Electrochemistry: Cyclic Voltammetry Study of Acidic Solutions from Electronic Wastes (Printed Circuit Boards) for Recovery of Cu, Zn, and Ni. *Metals* **2017**, *7*, 55. [CrossRef]
6. Zhang, Y.; Liu, S.; Xie, H.; Zeng, X.; Li, J. Current status on leaching precious metals from waste printed circuit boards. *Procedia Environ. Sci.* **2012**, *16*, 560–568. [CrossRef]
7. Hagelüken, C. Recycling of electronic scrap at Umicore's integrated metals smelter and refinery. *Erzmetall* **2006**, *59*, 152–161.
8. Lennartsson, A.; Engström, F.; Samuelsson, C.; Björkman, B.; Pettersson, J. Large-Scale WEEE Recycling Integrated in an Ore-Based Cu-Extraction System. *J. Sustain. Metall.* **2018**, *4*, 222–232. [CrossRef]
9. Ariizumi, M.; Takagi, M.; Inoue, O.; Oguma, N. Integrated processing of e-scrap at Naoshima smelter and refinery. In Proceedings of the Copper 2016, Kobe, Japan, 13–16 November 2016; Volume 6, p. RW 1-2.
10. Wood, J.; Creedy, S.; Matusewicz, R.; Reuter, M. Secondary copper processing using Outotec Ausmelt TSL technology. In Proceedings of the MetPlant 2011, Perth, Australia, 8–9 August 2011; pp. 460–467.
11. Anindya, A. Minor Elements Distribution during the Smelting of WEEE with Copper Scrap. Ph.D. Thesis, School of Civil, Environmental & Chemical Engineering, RMIT University, Melbourne, Australia, 30 July 2012.
12. Anindya, A.; Swinbourne, D.; Reuter, M.A.; Matusewicz, R. Indium distribution during smelting of WEEE with copper scrap. In Proceedings of the European Metallurgical Conference (EMC 2011), Düsseldorf, Germany, 26–29 June 2011; Volume 1, pp. 3–14.
13. Tran, T.; Wright, S.; Sun, S. Activity of lead in copper matte at very low lead concentration. *High Temp. Mater. Process.* **2013**, *32*, 197–206. [CrossRef]
14. Fan, Y.; Gu, Y.; Shi, Q.; Xiao, S.; Jiang, F. Experimental Study on Smelting of Waste Smartphone PCBs Based on Al_2O_3-FeO_x-SiO_2 Slag System. In Proceedings of the 10th International Conference on Molten Slags, Fluxes and Salts, Seattle, WA, USA, 22–25 May 2016; Springer: Cham, Switzerland, 2016; pp. 203–210.
15. Klemettinen, L.; Avarmaa, K.; Taskinen, P.; Jokilaakso, A. Behavior of nickel as a trace element and time-dependent formation of spinels in WEEE smelting. In Proceedings of the Extraction 2018, Ottawa, ON, Canada, 26–29 August 2018; pp. 1073–1082.
16. Avarmaa, K.; O'Brien, H.; Taskinen, P. Equilibria of Gold and Silver between Molten Copper and FeO_x-SiO_2-Al_2O_3 Slag in WEEE Smelting at 1300 °C. In Proceedings of the 10th International Conference on Molten Slags, Fluxes and Salts, Seattle, WA, USA, 22–25 May 2016; Springer: Cham, Switzerland; pp. 193–202.
17. Avarmaa, K.; Yliaho, S.; Taskinen, P. Recoveries of rare elements Ga, Ge, In and Sn from waste electric and electronic equipment through secondary copper smelting. *Waste Manag.* **2018**, *71*, 400–410. [CrossRef] [PubMed]
18. Klemettinen, L.; Avarmaa, K.; Taskinen, P. Trace Element Distributions in Black Copper Smelting. *Erzmetall* **2017**, *70*, 257–264.
19. Avarmaa, K.; Johto, H.; Taskinen, P. Distribution of precious metals (Ag, Au, Pd, Pt, and Rh) between copper matte and iron silicate slag. *Metall. Mater. Trans. B* **2016**, *47*, 244–255. [CrossRef]
20. Fan, Y.; Gu, Y. Prediction of the distribution of minor elements in sulfide smelting of waste PCBs. *Metal. Int.* **2014**, *19*, 85.
21. Jorgensen, F.R.A. Single particle combustion of chalcopyrite. *Proc. Australas Inst. Min. Metall.* **1983**, *288*, 37–46.

22. Chaubal, P.C.; Sohn, H.Y. Intrinsic kinetics of the oxidation of chalcopyrite particles under isothermal and nonisothermal conditions. *Metall. Trans. B* **1986**, *17*, 51–60. [CrossRef]
23. Jokilaakso, A.; Suominen, R.; Taskinen, P.; Lilius, K. Oxidation of chalcopyrite in simulated suspension smelting. *Trans IMM Sect. C* **1991**, *100*, C79–C90.
24. Jokilaakso, A. Removal of antimony and arsenic from impure copper concentrates under simulated flash smelting reaction shaft conditions. *Acta Polytech. Scand. Chem. Technol. Metall. Ser.* **1992**, *205*, 5–55.
25. Peuraniemi, E.; Jokilaakso, A. Reaction sequences in sulphide particle oxidation. In Proceedings of the TMS Annual Meeting 2000, Nashville, TN, USA, 12–16 March 2000; pp. 173–187.
26. Peuraniemi, E.; Järvi, J.; Jokilaakso, A. Behaviour of copper matte particles in suspension oxidation. In Proceedings of the Copper 99, Phoenix, AZ, USA, 10–13 October 1999; pp. 463–476.
27. Perez-Tello, M.; Sohn, H.Y.; St Marie, K.; Jokilaakso, A. Experimental investigation and three-dimensional computational fluid-dynamics modeling of the flash-converting furnace shaft: Part I. Experimental observation of copper converting reactions in terms of converting rate, converting quality, changes in particle size, morphology, and mineralogy. *Metall. Mater. Trans. B* **2001**, *32*, 847–868.
28. Guntoro, P.; Jokilaakso, A.; Hellstén, N.; Taskinen, P. Copper Matte—Slag Reaction Sequences and Separation Processes in Matte Smelting. *J. Min. Metall. Sect. B Metall.* **2018**, in press.
29. Kim, Y.H.; Themelis, N.J. Effect of phase transformation and particle fragmentation on the flash reaction of complex metal sulphides. In Proceedings of the Reinhardt Schuhmann International Symposium on Innovative Technology and Reactor Design in Extraction Metallurgy, Warrendale, PA, USA, 9–12 November 1986; pp. 349–369.
30. Ahokainen, T.; Jokilaakso, A. Numerical Simulation of the Outokumpu Flash Smelting Furnace Reaction Shaft. *Can. Metall. Q.* **1998**, *37*, 275–283. [CrossRef]
31. Fellman, J. Printed Circuit Board (PCB) Scrap Melting and Mixing with Molten Fayalite Slag. Master's Thesis, School of Chemical Engineering, Aalto University, Espoo, Finland, 10 September 2018.

© 2018 by the authors. Licensee MDPI, Basel, Switzerland. This article is an open access article distributed under the terms and conditions of the Creative Commons Attribution (CC BY) license (http://creativecommons.org/licenses/by/4.0/).

Article

Thermal Processing of Jarosite Leach Residue for a Safe Disposable Slag and Valuable Metals Recovery

Minna Rämä [1,*], Samu Nurmi [1], Ari Jokilaakso [1], Lassi Klemettinen [1], Pekka Taskinen [1] and Justin Salminen [2]

[1] Department of Chemical and Metallurgical Engineering, School of Chemical Engineering, Aalto University, Kemistintie 1, P.O. Box 16100, FI-00076 Aalto, Finland; samu.nurmi@aalto.fi (S.N.); ari.jokilaakso@aalto.fi (A.J.); lassi.klemettinen@aalto.fi (L.K.); pekka.taskinen@aalto.fi (P.T.)
[2] Boliden Kokkola Oy, P.O. Box 26, 67101 Kokkola, Finland; justin.salminen@boliden.com
* Correspondence: minna.rama@aalto.fi; Tel.: +358-50-526-4301

Received: 4 September 2018; Accepted: 20 September 2018; Published: 21 September 2018

Abstract: In electrolytic production of zinc, the iron levels in the solutions are controlled by precipitation of jarosite or goethite. These precipitates also co-precipitate unrecovered valuable metals (Zn, Pb, Cu, Ag) as well as elements of concern (As, Cd, Hg). After stabilization, the residues are traditionally landfilled. This work investigates pyrometallurgical treatment of jarosite residue to convert the material into reusable clean slag and to recover the valuable metals within the residue. The pyrometallurgical treatment is divided into two functional steps. First, the material is melted in an oxidizing atmosphere, after which the oxide melt is reduced to produce an inert, clean slag. Then, a liquid metal or speiss phase collects the valuable metals, such as silver. The aim was to examine the optimal process conditions for reaching the target values for remaining metals in the slag; Pb < 0.03 wt %, Zn < 1 wt %. As a conclusion, the limiting factor in sulfur, lead, and zinc removal is the contact between the oxidizing or reducing gas and the molten sample. The mass transfer and volatile metals removal were significantly improved with a gas lance installation. The improved gas-liquid interaction enabled the first steps of gas flow rate optimization and ensured the sufficiently low end-concentrations of the aforementioned elements.

Keywords: jarosite residue; pyrometallurgy; circular economy; slag valorization; metal recovery

1. Introduction

Approximately 85% of zinc worldwide is currently produced via RLE (Roasting-Leaching-Electrowinning) processes, which generate 0.5–0.9 tonnes of dry jarosite residue for every tonne of zinc produced, as iron level in the leaching solution is controlled by precipitation [1–3]. The iron precipitation is also possible by producing goethite, paragoethite, or hematite; however, the jarosite process is used most commonly worldwide [4]. The chemical formula of jarosite is $ZFe_3(SO_4)_2(OH)_6$, where Z represents either Ag^+, H_3O^+, K^+, Li^+, Na^+, NH_4^+, or $\frac{1}{2} Pb^{2+}$ [5]. On an industrial scale, typical cations used for precipitation are Na^+, K^+, and NH_4^+. Jarosite leach residues may also contain unrecovered base metals (Zn, Pb, Cu), critical elements (In, Ga, Ge, Sb), precious metals (Ag and Au), and elements of concern (As, Cd, Hg) [2,6,7]. These iron-rich leach residues are not suitable for use because of their classification as hazardous waste due to the heavy metals they contain [8].

Landfilling of iron rich sludge is still allowed in many countries; hence, stockpiling the waste is still commonly practiced [2]. However, the overall goal to improve yields and resource efficiency, legislative matters, as well as the aim towards a circular economy have increased the interest in processing these residues into a usable form. The goal is to produce an inert slag that could be used for construction purposes and, simultaneously, to recover the valuable metals from the residue [9,10].

Some production sites have already started to treat the generated residues with different pyrometallurgical technologies [2,11,12].

Several driving forces for process development, including both economic and environmental aspects, have been recognized. First of all, large amounts of already stockpiled or continuously produced iron-rich residues could be utilized, for example in road building or other construction material applications, instead of being disposed of [11,13]. The availability of land for this kind of hazardous waste landfilling is also limited, and causes additional costs [10,13,14]. In addition, recycling of the valuable metals within the residues is hindered due to the present practice, which results in these metals ending up in the landfills. However, the global aim for more efficient utilization of natural resources necessitates the use of residues as secondary raw material sources [2,10].

Pyrometallurgical treatment of jarosite leach residue has proved to be the most promising processing route due to its ability to both produce a clean slag and to recover most of the valuable metals. However, more detailed knowledge of the processing possibilities, as well as the effects of different processing parameters and variables, are required for process development and optimization.

A computational equilibrium model [15] created with MTDATA software (National Physical Laboratory, Teddington, UK) [16] and Mtox database (Version 8.2) [17] was used to do a preliminary scope for suitable conditions for the thermal treatment of jarosite residue. The model [15], earlier studies [18], and industrial processes [12] have proved that the pyrometallurgical treatment of this kind of residue can be divided into two functional steps. The first step of the treatment is conducted under oxidizing conditions to thermally decompose the material, releasing the sulfates and OH-groups, and to melt the material. During the second step, the formed oxide melt is reduced and the volatile elements are removed to the gas phase. Thereby, a clean slag and a liquid metal alloy are formed. The formation of the metal alloy during the reduction stage plays an important role due to its ability to collect various low-concentration metals from the material, such as Ag, As, and Sb [19].

High-temperature (1400 °C) experiments were conducted in a laboratory with industrial jarosite samples to investigate how to convert the jarosite residue into a clean (Pb < 0.03 wt %, Zn < 1 wt %) and inert slag product and to determine in what conditions the volatile metals and elements are extracted from the residue. The pyrometallurgical treatment in laboratory-scale showed that the targets can be reached with selected process conditions [20]. However, due to the low sample masses in the earlier study, it was not possible to conduct detailed elemental analyses for the samples. Thus, in this work, larger sample sizes will be used, allowing full chemical analysis to be performed as well. Different process parameter (treatment time, gas flow rates, gas lance/no gas lance) values were tested to determine the optimal conditions for reaching the targeted results.

The microstructures and phase compositions were determined using a scanning electron microscope equipped with an energy dispersive spectrometer (SEM-EDS). The inductively coupled plasma-optical emission spectroscopy (ICP-OES) method was used to determine the chemical composition of the bulk samples.

2. Materials and Methods

High-temperature (1400 °C) laboratory-scale experiments were conducted with industrial jarosite samples (about 15 wt % Fe, 4 wt % Zn, and 3 wt % Pb). The exact initial composition of the test material is omitted, and it may also have some variation due to the heterogeneous nature of industrial residue.

The test material was dried at 80 °C for 18 h to reduce its moisture content. After drying, the material was homogenized by grinding it manually in a mortar. Pre-treatment of the material was conducted at 700 °C for 15 min in air (65 mL/min) to release some of the OH-groups and sulfates by thermal decomposition. Several batches of pre-treatments were mixed to obtain a homogenous material for the experiments. The measured mass losses obtained in the pre-treatments were compared to previous TGA (thermogravimetric analysis) measurements [15] to verify the sufficiency of the treatment.

Figure 1 shows a schematic of the experimental arrangement. Pre-treatments as well as oxidation and oxidation-reduction experiments were all conducted in a vertical tube furnace (LTF 16/-/450, Lenton, Nottingham, UK) equipped with an alumina working tube (impervious pure alumina, 45/38 mm OD/ID). Temperature was monitored with a calibrated Pt/Pt10Rh thermocouple (Johnson Matthey, London, UK, uncertainty ± 3 °C) inside an alumina sheath, connected to multimeters (models 2000 and 2010, Keithley, Solon, OH, USA). Experimental gases (air: 80% N_2, purity 5.0 and 20% O_2, purity 5.0, O_2: purity 5.0, Ar: purity 5.0, CO: purity 3.7, CO_2: purity 5.3, all from AGA Linde, Riihimäki, Finland) were injected from the top of the furnace either into the working tube or directly onto the sample through an alumina rod acting as a gas lance (3 mm ID), see Figure 1. Either rotameters (air, O_2, Ar; Kytola Instruments, Muurame, Finland) or mass flow controllers (CO, CO_2; Aalborg DFC26, USA) were used for regulating the gas flow rates.

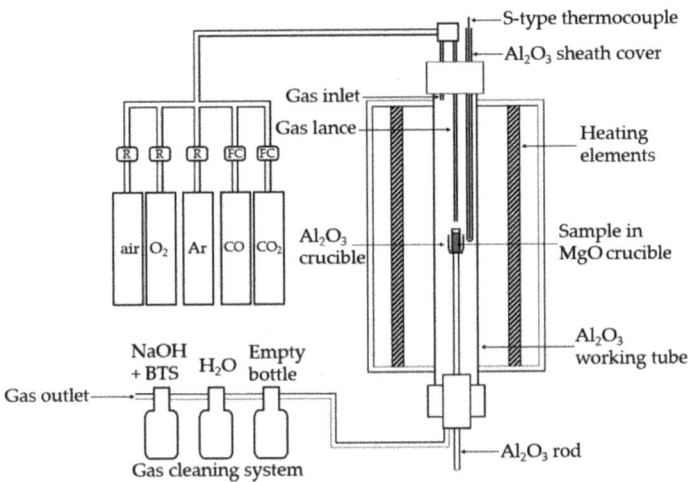

Figure 1. Schematic of the experimental arrangement including the vertical tube furnace, gas train, sample, and gas cleaning system. R = rotameter, FC = mass flow controller, BTS = bromothymol blue.

A gas cleaning system connected to the furnace ensured the safe neutralization or recovery of the off-gases and volatilized elements. An underpressure pump (LabVac LVH40, Piab, Sweden) enhanced the gas outflow from the furnace into the gas cleaning system. The first bottle of the system was empty and was used to prevent the escape of the liquids into the furnace in case of a blockage in the gas train. The second and third bottles contained H_2O and a NaOH solution (6.5 M, 98% NaOH pellets from J.T.Baker, Phillipsburg, NJ, USA), respectively. BTS indicator (Bromothymol blue, $C_{27}H_{28}Br_2O_5S$, Sigma-Aldrich, St. Louis, MO, USA) was dissolved into the NaOH solution to indicate the point where NaOH was consumed from the scrubber solution.

The test material was held in dense MgO crucibles (SC10030, 20/75 mm ID/H, Tateho Ozark Technical Ceramics, Webb City, MO, USA) during the pre-treatments and experiments. The sample was introduced into the hot zone of the furnace by elevating the alumina rod entering the furnace through a leak-proof plug at the bottom of the furnace. The sample was lifted to the hot zone under inert atmosphere (argon) gradually to prevent possible cracking of the crucible, after which the atmosphere was changed to oxidizing. After an oxidation experiment, the atmosphere was changed back to inert, and the sample was lowered to the colder section of the furnace for cooling. In the oxidation-reduction experiments, the atmosphere was directly changed from oxidizing to reducing, and again back to inert after the reduction stage was completed, followed by lowering of the sample for cooling.

Table 1 shows the experimental parameters used in the experiments. All the experiments were conducted at 1400 °C, as that temperature was previously shown to be required for melting the test material with smaller samples [20]. The target was to investigate the optimal conditions for reaching the desired results and, furthermore, how to enhance the process, for example, by adding a gas lance to the experimental arrangement.

Table 1. Experimental parameters used in the oxidation and oxidation-reduction experiments.

	Sample ID	Temperature (°C)	Oxidation Time (min)	Oxidizing Gas/Flow Rate (mL/min)	Reduction Time (min)	Reducing Gas/Flow Rate (mL/min)
Oxidation	O1	1400	30	O_2/65	-	-
	O2	1400	60	O_2/65	-	-
	O3	1400	60	O_2/130	-	-
	O4	1400	60	O_2/260	-	-
	O5 [1]	1400	30	O_2/65 [1]	-	-
	O6 [1]	1400	60	O_2/65 [1]	-	-
Oxidation + Reduction	O-R1	1400	60	O_2/130	120	$CO-CO_2$ (50:50)/260
	O-R2	1400	60	O_2/260	120	$CO-CO_2$ (50:50)/260
	O-R3 [1]	1400	60	O_2/130 [1]	120	$CO-CO_2$ (50:50)/260 [1]
	O-R4 [1]	1400	60	O_2/260 [1]	120	$CO-CO_2$ (50:50)/260 [1]

[1] Gas flow through a gas injection lance.

The first oxidation experiments (O1 and O2) conducted with 65 mL/min oxygen flow rate (STP) were for testing if 30 or 60 min treating time is needed for efficient sulfur removal. Based on the analysis results, the next experiments (O3 and O4) were conducted with doubled and quadrupled oxygen flow rates, respectively, to enhance the removal of sulfur from the material. The 60 min treating time was selected based on the results, which showed that 30 min treating time is not enough for sulfur removal. For the experiments O5 and O6, an oxygen gas lance straight above the sample was used to enhance the contact between oxygen and the sample and, thus, to perform the oxidation stage faster and with less oxygen.

Desulfurization during the oxidation stage occurs when sulfur in the material combusts and sulfur dioxide gas is formed as follows:

$$S + O_2 (g) \rightarrow SO_2 (g) \tag{1}$$

Based on this reaction, the theoretical need of oxygen for complete sulfur removal was calculated. The 30 min experiments O1 and O5, conducted with oxygen flow rate of 65 mL/min, were performed with the stoichiometric amount of oxygen. The 60 min experiments with the same oxygen flow rate (O2 and O6) had oxygen approximately 2 times the stoichiometric requirement. Experiments O3 and O4 had oxygen about 4 and 8 times the stoichiometric need, respectively. The oxidation stage of the oxidation-reduction experiments was conducted with about 4 times the amount of oxygen compared to the theoretical need in the experiments O-R1 and O-R3, and with about 8 times the amount of oxygen in the experiments O-R2 and O-R4. The ratio between the used and stoichiometrically needed oxygen in each experiment can be expressed as O_2 excess factor, and its effect on the sulfur removal efficiency during the oxidation experiments was investigated.

Oxidation time of 60 min and temperature of 1400 °C were used in all oxidation-reduction experiments. Based on the analysis results of the oxidation experiments, the oxidation stage in these experiments was selected to be conducted with two different oxygen flow rates; 130 and 260 mL/min (O-R1 and O-R2, respectively). The reduction time of 120 min was selected based on the earlier study conducted with smaller samples [20]. The reducing gas used was a 50:50 mixture of CO and CO_2 with a total flow rate of 260 mL/min. For enhancing the removal of zinc and lead from the material to the gas phase, the experiments O-R3 and O-R4 were conducted with a gas injection lance. Otherwise, the same experimental conditions were used as in the first oxidation-reduction experiments to demonstrate the effect of the lance.

The samples casted in epoxy resin were prepared for SEM-EDS analysis using traditional dry metallographic methods. The polished sections were cleaned in an ultrasonic bath for 15 min in ethanol. To ensure sufficient electrical conductivity during the analysis, a thin carbon layer was evaporated on the sample surfaces with a sputtering device (Leica EM SCD050, Leica microsystems, Wetzlar, Germany). The microstructure and phase composition were examined with an SEM (Zeiss LEO 1450 W-filament SEM, Carl Zeiss AG, Oberkochen, Germany) equipped with an EDS (Oxford X-Max 50 mm^2, Oxford Instruments, Oxford, UK). The acceleration voltage was set to 15 kV, and the standards employed for the main elements were albite (Al and Na, K-series peaks), apatite (Ca, K-series), barite (Ba, L-series), copper (Cu, K-series), hematite (Fe, K-series), lead (Pb, M-series), olivine (Mg and O, K-series), pyrite (S, K-series), quartz (Si, K-series), sanidine (K, K-series), and zinc sulfide (Zn, K-series). XPP-ZAF matrix correction procedure was applied for processing the raw data [21].

The bulk chemical composition of the samples was determined with an ICP-OES apparatus (iCAP 6000 series, Thermo Fisher Scientific, Pittsburgh, PA, USA). Microwave-assisted digestion of samples was performed with HNO_3, HCl, and HBF_4 using a MARS 6 microwave digestion system (CEM Corporation, Matthews, NC, USA).

3. Results and Discussion

To reach the target S < 1 wt % in the slag after oxidation, sulfur content of the material needs to decrease 97.2% compared to the dried material. During the oxidation experiments O1–O4, 48.5%, 93.3%, 99.3%, and 99.7% decreases in sulfur content were observed, respectively. In the oxidations O1 and O2, performed with 65 mL/min oxygen flow rate and treating times of 30 and 60 min, respectively, the targeted sulfur level after the treatment was not reached. Using the amount of oxygen needed based on the theory, or double that amount, did not result in efficient sulfur removal. In both O3 and O4, the sulfur removal was more efficient, and the target was reached. Thus, four times the amount of theoretically needed oxygen was enough to remove sulfur to an adequate level.

During the experiments O5 and O6, conducted with the gas lance, the desulfurization was 80.2% and 97.9%, respectively, compared to that of the dried material. Therefore, the oxygen flow rate of 65 mL/min was not enough with a treating time of 30 min (O5), but with 60 min treatment (O6), the sulfur concentration decreased below the target value. Thereby, when the gas lance was used, the amount of theoretically needed oxygen was insufficient for efficient sulfur removal, but twice that amount resulted in an acceptable level of sulfur in the sample. Evidently, adding a gas lance to the experimental arrangement considerably enhanced the sulfur removal from the sample. This can also be seen in Figure 2, where the dependence between the reached decrease in the sulfur concentration and the O_2 factor is depicted. The O_2 factor describes how many times the theoretically needed amount of oxygen (Equation (1)) for efficient sulfur removal was used in an experiment.

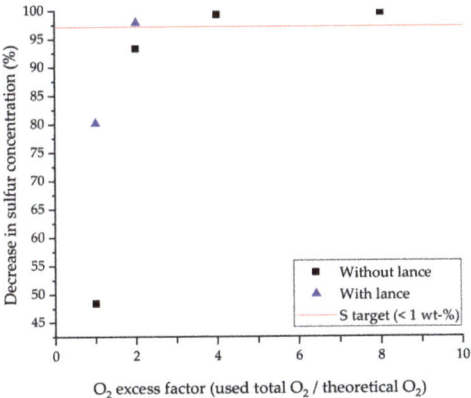

Figure 2. The effect of O_2 excess factor on the reached decrease in sulfur concentration of a sample. O_2 excess factor describes how many times the theoretically needed amount of oxygen was used. The target for achieving S < 1 wt % is at 97.2%.

Table 2 shows the results of the chemical analysis of the bulk compositions conducted with the ICP-OES method. The results show that most of the lead in the jarosite residue is volatilized into the off-gases already during the oxidation stage, and silver is fumed into the gas phase during the reduction stage. In the oxidation-reduction experiments conducted without the gas lance (O-R1 and O-R2), zinc and the remaining lead are quite effectively fumed during the reduction stage, but the target levels (Zn < 1 wt %, Pb < 0.03 wt %) in the bulk were not reached. During the experiments O-R3 and O-R4, conducted with the gas lance, due to the improved gas-liquid interaction, the removal of zinc and lead was more efficient and the target concentrations were reached. The increasing Mg content in the slag is mostly due to dissolution of the crucible material.

Table 2. Bulk compositions of the samples based on the chemical analysis results (wt %, rounded to two decimals). Oxygen concentration of the samples was not determined, and the trace elements have been omitted.

	Sample ID	Ag	Al	Ca	Cu	Fe	K	Mg	Na	Pb	Si	Zn
	Pre-treated	0.02	0.95	4.20	0.28	26.40	0.45	0.23	2.20	5.30	3.60	5.90
Oxidation	O1	0.04	0.67	3.10	0.44	38.70	0.60	1.90	2.70	2.30	2.60	7.10
	O2	0.02	1.30	5.70	0.39	29.70	0.53	4.50	3.00	0.11	5.90	4.20
	O3	0.02	1.30	5.90	0.39	28.80	0.49	4.60	2.90	0.81	5.50	4.80
	O4	0.02	1.30	5.60	0.37	29.20	0.45	4.70	2.70	0.77	5.30	5.20
	O5 [1]	0.03	1.20	5.40	0.39	31.30	0.51	4.10	2.90	0.25	5.50	5.40
	O6 [1]	0.02	1.30	6.10	0.38	28.90	0.48	4.70	3.10	0.29	6.40	3.80
Oxidation + Reduction	O-R1	0.00	1.40	5.80	0.39	28.90	0.54	6.20	3.00	0.05	5.90	1.40
	O-R2	0.00	1.40	5.60	0.38	29.10	0.52	6.50	2.90	0.04	5.80	1.50
	O-R3 [1]	0.00	1.40	5.90	0.35	28.00	0.28	6.80	2.40	0.00	6.50	0.15
	O-R4 [1]	0.00	1.30	5.70	0.35	30.50	0.23	6.80	2.20	0.01	6.20	0.15

[1] Gas flow through a gas injection lance.

SEM-microstructure images in Figure 3 show the two main phases formed during the oxidation stage in oxidation experiments O4 and O6; liquid slag (darker matrix) and solid iron-magnesia-zinc spinels (medium grey areas). Figure 3B also shows some bright metal phase areas consisting mainly of lead with some copper, antimony, and silver.

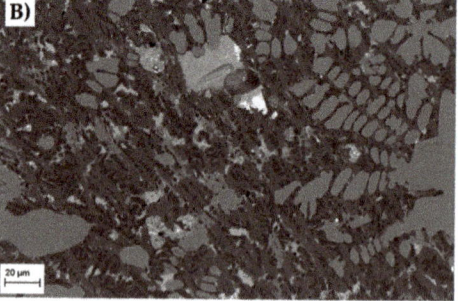

Figure 3. Microstructures of oxidation experiments (**A**) O4 and (**B**) O6. The liquid slag phase is the darker matrix and the solid spinels are the lighter phase areas. The brightest area in (**B**) is a lead-rich metal phase.

Since the sample was slowly cooled to room temperature, there is a possibility that the spinels are formed during cooling. However, the angular shape of the spinels shown in Figure 3A suggests that the spinels are present at 1400 °C in solid form. When the treatment is conducted using the gas lance, the spinel structure has mostly lost its angular shape (Figure 3B). One possible explanation is the better mixing of sample material caused by the injection of gases directly onto the sample surface with the lance.

Figure 4A,B show the microstructure images of oxidation-reduction experiments O-R1 and O-R3, respectively. Three phases; liquid slag (dark grey matrix), spinels (medium grey areas), and metal or speiss droplets (light grey and white spots) are clearly distinguishable. The metallic droplets seem to be mostly attached to the spinels, which apparently has hindered the formation of larger metallic phase areas during the treatment. The same phenomenon has been reported and investigated, for example, in copper making conditions [22,23].

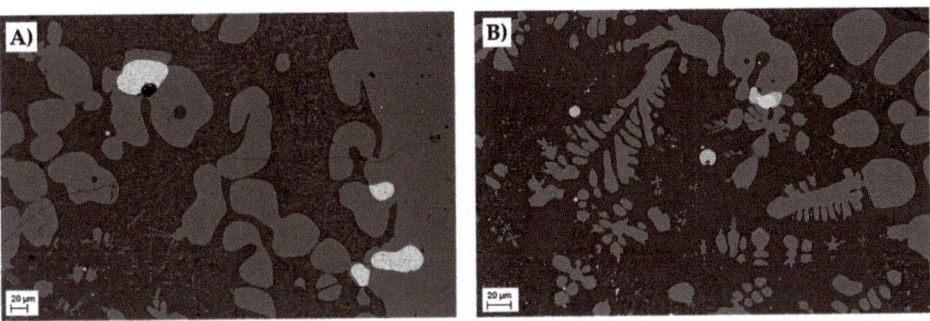

Figure 4. Microstructures of oxidation-reduction experiments (**A**) O-R1 and (**B**) O-R3. The darker matrix is the liquid slag phase and the medium grey areas are the spinels. Light grey and white spots are metal or speiss droplets formed during the reduction stage.

Figure 4B shows that the core and rim of the largest droplet consist of two different phases. Based on SEM-EDS analyses, the core of the droplet is mainly copper (>90 wt %), whereas the rim consists mainly of copper sulfide.

The bulk lead and zinc concentrations after oxidation-reduction experiments were plotted in Figure 5. Two experiments were conducted both with and without the gas lance, employing different oxygen flow rates (130 and 260 mL/min) during the oxidation stage. The parameters during the reduction stage were kept constant. The effect of oxygen flow rate is not very dramatic compared to the effect of the gas lance used.

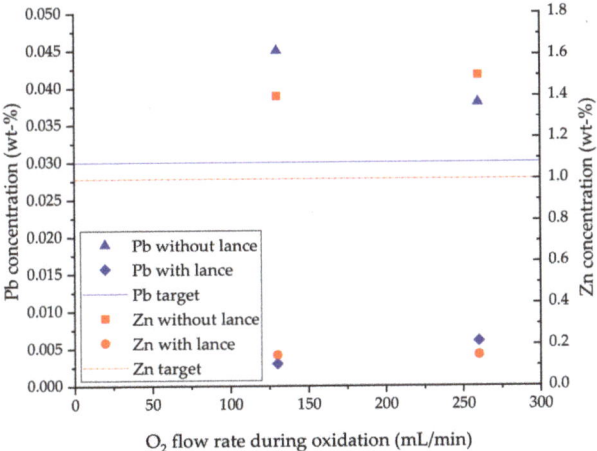

Figure 5. Lead and zinc concentrations in the bulk samples after oxidation-reduction experiments, as a function of oxygen flow rate (130 or 260 mL/min) during the oxidation stage. The effect of the gas lance is quite dramatic.

The target lead and zinc concentrations after the pyrometallurgical oxidation-reduction treatment were <0.03 wt % and <1 wt %, respectively. As Table 2 and Figure 5 clearly show, these targets were not reached with conventional gas introduction into the furnace work tube. The results also indicate that the effect of oxygen flow rate is quite insignificant at this stage. By far the most significant improvements were reached with the installation of the gas injection lance, which improved the interaction between gas and liquid sample material, thus enabling the first steps of experimental parameter optimization regarding the gas flow rates.

4. Conclusions

Pyrometallurgical treatment as a processing option for jarosite leach residue from the hydrometallurgical zinc production was investigated with high-temperature laboratory-scale experiments. Guidelines for the study were obtained from a thermodynamic model [15] and earlier experiments conducted with small sample sizes [20]. The aim was to study how to conduct the pyrometallurgical treatment for obtaining a clean slag (Pb < 0.03, Zn < 1, S < 1 wt %) as a product and to recover the valuable metals from the residue.

Based on the results, it can be concluded that the installation of a top blowing gas injection lance drastically improves the contact between gas and molten sample material, thus accelerating the mass transfer of O_2. Consequently, the assumption of the mass transfer of O_2 being the rate-limiting step in the oxidation step of the process [20] was confirmed, as the lance enabled more efficient oxidation and allowed the use of lower gas flow rates. Significantly more efficient reduction was also achieved by the gas lance installation, as the removal of zinc and lead from the slag was drastically improved.

During this study, the target levels regarding lead, zinc, and sulfur concentrations in the final slag product were reached. However, the spinel phase appears to hinder the formation of larger metal or speiss phase areas, which would collect the valuable metals, as the small individual metal droplets attach preferentially to the spinels. This issue will be investigated in the future by changing the Fe-Si-Ca ratio of the starting material, and thus the composition of the spinels, or by adding more copper-containing material into the jarosite residue before the pyrometallurgical treatment. Changing the composition of the initial feed mixture may also have a reducing effect to the melting temperature, which would be a step further in optimizing the process conditions.

Author Contributions: M.R. designed as well as supervised the experiments, analyzed the data, and wrote most of the manuscript; S.N. performed the experiments and prepared the samples for analysis; L.K. conducted the SEM-EDS analyses, helped in analyzing the data, and wrote parts of the manuscript; A.J. supervised the study and helped writing the manuscript; J.S. initiated the work, provided the experimental targets, and revised the manuscript; P.T. helped in designing the experimental setup and provided insights regarding the study.

Funding: Financial support for this study is provided by Tekes nationally funded CMEco project.

Conflicts of Interest: The authors declare no conflicts of interest.

References

1. Souza, A.D.D.; Pina, P.S.; Leão, V.A. Bioleaching and chemical leaching as an integrated process in the zinc industry. *Miner. Eng.* **2007**, *33*, 591–599. [CrossRef]
2. Glinin, A.; Creedy, S.; Matusewicz, R.; Hughes, S.; Reuter, M. Outotec® ausmelt technology for treating zinc residues. In Proceedings of the EMC, Weimar, Germany, 23–26 June 2013; GDMB Verlag GmbH: Clausthal-Zellerfeld, Germany, 2013; pp. 485–494.
3. Pappu, A.; Saxena, A.; Asolekar, S.R. Jarosite characteristics and its utilization potentials. *Sci. Total Environ.* **2006**, *359*, 232–243. [CrossRef] [PubMed]
4. Ismael, M.R.C.; Carvalho, J.M.R. Iron recovery from sulphate leach liquors in zinc hydrometallurgy. *Miner. Eng.* **2003**, *16*, 31–39. [CrossRef]
5. Han, H.; Sun, W.; Hu, Y.; Jia, B.; Tang, H. Anglesite and silver recovery from jarosite residues through roasting and sulfidization-flotation in zinc hydrometallurgy. *J. Hazard. Mater.* **2014**, *278*, 49–54. [CrossRef] [PubMed]

6. Kerolli-Mustafa, M.; Ćurković, L.; Fajković, H.; Rončević, S. Ecological risk assessment of jarosite waste disposal. *Croat. Chem. Acta* **2015**, *88*, 189–196. [CrossRef]
7. Mäkinen, J.; Salo, M.; Hassinen, H.; Kinnunen, P. Comparison of reductive and oxidative bioleaching of jarosite for valuable metals recovery. *Solid State Phenom.* **2017**, *262*, 24–27. [CrossRef]
8. Kerolli-Mustafa, M.; Fajković, H.; Rončević, S.; Ćurković, L. Assessment of metal risks from different depths of jarosite tailing waste of Trepça Zinc Industry, Kosovo based on BCR procedure. *J. Geochem. Explor.* **2015**, *145*, 161–168. [CrossRef]
9. Jha, M.K.; Kumar, V.; Singh, R.J. Review of hydrometallurgical recovery of zinc from industrial wastes. *Resour. Conserv. Recycl.* **2001**, *33*, 1–22. [CrossRef]
10. Wegscheider, S.; Steinlechner, S.; Leuchtenmüller, M. Innovative concept for the recovery of silver and indium by a combined treatment of jarosite and electric arc furnace dust. *JOM* **2016**, *69*, 388–394. [CrossRef]
11. Hughes, S.; Reuter, M.A.; Baxter, R.; Kaye, A. Ausmelt technology for lead and zinc processing. In Proceedings of the Lead & Zinc, Johannesburg, South Africa, 25–29 February 2008; pp. 141–162.
12. Choi, C.Y.; Lee, Y.H. Treatment of zinc residues by Ausmelt technology at Onsan zinc refinery. In Proceedings of the REWAS '99: Global Symposium on Recycling Waste Treatment and Clean Technology, San Sebastian, Spain, 5–9 September 1999; pp. 1613–1622.
13. Wood, J.; Coveney, J.; Helin, G.; Xu, L.; Xincheng, S. The Outotec® Direct zinc smelting process. In Proceedings of the EMC 2015: European Metallurgical Conference, Düsseldorf, Germany, 14–17 June 2015; GDMB Verlag GmbH: Clausthal-Zellerfeld, Germany; pp. 537–548.
14. Mombelli, D.; Mapelli, C.; Di Cecca, C.; Barella, S.; Gruttadauria, A.; Ragona, M.; Pisu, M.; Viola, A. Characterization of cast iron and slag produced by jarosite sludges reduction via Arc Transferred Plasma (ATP) reactor. *J. Environ. Eng.* **2018**, *6*, 773–783. [CrossRef]
15. Toropainen, A. A Computational Thermodynamic Model for Conversion of Jarosite Residue into an Inert Slag via a Pyrometallurgical Process. Master's Thesis, Aalto University, Espoo, Finland, 2016.
16. Davies, R.H.; Dinsdale, A.T.; Gisby, J.A.; Robinson, J.A.J.; Martin, S.M. MTDATA-thermodynamic and phase equilibrium software from the national physical laboratory. *Calphad* **2002**, *26*, 229–271. [CrossRef]
17. Gisby, J.; Taskinen, P.; Pihlasalo, J.; Li, Z.; Tyrer, M.; Pearce, J.; Avarmaa, K.; Björklund, P.; Davies, H.; Korpi, M.; et al. MTDATA and the prediction of phase equilibria in oxide systems: 30 years of industrial collaboration. *Metall. Mater. Trans. B* **2017**, *48*, 91–98. [CrossRef]
18. Hoang, J.; Reuter, M.A.; Matusewicz, R.; Hughes, S.; Piret, N. Top submerged lance direct zinc smelting. *Miner. Eng.* **2009**, *22*, 742–751. [CrossRef]
19. Chaidez-Felix, J.; Romero-Serrano, A.; Hernandez-Ramirez, A.; Perez-Labra, M.; Almaguer-Guzman, I.; Benavides-Perez, R.; Flores-Favela, M. Effect of copper, sulfur, arsenic and antimony on silver distribution in phases of lead blast furnace. *Trans. Nonferrous Met. Soc. China* **2014**, *24*, 1202–1209. [CrossRef]
20. Rämä, M.; Jokilaakso, A.; Klemettinen, L.; Salminen, J.; Taskinen, P. Experimental investigation of pyrometallurgical treatment of zinc residue. In Proceedings of the Extraction 2018, Ottawa, ON, Canada, 26–29 August 2018; pp. 981–992.
21. Lavrent'Ev, Y.G.; Korolyuk, V.N.; Usova, L.V. Second generation of correction methods in electron probe x-ray microanalysis: Approximation models for emission depth distribution functions. *J. Anal. Chem.* **2004**, *59*, 600–616. [CrossRef]
22. Wilde, E.D.; Bellemans, I.; Zheng, L.; Campforts, M.; Guo, M.; Blanpain, B.; Moelans, N.; Verbeken, K. Origin and sedimentation of Cu-droplets sticking to spinel solids in pyrometallurgical slags. *Mater. Sci. Technol.* **2016**, *32*, 1911–1924. [CrossRef]
23. Wilde, E.D.; Bellemans, I.; Campforts, M.; Guo, M.; Vanmeensel, K.; Blanpain, B.; Moelans, N.; Verbeken, K. Study of the effect of spinel composition on metallic copper losses in slags. *J. Sustain. Metall.* **2017**, *3*, 416–427. [CrossRef]

© 2018 by the authors. Licensee MDPI, Basel, Switzerland. This article is an open access article distributed under the terms and conditions of the Creative Commons Attribution (CC BY) license (http://creativecommons.org/licenses/by/4.0/).

Article

Hydrometallurgical Process for Selective Metals Recovery from Waste-Printed Circuit Boards

Željko Kamberović [1], Milisav Ranitović [2,*], Marija Korać [1], Zoran Anđić [3], Nataša Gajić [2], Jovana Djokić [3] and Sanja Jevtić [1]

1. Faculty of Technology and Metallurgy, University of Belgrade, Karnegijeva 4, Belgrade 11000, Serbia; kamber@tmf.bg.ac.rs (Ž.K.); marijakorac@tmf.bg.ac.rs (M.K.); sanja@tmf.bg.ac.rs (S.J.)
2. Innovation Center of Faculty of Technology and Metallurgy in Belgrade Ltd., Karnegijeva 4, Belgrade 11000, Serbia; ngajic@tmf.bg.ac.rs
3. Innovation Center of the Faculty of Chemistry Ltd., Studentski trg 12-16, Belgrade 11000, Serbia; zoranandjic@yahoo.com (Z.A.); djokic@chem.bg.ac.rs (J.D.)
* Correspondence: mranitovic@tmf.bg.ac.rs; Tel.: +381-62-972-42-56

Received: 11 May 2018; Accepted: 29 May 2018; Published: 11 June 2018

Abstract: This paper presents an experimentally-proved hydrometallurgical process for selective metals recovery from the waste-printed circuit boards (WPCBs) using a combination of conventional and time-saving methods: leaching, cementation, precipitation, reduction and electrowinning. According to the results obtained in the laboratory tests, 92.4% Cu, 98.5% Pb, 96.8% Ag and over 99% Au could be selectively leached and recovered using mineral acids: sulfuric, nitric and aqua regia. Problematic tin recovery was addressed with comprehensive theoretical and experimental work, so 55.4% of Sn could be recovered through the novel physical method, which consists of two-step phase separation. Based on the results, an integral hydrometallurgical route for selective base and precious metals recovery though consecutive steps, (i) Cu, (ii) Sn, (iii) Pb and Ag, and (iv) Au, was developed. The route was tested at scaled-up laboratory level, confirming feasibility of the process and efficiencies of metals recovery. According to the obtained results, the proposed hydrometallurgical route represents an innovative and promising method for selective metals recovery from WPCBs, particularly applicable in small scale hydrometallurgical environments, focused on medium and high grade WPCBs recycling.

Keywords: gold; copper; WPCBs; leaching; physical separation; Tin recovery

1. Introduction

Due to high concentrations of the base and precious metals and their corresponding values, waste printed circuit boards (WPCBs) represent the most significant part of electronic waste. Hence, important for WPCBs recycling is increasing from the point of the resource efficiency and sustainable use of resources, and also from the point of environmental issue [1]. Nevertheless, materials and components variety makes recycling of WPCBs a very complex and demanding process, which consists of different pre-processing and end-processing operations.

Therefore, various mechanical operations are suggested as the first step, which helps that further operations, either pyro- or hydrometallurgical, can be carried out more effectively [2]. In pyrometallurgy, WPCBs recycling is integrated in the base metals production process in which valuable metals are concentrated in the metal carrier, copper or lead, and subsequently valorized through series of pyro-hydro-electro refining operations to produce high-purity metals [3,4]. Although these operations are well-established on the industrial scale, intense marketing activities hinder environmental risks, and their ecological acceptance is dependent on implementation of high-efficient off-gas treatment systems, which affects already high capital and operating costs. Therefore, pyrometallurgical processes are dependent on the high operating capacities and on the processing high-grade WPCBs. However, changes in the material composition and recent trends toward decreasing metals content in the electronic devices may represent serious drawbacks for these operations in the near future [4]. Recycling technologies lagging behind product designers' activities and distance is regrettably increasing.

In contrast, high flexibility and selectivity of the hydrometallurgical operations are offering possibility for separate metals recovery in early stages of processing. This makes hydrometallurgical recycling of electronic waste economically feasible even at lower operating capacities. From a technological perspective, due to morphological similarity between WPCBs and specific "refractory" copper ores, traditional hydrometallurgical methods are applied for WPCBs recycling as well [5,6]. Thus, in the beginning, hydrometallurgical recycling of WPCBs was focused only on gold and silver recovery, mainly through simplified variations of the cyanidation process. However, increased environmental awareness and need for sustainable material management has imposed necessity for development of more comprehensive metals recovery processes. Accordingly, in the scientific literature, various studies have been reported, but generally focused on the independent recovery of the base (BM) [7–9], precious (PM) [10–12], or solder metals [13–15]. However, investigations of compatibility and possible adverse interactions between consecutive leaching and recovery steps are scarce. For that reason, in recent years, significant efforts have been made to combine and incorporate the above mentioned methods in the unique multi-step processing route [16–18]. Since the major economic driver for WPCBs recycling relies on the efficient PM recovery, most of these processes are developed following the concept of metals concentration, in which BM are selectively removed in first stage of processing by mainly using oxidative sulfate solutions [19,20]. Even though the main objective of this step is leaching of copper, other BM like iron, zinc, nickel, or cobalt are leached as well, reducing the possibility for any inhibiting effect on the successive PM leaching [19]. Subsequently, leaching of the gold and silver is performed using cyanides, halides, aqua-regia (AR) or acidic thiosulfate or thiourea solutions [21]. Over the decades, cyanide leaching represented the dominant method for gold recovery from primary and secondary sources. In spite of low cost and consumption of cyanides, their high toxicity generally restricts their implementation in modern WPCBs recycling process. Instead, usage of the thio-based solutions is emphasized in many recent studies, due to their high selectivity for precious metals and non-toxicity. However, instability of the thio-based solutions, relatively slow leaching rates and high consumption makes them less attractive for industrial application. Also, gold surface passivation as well as insufficiently tested risks for human health additionally complicate their application. Consequently, in addition to promising results of extraction rate and yield for iodine leaching, favorable leaching kinetic and ability to effectively prevent any interfering effect caused by residual BM still make AR one of the main reagents for PM recovery [22].

Nevertheless, even with these great improvements done in the past period, shortfalls in comprehensive process overview obtained from scaled-up demonstrations are preventing better understanding of the full-scale system performances. Additionally, even though lead and tin quantitatively are next in-line to copper, a small number of studies considers these metals of interest [23].

In response, investigations presented in this study are related to Cu, Pb, Sn, Ag, and Au recoveries, which are based on quantities and economic importance identified as metals of major interest. According to the experimental investigation, by implementing a specially designed combination of simple and time-saving methods (leaching, cementation, precipitation, reduction and electrowinning), Cu, Pb, Ag, and Au could be successfully leached and recovered through three consecutive steps using mineral acids (sulfuric, nitric and aqua regia). In addition, the simple mechanical method of coagulation, gravitation concentration, and filtration for Sn recovery from WPCBs is introduced, improving the overall metals recovery rate. Obtained results were used for definition of an improved multi-step hydrometallurgical processing route tested at scaled-up laboratory level, confirming the process feasibility and the metals recovery efficiencies.

2. Materials and Methods

2.1. Materials

In the presented research, 100 kg of mechanically treated WPCBs were provided by a local electronic waste recycler. Material was obtained after primary crushing in the cross-flow QZ® chain mill equipped with the magnetic separator and secondary shredding in the Meccanoplastica® cutter. Chemicals used in the laboratory tests were p.a. (*pro analysi*) certified, while chemicals used in the scaled-up laboratory test were technical grade.

2.2. Analytical Methods

Chemical composition of WPCBs and solid products was determined by X-ray fluorescence spectrometry (XRF) using Thermo Scientific ARL Quant'x EDXRF Spectrometer. Chemical composition of pregnant leaching solutions (PLS) and solutions after metals recovery (MRS) was determined by atomic absorption spectrometry (AAS) using Perkin Elmer 4000 spectrometer. Phase composition and material morphology was analyzed by X-ray diffraction analysis (XRD) and scanning electron microscopy (SEM) using Philips PW-1710 difractometer and Mira3 Tescan microscope.

Chemical composition of WPCBs was determined after thermal treatment of three samples, each weighing 1.00 kg, in medium induction furnace (reduction conditions, at 1250 °C for 1 h) in which organic fraction was burned (loss on ignition), oxides together with other slag-forming compounds were transferred to slag phase, and metals were collected in melt ($M°$ state). Obtained fractions were measured and analyzed, and chemical composition was determined as average value of three analyses. For this analysis, the slag phase was grinded to a particle size <100 μm to support a representative measurement.

Due to high heterogeneity of investigated material, the leaching degree was determined by material balance method [19]:

$$Metal_{leaching\ efficiency}, \% = \frac{Metal_{extracted}\ \text{into leaching solution}}{Metal_{extracted}\ \text{in leaching solution} + Metal_{\text{in solid residue}}} \times 100 \qquad (1)$$

Concentration of metals in PLS and MRS was determined directly by AAS. Concentration of targeted metal in leaching residue was determined by AAS after leaching of residue in hot AR solution.

2.3. Experimental Methods

Development of the integral hydrometallurgical process for selective base and precious metals recovery from WPCBs was performed through theoretical and experimental investigations, which included simulation and modeling of the leaching reactions, laboratory tests, and scaled-up laboratory testing (Figure 1).

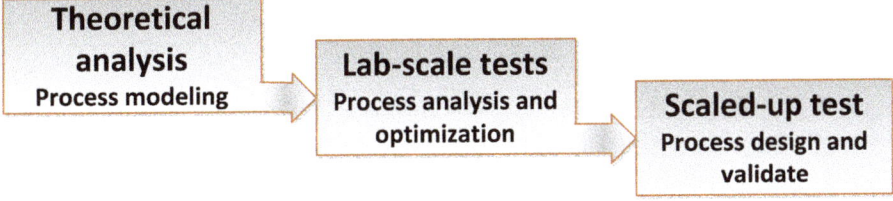

Figure 1. Process development methodology.

2.3.1. Simulation of Leaching Reactions

Theoretical analysis of the thermodynamic characteristics of selected leaching reactions was performed as the first step, to obtain more comprehensive overview of the highly-complex reaction system and to define major processing steps and their relative order. Therefore, analysis of phase stability of the metals and their possible chemical reactions (based on Pourbax, Eh-pH, diagrams) with common mineral acids, sulfuric, nitric and hydrochloric, in predefined temperature range, was performed using HSC Chemistry software [24].

2.3.2. Laboratory Tests

Leaching tests were performed in order to investigate influence of time, temperature, leaching and oxidizing agent concentration, solid:liquid (S/L) ratio, and stirring rate on leaching degree.

Cu leaching tests were performed through oxidative sulfate leaching, using 2 mol/dm^3 solution of H_2SO_4 with continuous H_2O_2 addition (30 vol. %) as oxidizing agent. Electrowinning tests (EW) for Cu recovery were performed to investigate influence of the process control, current density and temperature, with and without mixing. Ag cementation tests were performed using Cu powder (1.5 g, 100 wt. % excess, purity >99.5%, <125 μm,). Tests were performed in order to investigate influence of the temperature under fixed time (30 min) and pH (2.2–2.4). Adjustment of pH was performed by adding Na_2CO_3 into solution [25]. Sn recovery was performed by a two-step phase separation of micron-sized, coarse, suspended particles of stannic acid obtained during Cu leaching step. Extracted stannic acid is probably a mixture of α- (H_2SnO_3) and β- modifications ($SnO_2 \times xH_2O$). In this step, PLS after Cu leaching was filtered through coarse cellulose filter, pore size 38–75 μm, separating solid residue and solution containing suspended particles. Subsequently, fine suspended particles were extracted using laboratory centrifuge.

Pb and Ag leaching tests were performed using HNO_3 at fixed temperature (80 °C) and stirring rate (300 rpm) [23]. Tests were performed using sample of WPCBs after liberation of BM through oxidative sulfate leaching. Pb and Ag recovery from leaching solution was performed by chemical precipitation using excessive amount of 4 mol/dm^3 NaCl solution (100% molar excess), at room temperature.

Au leaching was performed using standard AR solution, a mixture of nitric and hydrochloric acid with 1:3 molar ratio. Knowing that an increase of temperature tends to give faster Au leaching rates, the test was performed at 60 °C, which is adopted as upper temperature level, above which decomposition reactions and evaporation of AR are intensified [22]. Au leaching kinetic was investigated by measuring the concentration of Au in solution, in intervals of 30 min for total of 180 min of leaching. Tests were performed using sample of the WPCBs liberated from the major metals presence, which is obtained after dissolution in 50 vol. % HNO_3. Due to rapid decrease in apparent density after dissolution of metals, S/L ratio was changed (increased) to 500 g/dm^3. Au recovery from leachate was performed by chemical reduction using an excessive amount of $FeSO_4$ (0.2 g, 100 wt. % excess).

Due to low initial content, other valuable metals like Ni, Co, and Pd were not of interest in this study, though chemical behavior is known.

Laboratory tests were performed in three-neck glass reactor, 2 dm^3 volume, equipped with condenser, stirrer, pH and temperature control, and chemicals addition system. Depending on the S/L ratio, a different amount of the WPCBs sample was leached with 1 dm^3 of the corresponding leaching agent. Laboratory galvanostatic polarization tests were performed using a Potentiostat/Galvanostat Bank STP 84. Two-electrodes configuration, Cu foil as cathode and PbSb$_7$ as anode with same surface area of 3 cm^2 were used. For enlarged laboratory EW tests, system of parallel-connected monopolar electrodes placed in rectangular electrolytic cell, with 2 dm^3 volume, equipped with heater and stirrer was used. A system of 10 cathodes (Cu sheet) and 11 anodes (PbSb$_7$) with the same surface area, 50 × 60 × 1 mm, and electrode distance 20 mm, was used. A rectifier with maximum capacity of 100 A and 40 V was used as DC source. For electric measurements, an HP 3466a multimeter was used.

2.3.3. Scaled-up Laboratory Test

Scaled-up test was performed using 5 kg of WPCBs sample, through consecutive reproduction of process parameters of each leaching and recovery step.

Leaching, chemical precipitation, and reduction were performed in glass-leaching rector, effective volume 30 dm^3, equipped with stirrer, chemicals addition system, temperature control, and condenser coupled with off-gas washing system. EW recovery of Cu was performed in polypropylene electrolytic cell—volume 15 dm^3, cathode surface 150 mm × 100 mm, electrodes distance 30 mm, equipped with heating tank, volume 80 dm^3, and electrolyte recirculation system. Same DC source, 100 A and 40 V, as in galvanostatic laboratory tests, was used. Phase separation was performed using acid-resistant trommel screen (<105 μm) and plate and frame filter press. Process efficiency of each consecutive leaching step was followed by AAS, analyzing concentration of targeted metal in leachate aliquot in time interval of 30 min, until constant value of concentration.

2.4. Development of the Process Flow Sheet

Based on the results obtained in previous stages, an integral multi-step hydrometallurgical processing route for selective base and precious metals recovery was developed. The process flow sheet with overall mass balance and efficiency of each processing step is presented. Also, improvements related to innovative two-step separation method for Sn recovery is presented.

3. Results

3.1. Material Characterization

Physical–chemical characterization included determination of chemical composition (Table 1), content of magnetic fraction (2.3 wt. %), and moisture (1.71 wt. %), bulk density (889 kg/m^3) and particle size distribution (Scheme 1).

Table 1. Chemical composition of mechanically treated WPCBs.

Material	Share	Unit
Metals		
Cu	25.51	wt. %
Sn	3.57	wt. %
Pb	2.47	wt. %
Zn	2.18	wt. %
Fe	0.85	wt. %
Ni	0.18	wt. %
Al	<0.1	wt. %
Sb	960	ppm
Co	620	ppm
Ag	6800	ppm
Au	203	ppm
Glass/Ceramic		
SiO$_2$	21.94	wt. %
Al$_2$O$_3$	6.16	wt. %
TiO$_2$	0.88	wt. %
Na$_2$O	0.14	wt. %
Polymer	~35	wt. %

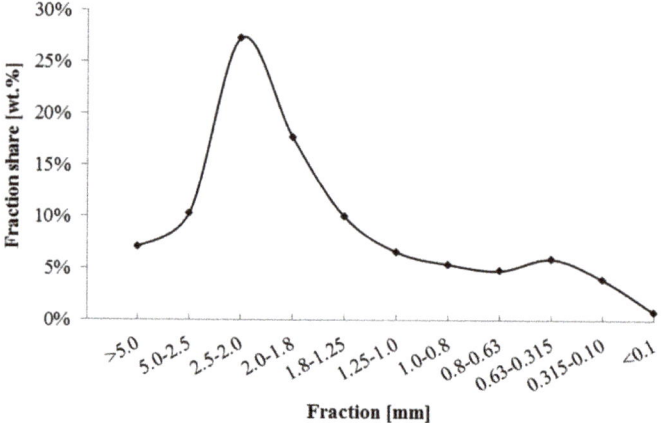

Scheme 1. Particle size distribution.

Results obtained by sieve analysis suggest that material is suitable for hydrometallurgical processing as only 7 wt. % of the complete sample consists of particles bigger than 5.00 mm (coarse fraction), whereas less than 10 wt. % are particles smaller than 0.63 mm (dusty fraction).

Considering the generally adopted methodology by which WPCBs are categorized into several groups depending on the gold content, it may be concluded that used material is on threshold for medium and high grade (minimum content 200 ppm) [1,26]. Additionally, very low amounts of Fe and Al suggest on efficient magnetic separation as well as that aluminum parts were most probably extracted prior to mechanical treatment, i.e., during the processor's dismantling and removal of cooling and housing parts.

3.2. Simulation of Leaching Reactions

Stability of the different phases of metals was determined as a function of the electrochemical potential (Eh) and the pH, which are major parameters for determination of the conditions required for the leaching process. Pourbaix (Eh-pH) diagrams are presented on Figure 2.

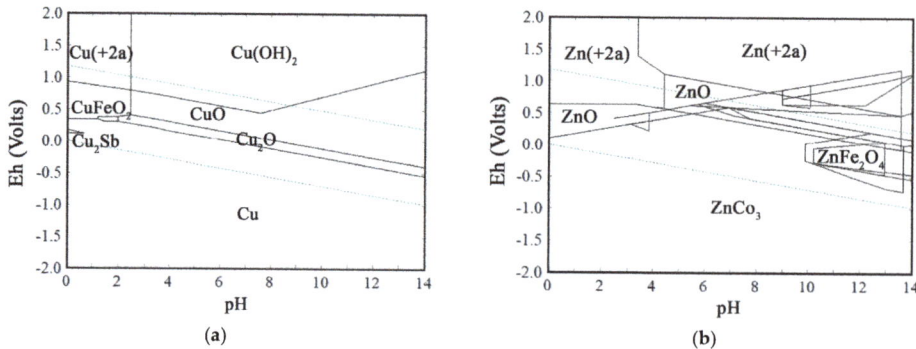

Figure 2. *Cont.*

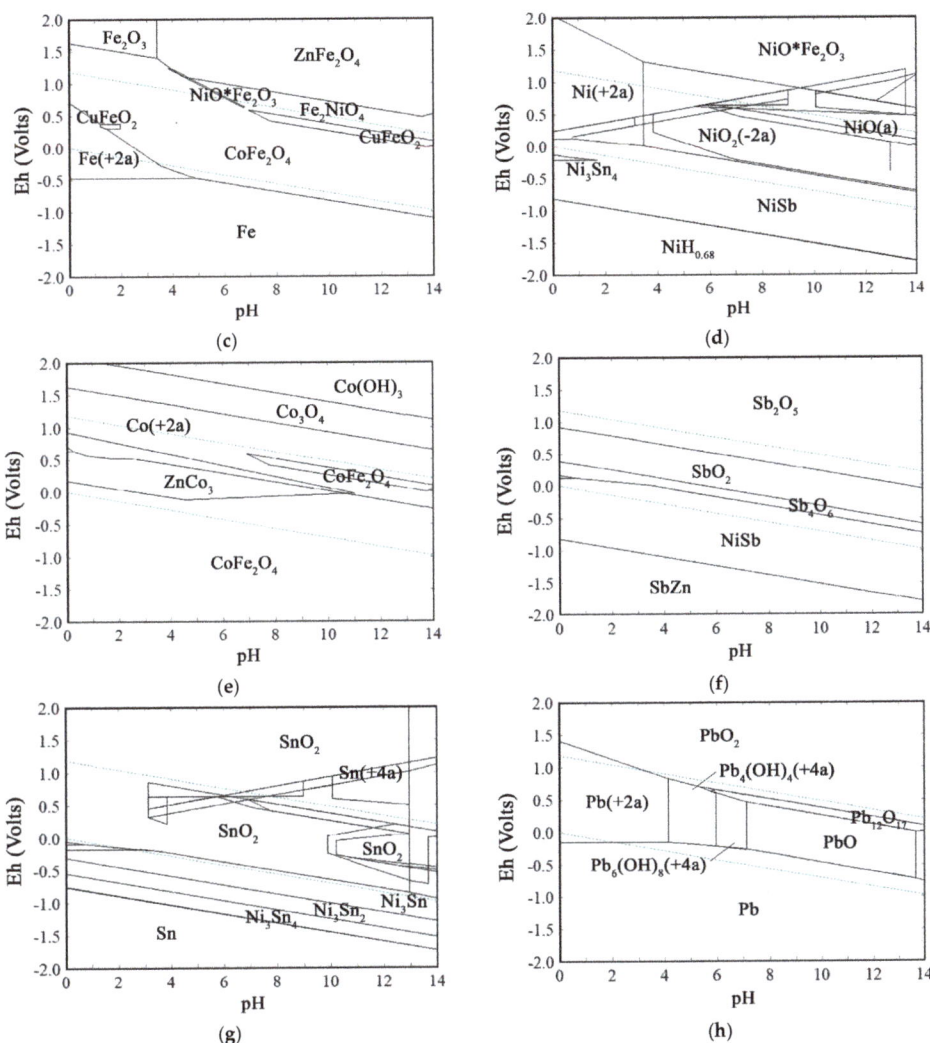

Figure 2. Pourbaix diagrams (Eh-pH) for 1 mol/dm^3 (metal)/kg H$_2$O and P$^\theta$: (**a**) Cu; (**b**) Zn; (**c**) Fe; (**d**) Ni; (**e**) Co; (**f**) Sb; (**g**) Sn and (**h**) Pb.

Figure 2a–d show the stability regions of Cu, Zn, Fe and Ni, respectively. As presented, the stability of soluble ionic phase, M$^{(2+a)}$, of said metals is influenced by both Eh (>0.0 V) and pH (<3). In negative regions of electrochemical potential (Eh < 0.0), metallic forms are stable, while at pH > 3, various combinations of oxides, hydroxides, and oxyhydroxides are stable. On the contrary, the soluble form of Co is highly influenced by electrochemical potential, where in only a small region of Eh in between 1.0–1.5 V of Co$^{(2+a)}$ is in stable phase, whereby this region narrows as pH increases(Figure 2e). A similar situation is observed in the case of Sb, Figure 2f, with the exception that in complete pH range and Eh in between 0.5–1.0 V, various oxides of Sb are stable.

According to the results presented in Figure 2g,h, both Sn and Pb have soluble forms within certain regions of pH (>6) and Eh in between 0.0–0.5 V. However, unlike Pb, which has stable soluble phase Pb$^{(2+a)}$ at pH < 4 and Eh > 0.0 V, for almost all values of pH and Eh > 0.0 V, the stable phase of Sn is SnO$_2$.

On contrary to Au, where the phase stability analysis confirmed its limited solubility, i.e., in most regions Au form is dominate, analysis of the phase stability of Ag, Figure 3a, showed that soluble form is influenced with highly oxidative conditions and may be achieved at Eh > 1.0 V and pH < 4.

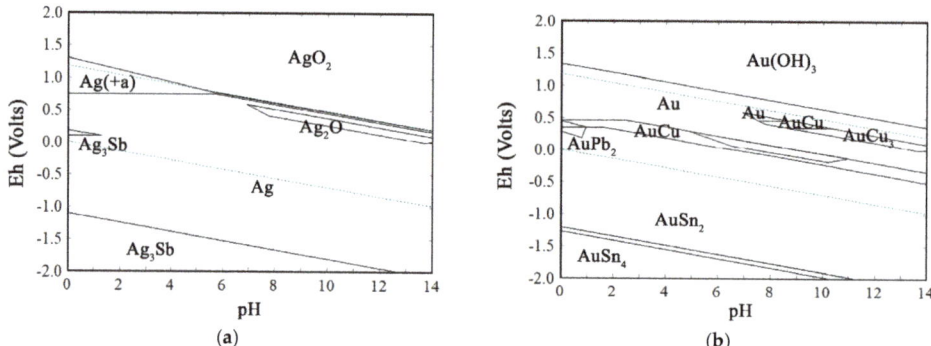

Figure 3. Pourbaix diagrams (Eh-pH) for 1 mol/dm^3 (metal)/kg H$_2$O and P$^\theta$: (**a**) Ag; (**b**) Au.

During the calculations, it was determined that temperature (if changed from 25 to 80 °C) and concentration (if changed from 1/10 to 1/100 mol/dm^3/kg H$_2$O) does not have significant influence on the type of phase in which stability prevails, but only on the stability region, shifting the regions insignificantly to higher or lower values of pH and Eh. This is the reason why figures for different temperatures and concentrations were excluded.

Since investigations of phase stability showed that the majority of investigated metals have their soluble forms in acidic (pH < 4) and oxidative (Eh > 0.0) regions, further theoretical investigations were focused on thermodynamic modeling of the possible chemical reactions of the metals with the selected acidic leaching agents: sulfuric, nitric, and hydrochloric acid.

Accordingly, values of standard-state free energy (ΔG^θ) for reactions of metals with H$_2$SO$_4$ and H$_2$O$_2$, within the predefined temperature range (25–80 °C), are presented in Table 2.

Table 2. ΔG^θ for reactions of metals with H$_2$SO$_4$ and H$_2$O$_2$.

Leaching Reaction	ΔG^θ_{298}	ΔG^θ_{353}	
	kJ·mol^{-1}		
Me * + H$_2$SO$_4$ + H$_2$O$_2$ → Me * SO$_4$ (a) + 2H$_2$O	<0	<0	(1)
Pb + H$_2$SO$_4$ + H$_2$O$_2$ → PbSO$_4$ (s) + 2H$_2$O	−477.19	−474.60	(2)
Sn + 2H$_2$O$_2$ → H$_2$SnO$_3$ (s)+ H$_2$O	−753.80	−749.11	(3)
Sn + H$_2$SO$_4$ + H$_2$O$_2$ → SnSO$_4$ (s) + 2H$_2$O	−572.90	−570.56	(4)
2Ag + H$_2$SO$_4$ → Ag$_2$SO$_4$ (a) + H$_2$(g)	71.02	65.89	(5)
2Ag + H$_2$O$_2$ → Ag$_2$O (s) + 2H$_2$O	−127.19	−127.70	(6)
Ag$_2$O + H$_2$SO$_4$ → Ag$_2$SO$_4$ (a) + H$_2$O	−155.42	−154.93	(7)

* Me = Sb, Zn, Fe, Co, Ni, Cu—in reaction priority order (decreasing negative value ΔG^θ); (a) = aqua; (s) = solid; (g) = gas.

Analyzing the results presented in Table 2, negative values of ΔG^θ for Equation (1) suggest that H$_2$SO$_4$-H$_2$O$_2$ could represent a suitable system for BM leaching. Also, a slight decrease of ΔG^θ suggests that in the investigated temperature range, its increase has a positive effect on BM leaching kinetic. However, according to values of ΔG^θ for Equations (3) and (4), in an oxidative sulfate system, the reaction of Sn with H$_2$O$_2$ is dominant, resulting in intensive oxidation of Sn in highly stable hydrolyzed tin (IV) oxide, H_2SnO_3, and/or $SnO_2 \times xH_2O$. In addition, although reaction between metallic Ag and H_2SO_4 is disfavored according to Equation (5), a reaction may be expected as a result of partial oxidation of Ag, since Ag_2O may be subsequently dissolved in H_2SO_4 (refer to Equations (6) and (7)).

Values of standard-state free energy of possible chemical reactions of metals with HNO_3 and HCl, within the predefined temperature range (25–80 °C) are presented in Tables 3 and 4.

Table 3. ΔG^θ possible chemical reactions of metals with HNO_3.

Leaching Reaction	ΔG^θ_{298}	ΔG^θ_{353}	
	kJ·mol^{-1}		
$aMe^* + bHNO_3 \rightarrow aMe^*(NO_3)_2 \text{(a)} + cNO_2 \text{(g)} + dH_2O$	<0	<0	(8)
$Sn + 4HNO_3 \rightarrow SnO_2 \text{(s)} + 4NO_2 \text{(g)} + 2H_2O$	−462.95	−492.19	(9)
$Sn + 4HNO_3 \rightarrow Sn(NO_3)_2 \text{(a)} + 2NO_2 \text{(g)} + 2H_2O$	−310.52	−319.70	(10)
$2Sb + 6HNO_3 \rightarrow Sb_2O_3 \text{(s)} + 6NO_2 \text{(g)} + 3H_2O$	−408.56	−445.91	(11)

* *Me* = Fe, Zn, Pb, Co, Ni, Cu, Ag—in reaction priority order (decreasing negative value ΔG^θ); a, b, c, d, x, y—stoichiometric coefficients; (a) = aqua; (s) = solid; (g) = gas.

Table 4. ΔG^θ possible chemical reactions of metals with HCl.

Leaching Reaction	ΔG^θ_{298}	ΔG^θ_{353}	
	kJ·mol^{-1}		
$Me^* + 2HCl \rightarrow Me^{**}Cl_2 \text{(a)} + H_2 \text{(g)}$	<0	<0	(12)
$Pb + 2HCl \rightarrow PbCl_2 \text{(s)} + H_2 \text{(g)}$	−58.82	−68.60	(13)
$2Ag + 2HCl \rightarrow 2AgCl \text{(s)} + H_2 \text{(g)}$	35.92	22.12	(14)
$2Sb + 6HCl \rightarrow 2SbCl_3 \text{(a)} + 3H_2 \text{(g)}$	58.70	41.63	(15)
$Cu + 2HCl \rightarrow CuCl_2 \text{(a)} + H_2 \text{(g)}$	81.59	71.60	(16)
$Au + 4HCl \rightarrow [AuCl_4]^- \text{(a)} + 4H^+ \text{(g)}$	420.42	420.85	(17)
$2Cu + 4HCl \text{(a)} + O_2 \text{(g)} = 2CuCl_2 \text{(a)} + 2H_2O$	−352.38	−341.75	(18)
$Au + 4HCl + O_2 \text{(g)} \rightarrow [AuCl_4]^- \text{(a)} + 2H_2O$	84.48	100.85	(19)

* *Me* = Zn, Sn, Ni, Fe, Co—in reaction priority order (decreasing negative value ΔG^θ); a, b, c, d, x, y—stoichiometric coefficients; (a) = aqua; (s) = solid; (g) = gas.

Analysis of possible chemical reactions of metals with HNO_3 and negative values of ΔG^θ for Equation (8) indicate spontaneous character of reactions which means that all BM including Ag can be easy dissolved in HNO_3. However, comparing the values of ΔG^θ for the reactions Equations (9) and (10), the dominant reaction of Sn with HNO_3 is intensive oxidation to highly stable $SnO_2 \times xH_2O$. An analogue situation is observed in the case of Sb, where oxidation to Sb_2O_3 represents the dominant reaction, as in Equation (11).

In case of HCl, negative values of the ΔG^θ calculated for the reactions defined with Equations (12) and (13) indicate that Sn along with other BM (Pb, Zn, Fe, Ni, Co) may be dissolved in HCl. However, comparing the values of the ΔG^θ, Equations (16)–(19), leaching of Au and Cu is dependent on the presence of an oxidizing agent. In addition, the low value of ΔG^θ for the reaction of Ag with HCl, as in Equation (14), points out that the transformation of Ag to stable solid salt AgCl may occur.

In line with the results obtained after theoretical reconsiderations, it may be concluded that the majority of investigated metals may be leached using the aforementioned mineral acids. However, the obtained results indicate that the main drawback is related to the limited selectivity in this highly complex system, which is the consequence of the different hydrometallurgical behaviors and properties of metals contained within WPCBs. This eventually may influence overall process selectivity, if priority of leaching operations is not designed properly.

Contrary to the high selectivity of H_2SO_4 for BM leaching, the ability of HNO_3 to dissolve the majority of metals can be seen as a disadvantage rather than an advantage, because in subsequent recovery steps, multi-metal composition of PLS highly complicates their selective recovery. On the contrary, usage of HCl showed better properties in terms of selectivity, i.e., possibility for selective recovery of Pb and Sn. However, the main disadvantage is related to the possible transformation of Ag to highly stable AgCl, which is why this leaching operation should be performed only after Ag leaching.

These findings were essential for the definition and relative order of hydrometallurgical operations. Accordingly, implementation of successive leaching steps using H_2SO_4, HNO_3 and AR may result in selective recovery of BM, Pb and Ag, and lastly, Au. Additionally, due to the fact that successive usage of H_2SO_4-HNO_3-AR does not require significant adjustments of pH, laboratory tests investigating possibility of selective metals recovery from mechanically treated WPCBs were organized as follows:

- Oxidative sulfate leaching of BM, specially referenced to Cu leaching and recovery, was investigated as the first processing step,
- Following the Cu recovery and removal of other BM, leaching and recovery of Pb and Ag using HNO_3 was investigated as a second processing step,
- Au recovery using AR solution was investigated as the last processing step.

3.3. Laboratory Tests

3.3.1. Cu Leaching and Recovery

Influence of various process parameters on Cu leaching efficiency is presented on Figure 4a–d.

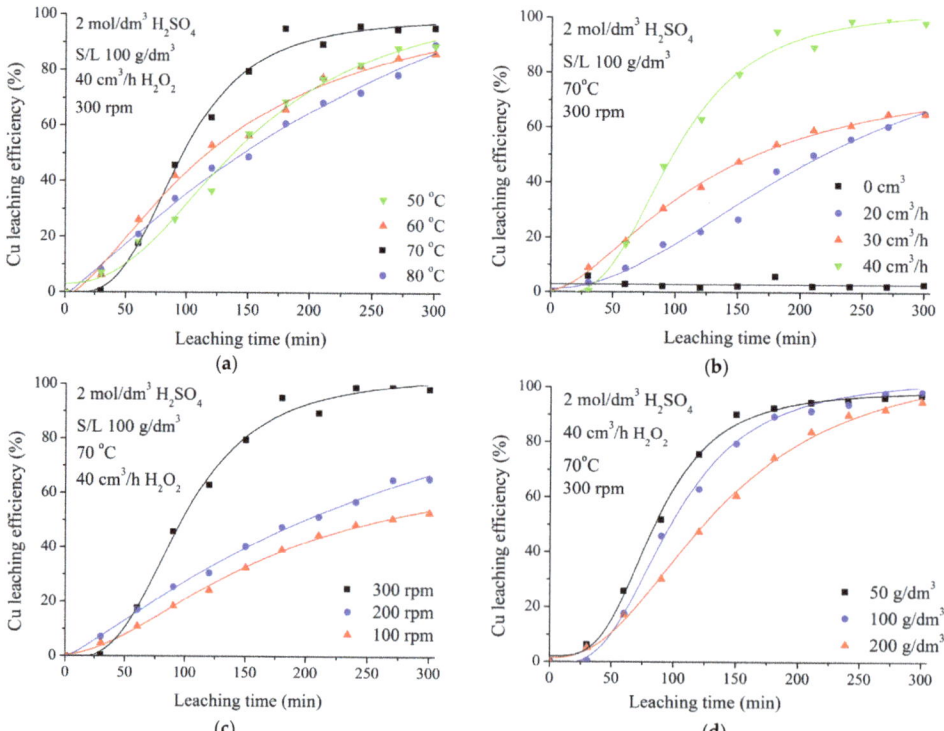

Figure 4. Influence of various process parameters on Cu leaching efficiency: (**a**) temperature; (**b**) H_2O_2 addition; (**c**) stirring rate; (**d**) S/L ratio.

Considering the positive effect on leaching kinetic determined in the theoretical analysis, influence of temperature was investigated in the range of 50–80 °C. As seen from results presented in Figure 4a, temperature has no significant influence on overall process efficiency, giving the high leaching efficiencies for all analyzed temperatures. However, in terms of leaching kinetic, the highest leaching rate was obtained at 70 °C. Due to rapid decomposition of H_2O_2, further temperature increase had no significant impact on the Cu leaching efficiency.

Investigating the influence of oxidizing agent addition, as presented in Figure 4b, Cu leaching efficiency increases with the increase of H_2O_2 concentration, as a result of oxygen produced from the decomposition of H_2O_2 and which acts as oxidant in the reaction with metallic Cu.

Figure 4c shows that an increase of the stirring rate has a positive impact on Cu leaching efficiency. This may be a consequence of the complex and composite structure of WPCBs, where in the case of inappropriate mixing of suspension, heavier, metal-bearing particles are remaining at the bottom of the reactor, covered with lighter particles of non-metallic materials, affecting the metal's accessibility to the leaching agent. Due to highly achieved leaching efficiency, the further increase of stirring rate was not investigated.

Investigation of S/L ratio influence on Cu leaching efficiency, presented in Figure 4d, revealed that S/L ratio has no significant effect on overall leaching efficiency, but only on leaching kinetic, meaning that lower S/L ratio allows achievement of the same Cu leaching efficiency in shorter leaching time.

From the presented experimental results, it can be summarized that optimum conditions for Cu leaching using 2 mol/dm^3 H_2SO_4 solution as leaching agent, are as follows: temperature of 70 °C, S/L ratio of 200 g/dm^3, H_2O_2 addition 40 cm^3/h, time of 5 h, and stirring rate of 300 rpm. By applying these conditions, achieved Cu leaching efficiency was 95.4%.

Sn Recovery

High concentration of fine suspended particles, present in most Cu leaching suspensions, was identified as the main reason for difficulties faced during the phase separation process. To further investigate this phenomenon, suspension obtained after the leaching test yielding highest Cu recovery was subjected to two-step phase separation, consisting of filtration and centrifugal separation. After washing and drying, morphology, phase composition, and chemical composition of extracted suspended particles was subjected to SEM, XRD, and XRF analysis. According to obtained results presented in Figures 5 and 6 and in Table 5, it was determined that the basis of extracted suspended particles represent Sn in the form of agglomerated stannic acid particles smaller than 1 μm.

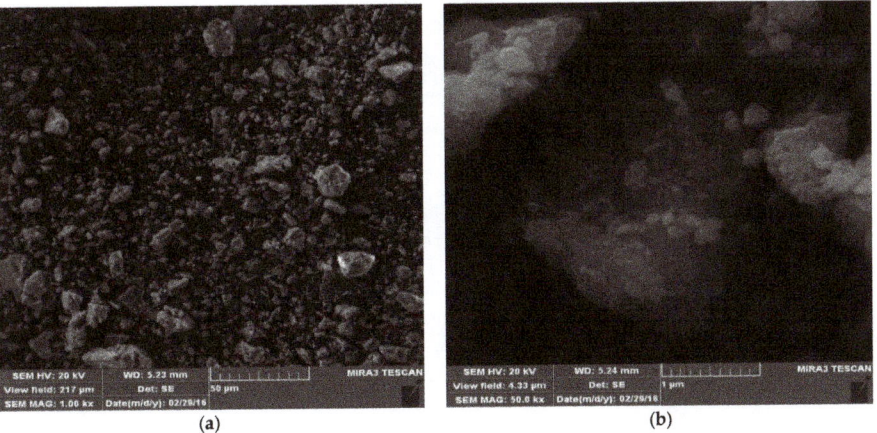

Figure 5. SEM micrographs of extracted suspended particles under different magnifications: (**a**) 1000×; (**b**) 50,000×.

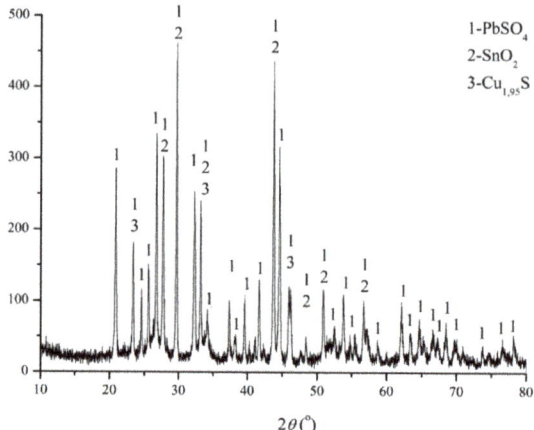

Figure 6. XRD pattern of extracted suspended particles.

Table 5. Chemical composition of extracted suspended particles.

Element	Fe	Cu	Zn	Sn *	Pb	Non-Metallic
Amount, %	0.31	0.24	0.36	82.53	3.46	balance

* as stannic acid.

Presented results confirm theoretical findings that in an oxidative sulfate system, transformation of Sn in highly stable stannic acid is the dominant reaction. Additionally, according to mass balance, it is calculated that 55.4 wt. % of total Sn contained in investigated WPCBs may be easily extracted in form of mixture of H_2SnO_3 and $SnO_2 \times xH_2O$.

Ag Cementation

Cementation of Ag was investigated using leaching solutions from the last three tests of Cu leaching (Figure 4d influence of S/L). Content of Ag in solutions prior and after cementation, as average value of three different measurements, as well efficiencies of cementation process at various temperatures, are presented in Table 6.

Table 6. Degree of Ag cementation by Cu powder.

C_{Ag} Prior Cementation, mg/dm^3	Temperature, °C	C_{Ag} after Cementation, mg/dm^3	Efficiency, %
11.85	20	<1	>95
26.01	30	4.57	82.43
29.91	40	12.10	59.55

As presented, the best results for Ag cementation were obtained at room temperature. Notable decrease in process efficiency on higher temperatures, 30 °C and 40 °C, most probably come from intensified dissolution of Cu powder. According to AAS analysis, Ag concentration in the leachates after Cu leaching step is in range of 11.85–29.91 mg/dm^3, indicating that, comparing to total content of Ag in WPCBs, on average 3.10 wt. % of Ag may be expected in solutions after Cu leaching.

EW Recovery of Cu

In the final phase of Cu leaching and recovery, laboratory tests for Cu electrowinning from leaching solution were performed. The tests were performed for 2 h. According to the results presented in Table 7, most adequate parameters for scaled-up Cu EW recovery, considering the Cu quality and current efficiency, were adopted as follows: 2.2 V, 250 A/m^2, 40 °C, and 100 rpm. Under these conditions, Cu concentration in electrolyte prior and after EW was 48.43 g/dm^3 and 32.25 g/dm^3, respectively. Concentration of Fe in electrolyte was 1.68 g/dm^3 and virtually unchanged during the process. According to results of XRF analysis, in obtained Cu deposits, only presence of Fe and Ni were detected. Ni concentration in all analysis was below 20 ppm. Copper recovery was enhanced by the second stage of the EW process. It was performed with i_c = 135 Am^{-2} until the Cu concentration of 2 g/L. The obtained cathode deposit has lower purity (<99% Cu) but can be used for anode casting and further refined up to >99.97 wt. % Cu by standard procedure for electrorefining [27].

Table 7. Investigated EW process parameters.

Cathodes	Cu Foil				Cu Sheet		
i_c, A/m^2	150	300	300	450	200	250	300
U (cell), V	1.93	2.12	2.04	2.38	2.1–2.2	2.2–2.3	2.2–2.4
Current efficiency, %	94	89	92	85	93	91	88
Stirring, rpm	No	No	100	100	100	100	100
Temperature, °C	Room	Room	40	40	40	40	40
Chemical composition of deposit/%							
Cu	99.98	99.77	99.92	99.78	99.93	99.91	99.87
Fe	n/a	0.153	0.056	0.192	0.043	0.059	0.078

3.3.2. Pb and Ag Leaching and Recovery

Figure 7a–d represents the influence of various process parameters on Pb and Ag leaching efficiency.

As presented in Figure 7a, acid concentration significantly impacts leaching efficiency, Ag in particular. Increasing the HNO$_3$ concentration to 8 mol/dm^3, after 60 min of leaching, achieved efficiency for Pb and Ag was over 98% and 87%, respectively.

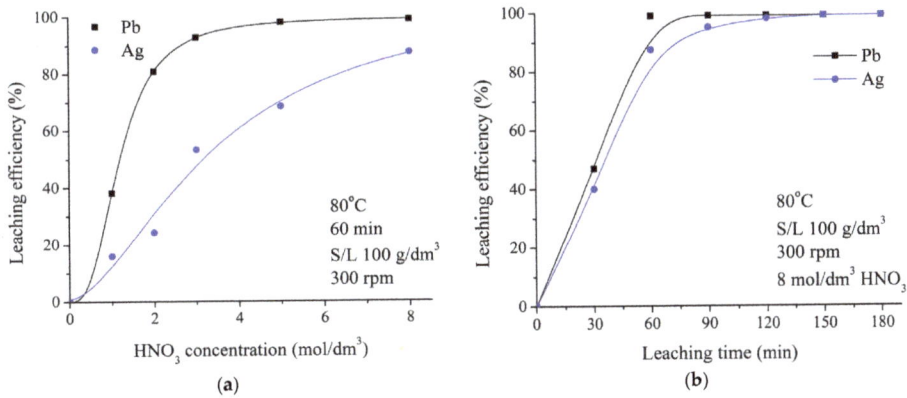

Figure 7. Cont.

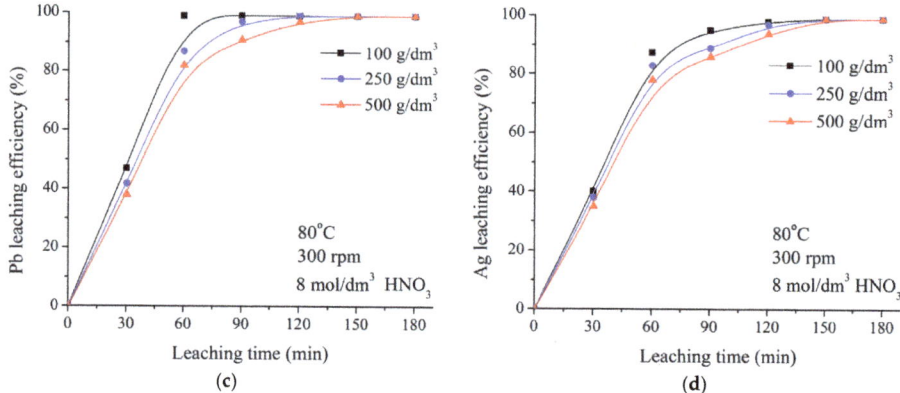

Figure 7. Influence of various process parameters on Pb and Ag leaching efficiency: (**a**) HNO$_3$ concentration; (**b**) leaching time; (**c**) S/L ratio on Pb leaching efficiency; (**d**) S/L ratio on Ag leaching efficiency.

Figure 7b shows influence of leaching time, revealing that prolonging the process duration for more than 120 min has no influence on already high Pb and Ag leaching efficiencies. Based on this conclusion and high achieved efficiencies for Pb and Ag leaching within first 120 min, leaching time of 2 h was adopted as optimal.

Figure 7c,d shows that S/L ratio has no significant influence on Pb and Ag leaching efficiency. The probable reason for obtained high efficiencies comes from the significant decrease of apparent density after removal of all base metals, from 889 to 520 kg/m^3, and which consequently allows better mixing of suspension.

From the presented experimental results, it can be summarized that best results for Pb and Ag leaching were achieved using 8 mol/dm^3 HNO$_3$, temperature of 80 °C, S/L ratio of 500 g/dm^3, time of 2 h, and stirring rate of 300 rpm. By applying these conditions, achieved Pb and Ag leaching efficiency was 98.5% and 96.8%, respectively.

Selective precipitation of PbCl$_2$ and AgCl was performed using solution after the last leaching test with highest S/L ratio (500 g/dm^3), by adding an excessive amount of 4 mol/dm^3 NaCl solution at room temperature. Comparing the concentration of these metals in PLS and MRS, presented in Table 8, efficiency of recovery process is calculated to over 99%.

Table 8. Pb and Ag concentration of PLS and MRS.

	PLS, g/dm^3	MRS, mg/dm^3	Efficiency, %
Pb	16.17	91.11	99.40
Ag	4.31	17.40	99.60

3.3.3. Au Leaching and Recovery

Figure 8 shows the relationship between Au leaching efficiency versus time, under constant temperature (60 °C) and stirring rate (300 rpm). As presented, no change of Au concentration in AR solution was detected when leaching time was extended more than 120 min, suggesting that under assessed process parameters leaching time has no significant influence on Au leaching efficiency. Moreover, AAS analysis, after repeated dissolution of the solid residue in freshly prepared hot AR solution, suggest that under applied process parameters (60 °C, 2 h, S/L ratio 500 g/dm^3, 300 rpm) full dissolution (exceeding 99%) of Au was achieved.

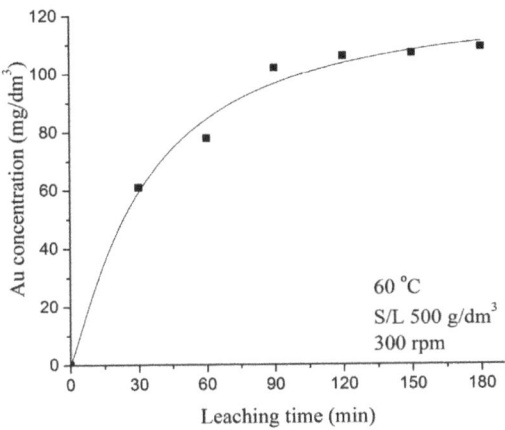

Figure 8. Influence of leaching time on Au leaching efficiency.

Gold recovery from AR solution was performed by chemical reduction using excessive amount of $FeSO_4$ (0.2 g, 100 wt. % excess), at room temperature. According to the AAS analysis of solution after reduction, particles settling and phase separation, Au concentration in solution was below detection limit (1 ppm), proving the expected high efficiency, over 99%. The SEM micrograph of the obtained Au powder, consisted from agglomerated fine particles, is presented on Figure 9.

Figure 9. SEM micrograph of reduced Au particles.

3.4. Scaled-up Laboratory Test

Integration of processing steps in selective multi-step hydrometallurgical route for comprehensive metals recovery from WPCBs was tested on scaled-up laboratory level, through consecutive steps, as follows:

(i) Cu leaching and recovery: Oxidative sulfate leaching followed by Ag cementation and EW recovery of Cu, reproducing previously defined process parameters, was performed as first leaching and recovery stage. Overall Cu leaching and recovery efficiency of 92.5% was achieved after 10 h of leaching and 12 h of EW recovery stage,

(ii) Partial Sn recovery: In this recovery step, solution after Cu leaching preceding the Ag cementation and EW, was primarily filtered using acid-resistant trommel screen, mesh 150 (<105 μm), separating solid residue and solution containing suspended particles of stannic acid. Afterwards, solution was subjected to coagulation, by adding 0.2 vol. % of commercial coagulant, Brenntamer A3322®, concentration 1 g/dm^3, in PLS at room temperature and constant stirring, 100 rpm. After settling of coagulated particles, solution was filtered using filter press, separating purified leachate from Sn-enriched residue,

(iii) Pb-Ag recovery: Leaching and recovery of Pb and Ag using HNO_3 followed by NaCl precipitation, was performed as the third processing step by applying process parameters defined in laboratory tests, using solid residue after Cu leaching,

(iv) Gold recovery: Finally, gold leaching using AR solution from solid residue after Pb–Ag recovery, by applying process parameters adopted as optimal, followed by reduction using ferrous sulfate, was performed as the last leaching and recovery step.

According to mass distribution and chemical composition of all obtained products and solutions after the scaled-up laboratory test, efficiency of each processing step was determined and presented in Table 9.

Table 9. Mass distribution, chemical composition of products, and process efficiencies.

Product	Mass	Efficiency	Cu	Pb	Sn *	Ag	Au
	g	%			wt. %		
Coagulated Sn	232.10	54	0.14	6.41	57.55	n.d.	n.d.
Ag cement	7.20	>95	81.93	0.02	0.02	16.89	n.d.
Cathode Cu	1180.00	>92	99.71	0.04	0.08	n.d.	n.d.
PbCl$_2$ and AgCl precipitate	155.10	>98	n.d.	79.33	0.07	20.60	n.d.
Au powder	0.92	>98	n.d.	n.d.	n.d.	n.d.	>99.9

n.d.—not detected; * as stannic acid.

3.5. Development of the Process Flow Sheet

Based on all obtained results, integral multi-step hydrometallurgical process for selective base and precious metals recovery was developed (Figure 10). Overall process efficiency as well as mass distribution is also presented.

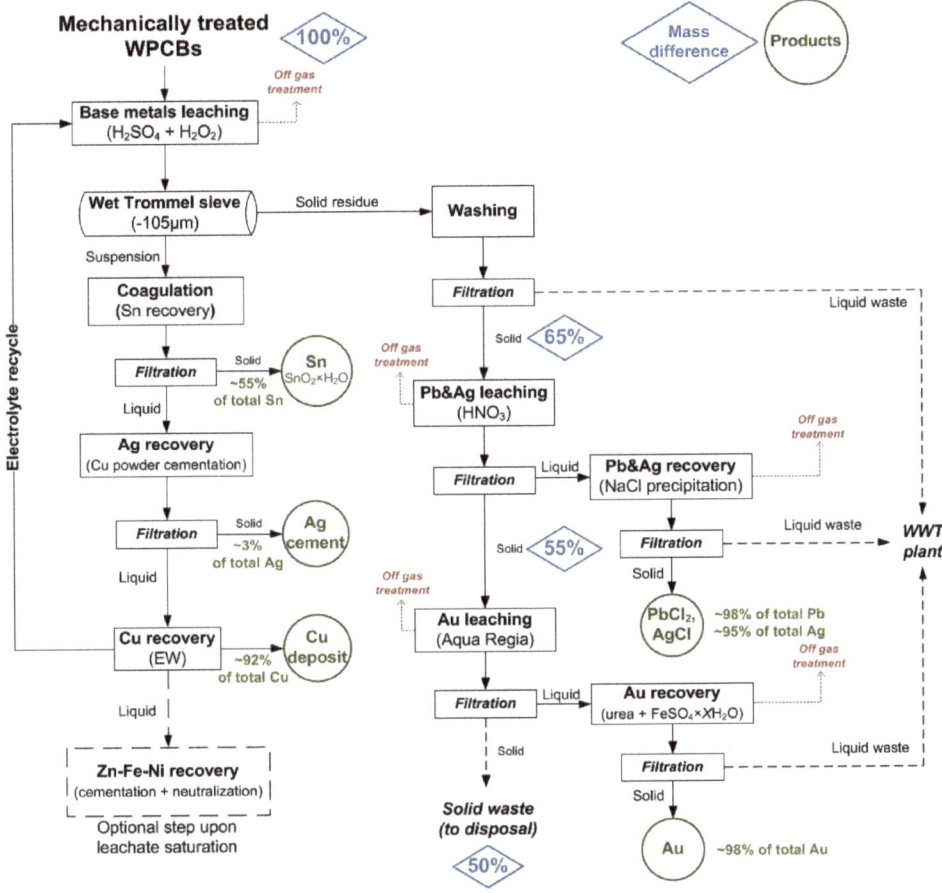

Figure 10. Process flow sheet for integral multi-step hydrometallurgical method for selective base and precious metals recovery from WPCBs.

4. Conclusions

In this study, presented integral multi-step technological route for the selective base and precious metals recovery from WPCBs is experimentally proved. As determined, by implementing simple and time saving methods (leaching, cementation, precipitation, reduction and electrowinning), Cu, Sn, Pb, Ag, and Au could be successfully recovered through the consecutive processing steps.

In the first stage, Cu was leached using 2 mol/dm^3 H_2SO_4 solution in the presence of H_2O_2 and subsequently recovered by EW, with the overall process efficiency of 92.4%. Additionally, around 3.10% of the total Ag leached during this leaching step was recovered by cementation with efficiency exceeding 95%. In the second stage, after 2 h, 98.5% Pb and 96.8% Ag were leached using 8 mol/dm^3 solution of HNO_3 at 80 °C, and recovered by precipitation using NaCl, with efficiency that exceeds 99%. In the third stage, over 99% of gold was leached using AR solution and completely recovered from the PLS by chemical reduction using $FeSO_4$. In addition, by implementing the innovative method, consisting of the two-step phase separation of suspension after Cu leaching, 55.4% of Sn may be recovered in the form of highly stable hydrolyzed tin (IV) oxide.

According to the results of the scaled-up laboratory test, in addition to the demonstrated high efficiencies of each consecutive step, which were over 92% for Cu and over 95% for Pb–Ag and Au recovery, implementation of the novel method for partial Sn recovery was tested as well. Although preliminary process efficiency was limited to 55%, technological simplicity of the novel method indicates possibility for further improvement of process efficiency, yielding the material with significantly increased Sn content.

Recovery of major beneficial metals, Cu, Sn, Pb, Ag and Au, followed by production of minimal waste materials with known character, represents great advantage of presented technological route which is of the highest importance for the lower operating capacities. Further analysis of the process economy could give a more comprehensive overview if such a process represents a promising alternative for the present status dominated by the pyrometallurgical sector.

Author Contributions: Conceptualization, Ž.K., M.R., M.K.; Methodology, Ž.K., M.R., Z.A.; Validation, Ž.K., M.R. and S.J.; Investigation, M.R.; N.G.; J.D.; Writing-Original, M.R.; Writing-Review & Editing, Ž.K., M.K.; Visualization, N.G.; J.D.; Supervision, Ž.K.

Acknowledgments: Presented research was supported by Ministry of Education, Science and Technological Development, Republic of Serbia, project "Innovative synergy of by-products, waste minimization and cleaner production in metallurgy", No. TR34033.

Conflicts of Interest: The authors declare no conflict of interest.

References

1. Reuter, M.; Hudson, C.; Van Schaik, A.; Heiskanen, K.; Meskers, C.; Hagelüken, C. *Metal Recycling: Opportunities, Limits, Infrastructure*; UNEP: Nairobi, Kenya, 2013; ISBN 978-92-807-3267-2.
2. Cui, J.; Forssberg, E. Mechanical recycling of waste electric and electronic equipment: A review. *J. Hazard. Mater.* **2003**, *99*, 243–263. [CrossRef]
3. Ghodrat, M.; Rhamdhani, M.; Brooks, G.; Masood, S.; Corder, G. Techno economic analysis of electronic waste processing through black copper smelting route. *J. Clean. Prod.* **2016**, *126*, 178–190. [CrossRef]
4. Yazici, E.Y.; Deveci, H. Extraction of metals from waste printed circuit boards (WPCBs) in H_2SO_4–$CuSO_4$–NaCl solutions. *Hydrometallurgy* **2013**, *139*, 30–38. [CrossRef]
5. Kamberović, Ž.; Sokić, M.; Korać, M. On the physicochemical problems of aqueous oxidation of polymetalic gold bearing sulphide ore in an autoclave. *Physicochem. Probl. Miner. Process.* **2003**, *37*, 107–114.
6. Lee, J.; Kim, S.; Kim, B.; Lee, J.-C. Effect of Mechanical Activation on the Kinetics of Copper Leaching from Copper Sulfide (CuS). *Metals* **2018**. [CrossRef]
7. Birloaga, I.I.; De Michelis, I.; Ferella, F.; Buzatu, M.; Vegliò, F. Study on the influence of various factors in the hydrometallurgical processing of waste printed circuit boards for copper and gold recovery. *Waste Manag.* **2013**, *33*, 935–941. [CrossRef] [PubMed]
8. Sun, Z.H.I.; Xiao, Y.; Sietsma, J.; Agterhuis, H.; Visser, G.; Yang, Y. Selective copper recovery from complex mixtures of end-of-life electronic products with ammonia-based solution. *Hydrometallurgy* **2015**, *152*, 91–99. [CrossRef]
9. Orac, D.; Havlik, T.; Maul, A.; Berwanger, M. Acidic leaching of copper and tin from used consumer equipment. *J. Min. Metall. Sect. B Metall.* **2015**, *51*, 153–161. [CrossRef]
10. Park, Y.J.; Fray, D.J. Recovery of high purity precious metals from printed circuit boards. *J. Hazard. Mater.* **2009**, *164*, 1152–1158. [CrossRef] [PubMed]
11. Kim, E.; Kim, M.; Lee, J.C.; Pandey, B. Selective recovery of gold from waste mobile phone PCBs by hydrometallurgical process. *J. Hazard. Mater.* **2011**, *198*, 206–215. [CrossRef] [PubMed]
12. Li, J.; Xu, X.; Liu, W. Thiourea leaching gold and silver from the printed circuit boards of waste mobile phones. *Waste Manag.* **2012**, *32*, 1209–1212. [CrossRef] [PubMed]
13. Jha, M.; Kumari, A.; Choubey, P.; Lee, J.; Kumar, V.; Jeong, J. Leaching of lead from solder material of waste printed circuit boards (PCBs). *Hydrometallurgy* **2012**, *121–124*, 28–34. [CrossRef]
14. Havlik, T.; Orac, D.; Petranikova, M.; Miskufova, A. Hydrometallurgical treatment of used printed circuit boards after thermal treatment. *Waste Manag.* **2011**, *31*, 1542–1546. [CrossRef] [PubMed]

15. Cheng, C.Q.; Yang, F.; Zhao, J.; Wang, L.H.; Li, X.G. Leaching of heavy metal elements in solder alloys. *Corros. Sci.* **2011**, *53*, 1738–1747. [CrossRef]
16. Kamberović, Ž.; Korać, M.; Ranitović, M. Hydrometallurgical process for extraction of metals from electronic waste-part II: Development of the processes for the recovery of copper from printed circuit boards (PCB). *Metalurgija* **2011**, *17*, 139–149.
17. Birloaga, I.; Coman, V.; Kopacek, B.; Vegliò, F. An advanced study on the hydrometallurgical processing of waste computer printed circuit boards to extract their valuable content of metals. *Waste Manag.* **2014**, *34*, 2581–2586. [CrossRef] [PubMed]
18. Kumari, A.; Jha, M.K.; Singh, R.P. Recovery of metals from pyrolysed PCBs by hydrometallurgical techniques. *Hydrometallurgy* **2016**, *165*, 97–105. [CrossRef]
19. Behnamfard, A.; Salarirad, M.; Veglio, F. Process development for recovery of copper and precious metals from waste printed circuit boards with emphasize on palladium and gold leaching and precipitation. *Waste Manag.* **2013**, *33*, 2354–2363. [CrossRef] [PubMed]
20. Yang, H.; Liu, J.; Yang, J. Leaching copper from shredded particles of waste printed circuit boards. *J. Hazard. Mater.* **2011**, *187*, 393–400. [CrossRef] [PubMed]
21. Syed, S. Recovery of gold from secondary sources—A review. *Hydrometallurgy* **2012**, *115–116*, 30–51. [CrossRef]
22. Birch, A.; Mohamed, S.R.; Friedrich, B. Screening of Non-cyanide Leaching Reagents for Gold Recovery from Waste Electric and Electronic Equipment. *J. Sustain. Metall.* **2018**. [CrossRef]
23. Ranitović, M.; Kamberović, Ž.; Korać, M.; Jovanović, N.; Mihjalović, A. Hydrometallurgical recovery of tin and lead from waste printed circuit boards (WPCBs): Limitations and opportunities. *Metalurgija* **2016**, *55*, 153–156.
24. Roine, A. *HSC Chemistry® v 6.12*; Outotec Research Oy Center: Pori, Finland, 2006.
25. Timur, S.; Cetinkaya, O.; Erturk, S. Investigating silver cementation from nitrate solutions by copper in forced convection systems. *Miner. Metall. Proc.* **2005**, *22*, 205–210.
26. Cucchiella, F.; D'Adamo, I.; Koh, S.C.L.; Rosa, P. A profitability assessment of European recycling processes treating printed circuit boards from waste electrical and electronic equipments. *Renew. Sustain. Energy Rev.* **2016**, *64*, 749–760. [CrossRef]
27. Dimitrijević, S.; Ivanović, A.; Simonović, D.; Kamberović, Ž.; Korać, M. Electrodepositon of copper and precious metals from waste sulfuric acid solution. In Proceedings of the 15th International Research TMT, Prague, Czech Republic, 12–18 September 2011; pp. 689–692.

© 2018 by the authors. Licensee MDPI, Basel, Switzerland. This article is an open access article distributed under the terms and conditions of the Creative Commons Attribution (CC BY) license (http://creativecommons.org/licenses/by/4.0/).

Article

Thermodynamic Considerations for a Pyrometallurgical Extraction of Indium and Silver from a Jarosite Residue

Stefan Steinlechner * and Jürgen Antrekowitsch

Chair of Nonferrous Metallurgy, Montanuniversitaet Leoben, Leoben 8700, Austria; juergen.antrekowitsch@unileoben.ac.at
* Correspondence: stefan.steinlechner@unileoben.ac.at; Tel.: +43-3842-402-5254

Received: 26 March 2018; Accepted: 3 May 2018; Published: 9 May 2018

Abstract: Indium and silver are technologically important, critical metals, and in the majority of cases, they are extracted as a by-product of another carrier metal. The importance of indium has seen recent growth, and for technological reasons, these metals can be found in industrial residues from primary zinc production, such as the iron precipitate—jarosite. To secure the supply of such metals in Europe, and with the idea of a circular economy and the sustainable use of raw materials, the recycling of such industrial residues is coming into focus. Due to the low value of jarosite, the focus must lie simultaneously on the recovery of valuable metals and the production of high-quality products in order to pursue an economical process. The objective of this article is to give the fundamentals for the development of a successful process to extract the minor elements from roasted jarosite. As such, we use thermodynamic calculations to show the behavior of indium and silver, leading to a recommendation for the required conditions for a successful extraction process. In summary, the formation of chlorine compounds shows high potential to meet the challenge of simultaneously recovering these metals together with zinc at the lowest possible energy input.

Keywords: indium; silver; jarosite; recycling; industrial residue; process development; selective extraction; simultaneous recovery; pyrometallurgy

1. Introduction

Recycling rates have increased in recent decades, but this is mainly due to the reutilization of end-of-life products and their better collection logistics. For economic reasons, a decrease in the level of easily minable primary resources, and stricter environmental legislation regulating, for instance, land filling, different industrial residues have also become a focal point of the recycling industry. As the valuable metal content when compared to various end-of-life products or primary ore concentrates is in most cases drastically lower, the requirements of newly-developed recycling processes are more demanding. This results in a strategy to develop concepts capable of the simultaneous extraction of metals and generation of products with an "added value".

The zinc industry is one of the base metal industries carrying a broad variety of accompanying side elements and is the focus of this research—in particular, its iron precipitate residue. As an example, zinc ore concentrate includes several metals, such as indium, defined as a critical metal by the European Commission [1]. Today, the hydrometallurgical zinc extraction from oxidic or sulfidic zinc concentrates dominates, with a ratio of more than 90% of primary zinc production. Even though several new technologies have been implemented in the last few decades (e.g., direct leaching of sulfidic ore and solvent extraction for solution purification), the typical flow sheet of a hydrometallurgical zinc plant still features the concentration, roasting, neutral- and hot-acid leaching of the zinc ore followed by electrowinning, as shown in Figure 1.

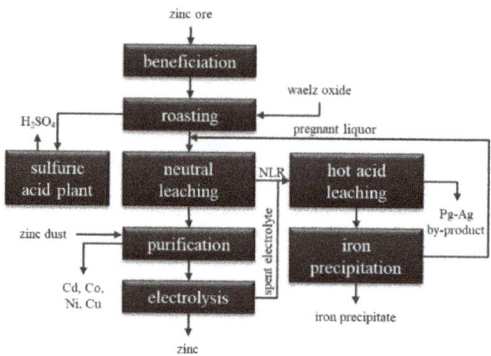

Figure 1. Flow sheet of zinc production by roasting, neutral- and hot-acid leaching, and electrowinning [2, 3]. NLR: neutral leaching residue.

To begin with, the zinc ore concentrate is oxidized in a roasting step, prior to the leaching of the calcine. During the roasting step, the contained iron reacts with zinc, forming a barely-soluble zinc ferrite spinel, rendering the hot-acid leaching step mandatory. To avoid a high volume of processed solution for the following iron precipitation step, the hot-acid leaching is only carried out for the solid neutral leaching residue (NLR). In the case of iron ores with low iron content, the hot-acid leaching step can be avoided. Over time, various iron removal technologies have evolved. An overview of the three common iron residues (hematite, goethite and jarosite), including the metal contents of iron and zinc, as well as their typical moisture contents, is shown in Table 1.

Table 1. Iron precipitation residues and their characteristics [4,5].

Residue	Content		Quantity Dry	Moisture			Quantity Moist
				Adherent	Chem. Bound	Total	
	Fe (wt. %)	Zn (wt. %)	(kg/t Zn)	(kg H_2O/t Zn)		(wt. %)	(kg/t Zn)
Hematite	57–60	1	245	27	0	10	272
Goethite	34–40	8.5–13	370	250	24	43	637
Jarosite	29	2–6	520	350	47	44	907

Furthermore, the typical contents of iron and zinc, and the relative amount of generated residue per ton of produced zinc are given. Since the material is dumped as sludge, the content of H_2O, either chemically bound or adherent to the sludge, is also shown. Although the jarosite precipitate has the lowest iron content and, with this, the highest level of moisture, it is the most commonly used source of iron removal among those listed, and offers the highest possibility of also finding silver and indium in the residue.

Equation (1) is the mineralogical structural formula of jarosite. It shows that it is a basic iron sulfate compound, where R is a placeholder for potential cations like H_3O^+, Na^+, K^+, NH_4^+, Ag^+, Li^+, Ti^+, Rb^+, and R_2 for Hg^{2+}, Pb^{2+} [2,6]. On an industrial scale, jarosite precipitation is based on the addition of ammonium or sodium for economical and ecological reasons [2,3,7,8].

$$R_2Fe_6(OH)_{12}(SO_4)_4 \tag{1}$$

Moreover, Zn, Cd, Ag, Cu^{2+}, and In^{3+} can be assimilated into the jarosite structure [9]. For that reason, this article describes the thermodynamic fundamentals for the extraction of minor elements like silver and indium from industrially generated jarosite over the course of primary zinc production.

The main focus is placed on the simultaneous extraction of a zinc oxide product also carrying compounds of indium and silver.

This offers the advantage of an easy reutilization of the "added value" product in the primary zinc production industry. The silver and indium form the additional value on top of the zinc oxide, and they can be recovered if the zinc oxide is produced in adequately equipped zinc plants, without further investment in new facilities. Specifically, the simultaneous recovery of an added value product is the basis for the development of an economical and sustainable process for the recycling of jarosite.

2. Materials and Methods

2.1. Formation of Jarosite

Jarosite precipitation is summarized in Equations (2)–(4), using the case of ammonium jarosite formation:

$$3Fe_2(SO_4)_3 + 6H_2O \rightarrow 6Fe(OH)SO_4 + 3H_2SO_4, \tag{2}$$

$$4Fe(OH)SO_4 + 4H_2O \rightarrow 2Fe_2(OH)_4SO_4 + 2H_2SO_4, \tag{3}$$

$$2Fe(OH)SO_4 + 2Fe_2(OH)_4SO_4 + 2NH_4OH \rightarrow (NH_4)_2Fe_6(SO_4)_4(OH)_{12}. \tag{4}$$

If the above reactions are combined, the overall reaction can be written as in Equation (5):

$$3Fe_2(SO_4)_3 + 10H_2O + 2NH_4OH \rightarrow (NH_4)_2Fe_6(SO_4)_4(OH)_{12} + 5H_2SO_4. \tag{5}$$

Although jarosite precipitation is a selective removal process of iron, small losses of other dissolved metals like zinc, lead, or silver cannot be avoided, leading to the aforementioned assimilation into the jarosite structure. Patiño mentioned the substitution of a cation with silver in a sodium jarosite sample in literature [10]—as in Equation (1)—with the ratio of elements as shown in Equation (6):

$$[Na_{0.675}Ag_{0.005}(H_3O)_{0.32}] \cdot Fe_3(SO_4)_2(OH)_6. \tag{6}$$

Salinas [11] reports a similar composition of an investigated industrial jarosite sample, with its molar ratio shown in Equation (7):

$$[Na_{0.07}Ag_{0.001}K_{0.02}(NH_4)_{0.59}(H_3O)_{0.31}] \cdot Fe_3(SO_4)_2(OH)_6. \tag{7}$$

Furthermore, it can be observed in Equation (5) that H_2SO_4 is formed during this precipitation, which leads to the necessity of the addition of neutralization agents to avoid resolubization. Commonly, calcined ore from the roasting step is utilized for neutralization during the precipitation of jarosite, leading to a certain amount of impurities not caused by the precipitation itself. Going hand in hand with this, elements or compounds in the ore concentrate are present in the precipitate as well, including lead, silver, zinc-ferrite, and indium, in addition to gangue compounds.

Table 2 shows the elemental analysis of a typical jarosite residue, with 27.1 wt. % of iron, approximately 6.5 wt. % of zinc and lead, and also sulfur, silver, and indium, in concentrations of 8.4 wt. %, 180 ppm, and 230 ppm, respectively.

The substitution of the cation placeholder in the jarosite and the utilization of calcine for neutralization are not the only possible sources of side elements in the residue. In some plants, no Pb–Ag–residue separation is performed, due to low contents of those elements in the concentrate. Still, a certain amount of this by-product is always generated, and in such a case, ends up together with the iron in the precipitation residue—jarosite—as well [4]. Aside from that, dissolved silver oxide precipitates to insoluble silver chloride in the presence of chlorine ions in the pregnant solution, resulting in up to 5 wt. % of the silver contained in the residue being present as a chloride compound [13,14].

Table 2. Typical elemental analysis of a jarosite sample [12].

Element	Concentration (wt. %)	Method
Ag	0.018	DIN EN ISO 11885
In	0.023	DIN EN ISO 11885
Fe	27.1	DIN EN ISO 11885
Zn	6.5	DIN EN ISO 11885
Pb	6.2	DIN EN ISO 11885
S	8.4	DIN EN ISO 11885
Cl	<0.1	DIN 38405 D1-2
F	0.01	DIN 38405 part 1

2.2. Behavior of Jarosite during Thermal Treatment

During a pyrometallurgical treatment under oxidizing atmosphere, the jarosite residue decomposes into separate compounds. The splitting of crystal water, the separation of the OH-group, and the separation of the sulfate group can be described for all common types of jarosite in the three characteristic steps shown, respectively, in Equations (8)–(10) for sodium jarosite:

$$NaFe_3(SO_4)_2(OH)_6 \cdot xH_2O \rightarrow NaFe_3(SO_4)_2(OH)_6 + x\{H_2O\} \quad (Temp. :> 230\,°C), \tag{8}$$

$$NaFe_3(SO_4)_2(OH)_6 \rightarrow NaFe(SO_4)_2 + Fe_2O_3 + 3\{H_2O\} \quad (Temp. :> 420\,°C), \tag{9}$$

$$NaFe(SO_4)_2 \rightarrow Na_2SO_4 + Fe_2O_3 + 3\{SO_3\} \quad (Temp. : 600–800\,°C). \tag{10}$$

Based on the stoichiometry of the jarosite compounds, a theoretical amount of 35.9 wt. % is separated via the gaseous reaction products, mainly leaving behind iron oxide. In the context of silver and indium, it can be stated that the possibly formed Ag_2O is not stable at an increased temperature, and consequently dissociates to metallic silver. Moreover, stable indium(III)oxide does not show any significant vapor pressure and is therefore also in the calcined residue.

2.3. Thermodynamic Considerations of the Behavior of Indium and Silver

Typically, the volatility of metallic zinc is exploited to extract the valuable zinc from secondary resources, as is done for steel mill dust in Waelz kilns, or for drosses in the French and American processes for the production of high-purity zinc oxide. Based on the previously described idea of an added value zinc oxide for the primary zinc industry, extracted from jarosite, the thermodynamic investigations in this article describe the possibility of forming volatile compounds of silver and indium, resulting in co-evaporation with zinc. To achieve this, the software package HSC Chemistry 8.1.4 (Outotec Oyj., Helsinki, Finland) was used to calculate two kinds of phase stability diagrams. The first module used for calculation showed stability regions (the Tpp diagram module) for the variation of $Cl_{2(g)}$ partial pressure as a function of temperature at a fixed $O_{2(g)}$ partial pressure. With this module, the required temperature for a certain reaction could easily be determined. The second module illustrated phase stability boundaries as lines (the Lpp diagram module) for the variation of $Cl_{2(g)}$ vapor pressure as a function of $O_{2(g)}$ partial pressure at a fixed temperature.

Additionally, vapor pressure curves for different indium and silver compounds were plotted to determine whether they were volatile or stayed in the solid phase. In these cases, the reaction equation module of the HSC software was used to determine the equilibrium constant of the general reaction applied to the relevant compound, as shown in Equation (11):

$$aMeX(s) \rightarrow bMeX(g) \quad K = \frac{p^b_{MeX(g)}}{a^a_{MeX(s)}}. \tag{11}$$

The formula of the equilibrium constant, K, for the general vaporization reaction above can be transformed according to the vapor pressure, p. Based on the assumption of pure substances, the activity can be set to 1, resulting in the simplified Equation (12) for the vapor pressure of a compound at the temperature assumed for the calculation of the equilibrium constant:

$$p_{MeX(g)}^b = \sqrt[b]{\left(a_{MeX(s)}^a \cdot K\right)} \xrightarrow{\text{simplified}} p_{MeX(g)}^b = \sqrt[b]{(K)}. \tag{12}$$

With a stoichiometric coefficient of 1, the partial pressure becomes equal to the equilibrium constant of the vaporization reaction.

3. Results and Discussion

The results of the thermodynamic investigation concerning the vapor pressure curves plotted with the software package HSC Chemistry 8.1.4 are shown in Figure 2. At a temperature of 900 °C, metallic indium (solid blue line) started to have a slowly-increasing vapor pressure, despite its theoretical boiling point being 2071 °C. In addition, metallic silver (solid red line) first increased in vapor pressure at slightly over 1000 °C, and with this, represented a possible partial transition to the gaseous state. The partial pressures of the metallic forms of indium and silver were very low, and were therefore not suitable for extraction via the vapor phase under moderate temperature regimes.

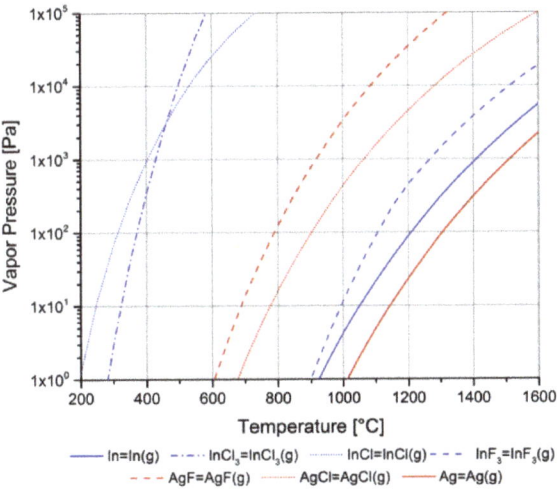

Figure 2. Vapor pressures for selected indium (blue lines) and silver compounds (red lines) as a function of temperature.

If the aim is to reach high extraction yields for silver and indium via a vaporization process, in addition to the recovery of zinc, the vapor pressure of the compounds should be as high as possible. Typically, zinc is vaporized above 907 °C, but it has significant vapor pressure at even lower temperatures. As such, the aim for silver and indium was to find volatile compounds below 907 °C. Consequently, oxides and halides were also evaluated, as they are typically known to be volatile compounds. Concerning the oxides, it can be stated that silver oxide was not stable and dissociated into its metallic form, while indium(II)oxide was not volatile in the range mentioned. Other indium oxides pointed out in the literature (e.g., In_4O_3, In_4O_6, and In_7O_9) could be formed during reduction processes, but reacted under oxidizing conditions once again forming indium(II)oxide. Indium(I)oxide, In_2O, could also be formed during a reduction process, but was only stable in the gaseous phase, which is the reason why it is not shown in Figure 2. In the case of a reduction potential that was too high, it was further reduced to the metallic state. Thermodynamic calculations also showed that the presence of a chlorine carrier such as $ZnCl_2$ immediately led to the formation of InCl. It could be seen that InCl and $InCl_3$ already had significant vapor pressures at very low temperatures (>300 °C), while AgCl started demonstrating vapor pressure at >700 °C. The fluorine compounds (dashed lines) of silver (AgF) and indium (InF_3) started showing vapor pressures at >900 °C and >600 °C, respectively.

In summary, the vapor pressure calculations showed that the volatility of halides was higher than that of the oxide compounds or the metallic forms of indium and silver. Moreover, the fact that chlorine compounds are less disturbing in the primary zinc industry than fluorine compounds, and can be removed more easily from a zinc oxide product before utilization, means that they are the preferred compounds for selective extraction via the vapor phase.

To determine the required conditions for the formation of volatile indium and silver compounds, phase stability diagrams of indium and silver, as well as their chlorides and oxides were calculated, as shown in Figure 3.

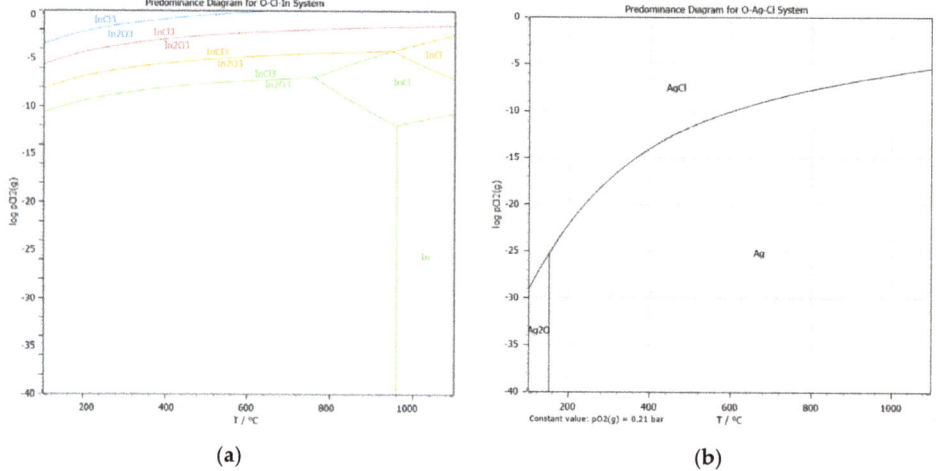

Figure 3. (**a**) Predominance diagram (Tpp) for the system O-Cl-In at four $O_{2(g)}$ partial pressures (green: 1×10^{-15} bar; yellow: 1×10^{-10} bar; red: 1×10^{-5} bar; blue: 0.21 bar;) (**b**) Predominance diagram (Tpp) for the system O-Cl-Ag at a $O_{2(g)}$ partial pressures of 0.21 bar.

Figure 3a shows, for four selected $O_{2(g)}$ partial pressures ranging from 0.21 bar (blue line) to 1×10^{-15} bar (green line), the predominance diagrams for metallic indium, In_2O_3, InCl, and $InCl_3$. Indium oxide had a wide range of stability in all cases, but was a non-volatile compound in the temperature range considered. It was also determined that a certain $Cl_{2(g)}$ partial pressure or any chlorine source was required, which could be ensured by, for instance, the addition of less-stable chlorine compounds, forming indium chloride. The stability region of In_2O_3 could be decreased by lowering the $O_{2(g)}$ partial pressure, which could be realized by purging with an inert gas or through the addition of reducing agents like carbon. Furthermore, it could be seen that at increased temperatures (>800 °C) and adequate $O_{2(g)}$ and $Cl_{2(g)}$ partial pressures, InCl showed a more stable region. The advantage of InCl when compared to $InCl_3$ is the ratio of indium to the consumed chlorine carrier for the formation of volatile indium chloride, which is one-third of its molar ratio in $InCl_3$. The predominantly-formed compound at lower $Cl_{2(g)}$ partial pressure in the temperature range considered was $InCl_3$.

As mentioned before, silver oxide was not stable when the temperature was increased above 180 °C, and reacted to form metallic silver. Figure 3b illustrates this fact on the left side of the diagram. The influence of $O_{2(g)}$ partial pressure on the stability regions concerning AgCl and Ag was not significant; the only change upon decrease was that the silver oxide already dissociated at a lower temperature. This was why only one $O_{2(g)}$ partial pressure is shown in the diagram. It can also be stated that, over the complete investigated temperature range, silver(I)chloride was formed so long as a source for chlorine was available, which was less stable than AgCl. As exhibited in Figure 2, silver(I)chloride had a higher vapor pressure than metallic silver, which positively influenced a targeted recovery.

Figure 4 shows the phase stability boundaries calculated for three temperatures, ranging from 700 °C (green line) to 1100 °C (red line), for indium and silver, and their chlorides and oxides.

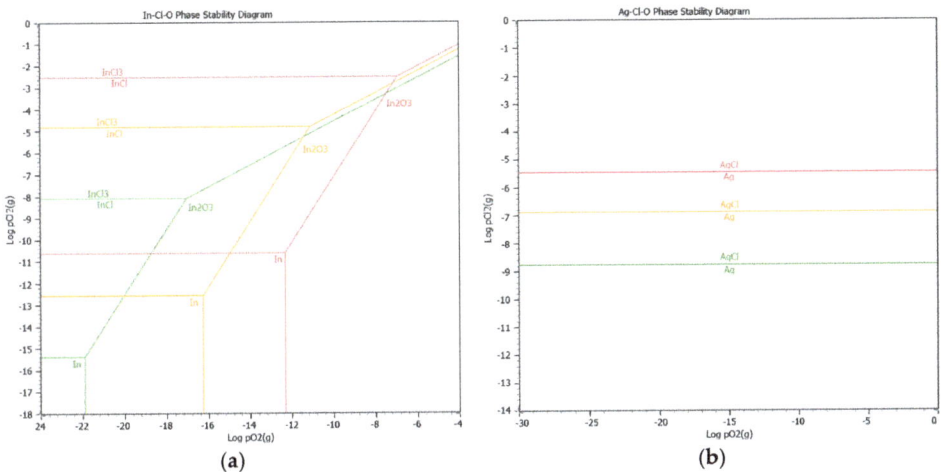

Figure 4. (a) Predominance diagram (Lpp) for the system O-Cl-In at three temperatures (green: 700 °C; yellow: 900 °C; red: 1100 °C) (b) Predominance diagram (Lpp) for the system O-Cl-Ag at three temperatures (green: 700 °C; yellow: 900 °C; red: 1100 °C).

Figure 4a shows again that the indium oxide was stable so long as the partial pressure of oxygen was high enough and the $Cl_{2(g)}$ partial pressure was low. In the case of decreased $O_{2(g)}$ partial pressure, metallic indium could be formed. The lower the temperature, the lower the partial pressure of $O_{2(g)}$ had to be to form metallic indium. If the $Cl_{2(g)}$ partial pressure was also taken into account, the diagram shows that the partial pressure itself had the main influence on $InCl_3$ formation, instead of temperature variation. For the formation of InCl from In_2O_3, the temperature could also significantly move the stability boundaries, but as described in the preliminary paragraph, the formation of $InCl_3$ was more likely in the investigated temperature region due to a requirement for lower $Cl_{2(g)}$ partial pressure, combined with an acceptably high $O_{2(g)}$ partial pressure. The increase in temperature shifted the stability region of InCl to a lower $Cl_{2(g)}$ partial pressure, and at a temperature of 1415 °C (not shown in Figure 4a), indium(I)chloride was formed instead of $InCl_3$.

The predominance diagram for silver shown in Figure 4b illustrates the $Cl_{2(g)}$ partial pressure as a function of the $O_{2(g)}$ partial pressure at constant temperatures. The horizontal lines for all three investigated temperatures highlight the aforementioned fact that the $O_{2(g)}$ partial pressure had no significant influence on the stability boundary of $AgCl_{(g)}$ and Ag. Silver(I)oxide was not stable at the investigated temperatures between 700 to 1100 °C, and therefore does not appear in the diagram.

In summary, it can be said that the volatilization of silver could be positively influenced by the presence of a chlorine carrier, leading to the formation of silver(I)chloride at any $O_{2(g)}$ partial pressure in the investigated range (below 1 bar) and with this, the required temperature shifted to lower values.

Regarding indium oxide, reducing conditions were required to bind the oxygen and reduce the $O_{2(g)}$ partial pressure, supporting the formation of $InCl_{3(g)}$ or $InCl_{(g)}$ for successful volatilization.

4. Conclusions

For the development of an economical and ecological process, the target is the simultaneous recovery of product with optimized quality. Especially in the case of industrial residues such as the iron precipitate of the zinc industry (which is typically low in value), elements like indium and silver contribute significantly to the overall economy of a potential recycling process if they are recovered in a meaningful way. Therefore, the investigated extraction method by the formation of volatile chlorine compounds of indium and silver supports the idea of generating an "added value" product through the simultaneous recovery of the elements zinc, indium, and silver in one metallurgical step. The fact that indium oxide requires a reducing agent is not a disadvantage, as it goes hand-in-hand with a potential reduction of zinc oxide for its vaporization as metallic zinc. Furthermore, this does not influence the silver extraction, where the oxygen partial pressure is of minor significance.

In the context of minimizing the operational costs of the process, the substitution of required additives (e.g., chlorine carriers in this case) is also a factor. This was investigated in [11], describing a possible combined treatment of two residues, one of them carrying chlorine as a constituent, along with electric arc furnace dust.

Author Contributions: S.S. conceived, designed and performed the experiments/calculations; J.A. contributed the calculation tool; S.S. and J.A. analyzed the data; S.S. wrote the paper.

Acknowledgments: The authors want to thank the Austrian Research Promotion Agency (FFG) and the Federal Ministry of Science, Research and Economy (BMWFW) for the financial support of this research activity (FFG project number: 844725).

Conflicts of Interest: The authors declare no conflict of interest.

References

1. European Commission: Report on Critical Materials for the EU. Available online: http://ec.europa.eu/DocsRoom/documents/10010/attachments/1/translations/en/renditions/native (accessed on 24 November 2017).
2. Pawlek, F. *Metallhüttenkunde*; Walter de Gruyter: Berlin, Germany, 1983; ISBN 3-11-007458-3.
3. Unger, A. Charakterisierung und Evaluierung von Rückständen aus der Primären Hydrometallurgischen Zinkgewinnung. Master's Thesis, Montanuniversität Leoben, Leoben, Austria, 2011.
4. Sinclair, R.J. *The Extractive Metallurgy of Zinc*; AusIMM, Carlton, Vic.: Carlton, Australia, 2005; ISBN 1-92080634-2.
5. Nitzert, M.R. Beitrag zur Thermischen Aufarbeitung von Jarosit und Weiteren Zinklaugungsrückstände im Gleichstromelektroreduktionsofen. Ph.D. Thesis, Technical University Aachen, Aachen, Germany, 1986.
6. Röpenack, A. Jarositfällung: Die Bedeutung der Eisenfällung für die hydrometallurgische Zinkgewinnung. *Erzmetall* **1978**, *32*, 272–276.
7. Frost, R.L.; Wills, R.A.; Weier, M.L.; Musumeci, A.W.; Martens, W. Thermal decomposition of natural and synthetic plumbojarosites: Importance in "archeochemistry". *Thermochim. Acta* **2005**, *432*, 30–35. [CrossRef]
8. Graf, G.; Ullmann, Z. *Ullmann's Encyclopedia of Industrial Chemistry*; Wiley-VCH: Weinheim, Germany, 2003; Volume A28, pp. 509–530.
9. Dutrizac, J.E.; Jambor, J.L. Jarosites and Their Application in Hydrometallurgy. *Rev. Mineral. Geochem.* **2000**, *40*, 405–452. [CrossRef]
10. Patiño, F.; Salinas, E.; Cruells, M.; Roca, A. Alkaline decomposition–cyanidation kinetics of argentian natrojarosite. *Hydrometallurgy* **1998**, *49*, 323–336. [CrossRef]

11. Salinas, E.; Roca, A.; Cruells, M.; Patiño, F.; Córdoba, D.A. Characterization and alkaline decomposition–cyanidation kinetics of industrial ammonium jarosite in NaOH media. *Hydrometallurgy* **2001**, *60*, 237–246. [CrossRef]
12. Wegscheider, S.; Steinlechner, S.; Leuchtenmüller, M. Innovative Concept for the Recovery of Silver and Indium by a Combined Treatment of Jarosite and Electric Arc Furnace Dust. *JOM* **2016**, *69*, 388–394. [CrossRef]
13. Wyslouzil, D.M.; Salter, R.S. Silver leaching fundamentals. In *Lead-Zinc '90 Proceedings of a World Symposium on Metallurgy and Environmental Control*; The Minerals, Metals & Materials Society: Warrendale, PA, USA, 1990; pp. 87–105.
14. Huang, Z. The recovery of silver and scare elements at Zhuzhou smelter. In *Lead-Zinc '90 Proceedings of a World Symposium on Metallurgy and Environmental Control*; The Minerals, Metals & Materials Society: Warrendale, PA, USA, 1990; pp. 239–250.

© 2018 by the authors. Licensee MDPI, Basel, Switzerland. This article is an open access article distributed under the terms and conditions of the Creative Commons Attribution (CC BY) license (http://creativecommons.org/licenses/by/4.0/).

MDPI
St. Alban-Anlage 66
4052 Basel
Switzerland
Tel. +41 61 683 77 34
Fax +41 61 302 89 18
www.mdpi.com

Metals Editorial Office
E-mail: metals@mdpi.com
www.mdpi.com/journal/metals

www.ingramcontent.com/pod-product-compliance
Lightning Source LLC
La Vergne TN
LVHW070245100526
838202LV00015B/2181